# ENVIRONMENTAL POLLUTANTS AND THEIR BIOREMEDIATION APPROACHES

# ENVIRONMENTAL POLLUTANTS AND THEIR BIOREMEDIATION APPROACHES

Edited by
Ram Naresh Bharagava

CRC Press
Taylor & Francis Group
Boca Raton London New York

CRC Press is an imprint of the
Taylor & Francis Group, an **informa** business

CRC Press
Taylor & Francis Group
6000 Broken Sound Parkway NW, Suite 300
Boca Raton, FL 33487-2742



Printed on acid-free paper

International Standard Book Number-13: 978-1-138-62889-2 (Hardback)



**Library of Congress Cataloging-in-Publication Data**

Names: Bharagava, Ram Naresh, editor.
Title: Environmental pollutants and their bioremediation approaches / editor, Ram Naresh Bharagava.
Description: Boca Raton : Taylor & Francis, CRC Press, 2017. | Includes bibliographical references.
Identifiers: LCCN 2016057862 | ISBN 9781138628892 (hardback : alk. paper)
Subjects: LCSH: Bioremediation.
Classification: LCC TD192.5 .E9565 2017 | DDC 628.5--dc23
LC record available at https://lccn.loc.gov/2016057862

**Visit the Taylor & Francis Web site at**
**http://www.taylorandfrancis.com**

**and the CRC Press Web site at**
**http://www.crcpress.com**

# Contents

Preface ........ vii
Editor ........ ix
Contributors ........ xi

**Chapter 1** Bioremediation: An Eco-Sustainable Green Technology: Its Applications and Limitations ........ 1

*Ram Naresh Bharagava, Pankaj Chowdhary, and Gaurav Saxena*

**Chapter 2** Organic and Inorganic Pollutants in Industrial Wastes: Ecotoxicological Effects, Health Hazards, and Bioremediation Approaches ........ 23

*Gaurav Saxena and Ram Naresh Bharagava*

**Chapter 3** Pesticides Contamination in the Environment: Toxicological Effects, and Biodegradation and Bioremediation Mechanisms for Environmental Safety ........ 57

*Namasivayam Vasudevan and Odukkathil Greeshma*

**Chapter 4** Uranium Radionuclide Contamination in the Environment: Ecotoxicological Effects, Health Hazards, and Bioremediation ........ 103

*Meenakshi Nandal and Mansi Rastogi*

**Chapter 5** Dyes Contamination in the Environment: Ecotoxicological Effects, Health Hazards, and Biodegradation and Bioremediation Mechanisms for Environmental Cleanup ........ 127

*Uma Shankar Prasad Uday, Nibedita Mahata, Soumya Sasmal, Tarun Kanti Bandyopadhyay, Abhijit Mondal, and Biswanath Bhunia*

**Chapter 6** Lindane Contamination in the Environment: Toxicological Effects and Bioremediation Approaches ........ 177

*Siddharth Boudh and Jay Shankar Singh*

**Chapter 7** Toxic Metals Contamination in the Environment: Toxicological Effects and Bioremediation Approaches for Environmental Cleanup ........ 209

*Janardhan Reddy Koduru, Shiv Shankar, Nandkishor S. More, Shikha, Lakshmi Prasanna Lingamdinne, and Jiwan Singh*

**Chapter 8** Bioremediation of Nonaqueous Phase Liquids (NAPLs)-Polluted Soil-Water Resources ........ 241

*Pankaj Kumar Gupta and Brijesh Kumar Yadav*

**Chapter 9** Microbial Strategies for Enhanced Phytoremediation of Heavy Metal-Contaminated Soils ........ 257

*Jyoti Singh and Ajay Veer Singh*

**Chapter 10** Microbial Cellulases and Their Applications in the Pulp and Paper Industry: An Emerging Paradigm ........ 273

*Shiv Shankar, M. Gupta, and Shikha*

**Chapter 11** Phytoremediation: A Green Technology for Remediation of Metal-Contaminated Sites ........ 305

*Manoj Kumar and Rajesh Singh*

**Chapter 12** Solid Waste: Environmental Threats and Management ........ 337

*V. Sudharsan Varma and Ajay S. Kalamdhad*

**Chapter 13** Pollutants in Tannery Wastewater: Pharmacological Effects, and Bioremediation Approaches for Human Health Protection and Environmental Safety ........ 369

*Swetlana Gautam, Gaurav Kaithwas, Ram Naresh Bharagava, and Gaurav Saxena*

**Chapter 14** Constructed Wetlands: An Emerging Phytotechnology for Degradation and Detoxification of Industrial Wastewaters ........ 397

*Ram Naresh Bharagava, Gaurav Saxena, and Pankaj Chowdhary*

**Chapter 15** Biomedical Wastes: Environmental Threats and Management ........ 427

*Sahar Mohamed Soliman and Amel Ibrahim Ahmed*

**Index** ........ 443

# Preface

The contamination of environments (soil/water/air) with toxic environmental pollutants released from various sources, their ecotoxicological effects, and health hazards is a serious concern worldwide. The degradation and detoxification of toxic environmental pollutants for environmental safety are global challenges. The wastes/pollutants released from various natural and anthropogenic sources are widespread in most of the countries, causing either the direct pollution in soil and aquatic resources or indirect pollution of the surface water and groundwater. Governments around the globe are strictly advocating for the mitigation of environmental pollution. Hence, the removal/elimination of toxic pollutants from contaminated environments is of utmost importance to promote the sustainable development of our society with low environmental impact. Conventional remediation technologies are expensive, environmentally destructive, and not aesthetically pleasing. In this regard, *bioremediation* has recently emerged as an "eco-friendly" and "cost-effective" sustainable alternative for environmental cleanup.

Bioremediation is an environment-friendly and cost-effective approach, which uses microbes such as bacterial/fungi/algae/yeast/cyanobacteria etc., green plants, or their enzymes, to degrade/detoxify toxic environmental pollutants from polluted soils and water resources. Thus, detailed knowledge of the various types of environmental pollutants, their toxicological effects in environments, in humans, in animals, and in plants as well as various bioremediation approaches for their degradation/detoxification are urgently required for environmental cleanup and safety of human and animals. However, the degradation and detoxification potential of microbes and their interactions with environmental pollutants has been studied in past few years, but detailed knowledge of the various types of environmental pollutants, their toxicological effects in environments, humans, and animals, as well as various bioremediation approaches used for their degradation/detoxification is scanty and not available as a collection.

This book will provide detailed, comprehensive knowledge of various types of toxic environmental pollutants, such as toxic metals, pesticides, azo dyes, petroleum hydrocarbons, plastics wastes, phenols, chlorophenols, etc., discharged from natural as well as anthropogenic sources; their toxicological effects in environments, humans, and animals; and their biodegradation and bioremediation through microbes/plants and/or by their enzymes as well as by constructed wetland treatment approaches for environmental cleanup. A number of experts from colleges, universities, and research laboratories have contributed their valuable knowledge for this book in the areas of biodegradation and bioremediation of environmental pollutants.

This book will also serve as an invaluable source of basic knowledge and recent advances in biodegradation and bioremediation of toxic environmental pollutants that widen its scope for a diverse community of professionals having interests in bioremediation technologies. All the chapters compiled in this book cover different aspects of environmental problems and their remedies with up-to-date developments in the field of biodegradation and bioremediation. This book will be of great value to researchers, environmental chemists and scientists, microbiologists and biotechnologists, eco-toxicologists, waste treatment engineers and managers, environmental science managers, administrators and policy makers, industry persons, and students at bachelor's, master's, and doctoral level in the relevant field. At last, I sincerely extend my heartfelt thanks to all the leading experts for contributing valuable chapters to this book and to my research scholar Mr. Gaurav Saxena for helping me in various ways during this book project. Thus, in this book, readers will find the updated information as well as the future direction for research in the field of biodegradation and bioremediation of environmental pollutants.

**Ram Naresh Bharagava**
*Babasaheb Bhimrao Ambedkar Central University*
*Lucknow, Uttar Pradesh, India*

# Editor

**Dr. Ram Naresh Bharagava** was born in 1977 and completed his education in Government Schools at Lakhimpur Kheri, India. He received a BSc degree (1998) in zoology, botany, and chemistry from the University of Lucknow, India, and an MSc degree (2004) in molecular biology and biotechnology from Govind Ballabh Pant University of Agriculture and Technology (GBPUAT), Pantnagar, India. He received a PhD (2010) in the area of microbiology jointly from the environmental microbiology division, Indian Institute of Toxicology Research (IITR) Council of Scientific and Industrial Research (CSIR), Lucknow, India, and Pt. Ravishankar Shukla University, Raipur, India. He was the Junior Research Fellow (JRF) during his PhD, qualified twice (2002 & 2003) on the CSIR-National Eligibility Test (NET) and Graduate Aptitude Test in Engineering (GATE) in 2003. His major research work during his PhD focused on the bacterial degradation of recalcitrant melanoidin from distillery wastewater. He has authored one book and edited two books. He has also written many chapters for national and international books. He has published many popular science articles in national newspapers, and national and international magazines. Dr. Bharagava has presented many research papers relevant to his research areas in national and international conferences. He has also served as a potential reviewer for various national and international journals in his respective areas of the research. Dr. Bharagava was awarded a postdoctoral appointment at CSIR-IITR, Lucknow, India, but left and subsequently joined (2011) Babasaheb Bhimrao Ambedkar (Central) University, Lucknow, India. He is currently working as an assistant professor of environmental microbiology and actively engaged in teaching at the postgraduate and doctoral levels and conducting research on the various Government of India (GOI)–sponsored projects in the area of environmental toxicology and bioremediation at Laboratory for Bioremediation and Metagenomics Research (LBMR) under the Department of Environmental Microbiology (DEM). His research has been supported by the University Grants Commission (UGC) and Department of Science and Technology (DST), India. He has been the advisor to more than 40 postgraduate students and currently, the mentor of 1 project fellow and 6 doctoral students. His major areas of research include the environmental toxicology, biodegradation, and bioremediation of environmental pollutants/industrial wastewaters, metagenomics, and wastewater microbiology. He is a life member of the Academy of Environmental Biology (AEB), Association of Microbiologists of India (AMI), and Biotech Research Society, India (BRSI),

Indian Science Congress Association (ISCA). In his spare time, he enjoys walking in peaceful environments and spending time with his family. He lives in south Lucknow with his wife (Ranjana) and three kids (Shweta, Abhay, and Shivani). He can be reached at bharagavarnbbau11@gmail.com, ramnaresh_dem@bbau.ac.in.

# Contributors

**Amel Ibrahim Ahmed**
Department of Community Health Nursing
Mansoura University
Mansoura, Egypt

**Tarun Kanti Bandyopadhyay**
Department of Chemical Engineering
National Institute of Technology
Agartala, India

**Ram Naresh Bharagava**
Laboratory for Bioremediation and Metagenomics Research
Department of Environmental Microbiology
Babasaheb Bhimrao Ambedkar University
Lucknow, India

**Biswanath Bhunia**
Department of Bio Engineering
National Institute of Technology
Agartala, India

**Siddharth Boudh**
Department of Environmental Microbiology
Babasaheb Bhimrao Ambedkar University
Lucknow, India

**Pankaj Chowdhary**
Laboratory for Bioremediation and Metagenomics Research
Department of Environmental Microbiology
Babasaheb Bhimrao Ambedkar University
Lucknow, India

**Swetlana Gautam**
Department of Pharmaceutical Sciences
School of Biosciences and Biotechnology
Babasaheb Bhimrao Ambedkar University
Lucknow, India

**Odukkathil Greeshma**
Centre for Environmental Studies
Anna University
Chennai, India

**M. Gupta**
Department of Environmental Science
Babasaheb Bhimrao Ambedkar University
Lucknow, India

**Pankaj Kumar Gupta**
Department of Hydrology
Indian Institute of Technology
Roorkee, India

**Gaurav Kaithwas**
Department of Pharmaceutical Sciences
School of Biosciences and Biotechnology
Babasaheb Bhimrao Ambedkar University
Lucknow, India

**Ajay S. Kalamdhad**
Department of Civil Engineering
Indian Institute of Technology Guwahati
Guwahati, India

**Janardhan Reddy Koduru**
Graduate School of Environmental Studies
Kwangwoon University
Seoul, Republic of Korea

**Manoj Kumar**
School of Environment and Sustainable Development
Central University of Gujarat
Gandhinagar, India

**Lakshmi Prasanna Lingamdinne**
Department of Environmental Engineering
Kwangwoon University
Seoul, Republic of Korea

**Nibedita Mahata**
Department of Biotechnology
National Institute of Technology
Durgapur, India

**Abhijit Mondal**
Department of Chemical Engineering
National Institute of Technology
Agartala, India

**Nandkishor S. More**
Department of Environmental Sciences
School for Environmental Sciences
Babasaheb Bhimrao Ambedkar University
Lucknow, India

**Meenakshi Nandal**
Department of Environmental Science
Maharshi Dayanand University
Rohtak, India

**Mansi Rastogi**
Department of Environmental Science
Maharshi Dayanand University
Rohtak, India

**Soumya Sasmal**
Division of Bio Technology
Netaji Subhas Institute of Technology
Delhi, India

**Gaurav Saxena**
Laboratory for Bioremediation and Metagenomics Research
Department of Environmental Microbiology
Babasaheb Bhimrao Ambedkar University
Lucknow, India

**Shiv Shankar**
Department of Environmental Sciences
School for Environmental Sciences
Babasaheb Bhimrao Ambedkar University
Lucknow, India

**Shikha**
Department of Environmental Sciences
School for Environmental Sciences
Babasaheb Bhimrao Ambedkar University
Lucknow, India

**Ajay Veer Singh**
Department of Microbiology
College of Basic Sciences and Humanities
G. B. Pant University of Agriculture and Technology
Pantnagar, India

**Jay Shankar Singh**
Department of Environmental Microbiology
Babasaheb Bhimrao Ambedkar University
Lucknow, India

**Jiwan Singh**
Department of Environmental Sciences
School for Environmental Sciences
Babasaheb Bhimrao Ambedkar University
Lucknow, India

**Jyoti Singh**
Department of Microbiology
College of Basic Sciences and Humanities
G.B. Pant University of Agriculture and Technology
Pantnagar, India

**Rajesh Singh**
School of Environment and Sustainable Development
Central University of Gujarat
Gandhinagar, India

**Sahar Mohamed Soliman**
Department of Community Health Nursing
Mansoura University
Mansoura, Egypt

**Uma Shankar Prasad Uday**
Department of Chemical Engineering
National Institute of Technology
Agartala, India

**V. Sudharsan Varma**
Department of Civil Engineering
Indian Institute of Technology Guwahati
Guwahati, India

**Namasivayam Vasudevan**
Centre for Environmental Studies
Anna University
Chennai, India

**Brijesh Kumar Yadav**
Department of Hydrology
Indian Institute of Technology
Roorkee, India

# 1 Bioremediation
## *An Eco-Sustainable Green Technology: Its Applications and Limitations*

*Ram Naresh Bharagava, Pankaj Chowdhary, and Gaurav Saxena*

**CONTENTS**

1.1 Introduction ..... 2
1.2 Bioremediation ..... 3
1.3 Types of Bioremediation ..... 4
 1.3.1 *In Situ* Bioremediation ..... 4
  1.3.1.1 Bioattenuation ..... 5
  1.3.1.2 Biostimulation ..... 5
  1.3.1.3 Bioaugmentation ..... 5
  1.3.1.4 Some Other *In Situ* Bioremediation Technologies ..... 6
 1.3.2 *Ex Situ* Bioremediation ..... 9
  1.3.2.1 Landfarming ..... 9
  1.3.2.2 Composting ..... 9
  1.3.2.3 Biopiles ..... 10
  1.3.2.4 Bioreactors ..... 10
1.4 Various Factors Affecting the Bioremediation Process ..... 11
1.5 How We Can Enhance Bioremediation ..... 13
 1.5.1 Genetic Engineering in Bioremediation ..... 13
 1.5.2 Role of Nano-Biotechnology in Bioremediation ..... 16
1.6 Merits and Demerits of Bioremediation ..... 16
 1.6.1 Advantages and Disadvantages of *In Situ* Bioremediation ..... 16
  1.6.1.1 Advantages ..... 17
  1.6.1.2 Disadvantages ..... 17
 1.6.2 Advantages and Disadvantages of *Ex Situ* Bioremediation ..... 18
  1.6.2.1 Solid-Phase Treatment ..... 18
  1.6.2.2 Slurry-Phase Treatment ..... 18
1.7 Conclusion ..... 18
Acknowledgments ..... 19
References ..... 19

## 1.1 INTRODUCTION

Increasing global consumption of several man-made products (paints, adhesives, gasoline, plastics, etc.) generates many toxic compounds, which is responsible for several adverse effects on living systems. Unfortunately, the production of these man-made compounds and changed agricultural practices have led to the widespread use of hazardous organic compounds and the subsequent increase in contaminant concentrations in the environment (Collins et al. 2002; Latha and Reddy 2013; Kang 2014; Saxena and Bharagava 2015). If these pollutants are released into the environment without adequate treatment, they could cause severe health problems and also destroy many fauna and flora. Various types of pollutants are spread through air and water into the ground, which has led to a variety of serious environmental and health problems throughout the world (Boopathy 2000).

The effects of water pollution are not only devastating to aquatic organisms but also to terrestrial animals and birds. More seriously, contaminated water destroys aquatic life and reduces reproductive ability. Ultimately, the water becomes unfit for human consumption or domestic exercise; in severe cases water even becomes a hazard to human health (Kaiser 2000). Large amounts of toxic chemicals are released into the environment, either deliberately as in the application of pesticides or accidentally as in the case of oil spills. Moreover, in the bioremediation process, various types of pollutants are removed, such as organic compounds, heavy metals, and various types of xenobiotic compounds, which are generated by several activities. Organic compounds, mostly human generated, have been widely used as industrial solvents, fuel components, and intermediates. Many manufactured products such as paints, adhesives, gasoline, and plastics contain harmful organic compounds (Collins et al. 2002; Brar et al. 2006; Kang 2014).

Many researchers have stepped up efforts to find more sustainable and cost-effective alternatives to using hazardous chemicals and treatments to remove existing harmful pollutants. However, bioremediation is a process in which beneficial microbiological agents, such as yeast, fungi, or bacteria, are used to clean up contaminated soil and water (Strong and Burgess 2008). It is defined as the elimination, attenuation, or transformation of polluting or contaminating substances by the application of biological processes. Contamination of soils, groundwater, sediments, surface water, and air with hazardous and toxic chemicals are serious problems, which have been faced by our world today (Boopathy 2000). Several microbial communities present in the environment efficiently degrade many toxic compounds. Most of them are degraded slowly (recalcitrant) and thus tend to accumulate in the environment. In some cases, this accumulation can constitute a severe hazard. Various elements of the chemical structures of many of these pollutants are beyond the biodegradation capabilities of microorganisms (Dejonghe et al. 2000).

In addition, bioremediation involves enhanced degradation of toxic compounds by transforming them into innocuous substances, specifically carbon dioxide and water. The process can be carried out either on-site (*in situ* bioremediation) by taking advantage of indigenous microorganisms or the introduction of bacterial or fungal strains or off-site in bioreactors (*ex situ* bioremediation) to achieve complete detoxification of toxic pollutants (Brar et al. 2006; Kumar et al. 2011; Orji et al. 2012; Hamzah et al. 2014).

A variety of microorganisms capable of efficiently degrading toxic compounds and xenobiotics in the environment have been either isolated or engineered. However, the actual application of such microorganisms in bioremediation has not progressed with the same momentum as their invention, or as other innovations in the biotechnology arena. Bioremediation is a general concept that includes all those processes and actions that take place in order to biotransform an environment, already altered by contaminants, to its original status (Thassitou and Aryanitoyannis 2001).

The agricultural field has seen major progress with the use of genetically engineered organisms, which seems to be a very important tool for producing food for those in undeveloped countries using nonhosted areas. Industrial and environmental biotechnology are taking new paths, resulting in processes with clean technologies, with maximum production and fewer residues (Soccol et al. 2003; Ezezika and Singer 2010).

The field of biotechnology can be divided into four divisions: biomedical, agricultural, environmental, and industrial. Several works have been published that define the applications of bioremediation and physical, chemical, and biological conditions necessary to facilitate contaminant biodegradation (Romantschuk et al. 2000). Plant genetic engineering through inserting or overexpressing specific genes in the genome of the plants provides an efficient method to enhance the phytoremediation capacity of plants (James and Strand 2009; Ezezika and Singer 2010). Bioremediation is a function of various factors: the existence of a microbial population capable of degrading contaminants; contaminant bioavailability to microbes; environmental factors such as temperature, pH, and nutrients (organic, inorganic, and their availability); electron acceptor(s); redox potential; water activity; osmotic pressure; and concentration of contaminants (Thakur 2004; Al-Sulaimani et al. 2010; Kang 2014). This chapter intends to present a brief review of soil bioremediation, its development, and the main factors limiting its use; the characteristics of principal methods of bioremediation; and its application in soil contamination with pesticides, herbicides, oil (hydrocarbons), food waste, and heavy metals.

## 1.2 BIOREMEDIATION

Bioremediation is the use of biological organisms to solve an environmental problem such as contaminated soil or groundwater. Bioremediation is the use of living microorganisms to degrade environmental pollutants or to prevent pollution (Brar et al. 2006; Antizar-Ladislao 2010; Latha and Reddy 2013). In other words, it is a technology for removing pollutants from the environment and, thus, restoring the original natural surroundings and preventing further pollution. In bioremediation processes, various organisms and their products are involved with several strategies, such as bacteria (bacterial bioremediation), fungi (mycoremediation), algae (phycoremediation), plant (phytoremediation), rhizosphere (rhizoremediation), and biomolecules derived from organisms (derivative bioremediation). Bioremediation is basically a technique in which microorganisms are utilized for the management of biological waste. Their metabolism is utilized for the removal of pollutants from contaminated environments.

Bioremediation can occur on its own, which is natural attenuation, or can occur artificially by adding chemicals spurred on by microbes, which is termed “biostimulation.”

Note that not all heavy metals can be removed by using bioremediation processes. Bioremediation is helping humankind and society deal with toxic and hazardous wastes, which can be otherwise harmful to health and surroundings, if not removed or made safe.

## 1.3 TYPES OF BIOREMEDIATION

Bioremediation is an eco-friendly sustainable technique for the remediation of environmental pollutants (Figure 1.1). Based on the removal and nature of treatment strategies, there are basically two methods.

### 1.3.1 *In Situ* Bioremediation

*In situ* bioremediation can be described as the process whereby pollutants are biologically degraded under natural conditions to either carbon dioxide or water or an attenuated transformation product. It is a low-cost, low-maintenance, environment-friendly, and sustainable approach for the cleanup of polluted sites (Aggarwal et al. 1990; Jorgensen 2007; Megharaj et al. 2011; Latha and Reddy 2013). In addition, *in situ* bioremediation treats the contaminated material on-site. Most often, *in situ* bioremediation is applied to the degradation of contaminants in saturated soils and groundwater. It is a superior method to cleaning contaminated environments

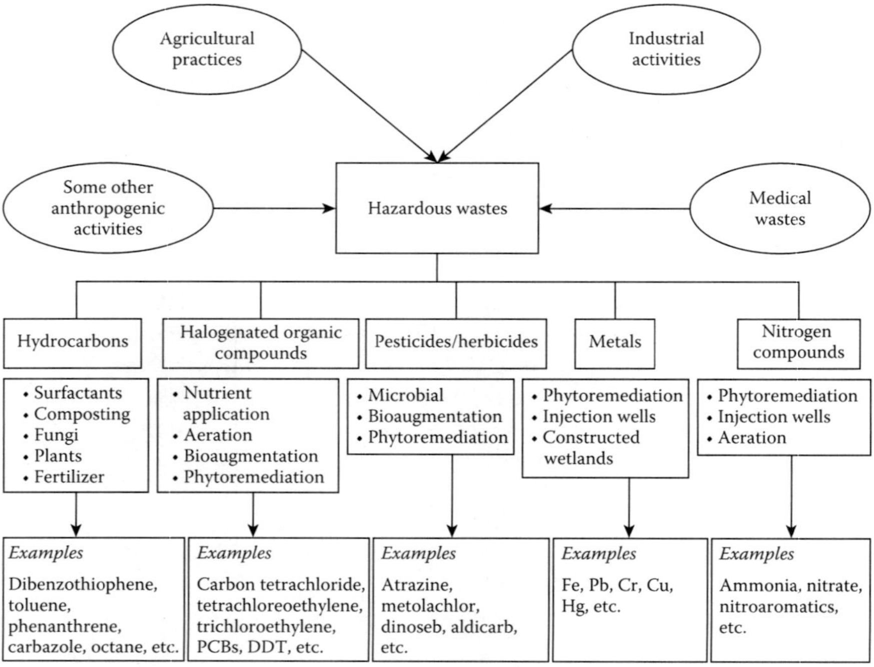


**FIGURE 1.1** Various bioremediation approaches for hazardous wastes.

because it is cheaper and uses harmless microbial organisms to degrade the chemicals. Chemotaxis is important to the study of *in situ* bioremediation because microbial organisms with chemotactic abilities can move into an area containing contaminants. Therefore, by enhancing the cells' chemotactic abilities, *in situ* bioremediation will become a safer method in degrading harmful compounds. There are generally three different types of *in situ* bioremediation processes: bioattenuation, biostimulation, and bioaugmentation.

#### 1.3.1.1 Bioattenuation

Bioattenuation depends on the natural process of degradation. This is the method of monitoring the natural progress of degradation to ensure that contaminant concentration decreases with time at relevant sampling points (Mulligan and Yong 2004; Latha and Reddy 2013). Bioattenuation is widely used as a cleanup method for underground storage tank sites with petroleum-contaminated soil and groundwater in the United States.

#### 1.3.1.2 Biostimulation

This is an intentional stimulation of degradation of chemicals achieved by adding water, nutrients, and electron donors or acceptors. If natural degradation does not occur or if the degradation is too slow, the environment has to be manipulated in such a way that biodegradation is stimulated and reaction rates are increased. The measures to be taken, called biostimulation, include supplying the environment with nutrients such as nitrogen and phosphorus, with electron acceptors such as oxygen, and with substrates such as methane, phenol, and toluene. The chemical additives used as substrates, phenol and toluene, are well-known toxic chemicals. Thus, the concentrations of these chemicals during biostimulation should be carefully monitored. In Japan, the effectiveness of *in situ* biostimulation by methane injection into trichloroethylene (TCE)-contaminated groundwater was demonstrated by small-scale field experiments funded separately by the Environment Agency (Peer et al. 2006; Latha and Reddy 2013) and by the Ministry of International Trade and Industry. By accumulating scientific evidence through these kinds of field experiments, *in situ* biostimulation is expected to become a reliable and safe cleanup technology.

#### 1.3.1.3 Bioaugmentation

In this process, microbial members with proven capabilities of degrading or transforming the chemical pollutants are added. The third choice in the treatment hierarchy is bioaugmentation, which is a way to enhance the biodegradation capacities of contaminated sites by inoculating bacteria with desired catalytic capabilities. This is considered to be an effective approach in the case of recalcitrant chemicals where bioattenuation or biostimulation does not work. However, we have to pay more attention to the application of bioaugmentation because of its unknown effects on the ecosystem. Because large amounts of degradative bacteria are added to contaminated sites, the effect of the bacteria on both human and environment must be clarified in advance. Moreover, it needs to be confirmed that the injected bacteria have perished after the remediation and thus do not affect the indigenous microbial community for a long period (Lee et al. 2006; Latha and Reddy 2013).

#### 1.3.1.4 Some Other *In Situ* Bioremediation Technologies

##### *1.3.1.4.1 Air Sparging or Biosparging*

*In situ* air sparging technology has been used since about 1985. This remediation technology is used on volatile organic compounds (VOCs) that have contaminated groundwater or soils in the saturated zone. The technology injects contaminant-free air into the subsurface saturated zone. This process causes a phase transfer of contaminants from a dissolved state to a vapor phase. The phenomenon can be described best as "controlled disequilibrium." In other words, air sparging involves injecting a gas (usually air/oxygen) under pressure into the saturated zone to volatilize groundwater contaminants and to promote biodegradation in saturated and unsaturated soils by increasing subsurface oxygen concentrations. Volatilized vapors migrate into the vadose zone where they are extracted via vacuum, generally by a soil vapor extraction system (Hardisty and Ozdemiroglu 2005; Neilson and Allard 2008).

##### *1.3.1.4.2 Bioventing*

This *in situ* remediation approach employs the indigenous microorganisms to biodegrade organic constituents adsorbed to soils in the unsaturated (vadose) zone. Bioventing uses low airflow rates to provide only enough oxygen to sustain microbial activity and prevent contaminant volatilization.

Bioventing is a promising technology that stimulates the natural *in situ* biodegradation of any aerobically degradable compounds in soil by providing oxygen to existing soil microorganisms. It typically uses low airflow rates to provide only enough $O_2$ to sustain microbial activity. Oxygen is most commonly supplied through direct air injection into residual contamination in soil. In addition to degradation of adsorbed fuel residuals, volatile compounds are biodegraded as vapors move slowly through biologically active soil. Bioventing techniques have been successfully used to remediate soils contaminated by petroleum hydrocarbons, nonchlorinated solvents, some pesticides, wood preservatives, and other organic chemicals. It is most often used at sites with mid-weight petroleum products (i.e., diesel fuel and jet fuel) and other volatile contaminants such as TCE, trichloroethane, ethylene dibromide, and dichloroethylene (Lee et al. 2006; Latha and Reddy 2013).

##### *1.3.1.4.3 Phytoremediation*

The process in which green plants are used to degrade/detoxify the pollutants from soil, sludge, sediments, surface water, and groundwater is known as phytoremediation. The principal application of phytoremediation is for lightly contaminated soils and waters where the material to be treated is at a shallow or medium depth and the area to be treated is large (Lee et al. 2006). This will make agronomic techniques economical and applicable for both planting and harvesting. In addition, the site owner must be prepared to accept a longer remediation period. Plants that are able to decontaminate soils do one or more of the following:

- Uptake contaminant from soil particles or soil liquid into their roots
- Bind the contaminant into their root tissue, physically or chemically
- Transport the contaminant from their roots into growing shoots and prevent or inhibit the contaminant from leaching out of the soil

Phytoremediation approaches can be applied to both inorganic and organic pollutants present in solid and liquid substrates (Salt et al. 1998; Karen et al. 2009; Kang 2014). Generally, phytoremediation of contaminants by a plant involves the following steps: uptake, translocation, transformation, compartmentalization, and sometimes mineralization (Schnoor et al. 1995). Factors affecting the uptake, distribution, and transformation of organic compounds by plants are mainly related to the physical and chemical properties of compounds (e.g., water solubility, molecular weight, and octanol–water partition coefficient), environmental conditions (e.g., temperature, pH, organic matter, and soil moisture content), as well as plant characteristics (e.g., shoot/root system, enzymes) (Susarla et al. 2002; Suresh and Ravishankar 2004).

In addition, phytoremediation technology can be divided into two main broad categories depending on whether the removal of contaminants or stabilization of geochemical conditions in the soil is accomplished. Many different mechanisms for pollutant uptake, removal, or stabilization have been identified (Table 1.1) (Alvarez and Illman 2006; Peer et al. 2006).

In phytodegradation, organic pollutants are converted by internal or secreted enzymes into compounds with reduced toxicity. Like phytodegradation, rhizosphere degradation or rhizodegradation involves the enzymatic breakdown of organic pollutants, but through microbial enzymatic activity (Chandra et al. 2008; Chandra and Chowdhary 2015). These breakdowns of products are either volatilized or incorporated into the microorganisms and soil matrix of the rhizosphere. The types of plants growing in the contaminated area influence the amount, diversity, and activity of microbial populations;

**TABLE 1.1**
**Different Phytoremediation Approaches and Their Applications to Various Contaminated Sites**

| Techniques | Plant Mechanism | Contaminated Sites |
|---|---|---|
| Phytoextraction | Uptake and concentration of metals via direct uptake into the plant tissue with subsequent removal of the plants (hyperaccumulation) | Soil |
| Phytostabilization | Root exudates cause metal to precipitate and become less available (complexation) | Soil, groundwater, mine tailing |
| Phytotransformation | Plant uptake and degradation of organic and inorganic compounds | Surface water and groundwater |
| Rhizofiltration | Uptake of metals into plant roots (rhizosphere accumulation) | Organic and inorganic contaminated soil |
| Phytovolatilization | Plants evaporate and transpirate selenium, mercury, and volatile hydrocarbons | Soil and groundwater |
| Phytodegradation | Enhances microbial degradation in rhizosphere | Soil and groundwater within rhizosphere |

*Source*: Dixit, R., et al., *Sustainability,* 7, 2189–2212, 2015; Ghosh and Singh, *Appl. Ecoland. Environ. Res.*, 3(1), 1–18, 2005; Vidali, *Pure Appl. Chem.*, 73(7), 1163–1172, 2001; Yadav et al., *Handbook of Metal-Microbe Interactions and Bioremediation*, CRC Press, Boca Raton, FL. 2017.

thus, there is a direct independence between the planted vegetation and the dominant microbial species in the soil matrix of the rhizosphere. Phytoextraction involves the removal of toxins, especially heavy metals (e.g., Cd, Ni, Hg), metalloids (e.g., Se), and radionuclides, by the roots of the plants with subsequent transport to aerial plant organs. Plants can also remove toxic substances, such as organics, from the soil through phytovolatilization. In this process, the soluble contaminants are taken up with water by the roots, transported to the leaves, and volatized into the atmosphere through the stomata.

Rhizofiltration removes contaminants from water and aqueous waste streams, such as agricultural runoff, industrial discharges, and nuclear material processing wastes. Absorption and adsorption by plant roots play a key role in this mechanism, and consequently, large root surface areas are usually required. Finally, phytostabilization can be used to reduce the erosion, leaching, and mobilization of soil contaminants, which result in aerial or waterborne pollution of additional sites. In phytostabilization, accumulation by plant roots or precipitation in the soil by root exudates immobilizes and reduces the availability of soil contaminants. Plants growing on polluted sites also stabilize the soil and can serve as a groundcover, thereby reducing wind and water erosion and direct contact of the contaminants with animals (Karen et al. 2009). An example of a simple phytoremediation system in use for many years is the constructed wetland, in which aquatic plants such as water hyacinths are cultivated to remove contaminants (metals, nitrate, etc.) from municipal or industrial wastewater.

*1.3.1.4.3.1 Merits of Phytoremediation* The merits of phytoremediation include the following (Mitsch et al. 2012; Ligi et al. 2014):

- It is a low-cost (approximately 20%–30% of costs associated with mechanical treatments), solar energy–driven cleanup technique.
- It is publicly acceptable.
- It is most useful at sites with shallow, low levels of contamination.
- It is useful for treating a large range of environmental pollutants.
- It is more effective as compared to mechanical cleanup methods.
- It is applicable to many remediation scenarios including large contaminated surface areas.

*1.3.1.4.3.2 Demerits of Phytoremediation* Demerits of phytoremediation include the following (Nedunuri et al. 2000):

- Treatment is generally limited to shallow soils within 90 cm from the surface and groundwater within 3 m from the surface.
- A frequently cited disadvantage of phytoremediation, compared to strategies such as excavation and *ex situ* treatment, is that the rate of remediation is much slower.
- In the phytoremediation approach, there are stressors that affect this technique in the field that are not encountered in a laboratory or greenhouse. These include variations in temperature, nutrients, and precipitation; herbivores (insects and/or animals); plant pathogens; and competition by weed species that are better adapted to the site.

- Hazardous materials when present in high concentration can be toxic to plants.
- It involves the same mass transfer limitations as other biotreatment technologies.
- It is a long-time treatment approach.
- It may be seasonal, depending on location.
- Climatic or hydrologic conditions may restrict the growth rate of certain plants.
- Contaminants may enter the food chain via animals (herbivores) or insects that consume plant material–containing contaminants.
- Degradation products may be mobilized into groundwater or bioaccumulated in animals.
- It can transfer contamination across media (e.g., from soil to air).
- It is not effective for strongly sorbed (e.g., polychlorinated biphenyls [PCBs]) and weakly sorbed contaminants.
- Disposal of secondary waste arising from the harvest of plants is problematic.
- The toxicity and bioavailability of biodegradation products are not always known.
- It is still in the demonstration stage.
- It is unfamiliar to regulators.

### 1.3.2 *Ex Situ* Bioremediation

*Ex situ* bioremediation processes require the excavation of contaminated soil or pumping of groundwater to facilitate microbial degradation. This technique has more disadvantages than advantages (Megharaj et al. 2011; Latha and Reddy 2013).

#### 1.3.2.1 Landfarming

This is a solid-phase treatment system for contaminated soil where tilling and soil amendment techniques are used to encourage the growth of beneficial microorganisms in contaminated areas (Thassitou and Arvanitoyannis 2001). Different conditions that are controlled during landfarming are as follows:

- Moisture content (usually by irrigation or spraying)
- Aeration (by tilling the soil with a predetermined frequency)
- pH (buffered near neutral pH by adding crushed limestone or agricultural lime)
- Other amendments (e.g., soil bulking agents and nutrients)

#### 1.3.2.2 Composting

Composting is an aerobic, thermophilic treatment process in which contaminated material is mixed with a bulking agent (compost rich in bioremediation microorganisms). This is a controlled biological process by which organic contaminants (e.g., polycyclic aromatic hydrocarbons [PAHs]) are converted by microorganisms to safe, stabilized by-products. Typically, thermophilic conditions (54°C–65°C) must be maintained to properly compost soil contaminated with hazardous organic contaminants; in most cases, this is achieved by the use of indigenous microorganisms. Soils are excavated and mixed with bulking agents and organic amendments such as wood chips, animal and vegetative wastes, and so on, to enhance the porosity of the mixture to be decomposed. Maximum degradation efficiency is achieved by maintaining the aeration and moisture as necessary, and closely monitoring moisture content and temperature (Atlas 1981; Atlas 1991; Lee et al. 2006; Latha and Reddy 2013; Kang 2014).

Basically, three different processes are designed for composting:

- Aerated static pile composting where compost is formed into piles and aerated with blowers or vacuum pumps.
- Mechanically agitated in-vessel composting where compost is placed in a reactor vessel, mixed, and aerated.
- Windrow composting where compost is placed in long piles known as windrows and periodically mixed with mobile equipment.

Windrow composting is usually considered to be the most cost-effective composting alternative, but it may also have the highest fugitive emissions. Pilot and full-scale projects have demonstrated that aerobic, thermophilic composting is able to reduce the concentration of explosives (trinitrotoluene [TNT], Research Department Explosive [RDX], and High-velocity Military Explosive or High-Molecular-weight RDX [HMX], ammonium picrate (or yellow-D), and associated toxicity to acceptable levels and is also applicable to PAH-contaminated soil. The substantial requirements of space and aeration coupled with the need for excavation of contaminated soil limit the application of composting. If VOC or SVOC (semivolatile organic compound) contaminants are present in soils, off-gas control may be required. Additionally, this method cannot treat metals and due to the addition of amendments ultimately leads to volumetric increase in the amount.

#### 1.3.2.3 Biopiles

A biopile is one of the many bioremediation techniques used to treat hydrocarbon-contaminated soil where the soil is piled over an air distribution system and aerated. It is a hybrid of landfarming and composting in which the engineered cells are constructed as an aerated composted pile and typically used for the treatment of surfaces contaminated with petroleum hydrocarbons. It is a refined version of landfarming that tends to control the physical losses of contaminants by leaching and volatilization. Biopiles provide a favorable environment for the indigenous aerobic and anaerobic microorganisms to perform the bioremediation process at optimum level (Lebkowska et al. 2011; Abhilash et al. 2012; Latha and Reddy 2013; Dixit et al. 2015).

#### 1.3.2.4 Bioreactors

Bioreactors can be used to treat liquid effluents/slurries or contaminated solid waste/soil. Biological treatment in bioreactors offers the benefits of degradation under controlled parameters with a continuous monitoring system and is an *ex situ* method of bioremediation. Several types of bioreactor are available worldwide: batch, continuous, sequential batch biofilm, membrane, fluidized bed, biofilm, and airlift bioreactors, and so on. Moreover, bioremediation in reactors involves the processing of contaminated solid material (soil, sediment, sludge) or water through an engineered system. A slurry bioreactor may be defined as a containment vessel apparatus used to create a three-phase (solid, liquid, and gas) mixing condition to increase the bioremediation of soil-bound and water-soluble pollutants as a water slurry of the contaminated soil and biomass (usually indigenous microorganisms) capable of degrading target contaminants. In general, the rate and extent of biodegradation are greater in a bioreactor system than *in situ* or in solid-phase systems because the contained environment is more manageable and, hence, more controllable and predictable. The contaminated soil requires pretreatment (e.g., excavation) or alternatively the contaminant can be stripped from the soil via soil washing or physical extraction (e.g., vacuum extraction) before being placed in a bioreactor. These are employed in the treatment of a wide array of organics (Brar et al. 2006; Latha and Reddy 2013).

## 1.4 VARIOUS FACTORS AFFECTING THE BIOREMEDIATION PROCESS

Bioremediation is an essential process for sustainable development. The suitability of a particular bioremediation technology is determined by several factors, such as site conditions, microbial population, oxygen, water, nutrients availability, temperature, pH, and the type, quantity, and toxicity of pollutant chemical species present (Figure 1.2).

Two more factors are worth mentioning: co-metabolism and consortia conditions. Some biodegraders need other substrates to degrade pollutants (Caplan 1993; Boopathy 2000). This phenomenon is called co-metabolism and is especially required for organochlorine compounds. In contrast, it has been shown that the presence of other carbon sources decreases organophosphate biodegradation.

For *in situ* and *ex situ* bioremediation:

- Oxygen, water, and nutrients are added, and the temperature and pH are controlled.
- The rate microorganisms degrade the contaminants is influenced by the specific contaminants present, their concentrations, the oxygen supply, moisture, temperature, pH, nutrient supply, bioaugmentation, and co-metabolism.

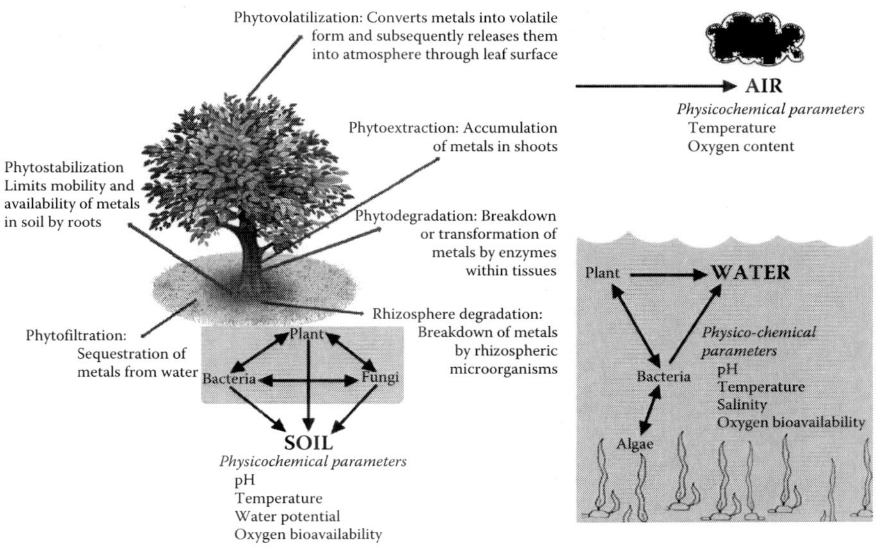


**FIGURE 1.2** Factors affecting biodegradation and bioremediation in soil, water, or air.

Main factors affecting *in situ* bioremediation of contaminated groundwater have been widely described in the literature and are shown in Table 1.2 (Boopathy 2000; McGuire et al. 2005; Farhadian et al. 2006; Andreoni and Gianfreda 2007).

There are some major points including:

- Source and concentration of pollutant
- Chemistry and toxicity of contamination
- Solubility, transport, adsorption, dispersion, and volatility of pollutant compounds
- Detection, determination, and monitoring of pollutants
- Chemistry and mechanics of soil at the contaminated site
- Hydrogeology and hydrology of the contaminated site
- Limitations of environmental standards for water and soil
- Environment conditions, nutrient sources, and the presence of electron acceptors
- Biodegradability of contaminants, and the presence of a competent biodegrading population of microorganisms
- Microorganisms that can be adapted to degrade specific contaminants or enhance the process

**TABLE 1.2**
**Factors Affecting the Bioremediation Process**

| Factors | Affected Functions |
|---|---|
| Microbial community | Growth until critical biomass is reached |
| | Mutation and horizontal gene transfer |
| | Enzyme induction |
| | Enrichment of the capable microbial populations |
| | Production of toxic metabolites |
| Environmental condition | Depletion of preferential substrates |
| | Lack of nutrients |
| | Inhibitory environmental conditions |
| Substrate | Too low concentration of contaminants |
| | Chemical structure of contaminant |
| | Toxicity of contaminants |
| | Solubility of contaminants |
| Biological, aerobic vs. anaerobic | Oxidation/reduction potential |
| | Availability of electron acceptors |
| | Microbial population present in the site |
| Growth substrate vs. co-metabolism | Type of contaminants |
| | Concentration |
| | Alternate carbon source present |
| | Microbial interaction (competition succession and predation) |
| Physicochemical bioavailability of pollutants | Equilibrium sorption |
| | Irreversible sorption |
| | Incorporation into humic matters |
| Mass transfer limitation | Oxygen diffusion of solubility |
| | Diffusion of nutrients |
| | Solubility/miscibility in/with water |

Moreover, microorganisms, through their enzymatic pathways, act as biocatalysts and facilitate the progress of biochemical reactions that degrade the targeted contaminants (Brar et al. 2006). As a result, bioremediation techniques are only applicable in environments that can sustain life. The microorganisms act against the contaminants only when they have access to a variety of materials—compounds to help them generate energy and nutrients to build more cells. In very few cases, the natural conditions at the contaminated site provide all the essential materials in large enough quantities that bioremediation can occur without human intervention—a process called "intrinsic bioremediation." More often, bioremediation requires the construction of engineered systems to supply microbe-stimulating materials—a process called "engineered bioremediation." Engineered bioremediation relies on accelerating the desired biodegradation reactions by encouraging the growth of more organisms, as well as by optimizing the environment, in which the organisms must carry out the detoxification reactions (Hatzikioseyian 2010).

## 1.5 HOW WE CAN ENHANCE BIOREMEDIATION

Various potential bioremediation processes are currently most popular for enhancing/improving the mechanism of remediation (Figure 1.3).

Bioremediation over the past two decades has been mainly gained from cultivable microbes that are at hand and have specific catalytic activities. However, much of the catabolic potential in nature remains undiscovered because our ability to culture microbes in the laboratory is less than 1% of all microbes present in the sample. Current genome-wide ("omics") technologies are being used to access hidden microbes without the need to cultivate them and with the aim of "rescuing" enzymatic activities (oxygenases, hydrolases, esterases) that are relevant for the degradation of pollutants in the environment.

### 1.5.1 Genetic Engineering in Bioremediation

The first genetically engineered microbe was created by Ananda Chakrabarty, an Indian-born microbiologist and genetic engineer, in 1971. The patent was approved in 1980 by the U.S. Supreme Court. The microbe was a variant of the genus *Pseudomonas* and was capable of breaking down the constituents of crude oil. Chakrabarty showed that four strains of the common *Pseudomonas* bacteria contained enzymes that enabled them to break down different hydrocarbons. He first determined that the genes for oil-degrading enzymes were carried not on the microorganism's chromosome, but rather on other extra-chromosomal elements known as "plasmids." He combined these plasmids into a strain of *Pseudomonas*. Unfortunately, due to regulations and public concerns of using the microbe for bioremediation, Chakrabarty's breakthrough microbe still sits on a shelf, unused. At the time, the new superbug created by Chakrabarty was said to have the potential to degrade oil 10–100 times faster than other nongenetically engineered independent strains.

In addition, in biotechnology and genetic engineering, the symbiotic relationship between bacteria and plants has been exploited for *in situ* bioremediation of a wide range of organic pollutants such as TCE (Kang 2014), toluene, PCBs,

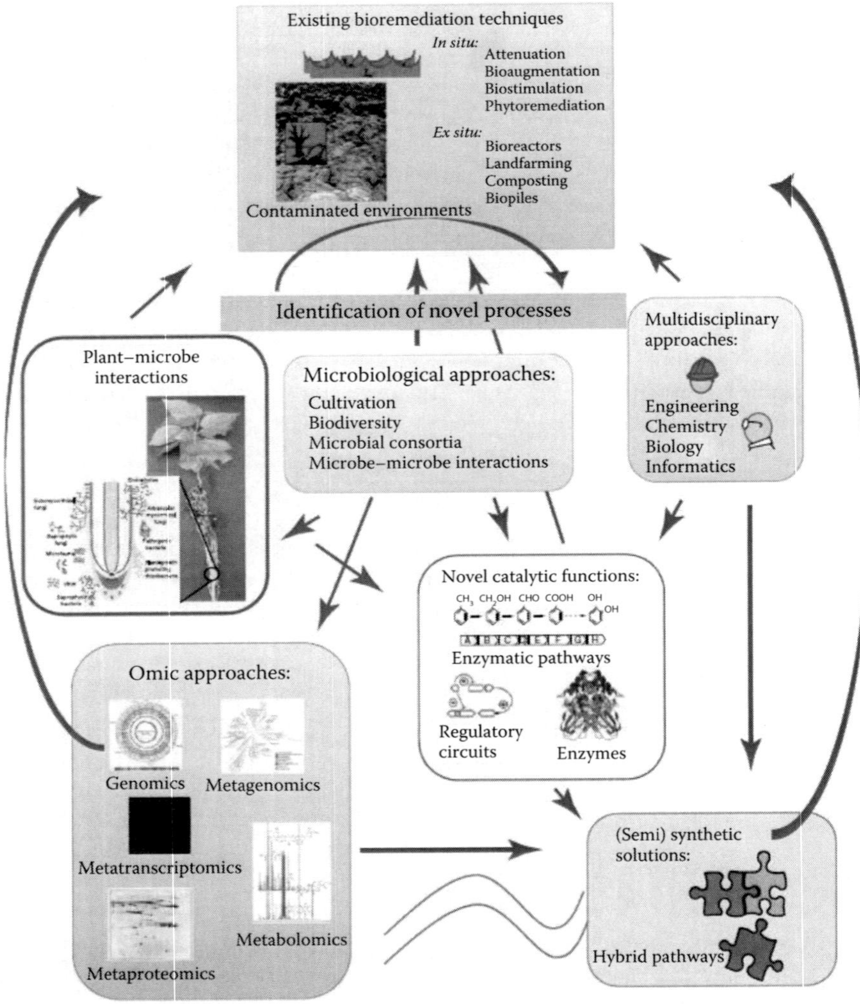


**FIGURE 1.3** New processes to improve/enhance the bioremediation of polluted sites. (From Ramos et al., *Trends in Biotechnology*, 29(12): 641–647, 2011. With permission.)

and so on. Further, scientists are currently looking into certain genetically engineered microorganisms to increase their ability to metabolize specific chemicals such as hydrocarbons and pesticides. The possibilities of using genetic engineering for improvement of bioremediation process had an early boost in the late 1980s. Recombinant DNA techniques have been studied intensively to improve the degradation of hazardous waste under laboratory conditions. The genetically engineered microorganisms have higher degradative capacity and have been demonstrated successfully for the degradation of various pollutants under defined conditions. Genetic modification technology has resulted often in a wide variety of current and potential applications for use in the process of bioremediation. Bioremediation explores gene diversity and metabolic versatility of microorganisms (Rittmann and McCarty 2001; Brar et al. 2006; Ezezika and Singer 2010).

The genetic architecture of these organisms makes them valuable in biodegradation, biotransformation, biosorption, and bioaccumulation. The necessary blueprint of genes encoding for the biodegradative enzymes is present in chromosomal and extra-chromosomal DNA of such microbes. Recombinant DNA techniques facilitate evolution of the ability of an organism to metabolize a xenobiotic by detecting such degradative genes and transforming them into appropriate hosts via suitable vectors under the tight control of appropriate promoters. This depends on their susceptibility to alteration and exchange of genetic information. The recombinant DNA technology explores polymerase chain reaction (PCR), anti-sense RNA technique, site-directed mutagenesis, electroporation, and particle bombardment techniques. The biotechnology armed with recombinant DNA technology is now fine-tuning the bioremediation technology by improving pollutant-degrading microbes through strain improvement and genetic modification of specific regulatory and metabolic genes that are crucial in developing effective, safe, and economical techniques for bioremediation (Iwamoto and Nasu 2001; Ezezika and Singer 2010).

Microbiological and ecological knowledge, biochemical mechanisms, and field engineering designs are essential elements for successful *in situ* bioremediation using genetically modified bacteria and microalgae (Brar et al. 2006). Recently, genetic engineering has also been used to improve the bioremediation of organic pollutants. The transgenic approach for engineering the plants either with bacterial or animal xenobiotic degrading genes has been successfully tried for *in situ* bioremediation and accumulation of heavy metal as well as degradation of various xenobiotics like explosives and hydrocarbons (Suresh and Ravishankar 2004; Ezezika and Singer 2010).

Transgenic plants and associated bacteria constitute a new generation of genetically modified organisms for efficient and environment-friendly treatment of soil and water contaminated with organic pollutants like PCBs. Genetic engineering can create genetically modified organisms, which can potentially degrade various types of PAHs, and the modern biotechnology could be applied for the removal of various toxic compounds. Genetically modified organisms have been developed to degrade or modify many different compounds including carbazole, a petroleum component that inhibits catalysts used in refining, pesticides, explosives, and aromatic compounds.

Mixotrophy in cyanobacteria and microalgae can provide many competitive advantages over bacteria and fungi in degrading organic pollutants, and molecular methods and metabolic and genomic information will help not only in identification and selection of mixotrophic species of cyanobacteria and microalgae with capabilities to degrade organic pollutants, but also in monitoring the efficiency of remediation efforts under field conditions. Scientists reported that microalgae are capable of producing lipids and hydrocarbons quickly, and their photosynthetic abilities make them promising candidates for wastewater treatment (bioremediation) and as an alternative energy source (biodiesel). Some microorganisms could be used in the biomonitoring of organic pollution by constructing these organisms using bioluminescence genes.

Expression of the catabolic genes of PCB-degrading microorganisms is a key factor for PCB biodegradation in contaminated soils. Transgenic plants expressing the bacterial xenobiotic degradation genes combine the advantages of both systems (i.e., the greater ability of biodegradation by bacteria and the high biomass and stability of the plants to have an ideal system for *in situ* bioremediation of contaminants). However, there are strict regulations for the introduction of genetically engineered organisms in the environment.

### 1.5.2 Role of Nano-Biotechnology in Bioremediation

Nanoparticles or nanomaterials are found to be very effective for the removal of a wide range of toxic metals from the environment as compared to conventional methods. A nanoparticles-enhancing microbial activity to remove toxic pollutants is called "nano-bioremediation." Nano-based technologies not only reduce the costs of cleaning up contaminated sites at a large scale but also reduce the process time as well.

However, bio-nanotechnology or nanotechnology through biotechnology is the biofabrication of nano-objects or bifunctional macromolecules used as tools to construct or manipulate nano-objects. Wide physiological diversity, small size, genetic manipulability, and controlled culturability make microbial cells ideal producers of nanostructures ranging from natural products, such as polymers and magnetosomes, to engineered proteins or protein constructs, such as virus-like proteins and tailored metal particles (Sarikaya et al. 2003). *Deinococcus radiodurans*, a radioactive-resistant organism, has the ability to withstand radiation well beyond the naturally occurring levels, and thus its application in radioactive waste cleanup initiatives funded by the U.S. Department of Energy (DOE) (Brim et al. 2000). Metal chelating polymers require toxic solvents for their synthesis and ultrafiltration for their separation, and this can be solved by developing metal binding materials that can be recovered by changing the environment surrounding them, including pH and temperature. One such material is nanoscale-modified biopolymers that are produced by the genetic and protein engineering of microorganisms, and their size can be controlled at the molecular level. This innovative technique would be a promising tool to address the escalating problem of heavy metals as well as organic contaminants in the environment.

## 1.6 MERITS AND DEMERITS OF BIOREMEDIATION

There are several merits and demerits compared to conventional techniques for removing the hazardous pollutants from the environment (Table 1.3):

- Lower cost for implementation.
- No residual treatment is required (e.g., microorganisms [in biological treatment] completely breakdown [biodegrade] the contaminants, whereas in physical/chemical treatment, the contaminants are often just separated, e.g., not actually remediating the contamination).

### 1.6.1 Advantages and Disadvantages of *In Situ* Bioremediation

*In situ* bioremediation has many advantages and disadvantages (Boopathy 2000; Farhadian et al. 2008).

**TABLE 1.3**
**Merits and Demerits of Various Types of Bioremediation Techniques**

| Bioremediation Strategies | Example | Merits | Demerits |
|---|---|---|---|
| *In situ* | *In situ* bioremediation<br>Bioventing<br>Biostimulation<br>Bioaugmentation | Most cost-efficient<br>Noninvasive<br>Relatively passive<br>Natural attenuation<br>Treats soil and water | Environmental constraints<br>Extended treatment times<br>Monitoring difficulties |
| *Ex situ* | Land farming<br>Composting<br>Biopiles/biocells | Cost-efficient<br>Can be done on-site | Extended treatment time<br>Need to control abiotic loss<br>Mass transfer problem<br>Bioavailability limitation |
| Bioreactors | Slurry/aqueous reactors | Rapid degradation kinetics<br>Optimized environmental parameters<br>Enhances mass transfer<br>Effective use of inoculants and surfactants | Requires excavation<br>Relatively high capital cost and operating cost |

#### 1.6.1.1 Advantages

- It can be done on-site.
- Transportation costs can be reduced.
- Waste can be removed permanently.
- Site disruption can be minimized.
- It is applicable to diluted and widely diffused contaminants.
- It is affordable.
- There is no need to excavate and transport soils (i.e., typically less expensive).
- It can treat a large volume of soil at once.
- It causes less contaminants to be released than *ex situ* techniques.
- Less dust is generated.
- It is most effective if there is permeable sandy soil (uncompacted).
- It is least effective in clays/highly layered subsurface environments because oxygen cannot be evenly distributed throughout the treatment area.
- It may take longer to reach the cleanup goal (if less easily degradable contaminant, requires years).
- It may be more difficult to manage than *ex situ* techniques.

#### 1.6.1.2 Disadvantages

*In situ* bioremediation takes more time compared to the other remediation methods, there is seasonal variation of the microbial activity due to direct exposure to changes in environmental factors that cannot be controlled, and application of treatment additives may be problematic. Microorganisms act well only when the waste materials present allow them to produce nutrients and energy for the development of more cells. When these conditions are not favorable, then their capacity to degrade is reduced. In such cases, genetically engineered microorganisms have to be used, although stimulating indigenous microorganisms is preferred (Farhadian et al. 2008).

### 1.6.2 Advantages and Disadvantages of *Ex Situ* Bioremediation

- *Ex situ* techniques can be faster, easier to control, and used to treat a wider range of contaminants and soil types than *in situ* techniques.
- There is more certainty about the uniformity of treatment because of the ability to homogenize, screen, and continuously mix the soil.
- There is more risk of material handling/worker exposure conditions.
- It usually requires treatment of the contaminated soil before and, sometimes, after the actual bioremediation step.
- Solid-phase soil treatment processes include landfarming, soil biopiles, and composting.
- Slurry-phase soil treatment processes include the slurry-phase bioreactor.

#### 1.6.2.1 Solid-Phase Treatment

- This treats soils in an aboveground treatment area.
- This type of system is equipped with collection systems to prevent any contaminant from escaping the treatment.
- Moisture, heat, nutrients, and oxygen are controlled to enhance biodegradation rates.
- Solid-phase systems are relatively simple to operate and maintain but require a large amount of space and require a longer time of treatment than slurry-phase processes.

#### 1.6.2.2 Slurry-Phase Treatment

- Contaminated soil is combined with water and other additives in a large bioreactor and mixed to keep the indigenous (native) microorganisms in contact with the contaminants.
- Nutrients and oxygen are added and the conditions in the bioreactor are controlled (to ensure the optimum environment for the microorganisms to degrade the contaminants.
- Slurry-phase treatment is a relatively rapid process (compared to other biological treatment processes), particularly for contaminated clays.

## 1.7 CONCLUSION

- It is well known that bioremediation is an emerging and interdisciplinary technology that involves knowledge of chemical and biochemical engineering, ecology, statistics, microbiology, chemical and biochemical properties, and geology. It is considered to be still in development.
- Bioremediation approaches mainly based on the principles of biostimulation and bioaugmentation include bioventing, landfarming, bioreactors, and composting.

- Regulations that limit the disposal of chemicals and escalation in the costs of physical and chemical treatments make bioremediation technologies more attractive.
- Several factors that limit the efficiency of microbial degradation of organic pollutants are numerous. Besides the bioavailability of the pollutant itself, low temperature, anaerobic conditions, low levels of nutrients and co-substrates, the presence of toxic substances, and the physiological potential of microorganisms are particularly important in the polluted sites (Romantschuk et al. 2000).
- The bioremediation process is influenced by various factors—existence of a specific microbial population, bioavailability of contaminants, and environmental factors (soil type, temperature, pH, nutrients, and presence of oxygen or other electron acceptors).
- The choice of methods in each technology requires careful consideration.
- The reluctance among the public to accept genetically modified organisms for bioremediation also needs to be considered in future studies, and they must prove nontoxic to the environment.
- Bioremediation technologies have been successfully employed in the field and are gaining more and more importance with increased acceptance of eco-friendly remediation solutions.

## ACKNOWLEDGMENTS

The Rajiv Gandhi National Fellowship (Letter No. RGNF 2015-17SC-UTT-20334) awarded to Mr. Pankaj Chowdhary for PhD work from University Grant Commission (UGC), Government of India (GOI), New Delhi, is duly acknowledged.

## REFERENCES

Abhilash, P.C., Powell, J.R., Singh, H.B., and Singh, B.K. 2012. Plant-microbe interactions: Novel applications for exploitation in multipurpose remediation technologies. *Trends in Biotechnology* 30: 416–420.

Aggarwal, P.K., Means, J.L., Hinchee, R.E., Headington, G.L., and Gavaskar, A.R. 1990. *Methods to select chemicals for in-situ biodegradation of fuel hydrocarbons*, Tyndall AFB, Air Force Engineering and Services Center, Panama City, FL.

Al-Sulaimani, H., Al-Wahaibi, Y.M., Al-Bahry, S., and Zaragari, S. 2010. Experimental investigation of biosurfactants produced by Bacillus species and their potential for MEOR in Omani oil field. SPE EOR Conference at Oil & Gas West Asia, 11–13 April, Muscat, Oman. doi.org/10.2118/129228-MS.

Alvarez, P.J.J., and Illman, W.A. 2006. Bioremediation and natural attenuation: Process fundamentals and mathematical models, Wiley.

Andreoni, V., and Gianfreda, L., 2007. Bioremediation and monitoring of aromatic-polluted habitats. *Applied Microbiology Biotechnology* 76: 287–308.

Antizar-Ladislao, B. 2010. Bioremediation: Working with bacteria. *Elements* 6: 389–94.

Atlas, R.M. 1981. Microbial degradation of petroleum hydrocarbons: An environmental perspective. *Microbiological Reviews* 45: 180.

Atlas, R.M. 1991. Microbial hydrocarbon degradation-bioremediation of oil spills. *Journal of Chemical Technology and Biotechnology* 52: 149.

Boopathy, R. 2000. Factors limiting bioremediation technologies. *Bioresource Technology* 74: 63.

Brar, S.K., Verma, M., Surampalli, R.Y., Misra, K., Tyagi, R.D., Meunier, N., and Blais, J.F. 2006. Bioremediation of hazardous wastes—A review. *Practice Periodical of Hazardous, Toxic, and Radioactive Waste Management* 10: 59–72.

Brim, H., McFarlan, S.C., Fredrickson, J.K., Minton, K.W., Zhai, M., Wackett, L.P., and Daly, M.J. 2000. Engineering *Deinococcus radiodurans* for metal remediation in radioactive mixed waste environments. *Nature Biotechnology* 18: 85–90.

Caplan, J.A. 1993. The worldwide bioremediation industry: Prospects for profit. *TIB Tech* 11: 320–323.

Chandra, R., Bharagava, R.N., and Rai, V. 2008. Melanoidins as major colourant in sugarcane molasses based distillery effluent and its degradation. *Bioresource Technology* 99: 4648–4660.

Chandra, R., and Chowdhary, P. 2015. Properties of bacterial laccases and their application in bioremediation of industrial wastes. *Environmental Science Processes and Impacts* 17: 326–342.

Collins, C., Laturnus, F., and Nepovim, A. 2002. Remediation of BTEX and trichloroethene. *Environmental Science Pollution and Research International* 9: 86–94.

Dejonghe, W., Goris, J., Fantroussi, S.E., Hofte, M., De Vos, P., Verstraete, W., and Top, E.M. 2000. Effect of dissemination of 2,4- dichlorophenoxyacetic acid (2,4-D) degradation plasmids on 2,4-D degradation and on bacterial community structure in two different soil horizons. *Applied and Environmental Microbiology* 66: 3297–3304.

Dixit, R., Wasiullah, Malaviya, D., et al. 2015. Bioremediation of heavy metals from soil and aquatic environment: An overview of principles and criteria of fundamental processes. *Sustainability* 7: 2189–2212.

Ezezika, O.C., and Singer, P.A. 2010. Genetically engineered oil-eating microbes for bioremediation: Prospects and regulatory challenges. *Technology in Society* 32: 331–335.

Farhadian, M., Vachelard, C., Duchez, D., and Larroche, C. 2008. In situ bioremediation of monoaromatic pollutants in groundwater: A review. *Bioresource Technology* 99: 5296–5308.

Farhadian, M., Larroche, C., Borghei, M., Troquet, J., and Vachelard, C. 2006. Bioremediation of BTEX-contaminated groundwater through bioreactors, 4me colloque Franco-Roumain de chimie appliquee, Université Blaise Pascal, Clermont-Ferrand, France, 28 June–2 July, p. 438.

Ghosh, M., and Singh, S.P. 2005. A review on phytoremediation of heavy metals and utilization of its byproducts. *Appl. Ecoland. Environ. Res.* 3(1): 1–18.

Hamzah, A., Chia-Wei, P., Pek-Hoon, Y., and Nurul, H. 2014. Oil palm empty fruit bunch and sugarcane bagasse enhance the bioremediation of soil artificially polluted by crude oil. Soil and Sediment Contamination: *An International Journal* 23(7): 751–762.

Hardisty, P.E., and Ozdemiroglu, E. 2005. *The economics of groundwater remediation and Protection*, CRC Press, Boca Raton, FL.

Hatzikioseyian, A. 2010. Principles of bioremediation processes. In: Grażyna Płaza (ed), Trends in Bioremediation and Phytoremediation. Research Signpost: Trivandrum, India, 23–54.

Iwamoto, T., and Nasu, M. 2001. Current bioremediation practice and perspective. *Journal of Bioscience and Bioengineering* 92(1): 1–8.

James, C.A., and Strand, S.E. 2009. Phytoremediation of small organic contaminants using transgenic plants. *Current Opinion in Biotechnology* 20: 237–241.

Jorgensen, K.S. 2007. In situ bioremediation. *Advanced Applied Microbiology* 61: 285–305.

Kaiser, J. 2000. Just how bad is dioxin? *Science* 288: 1941–4.

Kang, J.W. 2014. Removing environmental organic pollutants with bioremediation and phytoremediation. *Biotechnology Letter* 36: 1129–1139.

Karen, E.G., Xiao-Dong, H., Bernard, R.G., and Bruce, M.G. 2009. Phytoremediation and rhizoremediation of organic soil contaminants: Potential and challenges. *Plant Science* 176: 20–30.

Kumar, A., Bisht, B.S., Joshi, V.D., and Dhewa, T. 2011. Review on bioremediation of polluted environment: A management tool. *International Journal of Environmental Sciences* 1: 1079.

Latha, A.P., and Reddy, S.S. 2013. Review on bioremediation—Potential tool for removing environmental pollution. *International Journal of Basic and Applied Chemical Sciences* 3(3): 21–33.

Lebkowska, M., Ewa, Z., Ewa, K, Ewa, M., Muszynski, A., Tabernacka, A., Naumczyk, J., and Jeczalikc, M. 2011. Bioremediation of soil polluted with fuels by sequential multiple injection of native microorganisms: Field-scale processes in Poland. *Ecological Engineering* 37: 1895–1900.

Lee,T.H., Byun, I.G., Kim, Y.O., Hwang, I.S., and Park, T.J. 2006. Monitoring biodegradation of diesel fuel in Bioventing processes using in situ respiration rate. *Water Science and Technology* 53(4–5): 263–72.

Ligi, T., Oopkaup, K., Truu, M., et al. 2014. Characterisation of bacterial communities in soil and sediment of a created riverine wetland complex using high-throughput 16S rRNA amplicon sequencing. *Ecological Engineering* 72: 56–66.

McGuire, J.T., Long, D.T., and Hyndman, D.W., 2005. Analysis of recharge induced geochemical change in a contaminated aquifer. *Ground Water* 43: 518–530.

Megharaj, M., Ramakrishnan, B., Sethunathan, K.V.N., and Naidu, R. 2011. Bioremediation approaches for organic pollutants: A critical perspective. *Environment International* 37: 1362–1375.

Mitsch, W.J., Zhang, L., Stefanik, K.C., et al. 2012. Creating wetlands: Primary succession, water quality changes, and self-design over 15 years. *Bio Science* 62(3): 237–50.

Mulligan, C.N., and Yong, R.N. 2004. Natural attenuation of contaminated soils. *Environment International* 30: 587–601.

Nedunuri, K.V., Govinda Raju, R.S., Banks, M.K., Schwab, A.P., and Chen, Z. 2000. Evaluation of phytoremediation for field-scale degradation of total petroleum hydrocarbons. *Journal of Environmental Engineering* 126: 483–490.

Neilson, A.H., and Allard, A.S. 2008. *Environmental degradation and transformation of organic chemicals*, CRC Press, Boca Raton, FL.

Orji, F.A., Abiye, A.I., and Dike, E.N. 2012. Laboratory scale bioremediation of petroleum hydrocarbon-polluted mangrove swamps in the Niger Delta using cow dung. *Malaysian Journal of Microbiology* 8(4): 219–228.

Peer, W.A., Baxter, I.R., Richards, E.L., Freeman, J.L., and Murphy, A.S. 2006. Phytoremediation and hyperaccumulator plants. In *Molecular biology of metal homeostasis and detoxification: From microbes to man*, Tamas, M.J., and Martinoia, E. (eds.), Springer, Berlin Heidelberg, pp. 299–341.

Ramos, J.L., Marques, S., Dillewijn, P. et al. 2011. Laboratory research aimed at closing the gaps in microbial bioremediation. *Trends in Biotechnology* 29(12): 641–647.

Rittmann, B.E., and McCarty, P.L. 2001. *Environmental biotechnology: Principles and applications*, McGraw Hill, New York.

Romantschuk, M., Sarand, I., Petanen, T., Peltola, R., Jonsson-Vihanne, M., Koivula, T., Yrjala, K., and Haahtela, K. 2000. Means to improve the effect of in situ bioremediation of contaminated soil: An overview of novel approaches. *Environmental Pollution* 107(2): 179–185.

Salt, D.E., Smith, R.D., and Raskin, I. 1998. Phytoremediation. *Annual Review of Plant Physiology and Plant Molecular Biology* 49: 643–68.

Sarikaya, M., Tamerler, C., Jen, A.K., Schulten, K., and Baneyx, F. 2003. Molecular biomimetics: Nanotechnology through biology. *Nature Materials* 2: 577–585.

Saxena, G., and Bharagava, R.N. 2015. Persistent organic pollutants and bacterial communities present during the treatment of tannery wastewater. In *Environmental waste management* Chandra, R. (ed.), CRC Press, Boca Raton, FL. pp. 217–247.

Schnoor, J.L., Licht, L.A., McCutcheon, S.C., Wolfe, N.L., and Carreira, L.H. 1995. Phytoremediation of organic and nutrient contaminants. *Environmental Science and Technology* 29: 318–23.

Soccol, C.R., Vandenberghe, L.P., Woiciechowski, A.L., Thomaz-Soccol, V., Correia, C.T., and Pandey, A. 2003. Bioremediation: An important alternative for soil and industrial wastes clean-up. *Indian Journal of Experimental Biology* 41(9): 1030–45.

Strong, P.J., and Burgess, J.E. 2008. Treatment methods for wine related ad distillery wastewaters: A review. *Bioremediation Journal* 12: 7087.

Suresh, B., and Ravishankar, G.A. 2004. Phytoremediation—A novel and promising approach for environmental clean-up. *Critical Reviews in Biotechnology* 24: 97–124.

Susarla, S., Medina, V.F., and McCutcheon, S.C. 2002. Phytoremediation: An ecological solution to organic chemical contamination. *Ecological Engineering* 18: 647–58.

Thakur, I.S. 2004. Screening and identification of microbial strains for removal of color and adsorbable organic halogens in pulp and paper mill effluent. *Process Biochemistry* 39: 1693–1699.

Thassitou, P.K., and Arvanitoyannis, I.S. 2001. Bioremediation: A novel approach to food waste management. *Trends Food Science and Technology* 12: 185.

Vidali, M. 2001. Bioremediation. An overview. *Pure Appl. Chem.* 73(7): 1163–1172.

Yadav, A., Chowdhary, P., Kaithwas, G., and Bharagava, R.N. 2017. Toxic Metals in the Environment Threats on Ecosystem and Bioremediation Approaches. In *Handbook of Metal-Microbe Interactions and Bioremediation*, Das, S and Dash, H.R. (eds.), CRC Press, Boca Raton, FL. pp. 127–141.

# 2 Organic and Inorganic Pollutants in Industrial Wastes

## *Ecotoxicological Effects, Health Hazards, and Bioremediation Approaches*

*Gaurav Saxena and Ram Naresh Bharagava*

**CONTENTS**

2.1 Introduction ........ 24
2.2 Industrial Wastes: Nature and Characteristics ........ 25
2.3 Environmental Pollutants and Their Toxicity in Environment ........ 27
2.3.1 Organic Pollutants ........ 27
2.3.1.1 Phenol Toxicity ........ 28
2.3.1.2 Endocrine Disrupting Chemicals and Toxicity ........ 28
2.3.1.3 Chlorinated Phenols and Toxicity ........ 29
2.3.1.4 Azo Dye and Toxicity ........ 30
2.3.1.5 Petroleum Hydrocarbons and Toxicity ........ 31
2.3.1.6 Melanoidins and Toxicity ........ 32
2.3.1.7 Pesticides and Toxicity ........ 32
2.3.1.8 Persistent Organic Pollutants and Their Toxicity ........ 33
2.3.2 Inorganic Pollutants ........ 34
2.3.2.1 Cadmium and Its Toxicity ........ 34
2.3.2.2 Chromium and Its Toxicity ........ 35
2.3.2.3 Arsenic and Its Toxicity ........ 36
2.3.2.4 Lead and Its Toxicity ........ 36
2.3.2.5 Mercury and Its Toxicity ........ 37
2.4 Bioremediation Approaches for Environmental Pollutants ........ 37
2.4.1 Composting ........ 37
2.4.2 Bioremediation ........ 38
2.4.3 Phytoremediation ........ 38
2.4.4 Bacterial-Assisted Phytoremediation ........ 41
2.4.5 Electro-Bioremediation ........ 41

2.5 Future Perspectives and Challenges ..... 44
2.6 Conclusion ..... 44
Acknowledgments ..... 44
References ..... 45

## 2.1 INTRODUCTION

Rapid industrialization and urbanization around the world has led to the recognition and understanding of the relationship between environmental contamination and public health. Industries are the key players in the national economies of many developing countries; however, unfortunately they are also the major polluters of the environment. Among the different sources of environmental pollution, industrial wastewater discharged from different industries is considered the major source of environmental pollution (soil and water). Industries use a variety of chemicals for the processing of raw materials to obtain good-quality products within a short period of time and economically. To obtain good-quality products within a short period of time, industries generally use cheap and poorly or nonbiodegradable chemicals, and their toxicity is usually ignored. However, in the public domain there are many reports available that confirm the presence of a variety of highly toxic chemicals in industrial wastewaters.

Industrial wastewaters contain a variety of organic and inorganic pollutants that cause serious environmental pollution and health hazards (Maszenan et al. 2011; Megharaj et al. 2011). During production processes, a variety of chemicals with large volumes of water are used to process raw materials in industries. This generates large volumes of high-strength wastewater, which is a major source of environmental pollution (Saxena et al. 2016). The wastewater generated from pollution-causing industries is characterized by high chemical oxygen demand (COD), biological oxygen demand (BOD), total dissolved solids (TDSs), total suspended solids (TSSs), and a variety of recalcitrant organic and inorganic pollutants. Organic pollutants include phenols, chlorinated phenols, endocrine disrupting chemicals (EDCs), azo dyes, polyaromatic hydrocarbons, polychlorinated biphenyls, and pesticides, whereas inorganic pollutants include a variety of toxic heavy metals such as cadmium (Cd), Chromium (Cr), arsenic (As), lead (Pb), and mercury (Hg).

The high concentration and poor biodegradability of recalcitrant organic pollutants and nonbiodegradable nature of inorganic metal pollutants in industrial wastewaters pose a major challenge for environmental safety and human health protection; thus, it is required to adequately treat industrial wastewater before its final disposal in the environment. Although a number of physicochemical methods are applied for the treatment of industrial wastewaters, all of these are costly, use a large amount of chemicals, and generate a large amount of sludge after treatment, which also acts as a secondary pollutant in the environment. Alternatively, biological treatment methods using an array of microorganisms have diverse metabolic pathways and, hence, are regarded as environmentally friendly, cost-effective methods for wastewater treatment with simple structural set-up, wider application, operational ease, and less sludge production compared to physicochemical methods (Mendez-Paz et al. 2005; Pandey et al. 2007). Biological methods using microbes are becoming much more popular for the treatment of industrial wastewaters in wastewater treatment plants.

Further, most chemical compounds are degraded by acclimated microorganisms during wastewater treatment at wastewater treatment plants; however, some of the chemicals compounds are not properly degraded/detoxified due to their recalcitrant nature during wastewater treatment and are discharged along with wastewaters, causing serious environmental pollution (Maszenan et al. 2011). Hence, the application of bioremediation technology using potential microorganisms and their consortia or of phytoremediation technology (use of green plants in constructed wetlands) is required for the degradation and detoxification of such types of recalcitrant industrial wastewaters prior to safe disposal in the environment. Therefore, this chapter provides a detailed account of the nature and characteristics of different industrial wastewaters, the toxicities of their organic and inorganic pollutants in living beings, and bioremediation approaches for their degradation and detoxification for environmental safety and human health protection.

## 2.2 INDUSTRIAL WASTES: NATURE AND CHARACTERISTICS

Industrial waste is the waste produced due to industrial activity and includes any material that is rendered useless during a product manufacturing process in industries. Industrial waste is of two types: solid waste and liquid waste. Liquid waste (i.e., wastewaters released from different industries) has become a major environmental concern worldwide because of its potentially toxic and hazardous nature to living beings. Many types of industrial wastewaters are produced based on different industries, such as distilleries, tanneries, textile, pulp and paper, pharmaceutical, electroplating, and so on. However, the nature and characteristics of industrial wastewater vary greatly from industry to industry, mainly depending on the production process adopted, raw materials used, and quality of product produced. Industries often discharge high-strength wastewaters that have potential to cause serious environmental hazards due to high BOD, COD, TDSs, and TSSs, and because they contain a variety of organic and inorganic pollutants (such as phenols, melanoidins, EDCs, azo dyes, polyaromatic hydrocarbons, polychlorinated biphenyls, etc.) and various toxic metals (such as cadmium [Cd], chromium [Cr], arsenic [As], mercury [Hg], etc.). Further, industrial wastewaters are mainly characterized by various physicochemical parameters and biological characteristics. Physicochemical parameters reflect the quality of wastewaters and are mainly used to define the pollution strength of wastewater. Physicochemical parameters mainly include color, pH, conductivity, temperature, BOD, COD, TDSs, TSSs, chloride, sulfate, phosphate, nitrate, total nitrogen, phenolics, and a variety of toxic metals. Biological characteristics include microbiological estimation of pathogenic and nonpathogenic bacteria, including aerobic and anaerobic bacteria, fungi, yeasts, actinomycetes, and viruses. The nature and characteristics of various types of wastewater are presented in Table 2.1.

Further, there has been increasing concern regarding the use of hazardous chemicals in industrial processes. From environmental safety and human health protection points of view, many regulations have been passed to ban the use of hazardous chemicals in industrial processes, such as the Integrated Pollution Prevention and Control Directive (96/61/EC 1996; 2008/1/EC 2008). Directive (REACH) EC 1907/2006 for European Regulatory Framework on chemicals, namely Registration, Evaluation,

**TABLE 2.1**
**Nature and Characteristics of Various Types of Industrial Wastewater**

| Type of Industrial Wastewater | Nature and Characteristics |
|---|---|
| Oil refinery wastewater | Poor biodegradability and contains oil, various aromatic hydrocarbons (highly toxic to natural environment) such as benzene, toluene, ethylbenzene and xylene (BTEX), persistent organic pollutants (POPs) such as dioxins (highly toxic in nature), metals/metalloids, phenolics, salts, and surfactants |
| Pulp and paper mill wastewater | Highly intense dark brown color, BOD, suspended solids (SS) and contains recalcitrant dioxins, furans, lignins, adsorbable organic halides (AOX), phenolic and chlorophenolic compounds, especially pentachlorophenol (highly toxic to living beings and hazardous to environment) |
| Textile wastewater | Alkaline in nature, highly colored, and often contains harmful residual dyes such as acidic, basic, reactive, disperse, azo, diazo, anthraquinone-based and metal complexes dyes (some carcinogenic in nature) |
| Tannery wastewater | Contains high organic loadings (BOD, COD, and TSS), salts (sodium, chloride, and sulfide), phenolic compounds, endocrine disrupting chemicals such as nonylphenols and phthalates, and other toxic metals, especially chromium (highly toxic proven carcinogen) |
| Distillery wastewater (spent wash) | High organic loading (BOD, COD, and TDS), contains phenolics and dark black color recalcitrant melanoidins (stop penetration of sunlight into water bodies and hence reduce photosynthesis) |
| Winery wastewater | Acidic in nature, variable flows and loadings, contains high content of organic matter, COD, and TSS and organic fraction consist of sugars, alcohols, acids, and high molecular weight recalcitrant compounds such as polyphenols, tannins, and lignins |
| Pharmaceutical wastewater | Acidic in nature, has high COD and TDS, and contains many organic solvents, formulations, disinfectants, and many generic drugs such as antibiotics, analgesics, etc. |
| Abattoir (slaughterhouse) wastewater | Contains high levels of organics (COD is mainly in colloidal form) and coarse suspended matter and heavy metals, nutrients, pathogenic and nonpathogenic microorganisms, detergents and disinfectants, and sometime pharmaceutical agents used for veterinary purposes |
| Agricultural wastewater | Alkaline in nature and contains high content of nitrogen, phosphorous, pesticides, and various toxic metals such as cadmium, lead, arsenic, etc. |
| Landfill leachate | Composition varies from landfill to landfill, generally colored, anoxic, and have high TDS, COD, BOD, and contains ammonia, phenols, benzene, toluene, chloride, iron, manganese, and other toxic metals such as lead, cadmium, zinc, arsenic, or chromium but little or no phosphorus |
| Acid mine drainage | Acidic in nature, contains high concentrations of iron, sulfate, copper, nickel, and toxic metals such as cadmium, lead, etc. (cause environmental damage) |

Note: BOD, biological oxygen demand; COD, chemical oxygen demand; TDS, total dissolved solid; TSS, total suspended solid.

Authorization and Restriction of Chemical substances, directed leather tanneries to restrict the use of those leather auxiliaries and basic chemicals that are not registered and listed in the Safety Data Sheet (Lofrano et al. 2013). Moreover, Directive 2003/53/EC restricted the marketing and use of products/product formulations containing more than 0.1% of nonyl ethoxyphenol (NPE) or nonylphenol (NP) and their use in making leather products in Europe (Lofrano et al. 2008). In addition, Directive 1999/815/EC has directed industries to label products if they contain more than 0.5% phthalates (benzyl butyl phthalate (BBP), dibutyl phthalate (DBP), and diethyl hexyl phthalate) due to the reproductive toxic potential of the phthalates (EU 2003). The use of o-phenyl phenol is restricted for leather finishing due to its carcinogenic potential (EPA 2007), and the use of formaldehyde (a cross-linker for casein top coats) is restricted due to its carcinogenic potential (EU 1998). Inorganic compounds such as cadmium sulfate and lead chromate (fastening agents) are also considered highly toxic in nature (IARC 2004; ATSDR 2008). Further, the EU Azo Colorants Directive (2002) has prioritized several azo dyes and restricted their use in leather tanneries due to higher toxicity. However, many chemicals that are considered to be hazardous for environment and human beings are still being used.

## 2.3 ENVIRONMENTAL POLLUTANTS AND THEIR TOXICITY IN ENVIRONMENT

Wastewaters discharged from different industries are considered the major sources of pollution in the environment. A variety of highly toxic and recalcitrant pollutants are being discharged along with industrial wastewaters in the environment after industrial operations. Environmental pollutants are of two types: organic and inorganic. Organic pollutants include phenols, NPs, chlorinated phenols, azo dyes, phthalic esters, petroleum hydrocarbons, pesticides, and a variety of persistant organic pollutants (POPs); however, inorganic pollutants comprise a variety of highly toxic nonbiodegradable metals such as arsenic (As), nickel (Ni), chromium (Cr), lead (Pb), mercury (Hg), and cadmium (Cd). Many of them have been regarded as priority pollutants by various environmental protection agencies such as the U.S. Environmental Protection Agency (USEPA) and the Agency for Toxic Substances and Disease Registry (ATSDR). Environmental pollutants are highly toxic in nature and are reported to cause pollution of soil and water and a variety of serious toxic effects in living beings (Chandra et al. 2008; Maszenan et al. 2011; Megharaj et al. 2011; Chandra et al. 2011a, 2011b, 2015; Saxena and Bharagava 2016; Saxena et al. 2016). The toxic effects caused by organic and inorganic pollutants in living beings are described in detail in the following sections.

### 2.3.1 ORGANIC POLLUTANTS

Various environment protection agencies have prioritized several organic chemicals as hazardous and restricted their use in industrial processes; however, many of these organic chemicals are being used in industries and discharged in wastewater.

These organic chemicals cause severe health hazards in living beings. Brief accounts of their toxicity in living beings are presented in the following sections.

#### 2.3.1.1 Phenol Toxicity

Among organic pollutants, phenol (hydroxybenzene) is one of the most common pollutants found in various industrial wastewaters as it is a basic structural unit of a variety of synthetic organic compounds, including agricultural chemicals as well as pesticides (Santos et al. 2009; Basha et al. 2010). At room temperature, phenol is a translucent, colorless, crystalline, white powder; at room temperature, it is a syrupy liquid on mixing with water; and crystals are hygroscopic and turn pink to red when exposed to air (Chandra et al. 2011b). Phenol has a sweet tar-like odor and is soluble in alcohol, glycerol, petroleum, and also in water to a lesser extent (Basha et al. 2010). The use of phenol has been limited by various environment protection agencies like USSEPA, ATSDR, and the International Agency for Research on Cancer (IARC), but it is still used by many industries. In the environment, phenol pollution is associated with distilleries, pulp and paper mills, coal mines, oil refineries, wood preservation plants, pharmaceuticals, coke-oven batteries, herbicides, and pesticides as well as their wastewaters (Sung et al. 2000; Gibson and Harwood 2002; Ahn et al. 2008). It is also used in preparation of several chemicals such as alkylphenols, cresols, xylenols, phenolic resins, aniline, pesticides, explosives, dyes, and other compounds (Michałowicz and Duda 2007). A variety of phenols such as chlorophenols (CPs), nitrophenols, and aminophenols have been reported to have severe toxic effects in living beings; however, toxic effects are also shown by catechol, chlorocatechols, methylphenols, bisphenols, and other phenolic compounds (Michałowicz and Duda 2007).

The presence of phenol in wastewaters imparts carbolic odor to receiving water bodies and causes toxic effects in aquatic fauna and flora. Moreover, phenols are highly toxic to human beings, affecting various biochemical functions. Acute exposure of phenol causes central nervous system (CNS) disorders leading to collapse and coma and sometimes muscular convulsions are also noted along with reduction in body temperature (hypothermia) (Chandra et al. 2011b). Inhalation and dermal contact of phenol cause cardiovascular diseases and skin blisters, respectively, while ingestion can cause serious gastrointestinal damage, and oral administration may result in muscle tremors and death (Manojlovic et al. 2007; Chakraborty et al. 2010; Chandra et al. 2011b). Further, the acute exposure of phenols to humans through ingestion causes dryness of throat and mouth, nausea, vomiting, and diarrhea, while chronic exposure causes methemoglobinemia, hemolytic anemia, profuse sweating, hypotension, arrhythmia, pulmonary edema, tachycardia, and dark-colored urine excreted due to lipid peroxidation (Gami et al. 2014).

#### 2.3.1.2 Endocrine Disrupting Chemicals and Toxicity

Endocrine disrupting chemicals (EDCs) are the organic chemical compounds that disturb the delicate hormonal balance (endocrine system) and compromise the reproductive fitness of living beings and ultimately may lead to carcinogenesis (Dixit et al. 2015; Saxena et al. 2016). There has been an increasing concern regarding the release of many EDCs along with industrial wastewaters into the environment (Ifelebuegu and Ezenwa 2011; Ye et al. 2012). There are many reports of the presence of EDCs in industrial wastewaters. Phthalic acid esters such as DBP, BBP, bis(2-ethylhexyl) phthalate (DEHP), and NP constitute a major group of EDCs. Kumar et al. (2008) have detected many EDCs like NP, 4-aminobiphenyl, hexachlorobenzene, and benzidine in tannery wastewater (TWW) collected from the northern region of India and tested their toxicity on the reproductive system of male rats. However, the presence of phthalates (EDCs) such as DEHP, DBP, bis(2-methoxyethyl)phthalate in TWW has also been reported (Alam et al. 2009, 2010).

Many regulations have been passed to restrict the use of toxic chemicals in industries (Saxena et al. 2016). Directive 2003/53/EC restricted the marketing and use of products/product formulations containing more than 0.1% of NPE or NP and their use in the making of the leather products in Europe (Lofrano et al. 2008). In addition, Directive 1999/815/EC has directed industries to label products if they contain more than 0.5% phthalates (BBP, DBP and diethyl hexyl phthalate) due to the reproductive toxic potential of the phthalates (EU 2003). Phthalic acid esters (benzene-1,2 dicarboxylic acid) are the aromatic compounds used widely as plasticizers in the manufacturing of plastic resins such as polyvinyl resins and cellulosic and polyurethane polymers used for the manufacturing of building materials, home furniture, transportation equipment, clothing, and packaging material for food and pharmaceutical industries; however, human and animal exposure to them causes carcinogenic and mutagenic effects (Maszenan et al. 2011). Phthalates cause irritation of the skin, conjunctiva, mucous membranes of oral and nasal cavities, testicular lesions, hypospadia, cryptorchidism in males, and in females cause obesity, prolongation of the estrous cycle, and anovulation due to a decrease in the serum estradiol level (Mikula et al. 2005; Kamrin 2009). Further, NP is reported to have strong endocrine disrupting potential, widely applied as surfactant for various industrial, institutional, and household applications and used for the synthesis of NPE (Graca et al. 2016). NPs act as xenoestrogen-like endocrine disruptors by binding to estrogen receptors and competitively inhibiting natural estrogens, and they have been shown to mimic the natural hormone 17β-estradiol and compete with the endogenous hormone for binding with estrogen receptors ERα and ER (Yang et al. 2015). Bisphenol also acts as EDC, belongs to the group of diphenylmethane derivatives and isphenols, and has been shown to cause negative health effects in many animal studies (Yang et al. 2015).

#### 2.3.1.3 Chlorinated Phenols and Toxicity

A CP is any organochloride of phenol that contains one or more covalently bonded chlorine atoms and is produced by electrophilic halogenation of phenol with chlorine. Most CPs are solid at room temperature. They have a strong, medicinal taste and smell and are commonly used as pesticides, herbicides, and disinfectants. CPs are considered as the major environmental pollutants discharged along with wastewaters from pulp and paper mills, tanneries, distilleries, dye and paint manufacturing, and pharmaceuticals industries (Crosby 1981; Chandra et al. 2006). The major source of CPs is as by-products when chlorine is used for the bleaching of pulp and for disinfection of drinking water. Therefore, large amounts of chlorinated aromatic compounds, including CPs, have been discharged into the environment (Steinle et al. 1998). IARC categorized

CPs into five groups—pentachlorophenol (PCP); 2,3,4,6-tetrachlorophenol (2,3,4,6-TeCP); 2,4,6-trichlorophenol (2,4,6-TCP); 2,4,5-TCP; and 2,4-dichlorophenol (2,4-DCP)—as belonging to the 2B group of potential human carcinogens; however, the World Health Organization (WHO) classified some chlorophenols (2,4,6-TCP, 2,4,5-TCP, and PCP) as compounds suspected of having carcinogenic properties (Igbinosa et al. 2013).

Among CPs, PCP is a wide-spectrum biocide with applications in agriculture, industry, and public health (Bevenue and Beckman 1967; Chandra et al. 2006). PCP is highly recalcitrant in nature and resistant to biotic biodegradation because it is highly chlorinated, and generally, aromatic compounds with higher amounts of chlorine are more resistant to biodegradation and lead to substantial environmental contamination (Anandarajah et al. 2000; Singh et al. 2009). PCP is widely applied as herbicides and fungicides and in wood protection, tanneries, distilleries, paint manufacturing, and pulp and paper mills (Chandra et al. 2006; Salmeron-Alcocer et al. 2007; Igbinosa et al. 2013). The effluents discharged from these industries are the main source of PCP contamination. In addition, owing to its molecular stability, sorption properties, and intensive use, PCP has progressively become a widespread contaminant in the environment (Mannisto et al. 2001; Singh et al. 2009). The USEPA regulates PCP as a priority pollutant and considers 1 mg/L of PCP hazardous for land disposal (Freeman 1989; Sittig 1981). It is highly carcinogenic, teratogenic, and mutagenic in nature and causes toxicity to living beings by inhibiting oxidative phosphorylation, inactivating respiratory enzymes, and damaging mitochondrial structure (Jain et al. 2005; Tripathi et al. 2011; Verma and Maurya 2013). The high concentration of PCP can also cause obstruction in the circulatory system of the lungs, heart failure, and damage to the CNS (U.S. Department of Health and Human Services [USDHHS] 2001; Tewari et al. 2011; Dixit et al. 2015).

#### 2.3.1.4 Azo Dye and Toxicity

Textile, leather, paint, acrylic, cosmetics, plastics, pharmaceutical, etc., industries use different dyes to color products and consume a substantial volume of water in the processing of the products (Ahmad 2009; Sujata and Bharagava 2016). However, during the treatment of wastewater containing dye, these dyes are not properly degraded/decolorized and are, hence, discharged into the environment. Dyes are also considered as serious environmental pollutants and cause a variety of toxic effects in living beings. These dyes affect the ecosystem by causing 45 severe health hazards, such as skin irritation, digestive tract irritation, nausea, vomiting, liver and kidney damage, etc., in humans and animals; reduced seed germination, root and shoot length in plants; and inhibition of the activity of soil microorganisms, therefore reducing soil fertility (Senthilkumaar et al. 2006; Mittal et al. 2010; Sujata and Bharagava 2016).

Aromatic compounds with one or more –N=N– groups constitute the largest class of synthetic dyes used in commercial applications. These are generally recalcitrant to biodegradation due to their xenobiotic nature (Pandey et al. 2007). TWW also contain azo dyes that are highly persistent in nature due to their complex chemical structure and xenobiotic nature leading to environmental pollution (Nachiyar and Rajkumar 2003; Gurulakshmi et al. 2008; Baccar et al. 2011; Patel et al. 2012; Mahmood et al. 2013; Preethi et al. 2013; Dixit et al. 2015; Saxena et al. 2016). Thus, the removal of azo dyes from TWW is essential because of their high mutagenicity, carcinogenicity, and intense coloration problems of contaminated aquatic resources (Osugi et al. 2009; Saratale et al. 2010). The discharge of azo dyes into surface water also leads to esthetic problems and obstructs light penetration and oxygen transport into water bodies, finally affecting the aquatic life (Khalid et al. 2008; Chen et al. 2011). In addition, these dyestuffs have been reported to cause other serious problems such as dermatitis, skin and eye irritation, and respiratory problems in human beings (Keharia and Madamwar 2003). Further, the EU Azo Colorants Directive (2002) has prioritized several azo dyes and restricted their use in industries due to associated toxic effects in living beings; however, azo dyes are still illegally used in industries and discharged along with industrial wastes.

#### 2.3.1.5 Petroleum Hydrocarbons and Toxicity

Hydrocarbons are organic compounds, containing primarily hydrogen and carbon, which are categorized as aliphatic (straight chain) and aromatic (cyclic). Petroleum hydrocarbons or total petroleum hydrocarbons (TPHs) is the common term used for any mixture of hydrocarbons that are found in crude oil. Crude oil is used to make petroleum products that contaminate the environment. It is very difficult to measure each chemical separately because there are many chemicals present in crude oil and in other petroleum products. However, it is useful to measure the total amount of TPHs at contaminated sites. TPHs include hexane, benzene, toluene, xylenes, naphthalene, and fluorine and other constituents of gasoline, jet fuels, mineral oils, and other petroleum products (Kumar et al. 2015). Refinery industry wastewaters are the major source of petroleum hydrocarbons in the environment. The most common petroleum hydrocarbons include aliphatic, branched, and cycloaliphatic alkanes, as well as monocyclic and polycyclic aromatic hydrocarbons (PAHs), which include naphthalene, fluorene, phenanthrene, anthracene, fluoranthene, pyrene, benzo[*a*]anthracene, and benzo[*a*]pyrene (Fuentes et al. 2014). The negative impact of hydrocarbons on the environment is due to the blocking effect of oil layer to water, nutrients, oxygen, and light access (Fuentes et al. 2014). Soil contamination with petroleum and its products is a serious worldwide concern, and petroleum-contaminated soil pollutes local groundwater, renders potable water unsafe, limits groundwater use, and causes ecological toxicity (Wang et al. 2008; Das and Chandran 2011; Hentati et al. 2013; Gargouri et al. 2014; Kumar et al. 2015). Moreover, petroleum hydrocarbons are poorly degraded and have thus become the most widespread environmental contaminant (Margesin et al. 2000; Kumar et al. 2015). Humans can be exposed to hydrocarbons in three main ways (Tormoehlen et al. 2014): (1) unintentional ingestion of household hydrocarbons by children, (2) dermal or inhalational occupational exposures to workers, and (3) intentional inhalation of hydrocarbons by adolescents and young adults. The toxicity of hydrocarbons mainly depends on the dose as well as the chemical characteristics of volatility, lipid solubility, viscosity, and surface tension (Tormoehlen et al. 2014).

Exposure to hydrocarbons occurs through one of the following routes (Tormoehlen et al. 2014): inhalation, ingestion with or without aspiration, or dermal. Inhalational abuse of hydrocarbons in humans can lead to criminal or violent behavior, development of memory and other cognitive deficits, cerebellar dysfunction, encephalopathy, weakness, dementia, depression of the CNS, metabolic acidosis, arrhythmia, or even a fatal malignant arrhythmia termed "sudden sniffing death" (Tormoehlen et al. 2014). In addition, the aspiration of hydrocarbons causes a potentially fatal pneumonitis characterized by cough, wheezing, respiratory distress, and hypoxia (Tormoehlen et al. 2014). Dermal exposure can cause dermatitis, chemical burns, and defatting injury, whereas oral exposure can cause local irritation as well as vomiting, diarrhea, and abdominal pain (Tormoehlen et al. 2014). Thus, acute hydrocarbon exposure can result in a wide array of pathologies, such as encephalopathy, pneumonitis, arrhythmia, acidosis, and dermatitis. Intentional inhalation and accidental ingestion exposures with aspiration also lead to the greatest morbidity and mortality (Tormoehlen et al. 2014).

#### 2.3.1.6 Melanoidins and Toxicity

Melanoidins (high molecular weight amino-carbonyl compounds) are dark brown to black-colored natural condensation products of sugars and amino acids produced by nonenzymatic browning reactions called Maillard reactions (Plavsic et al. 2006; Bharagava et al. 2009). Naturally melanoidins are widely distributed in food and drinks and widely discharged in huge amounts as environmental pollutants by various agro-based industries, especially from cane molasses–based distilleries and fermentation industries (Kumar and Chandra 2006). The wastewaters released from distilleries and fermentation industries are major sources of soil and aquatic pollution due to the presence of water-soluble recalcitrant melanoidins. These melanoidins are highly resistant to microbial attack, and conventional biological processes such as activated sludge treatment processes are insufficient to treat these melanoidins containing wastewater released from distilleries and fermentation industries (Chandra et al. 2008; Bharagava and Chandra 2010).

Melanoidins are the dark brown–colored compounds that pose a serious threat to soil and aquatic ecosystems; release of melanoidins causes an increased load of recalcitrant organic material to natural water bodies. Discharge of melanoidins containing distillery wastewater (DWW) into the environment causes several problems, such as reduction of sunlight penetration, decreased photosynthetic activity, and dissolved oxygen concentration, thereby posing deleterious effects to aquatic life. On land, it causes a reduction in soil alkalinity and inhibition of seed germination (Pant and Adholeya 2007; Chandra et al. 2008; Satyawali and Balakrishnan 2008). The melanoidins containing DWW have severe toxic effects on fishes and other aquatic organisms. The estimated $LC_{50}$ for distillery spent wash was found to be 0.5% using a biotoxicity study on freshwater fish *Cyprinus carpio* var. *communis* (Mahimaraja and Bolan 2004). Impacts of distillery effluent on carbohydrate metabolism of freshwater fish *Cyprinus carpio* were also studied by Ramakritinan et al. (2005).

#### 2.3.1.7 Pesticides and Toxicity

Pesticides are organic chemicals or organometallic compounds that are extensively used in agricultural applications to control pests and enhance crop productivity; they are low in cost, easy to produce, and readily available. However, their unregulated and indiscriminate use causes serious soil and water pollution and adverse effects in humans and animals, which depend on the degree of sensitivity of organisms and the toxicity of the pesticides (Ortiz-Hernandez et al. 2013). Pesticides can be divided into the following classes: insecticides (used to kill insects, chemicals such as DDT, pyrethroids, organophosphates, carbamates, etc.), fungicides (used to kill fungi, such as hexachlorobenzene, benzothiozole, PCP, etc.), herbicides (used to kill weeds, such as 2,4-D, atrazine, picloram, chlorophenoxy compounds, etc.), rodenticides (used to kill rodents, such as zinc phosphide, α-Napthyl thiourea [ANTU], 4-hydroxycoumarin, 1,3-indandiones, etc.), and fumigants (used to kill pests, such as phosphine, dibromochloropropane, etc.). The adverse impacts of pesticides on soil microbial diversity and activities have also been reported by many workers (Wang et al. 2006; Littlefield-Wyer et al. 2008). Pesticides also affect soil biochemical processes such as microbial mineralization of organic compounds, their biotransformations, nutrient dynamics, and bioavailability in soil, and reduction of soil enzymatic activities that act as a "biological index" of soil fertility and hence reduce the soil productivity (Hussain et al. 2009). Pesticides cause serious health hazards in living systems because of their rapid fat soluble and bioaccumulate nature in nontarget organisms (Ortiz-Hernandez et al. 2013). In the environment, pesticides cause surface and groundwater pollution, immune suppression, hormone disruption, diminished intelligence, reproductive abnormalities, cancer upon exposure to humans and animals, and also directly or indirectly affect the nontarget organisms (Aktar et al. 2009). The fate of pesticides in the environment is strongly governed by soil sorption processes, which regulate not only their transfer but also their bioavailability (Gavrilescu 2005; Ortiz-Hernandez et al. 2013). Further, pesticides are highly recalcitrant in nature and are resistant to microbial degradation; hence, they persist a long time in the environment and therefore cause severe toxic effects in living beings (Gavrilescu 2005).

#### 2.3.1.8 Persistent Organic Pollutants and Their Toxicity

Persistent organic pollutants (POPs) are the organic compounds that are resistant to environmental degradation through chemical, biological, and photolytic processes. POPs often have three common characteristics (i.e., presence of one or more cyclical ring structures of either aromatic or aliphatic nature, lack of polar functional groups, and variable amount of halogen, usually chlorine). POPs exhibit low water solubility but high lipid solubility and therefore bioaccumulate in fatty tissues. They exhibit greater stability that reflects the nonreactivity of C-Cl bonds toward hydrolysis and photolytic degradation (Chandra and Chaudhary 2013). POPs are highly toxic in nature, and if they enter the food chain, they bioaccumulate in the fat tissues of organisms at various trophic levels and biomagnify the physiological effects at higher levels resulting in severe health hazards (Saxena and Bharagava 2015, 2016). The contamination of the global environment with a complex mixture of POPs has resulted from deliberate discharges of industrial wastes and applications as well as from the inadvertent formation of by-products of incomplete combustion or industrial processes. They persist in the environment for a long time, are able to survive long-range transport, and have potential significant impacts on human health and the environment (Chandra and Chaudhary 2013).

POPs cause a variety of toxic effects in living beings, such as disruption of the endocrine system, neurobehavioral disorders, cancer (possibly including breast cancer), allergies, hypersensitivity, damage to the central and peripheral nervous systems, birth defects, learning disabilities, reproductive disorders, disruption of the immune system by altering the hormonal system, and mutagenicity and carcinogenicity (Chandra and Chaudhary 2013). The "Stockholm Convention" is a global treaty to protect human health and the environment from POPs, and listed are those compounds that have POP-like properties (Sharma et al. 2014). It focuses on eliminating or reducing releases of 12 POPs and is regarded as the "Dirty Dozen" (Chandra and Chaudhary 2013). These 12 POPs include eight pesticides (aldrin, chlordane, DDT, dieldrin, endrin, heptachlor, mirex, and toxaphene), two industrial chemicals (polychlorinated biphenyls and hexachlorobenzene), and two unintended by-products (dioxins and furans) (Gavrilescu 2005). The chemicals included to this convention must provide evidence of persistence in water, soil, or sediment. There are many chemicals with POP-like characteristics that are currently not listed in the "Stockholm Convention" or "Arhus Protocol." However, the candidate chemicals that are nominated by the POPs Review Committee as POPs are given preference for their inclusion in the convention.

## 2.3.2 Inorganic Pollutants

Inorganic pollutants mainly comprise the heavy metals, which are toxic in nature, such as Cd, Cr, As, Pb, Hg, and Ni. Various sources of heavy metals include soil erosion, natural weathering of the earth's crust, mining, industrial wastes, urban runoff, sewage discharge, insect or disease control agents applied to crops, and many others (Maszenan et al. 2011; Morais et al. 2012; Jaishankar et al. 2014). These toxic metals are often nonbiodegradable in nature and thus persist in the environment causing serious soil and water pollution and health threats in living beings (Jaishankar et al. 2014; Chandra et al. 2015). Therefore, stringent limits on heavy metal discharge concentrations have been imposed by various environmental protection agencies worldwide. A brief account of the toxicity of heavy metals is discussed below.

### 2.3.2.1 Cadmium and Its Toxicity

Cadmium (Cd) is a highly toxic metal and is used in rechargeable batteries, special alloys production, coatings, pigments, platings, as a plastic stabilizer, and it is also present in tobacco smoke (Jaishankar et al. 2014). Cd is a nonessential heavy metal that is well recognized for its adverse influence on the enzymatic systems of cells, for causing oxidative stress, and for inducing nutritional deficiency in plants (Irfan et al. 2013; Jaishankar et al. 2014). In plants, exposure to high levels of Cd causes reductions in photosynthesis, water uptake, and nutrient uptake; chlorosis; growth inhibition; browning of root tips; and finally death (Wójcik and Tukiendorf 2004; Mohanpuria et al. 2007; Yadav 2010). Human exposure to Cd is possible through a number of several sources including employment in primary metal industries, eating contaminated food, smoking cigarettes, and working in Cd-contaminated work places, with smoking being a major contributor (Jarup et al. 2003; Tchounwou et al. 2012). Cd is a severe pulmonary and gastrointestinal irritant and can be fatal if inhaled or ingested (Tchounwou et al. 2012; Jaishankar et al. 2014). Acute exposure to Cd causes abdominal pain, burning sensation, nausea, vomiting, salivation, muscle cramps, itai-itai disease (a combination of osteomalacia and osteoporosis), vertigo, and shock. Loss of consciousness and convulsions usually appear within 15 to 30 minutes, and gastrointestinal tract erosion;

pulmonary, hepatic, or renal injury; and coma may develop depending on the route of poisoning (Jarup et al. 2003; Singh et al. 2011; Tchounwou et al. 2012). However, chronic exposure of Cd causes a depressive effect on levels of norepinephrine, serotonin, and acetylcholine (Tchounwou et al. 2012). The mechanism of Cd toxicity is not well known but is assumed to cause cell damage through the generation of reactive oxygen species (ROS) that cause DNA damage (Aziz et al. 2014; Tchounwou et al. 2012). Further, the IARC has classified Cd as a human carcinogen (Group I) on the basis of sufficient evidence in both humans and experimental animals (Jarup et al. 2003; Tchounwou et al. 2012).

#### 2.3.2.2 Chromium and Its Toxicity

Chromium (Cr) is considered as one of the major environmental pollutants due to its toxicity for ecological, nutritional, and environmental reasons (Mishra and Bharagava 2016). Chromium occurs in several oxidation states in the environment ranging from $Cr^{+3}$ to $Cr^{+6}$ forms—both forms are toxic to animals, humans, and plants (Mohanty and Kumar 2013; Jaishankar et al. 2014). Industries such as metallurgical, chemical, refractory brick, leather, wood preservation, and pigments and dyes are the major consumers of chromium compounds for the processing of raw materials to obtain good-quality products within a short period of time and economically (Mishra and Bharagava 2016). $Cr^{+6}$ is more toxic than $Cr^{+3}$ because it is dominant, stable, highly soluble, and rapidly permeable through the sulfate transport system followed by interaction with proteins and nucleic acids, and has toxic, genotoxic, mutagenic, and carcinogenic effects on humans, animals, plants, and microbes (Mishra and Bharagava 2016; Saxena et al. 2016). $Cr^{+6}$ contamination in soil alters the structure of soil microbial communities and reduces their growth. If it enters the food chain, it causes severe health problems such as skin irritation, nasal irritation, ulceration, eardrum perforation, and lung carcinoma (Cheung and Gu 2007; Chandra et al. 2011a; Mishra and Bharagava 2016). However, $Cr^{+3}$ is less toxic, less mobile, and is actually an essential human nutrient mainly bound to organic matters in soil and the aquatic environment (Mishra and Bharagava 2016).

Tannery wastewater (TWW) is the major source of Cr contamination in the environment (Saxena and Bharagava 2015; Saxena et al. 2016). Generally, leather industries (LIs) discharge their wastewater into nearby canals/rivers, which are directly/indirectly being used by farmers for the irrigation of agricultural crops (Trujillo-Tapia et al. 2008; Gupta et al. 2012). This practice leads to the movement of potentially toxic metals like Cr from water to crop plants that ultimately reaches the human/animal body and causes toxicity (Sinha et al. 2008; Chandra et al. 2009; Saxena et al. 2016). However, Cr toxicity mainly depends on chemical speciation; thus, associated health effects are influenced by the chemical forms of exposure (Rameshraja and Suresh 2011). Thus, the use of Cr-loaded TWW for the irrigation of agricultural crops disrupts several physiological and cytological processes in cells leading to a reduction in root and shoot growth, biomass, seed germination, and seedling growth, and also induces chlorosis and photosynthetic impairment, finally leading to plant death (Shanker et al. 2005; Chidambaram et al. 2009; Hussain et al. 2010; Gupta et al. 2012; Saxena et al. 2016). $Cr^{+6}$ may enter the human and animal body by inhalation (occupational exposure), ingestion (of contaminated food and water), and dermal absorption. It is reported as a "human carcinogen" by WHO, IARC, USEPA, and the National Toxicological Programme (NTP) (Mishra and Bharagava 2016). However, the mechanism(s) of $Cr^{+6}$-induced carcinogenicity is not well understood but may be associated with the generation of free radicals, which causes DNA damage during the reduction of $Cr^{+6}$ to $Cr^{+3}$ (Mishra and Bharagava 2016).

#### 2.3.2.3 Arsenic and Its Toxicity

Arsenic (As) is a highly toxic, carcinogenic heavy metal, which is extensively available in the form of oxides or sulfides or as a salt of iron, sodium, calcium, copper, etc., and is reported as a widespread environmental pollutant in many parts of the world (Flora et al. 2005; Singh et al. 2007; Ventura-Lima et al. 2011; Jaishankar et al. 2014). It is commonly found in several chemical forms: inorganic forms of arsenic (arsenite and arsenate) are more toxic; methylated forms (methylarsonate [MMA] and dimethylarsenate [DMA]) and other arsenic species, like trimethylarsine oxide (TMAO) and tetramethylarsonium (TETRA) are moderately toxic; whereas arsenobetaine (AsB), arsenocholine (AsC), and other arsenosugars (AsS) are nontoxic in nature (Geiszinger et al. 2002; Fattorini and Regoli 2004; Fattorini et al. 2006; Ventura-Lima et al. 2011; Mirza et al. 2014). The major inorganic forms of arsenic include the trivalent arsenite, which is a more toxic form than pentavalent arsenate. Environmental contamination by arsenic occurs as a result of natural phenomena such as volcanic eruptions, soil erosion, and anthropogenic activities (Tchounwou et al. 2012). However, exposure to As in humans occurs via the oral route (ingestion), inhalation, dermal contact, and the parenteral route to some extent (ATSDR 2000; Tchounwou et al. 2012). Arsenic (As) causes severe disturbances of the cardiovascular and central nervous systems, bone marrow depression, hemolysis, hepatomegaly, melanosis, polyneuropathy, and encephalopathy, and exposure eventually leads to death. Ingestion may also cause black foot disease that is only reported in Taiwan (Jarup et al. 2003; Tchounwou et al. 2012). Drinking water has been considered as the major route of As exposure and is associated with the risks of cancer in lungs, kidney, bladder, and skin (Jarup et al. 2003; Mirza et al. 2014). Arsenate (As) also acts as an analog of phosphate (P) and competes for its uptake through carriers in the root plasmalemma in plants; thus, it causes suppression of a high-affinity P/As uptake system (Yadav 2010).

#### 2.3.2.4 Lead and Its Toxicity

Lead (Pb) is a highly toxic metal, rarely available in elemental form but found in association with zinc (Zn), copper (Co), iron (Fe) sulfide, as well as gold (Au), silver (Ag), bismuth (Bi), and antimony (Sb) minerals (Carocci et al. 2016). The contamination of soil and water with Pb mostly occurs from anthropogenic activities, industrial wastes, mining and smelting, and the past and present use of Pb in paints, batteries, gasoline, pesticides, and explosives (Kumar and Kumari 2015; Wani et al. 2015). In plants, Pb causes inhibition of enzyme activities at the cellular level by reacting with their sulfhydryl groups, causing water imbalance and a reduction in chlorophyll content, and thereby reduces photosynthesis, alters membrane permeability, and disturbs mineral nutrition as well as induces oxidative stress by increasing the production of ROS in plants (Reddy et al. 2005; Sharma and Dubey 2005; Yadav 2010; Jaishankar et al. 2014).

The inhalation of contaminated ambient air or ingestion of contaminated food and water are the major routes of Pb exposure to humans and animals; however, it is not biomagnified in aquatic or terrestrial food chains (Tchounwou et al. 2012; Carocci et al. 2016). Pb mainly causes injury to the CNS and causes headache, poor attention span, irritability, encephalopathy (characterized by sleeplessness and restlessness), loss of memory, acute psychosis, confusion, reduced consciousness, and dullness, and also adversely affects the kidneys, liver, hematopoietic system, endocrine system, and reproductive system (Jarup et al. 2003; Tchounwou et al. 2012). Further, Pb is a potential carcinogen and has been identified as a probable human carcinogen by the IARC (Jarup et al. 2003; Tchounwou et al. 2012).

#### 2.3.2.5 Mercury and Its Toxicity

Mercury (Hg) is a naturally occurring highly toxic metal that is exceedingly bioaccumulative in the aquatic system and mainly exists in three forms: metallic elements, inorganic salts, and organic compounds such as HgS, $Hg^{2+}$ (most common form in soil), Hg°, and methyl-Hg with different toxicities and bioavailabilities (Yadav 2010; Jaishankar et al. 2014). The major sources of Hg exposure include its use in dental amalgams, thermometers, sphygmomanometers, barometers, fossil fuel emissions, incandescent lights, batteries, ritualistic practices using mercury, and the incineration of medical waste (Guzzi and La Porta 2008; Rice et al. 2014). However, the major sources of chronic and low-level mercury exposure are dental amalgams and fish consumption (Tchounwou et al. 2012). In plants, the high concentration of $Hg^{2+}$ can induce visible injuries and several physiological disorders in plants as it binds to water channel proteins, thus inducing leaf stomata to close, physical obstruction of water flow in plants, interference with mitochondrial activity, and induction of oxidative stress by triggering the generation of ROS, ultimately leading to the disruption of biomembrane lipids and cellular metabolism in plants (Zhang and Tyerman 1999; Messer et al. 2005; Cargnelutti et al. 2006; Zhou et al. 2007; Yadav 2010). However, in humans and animals, it causes mental retardation, dysarthria, blindness, neurological deficits, loss of hearing, developmental defects, and abnormal muscle tone (Guzzi and La Porta 2008; Rice et al. 2014).

## 2.4 BIOREMEDIATION APPROACHES FOR ENVIRONMENTAL POLLUTANTS

Bioremediation approaches are eco-friendly in nature and can be applied for the degradation and detoxification of organic and inorganic pollutants for environmental safety and human health protection. These include the following methods.

### 2.4.1 Composting

Composting is a remediation method wherein the volume and water content of wastes are reduced, pathogens are destroyed, and microbes degrade highly toxic compounds into less toxic forms. Its efficiency depends on temperature and the soil/waste amendment ratio (Antizar-Ladislao et al. 2005; Maszenan et al. 2011). During composting, organic pollutants are degraded in the first phase of rapid decomposition, and heat is generated through microbial metabolism; however, maximum decomposition and microbial biomass formation occur during the thermophilic phase. Moreover, the mixing of composted material to noncomposted waste materials also increases the efficiency of composting because of the previously adapted microbes that enhance pollutant degradation during the composting process (Hwang et al. 2001; Maszenan et al. 2011). Further, composting as a remediation method has been reviewed by Maszenan et al. (2011).

### 2.4.2 Bioremediation

Bioremediation is an eco-friendly waste management method that utilizes the inherent potential of microbes such as algae, fungi, and bacteria to remove organic and inorganic pollutants from industrial wastes. Biodegradation is the degradation of highly toxic complex organic pollutants into less toxic forms. It involves detoxification and mineralization, where the waste is converted into inorganic compounds such as carbon dioxide, water, and methane (Reshma et al. 2011). But if the pollutants are persistent in nature, their biodegradation often proceeds through multiple steps utilizing different enzymes or microbial populations. The process of bioremediation depends on the metabolic potential of microbes to degrade/detoxify or transform the pollutants, which is also dependent on both the pollutant's accessibility and bioavailability (Antizar-Ladislao 2010). Bioremediation can be applied as both *in situ* (waste to be treated at the site) and *ex situ* (waste to be treated elsewhere) remediation technology. However, *in situ* bioremediation is usually preferred over *ex situ* bioremediation due to the high cost associated with excavation of contaminated sites to be treated. Further, *in situ* bioremediation involves following three strategies, including (Maszenan et al. 2011) bioattenuation (natural process of degradation and can be monitored by a decrease in pollutant concentration with increasing time), biostimulation (intentional stimulation of pollutants degradation by addition of water, nutrients, and electron donors or acceptors), and bioaugmentation (addition of laboratory-grown microbes with potential for degradation). Some of the applications of biodegradation and detoxification of organic and inorganic pollutants are summarized in Table 2.2. Further, a great deal of literature on the biodegradation and bioremediation of organic and inorganic pollutants has been reviewed (Maszenan et al. 2011; Megharaj et al. 2011).

### 2.4.3 Phytoremediation

Phytoremediation is an eco-sustainable green technology that makes use of green plants for the *in situ* remediation of environmental pollutants, whether organic or inorganic in nature. It is the set of phytotechnologies applied for decontamination of the environment, which includes (Chandra et al. 2015) phytoextraction (plants used to remove metal pollutants from contaminated sites via root absorption and their sequestration in the above-ground harvestable plant parts), rhizofiltration (constructed wetlands wherein plants are used to absorb metal pollutants from aquatic environments in their roots), phytostabilization (in-place inactivation of metal pollutants by plant roots),

**TABLE 2.2**
**Microbes Involved in the Degradation and Detoxification of Organic and Inorganic Pollutants from Industrial Wastewaters**

**Microorganisms Employed for the Bioremediation of Melanoidins Containing Distillery Wastewater**

| Microorganisms | Decolorization (%) | References |
|---|---|---|
| *Pseudomonas aeruginosa* | 67 | Mohana et al. (2007) |
| *Bacillus thuringiensis* | 22 | Kumar and Chandra (2006) |
| *Acetobacter acetii* | 76.4 | Sirianuntapiboon et al. (2004) |
| *Pleurotus florida* | 86.3 | Pant and Adholeya (2009) |
| *Xanthomonas fragariae* | 76 | Jain et al. (2002) |
| *Pseudomonas putida* | 60 | Ghosh et al. (2002) |
| *Coriolus versicolor* | 63 | Kahraman and Yesilada (2003) |

**Microorganisms Employed for the Bioremediation of Dye from Textile Wastewater**

| Microorganisms | Dyes and Decolorization (%) | References |
|---|---|---|
| *Pseudomonas* sp. | Reactive Blue 13 and 83.2 | Lin et al. (2010) |
| *Geobacillus stearothermophilus* (UCP 986) | Orange II and 96–98 | Evangelista-Barreto et al. (2009) |
| *Bacillus* sp. VUS | Navy Blue 2GL and 94 | Dawkar et al. (2009) |
| *Shewanella decolorationis S12* | Acid Red GR and 100 | Xu et al. (2007) |
| *Bacillus cereus, Pseudomonas putida, Pseudomonas fluorescence,* and *Stenotrophomonas acidaminiphila* | Acid Red 88 and 96 | Khehra et al. (2006) |
| *Aeromonas hydrophila* | Red RBN and 90 | Chen et al. (2003) |

**Microorganisms Employed for the Bioremediation of Tannery Wastewater**

| Microorganisms | Chromium Removal (%) | References |
|---|---|---|
| *Providencia vermicola* | 93.66 | El-Bestawy et al. (2013) |
| *Brachymonas denitrificans* | 88.5 | Kim et al. (2014) |
| *Thiobacillus ferrooxidans* | 5 | Mandal et al. (2010) |
| *Trichoderma* sp. | 97.93 | Vankar and Bajpai (2008) |
| *Spirogyra condensata* and *Rhizoclonium hieroglyphicum* | >75 | Onyancha et al. (2008) |
| *Acidithiobacillus thiooxidans* | 99.7 | Wang et al. (2007) |
| *Acenetobacter* sp. | 90 | Srivastava et al. (2007) |

phytovolatilization (uptake of pollutants through plant roots, their transformation into less toxic forms, and subsequent release through stomata), phytodegradation (breakdown of highly toxic organic pollutants into less toxic forms through the action of enzymes secreted within plant tissues), and rhizodegradation (breakdown of highly toxic organic pollutants into less toxic forms through the enzymatic activity of the plant rhizosphere). These phytotechnologies are applicable to a wide range of pollutants using several different mechanisms depending on their applications; however, all are not applicable for remediation of all pollutants (Marmiroli et al. 2006; Vangronsveld et al. 2009). Several applications of phytoremediation technology are presented in Table 2.3.

The selection of suitable plants for phytoremediation is not an easy task, but native plants are often preferred for phytoremediation due to less competition among them under certain environmental conditions. The ideal plants for phytoremediation should have the ability to hyperaccumulate heavy metals, have a fast growth rate, have the ability to tolerate high salt concentrations and pH, have high biomass, be easily harvestable, and must uptake and translocate metals to aerial parts efficiently (Vangronsveld et al. 2009; Sharma 2011). It can be a low-cost alternative to conventional remediation methods, which are environmentally destructive and require higher capital investments. The low biomass and slow growth rate are two major factors that limit the efficiency of phytoremediation; however, it can be enhanced by subjecting the plants to

**TABLE 2.3**
**Applications of Phytoremediation Technology**

| Mechanism | Pollutant | Media | Plant | Status | References |
|---|---|---|---|---|---|
| Phytoextraction | As | Soil | *Pteris vittata* L. | Soil | Hue (2013) |
| Phytoextraction | Zn, Cd, and As | Soil | *Datura stramonium* and *Chenopodium murale* | Applied | Varun et al. (2012) |
| Phytoextraction | Pb and Cu | Soil | *Zea mays* L. cv. LvBa Chaotian No. 15 | Soil pot | |
| Phytoextraction | Ni | Soil | *Ricinus communis* L. | Soil pot | Adhikari and Kumar (2012) |
| Phytoextraction | As | Soil | *Cassia fistula* | Applied | Preeti et al. (2011) |
| Phytoextraction | Cd | Soil | *Solanum nigrum* L. | Field demo | Ji et al. (2011) |
| Phytostabilization | Mn | Soil | *Chondrilla juncea* and *Chenopodium botrys* | Soil | Cheraghi et al. (2011) |
| Phytoextraction | $^{137}Cs$ | Soil | *Catharanthus roseus* | Applied | Fulekar et al. (2010) |
| Phytoextraction | Cr | Soil | *Anogeissus latifolia* | Applied | Mathur et al. (2010) |
| Phytoextraction | Zn and Cd | Soil | *Vetiveria, sesbania, Viola, sedum, Rumex* | Field demo | Mukhopadhyay and Maiti (2010) |
| Phytoextraction | Pb and Cd | Soil | *Jatropha curcas* L. | Applied | Mangkoedihardjo and Surahmaida (2008) |
| Phytostabilization | Cd | Soil | Sunflower | Applied | Zadeh et al. (2008) |
| Phytodegradation | U | Soil | *Brassica juncea* | Field demo | Huhle et al. (2008) |
| Phytoextraction | Pb and As | Soil | *Pteris vittata* and *Brassica juncea* | Soil | Salido et al. (2003) |

genetic manipulations using transgenic approaches, which confers the ability to tolerate and accumulate high metal concentrations from contaminated sites (Chandra et al. 2015). Moreover, phytoremediation efficiency can be enhanced by inoculating the plant growth-promoting rhizobacteria and endophytic bacteria exhibiting multiple traits, which are useful in the remediation of environmental pollutants (Saxena and Bharagava 2016). Further, a great deal of literature on the phytoremediation of heavy metals and organic pollutants can be found in the public domain (Vangronsveld et al. 2009; Chandra et al. 2015).

### 2.4.4 Bacterial-Assisted Phytoremediation

Phytoremediation is a low-cost but time-consuming clean-up approach, and toxicity in remediating plants is often associated with high metal concentrations; therefore, plant–microbes interaction can play a major role in ameliorating plant toxicity and assists them to tolerate and sequester high metal concentrations (Chandra et al. 2015; Saxena and Bharagava 2016). The term *rhizobacteria* is often used for those bacteria that colonize the rhizosphere of plants with high root exudate concentration and are technically termed *plant growth-promoting rhizobacteria* (PGPR). PGPR have numerous benefits, including production of growth-modulating hormones such as indole acetic acid (IAA), auxins, gibberellins, cytokinins, siderophores, ethylene, nitric oxide, organic acids, and antibiotics; all these help in the suppression of phytopathogens, improved mineral uptake, reduced ethylene production, and tolerance to high metal concentration, which ultimately lead to plant growth promotion (Ma et al. 2011). Further, PGPR can also be used in environmental remediation under stressful environments; however, phytoremediation is better depending on the integration among soil, bacteria, and plants. Moreover, endophytes (bacteria that colonize the internal tissues of plants) have been reported to tolerate high metal concentrations and, therefore, could be of potential interest for heavy metal phytoremediation (Ma et al. 2011; Li et al. 2012). Some of the applications of bacterial-assisted phytoremediation are presented in Table 2.4. Further, a great deal of literature on the bacterial-assisted phytoremediation of toxic metals can be found in the public domain (Zhuang et al. 2007; Glick 2010; Ma et al. 2011).

### 2.4.5 Electro-Bioremediation

Electro-bioremediation technology is becoming an increasingly popular hybrid technology that uses the combination of bioremediation and electrokinetics for the treatment of environmental pollutants (Maszenan et al. 2011). It involves the electrokinetics phenomena for the acceleration and orientation of transport of environmental pollutants and microbes for pollutants bioremediation (Li et al. 2010; Maszenan et al. 2011). Electrokinetics involves the use of several phenomenon like diffusion, electrolysis, electro-osmosis, electrophoresis, and electromigration, and uses weak electric currents of about 0.2 to 2 V $cm^{-1}$ (Saichek and Reddy 2005; Lukas Y.W. et al. 2007; Maszenan et al. 2011). A number of studies are available on the use of electro-bioremediation technology for contaminated soils (Lageman and Godschalk 2007; Martinez-Prado et al. 2014; Fei and Danny 2015). In addition, the applications, potentials, and limitations

**TABLE 2.4**
**Some Applications of Bacterial-Assisted Phytoremediation of Heavy Metals**

| Plant Growth-Promoting Rhizobacteria | Plant | Heavy Metal | Effects | References |
|---|---|---|---|---|
| *Bradyrhizobium* sp., *Pseudomonas* sp., *Ochrobactrum cytisi* | *Lupinus luteus* | Pb, Cu, and Cd | Decreased the metal accumulation; however, plant biomass increased. | Dary et al. (2010) |
| *Bacillus subtilis*, *Bacillus cereus*, *Pseudomonas aeruginosa*, *Flavobacterium* sp. | *Orychophragmus violaceus* | Zn | Increased shoot biomass and Zn accumulation. | He et al. (2010) |
| *Pseudomonas aeruginosa*, *Ralstonia metallidurans* | Maize | Cr, Pb, and Cr | Increased the uptake by shoot by a factor of 5.4 and 3.4, respectively.<br>Increased the accumulation of Cr in shoots by a factor of 5.2. | Braud et al. (2009) |
| *Achromobacter xylosoxidans* strain Ax10 | *Brassica juncea* | Cu | Increased the length of root and shoot, fresh and dry weight significantly, and extensively improved the Cu uptake of *B. juncea* plants as compared to the control. | Ma et al. (2009) |
| *Microbacterium* sp. G16, *Pseudomonas fluorescens* G10 | Rape | Pb | Increased root elongation of inoculated rape seedlings and total Pb accumulation as compared to the control plants. | Sheng et al. (2008) |
| *B. weihenstephanensis* SM3 | *Helianthus annuus* | Cu and Zn | Increased plant fresh and dry weight with increased Cu and Zn uptake during phytoextraction. | Rajkumar et al. (2008) |
| *Pseudomonas aeruginosa* | Black gram plants | Cd | Lessened the accumulation of Cd in plants; showed extensive rooting and enhanced plant growth. | Ganesan (2008) |
| *Pseudomonas aeruginosa* KUCd1 | *Cucurbita pepo* | Cd | Increased shoot length, root length, wet weight, dry weight, and chlorophyll, but reduced Cd uptake in shoots by 47.4%. | Sinha and Mukherjee (2008) |

*(Continued)*

**TABLE 2.4** *(CONTINUED)*
**Some Applications of Bacterial-Assisted Phytoremediation of Heavy Metals**

| Plant Growth-Promoting Rhizobacteria | Plant | Heavy Metal | Effects | References |
|---|---|---|---|---|
| *Burkholderia* sp. J62 | Maize and tomato | Cd and Pb | Increased the biomass of maize and tomato plant significantly; the increased Pb and Cd content in tissue varied from 38% to 192% and from 5% to 191%, respectively. | Jiang et al. (2008) |
| *Rhodococcus erythropolis* MtCC 7905 | *Pisum sativum* | Cr | Increased plant growth in the presence of $Cr^{+6}$ and its substantial amount reduced $Cr^{+3}$ at low temperature during phytostabilization. | Trivedi et al. (2007) |
| *Bradyrhizobium* sp. RM8 | Green gram var. K851 | Ni and Zn | Increased plant growth and decreased uptake of heavy metals by plant. | Wani et al. (2007) |
| *Brevibacillus* sp. B-I | *Trifolium repens* | Zn | Increased plant growth, N and P accumulation, nodule number, and mycorrhizal infection with reduced Zn uptake (phytostabilization). | Vivas et al. (2006) |

of electro-bioremediation technology have been reviewed by many authors (Wick et al. 2007; Maszenan et al. 2011; Gill et al. 2014).

## 2.5 FUTURE PERSPECTIVES AND CHALLENGES

Bioremediation has emerged as a low-cost alternative to conventional remediation technologies, which are environmentally destructive. However, it is limited by a number of reasons such as the nonavailability of potential microorganisms at the contaminated sites, persistent nature of organic and inorganic pollutants, and lack of enzymatic machinery in microbes for specific pollutants to be degraded or detoxified and insufficient biomass and slow growth rate of plants for phytoremediation. Molecular approaches have already proven their merits in modifying the plants and microbes for the enhanced degradation and detoxification of organic and inorganic pollutants; however, risks are also associated with the use of genetically modified plants and microbes for environmental cleanup, such as invasion of exotic plants and loss of biodiversity. Further, strict regulations in developed countries such as Europe and the United States also restrict the use of genetically modified organisms for environmental remediation for field applications. These limitations constitute major challenges in the use of environmental bioremediation technologies; however, continued efforts are required for improvement in existing bioremediation technologies.

## 2.6 CONCLUSION

- Industrial wastes are considered the major sources of environmental pollution and contain a variety of highly toxic organic and inorganic pollutants, which cause severe toxic effects in living beings.
- Physicochemical approaches are environmentally destructive and require high capital investment for environmental cleanup.
- Bioremediation technologies including phytoremediation can be eco friendly approaches for the treatment of industrial wastes for environmental safety and human health protection.
- There is a need to search for potential microbes for the degradation and detoxification of organic and inorganic pollutants from industrial wastes.
- A better understanding of plant–microbe interactions is required for the engineered phytoremediation of environmental pollutants.
- Risks associated with the use of genetically modified organisms should also be considered prior to release for field applications.
- Continued efforts are required to realize the economic feasibility of bioremediation technologies including phytoremediation for field applications.

## ACKNOWLEDGMENTS

The authors are extremely grateful to the "Science and Engineering Research Board" (SERB), Department of Science and Technology (DST), Government of India (GOI), New Delhi, India, for financial support as "Major Research Project" (Grant No.: SB/EMEQ-357/2013) for this work; the University Grant Commission (UGC) Fellowship received by Mr. Gaurav Saxena is also duly acknowledged.

## REFERENCES

Adhikari, T., and Kumar, A. 2012. Phytoaccumulation and tolerance of Riccinus communis L. to nickel. *International Journal of Phytoremediation* 14: 481–492.

Agency for Toxic Substances and Disease Registry (ATSDR). 2000. *Toxicological profile for arsenicTP-92/09*. Center for Disease Control, Atlanta, GA.

Ahmad, R. 2009. Studies on adsorption of crystal violet dye from aqueous solution onto coniferous pinus bark powder (CPBP). *Journal of Hazardous Material* 171: 767–773.

Ahn, S., Congeevaram, S., Choung, Y.K., and Park, J. 2008. Enhanced phenol removal by floating fungal populations in a high concentration phenol-fed membrane bioreactor. *Desalination* 221: 494–501.

Aktar, M.W., Sengupta, D., and Chowdhury, A. 2009. Impact of pesticides use in agriculture: Their benefits and hazards. *Interdisciplinary Toxicology* 2(1): 1–12.

Alam, M.Z., Ahmad, S., and Malik, A. 2009. Genotoxic and mutagenic potential of agricultural soil irrigated with tannery effluents at Jajmau (Kanpur), India. *Archives of Environmental Contamination and Toxicology* 57(3): 463–476.

Alam, M.Z., Ahmad, S., Malik, A., and Ahmad, M. 2010. Mutagenicity and genotoxicity of tannery effluents used for irrigation at Kanpur, India. *Ecotoxicology and Environmental Safety* 73(5): 1620–1628.

Anandarajah, K., Kiefer, P.M., Jr., Donohoe, B.S., and Copley, S.D. 2000. Recruitment of a double bond isomerase to serve as a reductive dehalogenase during biodegradation of pentachlorophenol. *Biochemistry* 39: 5303–5311.

Antizar-Ladislao, B. 2010. Bioremediation: Working with bacteria. *Elements* 6: 389–394.

Antizar-Ladislao, B., Lopez-Real, J., and Beck, A.J. 2005. In-vessel composting-bioremediation of aged coal tar soil: Effect of temperature and soil/green waste amendment ratio. *Environment International* 31: 173–178.

ATSDR. 2008. *Toxicological profile for cadmium*. Agency for Toxic Substances & Disease Register. ATSDR, Atlanta, GA.

Aziz, R., Rafiq, M.T., Yang, J., Liu, D., Lu, L., He, Z., Daud, M.K., Li, T., and Yang, X. 2014. Impact assessment of cadmium toxicity and its bioavailability in human cell lines (Caco-2 and HL-7702). *BioMed Research International* 2014: 839538.

Baccar, R., Blanquez, P., Bouzid, J., Feki, M., Attiya, H., and Sarra, M. 2011. Decolorization of a tannery dye: From fungal screening to bioreactor application. *Biochemical Engineering Journal* 56(3): 184–189.

Basha, K.M., Rajendran, A., and Thangavelu, V. 2010. Recent advances in the biodegradation of phenol: A review. *Asian Journal of Experimental Biological Science* 1(2): 219–234.

Bevenue, A., and Beckman, H. 1967. Pentachlorophenol: A discussion of its properties and its occurrence as a residue in human and animal tissues. *Residue Reviews* 19: 83–134.

Bharagava, R.N., Chandra, R., and Rai, V. 2009. Isolation and characterization of aerobic bacteria capable of the degradation of synthetic and natural melanoidins from distillery effluent. *World Journal of Microbiology and Biotechnology* 25: 737–744.

Braud, A., Jezequel, K., Bazot, S., and Lebeau, T. 2009. Enhanced phytoextraction of an agricultural Cr-and Pb contaminated soil by bioaugmentation with siderophore-producing bacteria. *Chemosphere* 74: 280–286.

Cargnelutti, D., Tabaldi, L.A., Spanevello, R.M., Jucoski, G.O., Battisti, V., Redin, M., Linares, C.E.B., et al. 2006. Mercury toxicity induces oxidative stress in growing cucumber seedlings. *Chemosphere* 65: 999–1006.

Carocci, A., Catalano, A., Lauria, G., Sinicropi, M.S., and Genchi, G. 2016. Lead toxicity, antioxidant defense and environment. *Reviews of Environmental Contamination and Toxicology* 238: 45–67.

Chakraborty, S., Bhattacharya, T., Patel, T.N., and Tiwari, K.K. 2010. Biodegradation of phenol by native microorganisms isolated from coke processing wastewater. *J. Environ. Biol.* 31: 293–296.

Chandra, R., Bharagava, R.N., and Rai, V. 2008. Melanoidins as major colourant in sugarcane molasses based distillery effluent and its degradation. *Bioresource Technology* 99: 4648–4660.

Chandra, R., Bharagava, R.N., Kapley, A., and Purohit, H.J. 2011a. Bacterial diversity, organic pollutants and their metabolites in two aeration lagoons of common effluent treatment plant (CETP) during the degradation and detoxification of tannery wastewater. *Bioresource Technology* 102(3): 2333–2341.

Chandra, R., Bharagava, R.N., Yadav, S., and Mohan, D. 2009. Accumulation and distribution of toxic metals in wheat (*Triticum aestivum* L.) and Indian mustard (*Brassica campestris* L.) irrigated with distillery and tannery effluents. *Journal of Hazardous Material* 162: 1514–1521.

Chandra, R., and Chaudhary, S. 2013. Persistent organic pollutants in environment and their health hazards. *International Journal of Bioassays* 02(09): 1232–1238.

Chandra, R., Ghosh, A., Jain, R.K., and Singh, S. 2006. Isolation and characterization of two potential pentachlorophenol degrading aerobic bacteria from pulp paper effluent sludge. *Journal of General and Applied Microbiology* 52: 125–130.

Chandra, R., Saxena, G., and Kumar, V. 2015. Phytoremediation of environmental pollutants: An eco-sustainable green technology to environmental management. In *Advances in biodegradation and bioremediation of industrial waste*, Chandra, R. (ed.), CRC Press, Boca Raton, FL, pp. 1–30.

Chandra, R., Yadav, S., Bharagava, R.N., and Rai, V. 2011b. Phenol degradation by *Paenibacillus thiaminolyticus* and *Bacillus cereus* in axenic and mixed conditions. *World Journal of Microbiology and Biotechnology* 27(12): 2939–2947.

Chen, G., Huang, M.H., Chen, L., and Chen, D.H. 2011. A batch decolorization and kinetic study of Reactive Black 5 by a bacterial strain Enterobacter sp. GY-1. *International Biodeterioration and Biodegradation* 65(6): 790–796.

Chen, K.C., Wu, J.Y., Liou, D.J., and Hwang, S.C.J. 2003. Decolorization of the textile dyes by newly isolated bacterial strains. *J. Biotechnol.* 101(1): 57–68.

Cheraghi, M., Lorestani, B., Khorasani, N., Yousefi, N., and Karami, M. 2011. Findings on the phytoextraction and phytostabilization of soils contaminated with Heavy Metals. *Biol. Trace Elem. Res* 144: 1133–1141.

Cheung, K.H., and Gu, J.D. 2007. Mechanism of hexavalent chromium detoxification by microorganisms and bioremediation application potential: A review. *International Biodeterioration and Biodegradation* 59(1): 8–15.

Chidambaram, A.P., Sundaramoorthy, A., Murugan, K., and Baskaran, S.G.L. 2009. Chromium induced cytotoxicity in black gram (*Vigna mungo* L). *Iranian Journal of Environmental Health Science and Engineering* 6(1): 17–22.

Crosby, D.G. 1981. Environmental chemistry of pentachlorophenol. *Pure and Applied Chemistry* 53: 1051–1080.

Dary, M., Chamber-Pérez, M.A., Palomares, A.J., and Pajuelo, E. 2010. In situ phytostabilisation of heavy metal polluted soils using Lupinus luteus inoculated with metal resistant plant-growth promoting rhizobacteria. *J. Hazard. Mater.* 177: 323–330.

Das, N., and Chandran, P. 2011. Microbial degradation of petroleum hydrocarbon contaminants: An overview. *Biotechnology Research International* 2011: 1–13. Article ID 941810.

Dawkar, V.V., Jadhav, U.U., Ghodake, G.S., and Govindwar, S.P. 2009. Effect of inducers on the decolorization and biodegradation of textile azo dye Navy Blue 2GL by *Bacillus* sp. VUS. *Biodegrad.* 20(6): 777–787.

Directive (1999/815/EC). http://www.tid.gov.hk/english/aboutus/tradecircular/cic/eu/2004/files/ci2132004a.pdf (accessed September 9, 2016).

Directive (2003/53/EC). http://www.tfl.com/web/files/Statement_NPE-surfactants.pdf (accessed September 9, 2016).

Directive (REACH (EC 1907/2006) for European Regulatory Framework on chemicals namely Registration, Evaluation, Authorization and Restriction of Chemical substances. www.hse.gov.uk/reach/reachtext.pdf (accessed September 9, 2016).

Dixit, S., Yadav, A., Dwivedi, P.D., and Das, M. 2015. Toxic hazards of leather industry and technologies to combat threat: A review. *Journal of Cleaner Production* 87: 39–49.

El-Bestawy, E., Al-Fassi, F., Amer, R., and Aburokba, R. 2013. Biological treatment of leather-tanning industrial wastewater using free living bacteria. *Adv. Life Sci. Technol.* 12: 46–65.

EPA. 2007. *Ortho-phenylphenol (OPP) & sodium ortho-phenylphenate (SOPP) risk characterization document.* Dietary Exposure Health Assessment Section, Medical Toxicology Branch, Department of Pesticide Regulation, California, Environmental Protection Agency, Sacramento, CA.

EU. 1998. Directive 98/8/EC of the European Parliament & of the Council of 16 February 1998 concerning the placing of biocidal products on the market.

EU. 2003. Commission Decision of 20 May 2003 Amending Decision 1999/815/EC concerning measures prohibiting the place on the market of toys and childcare articles intended to be placed in the mouth by children under three years of age made of soft PVC containing certain phthalates.

EU Azo Colorants Directive. 2002. http://www.tfl.com/web/files/eubanazodyes.pdf (accessed September 9, 2016).

Evangelista-Barreto, N.S., Albuquerque, C.D., Vieira, R.H.F.S., and CamposTakaki, G.M. 2009. Cometabolic decolorization of the reactive azo dye orange II by *Geobacillus stearothermophilus* UCP 986. *Textile Res. J.* 79(14): 1266–1273.

Fattorini, D., Notti, A., and Regoli, F. 2006. Characterization of arsenic content in marine organisms from temperate, tropical, and polar environments. *Chemistry and Ecology* 22: 405–414.

Fattorini, D., and Regoli, F. 2004. Arsenic speciation in tissues of the Mediterranean polychaete *Sabella spallanzanii*. *Environmental Toxicology and Chemistry* 23: 1881–1887.

Fei, Y., and Danny, R. 2015. Electro-bioremediation of contaminated sediment by electrode enhanced capping. *Journal of Environmental Management* 155: 154–161.

Flora, S.J.S., Bhadauria, S., Pant, S.C., and Dhaked, R.K. 2005. Arsenic induced blood and brain oxidative stress and its responses to some thiol chelators in rats. *Life Science* 77: 2324–2337.

Freeman, H.M. 1989. *Standard handbook of toxic and hazardous waste treatment and disposal.* McGraw-Hill Book Co., New York, NY.

Fuentes, S., Méndez, V., Aguila, P., and Seeger, M. 2014. Bioremediation of petroleum hydrocarbons: Catabolic genes, microbial communities, and applications. *Applied Microbiology and Biotechnology* 98: 4781–4794.

Fulekar, M.H., Singh, A., Thorat, V., Kaushik, C.P., and Eapen, S. 2010. Phytoremediation of 137 Cs from low level nuclear waste using Catharanthus roseus. *Indian Journal of Pure and Applied Physics* 48: 516–519.

Ganesan, V. 2008. Rhizoremediation of cadmium soil using a cadmium-resistant plant growth-promoting rhizopseudomonad. *Curr. Microbiol.* 56: 403–407.

Gami, A.A., Shukor, M.Y., Khalil, K.A., Dahalan, F.A., and Khalid, A. 2014. Phenol and phenolic compounds toxicity. *Journal of Environmental Microbiology and Toxicology* 2(1): 11–23.

Gargouri, B., Karray, F., Mhiri, N., Aloui, F., and Sayadis, S. 2014. Bioremediation of petroleum hydrocarbons-contaminated soil by bacterial consortium isolated from an industrial wastewater treatment plant. *Journal of Chemical Technology and Biotechnology* 89: 978–987.

Gavrilescu, M. 2005. Fate of pesticides in the environment and its bioremediation. *Engineering and Life Science* 5(6): 497–526.

Geiszinger, A.E., Goessler, W., and Francesconi, K.A. 2002. The marine polychaete *Arenicola marina*: Its unusual arsenic compound pattern and its uptake of arsenate from seawater. *Marine Environment Research* 53: 37–50.

Ghosh, M., Ganguli, A., and Tripathi, A.K. 2002. Treatment of anaerobically digested distillery spentwash in a two-stage bioreactor using *Pseudomonas putida* and *Aeromonas* sp. *Process Biochem.* 7: 857–862.

Gibson, J., and Harwood, C.S. 2002. Metabolic diversity in aromatic compound utilization by anaerobic microbes. *Annual Reviews of Microbiology* 56: 345–369.

Gill, R.T., Harbottle, M.J., Smith, J.W.N., and Thornton, S.F. 2014. Electrokinetic-enhanced bioremediation of organic contaminants: A review of processes and environmental applications. *Chemosphere* 107: 31–42.

Glick, B.R. 2010. Using soil bacteria to facilitate phytoremediation. *Biotechnology Advances* 28: 367–374.

Graca, B., Staniszewska, M., Zakrzewska, D., and Zalewska, T. 2016. Reconstruction of the pollution history of alkylphenols (4-tert-octylphenol, 4-nonylphenol) in the Baltic Sea. *Environmental Science and Pollution Research* 23: 11598–11610.

Gupta, K., Gaumat, S., and Mishra, K. 2012. Studies on phyto-genotoxic assessment of tannery effluent and chromium on *Allium cepa. Journal of Environmental Biology* 33(3): 557–563.

Gurulakshmi, M., Sudarmani, D.N.P., and Venba, R. 2008. Biodegradation of leather acid dye by Bacillus subtilis. *Advances in Biotechnology* 7: 12–19.

Guzzi, G., and La-Porta, C.A. 2008. Molecular mechanisms triggered by mercury. *Toxicology* 244(1): 1–12.

He, C.Q., Tan, G.E., Liang, X., Du, W., Chen, Y.L., Zhi, G.Y., et al. 2010. Effect of Zn-tolerant bacterial strains on growth and Zn accumulation in Orychophragmus violaceus. *Appl. Soil Ecol.* 44: 1–5.

Hentati, O., Lachhab, R., Ayadi, M., and Ksibi, M. 2013. Toxicity assessment for petroleum-contaminated soil using terrestrial invertebrates and plant bioassays. *Environmental Monitoring and Assessment* 185: 2989–2998.

Hue, N.V. 2013. Arsenic chemistry and remediation in Hawaiian soils. *Int. J. Phytoremediation.* 15(2): 105–116.

Huhle, B., Heilmeier, H., and Merkel, B. 2008. Potential of Brassica juncea and Helianthus annuus in phytoremediation for uranium. In *Uranium. Mine Hydrogeology*. pp. 307–318.

Hussain, F., Malik, S.A., Athar, M., Bashir, N., Younis, U., and Mahmood-ul-Hassan, M.S. 2010. Effect of tannery effluents on seed germination and growth of two sunflower cultivars. *African Journal of Biotechnology* 9(32): 5113–5120.

Hussain, S., Siddique, T., Saleem, M., Arshad, M., and Azeem, K. 2009. Impact of pesticides on soil microbial diversity, enzymes, and biochemical reactions. *Advances in Agronomy* 102: 159–200.

Hwang, E., Namkoong, W., and Park, J. 2001. Recycling of remediated soil for effective composting of diesel-contaminated soil. *Compost Science and Utilization* 9: 143–149.

IARC. 2004. *Monographs on the evaluation of carcinogenic risks to humans*. World Health Organization.

Igbinosa, E.O., Odjadjare, E.E., Chigor, V.N., Igbinosa, I.H., Emoghene, A.O., Ekhaise, F.O., Igiehon, N.O., and Idemudia, O.G. 2013. Toxicological profile of chlorophenols and their derivatives in the environment: The public health perspective. *The Scientific World Journal* 2013: 1–11. Article ID 460215.

Ifelebuegu, A.O., and Ezenwa, C.P. 2011. Removal of endocrine disrupting chemicals in wastewater treatment by fenton-like oxidation. *Water, Air, and Soil Pollution* 217(1): 213–220.

Integrated Pollution Prevention and Control Directive 96/61/EC. 1996. http://eur-lex.europa.eu/LexUriServ/LexUriServ.do?uri=CONSLEG:1996L0061:20060224:EN:PDF (accessed September 9, 2016).

Integrated Pollution Prevention and Control Directive 2008/1/EC. 2008. http://eur lex.europa.eu/LexUriServ/LexUriServ.do?uri=OJ:L:2008:024:0008:0029:en:PDF (accessed September 9, 2016).

Irfan, M., Hayat, S., Ahmad, A., and Alyemeni, M.N. 2013. Soil cadmium enrichment: Allocation and plant physiological manifestations. *Saudi Journal of Biological Science* 20(1): 1–10.

Jain, N., Minocha, A.K., and Verma, C.L. 2002. Degradation of predigested distillery effluent by isolated bacterial strains. *Ind. J. Exp. Bot.* 40: 101–105.

Jain, R.K., Kapur, M., Labana, S., Lal, B., Sarma, P.M., Bhattacharya, D., and Thakur, I.S. 2005. Microbial diversity: Application of microorganisms for the biodegradation of xenobiotics. *Current Science* 89(1): 101–112.

Jaishankar, M., Tseten, T., Anbalagan, N., Mathew, B.B., and Beeregowda, K.N. 2014. Toxicity, mechanism and health effects of some heavy metals. *Interdisciplinary Toxicology* 7(2): 60–72.

Jarup, L. 2003. Hazards of heavy metal contamination. *British Medical Bulletin* 68: 167–182.

Ji, P.H., Sun, T.H., Song, Y.F., Ackland, M.L., and Liu, Y. 2011. Strategies for enhancing the phytoremediation of cadmium-contaminated agricultural soils by *Solanum nigrum* L. *Environ. Pollut.* 159: 762–768.

Jiang, C.Y., Sheng, X.F., Qian, M., Wang, Q.Y. 2008. Isolation and characterization of a heavy metal resistant Burkholderia sp. from heavy metal-contaminated paddy field soil and its potential in promoting plant growth and heavy metal accumulation in metal polluted soil. *Chemosphere* 72: 157–164.

Kahraman, S., and Yesilada, O. 2003. Decolourization and bioremediation of molasses wastewater by white-rot fungi in a semi-solid state condition. *Folia Microbiol.* 48: 525–528.

Kamrin, M.A. 2009. Phthalate risks, phthalate regulation, and public health: A review. *Journal of Toxicology and Environmental Health, Part B* 12: 157–174.

Keharia, H., and Madamwar, D. 2003. Bioremediation concepts for treatment of dye containing wastewater: A review. *Indian Journal of Experimental Biology* 41(9): 1068–1075.

Khalid, A., Arshad, M., and Crowly, D.E. 2008. Accelerated dechlorination of structurally different azodyes by newly isolated bacterial strains. *Applied Microbiology and Biotechnology* 78(2): 361–369.

Khehra, M.S., Saini, H.S., Sharma, D.K., Chadha, B.S, and Chimni, S.S. 2006. Biodegradation of azo Dye C.I. acid red 88 by an anoxic-aerobic sequential bioreactor. *Dyes Pigments* 70(1): 1–7.

Khorasani, N., Yousefi, N., and Karami, M. 2011. Findings on the phytoextraction and phytostabilization of soils contaminated with heavy metals. *Biology of Trace Elements and Research* 144: 1133–1141.

Kim, I-S., Ekpeghere, K.I., Ha, S-Y., Kim, B-S., Song, B., Kim, J-T., Kim, H-G., and Koh, S-C. 2014. Full-scale biological treatment of tannery wastewater using the novel microbial consortium BM-S-1. *J. Environ. Sci. Health A Tox. Hazard Subst. Environ. Eng.* 49(3): 355–364.

Kumar, H., and Kumari, J.P. 2015. Heavy metal lead influative toxicity and its assessment in phytoremediating plants – A review. *Water, Air and Soil Pollution* 226: 324. doi: http://dx.doi.org/10.1007/398_2015_5003.

Kumar, P., and Chandra, R. 2006. Decolourisation and detoxification of synthetic molasses melanoidins by individual and mixed cultures of Bacillus spp. *Bioresource Technology* 7: 2096–2102.

Kumar, R., Das, A.J., and Lal, S. 2015. Petroleum hydrocarbon stress management in soil using microorganisms and their products. In *Environmental waste management,* Chandra, R. (ed.), CRC Press, Boca Raton, FL, pp. 525–550.

Kumar, V., Majumdar, C., and Roy, P. 2008. Effects of endocrine disrupting chemicals from leather industry effluents on male reproductive system. *Journal of Steroid Biochemistry and Molecular Biology* 111(3–5): 208–216.

Lageman, R., and Godschalk, M.S 2007. Electro-bioreclamation: a combination of in situ remediation techniques proves successful at a site in Zeist, the Netherlands. *Electrochim. Acta* 52: 3449–3453.

Li, H.-Y., Wei, D.-Q., Shen, M., and Zhou, Z.-P. 2012. Endophytes and their role in phytoremediation. *Fungal Diversity* 54: 11–18.

Li, T., Guo, S., Wu, B., Li, F., and Niu, Z. 2010. Effect of electric intensity on the microbial degradation of petroleum pollutants in soil. *Journal of Environmental Science* 22: 1381–1386.

Lin, J., Zhang, X., Li, Z., and Lei, L. 2010. Biodegradation of reactive blue 13 in a two-stage anaerobic/aerobic fluidized beds system with a *Pseudomonas* sp. isolate. *Bioresour. Technol.* 101(1): 34–40.

Littlefield-Wyer, J.G., Brooks, P., and Katouli, M. 2008. Application of biochemical fingerprinting and fatty acid methyl ester profiling to assess the effect of the pesticide Atradex on aquatic microbial communities. *Environmental Pollution* 153: 393–400.

Lofrano, G., Aydin, E., Russo, F., Guida, M., Belgiorno, V., and Meric, S. 2008. Characterization, fluxes and toxicity of leather tanning bath chemicals in a large tanning district area (IT). *Water Air Soil Pollut.* 8: 529–542.

Lofrano, G., Meric, S., Zengin, G.E., and Orhon, D. 2013. Chemical and biological treatment technologies for leather tannery chemicals and wastewaters: A review. *Science of the Total Environment* 461–462: 265–281.

Lukas, Y.W., Lei, S., and Hauke, H. 2007. Electro-bioremediation of hydrophobic organic soil-contaminants: A review of fundamental interactions. *Electrochimica Acta* 52: 3441–3448.

Ma, Y., Prasad, M.N.V., Rajkumar, M., and Freitas, H. 2011. Plant growth promoting rhizobacteria and endophytes accelerate phytoremediation of metalliferous soils. *Biotechnology Advances* 29: 248–258.

Ma, Y., Rajkumar, M., and Freitas, H. 2009. Inoculation of plant growth promoting bacterium Achromobacter xylosoxidans strain Ax10 for the improvement of copper phytoextraction by Brassica juncea. *J. Environ. Manage.* 90: 831–837.

Mahimaraja, S., and Bolan, N.S. 2004. Problems and prospects of agricultural use of distillery spentwash in India. Super Soil 3rd Australian New Zealand Soils Conference, 5–9 December 2004, University of Sydney, Sydney, New South Wales, Australia.

Mahmood, S., Khalid, A., Mahmood, T., Arshad, M., and Ahamad, R. 2013. Potential of newly isolated bacterial strains for simultaneous removal of hexavalent chromium and reactive Black-5 azodye from tannery effluent. *Journal of Chemical Technology and Biotechnology* 88(8): 1506–1513.

Mandal, T., Dasgupta, D., Mandal, S., and Datta, S. 2010. Treatment of leather industry by aerobic biological fenton oxidation process. *J. Hazard Mater* 180(1–3): 204–211.

Mangkoedihardjo, S., and Surahmaida. 2008. Jatropha curcas L. for phytoremediation of lead and cadmium polluted soil. *World Applied Science Journal* 4: 519–522.

Mannisto, M.K., Salinoja-Salonen, M.S., and Puhakka, J.A. 2001. In situ poly-chlorophenol bioremediation potential of the indigenous bacterial community of boreal ground water. *Water Research* 35: 2496–2504.

Manojlovic, D., Ostojic, D.R., Obradovic, B.M., Kuraica, M.M., Krsmanovic, V.D., and Puric, J. 2007. Removal of phenol and chlorophenols from water. *Desalination* 213): 116–122.

Margesin, R., Walder, G., and Schinner, F. 2000. The impact of hydrocarbon remediation (diesel oil and polycyclic aromatic hydrocarbons) on enzyme activities and microbial properties of soil. *Acta Biotechnologica* 20: 313–333.

Marmiroli, N., Marmiroli, M., and Elena, M. 2006. Phytoremediation and phytotechnologies: A review for the present and the future. In *Soil and Water Pollution Monitoring, Protection and Remediation*, Twardowska, I., et al. (eds.), Springer, The Netherlands, pp. 403–416.

Martınez-Prado, M.A., Unzueta-Medina, J., and Perez Lopez, M.E. 2014. Electrobioremediation as a hybrid technology to treat soil contaminated with total petroleum hydrocarbons. *Revista Mexicana de Ingeniería Química* 13(1): 113–127.

Maszenan, A.M., Liu, Y., and Ng, W.J. 2011. Bioremediation of wastewaters with recalcitrant organic compounds and metals by aerobic granules. *Biotechnology Advances* 29: 111–123.

Mathur, N., Singh, J., Bohra, S., Bohra, A., Mehboob Vyas, M., and Vyas, A. 2010. Phytoremediation potential of some multipurpose tree species of Indian Thar Desert in oil contaminated soil. *Advances in Environment Biology* 4(2): 131–137.

Megharaj, M., Ramakrishnan, B., Venkateswarlu, K., Sethunathan, N., and Naidu, R. 2011. Bioremediation approaches for organic pollutants: A critical perspective. *Environment International* 37: 1362–1375.

Mendez-Paz, D., Omil, F., and Lema, J.M. 2005. Anaerobic treatment of azo dye Acid Orange 7 under fed-batch and continuous. *Water Research* 39: 771–778.

Messer, R.L., Lockwood, P.E., Tseng, W.Y., Edwards, K., Shaw, M., Caughman, G.B., Lewis, J.B., and Wataha, J.C. 2005. Mercury (II) alters mitochondrial activity of monocytes at sublethal doses via oxidative stress mechanisms. *Journal of Biomedical Materials Research B* 75: 257–263.

Michałowicz, J., and Duda, W. 2007. Phenols-sources and toxicity. *Polish Journal of Environmental Studies* 16(3): 347–362.

Mikula, P., Svobodová, Z., and Smutná, M. 2005. Phthalates: toxicology and food safety – a review. *Czech Journal for Food Science* 23: 217–223.

Mishra, S., and Bharagava, R.N. 2016. Toxic and genotoxic effects of hexavalent chromium in environment and its bioremediation strategies. *Journal of Environmental Science and Health, Part. C Environmental Carcinogenesis and Ecotoxicology Reviews* 34(1): 1–32.

Mittal, A., Mittal, J., Malviya, A., Kaur, D., and Gupta, V.K. 2010. Adsorption of hazardous crystal violet from waste water by waste materials. *Journal of Colloid and Interface Science* 343: 463–473.

Mohana, S., Desai, C., and Madamwar, D. 2007. Biodegradation and decolourization of anaerobically treated distillery spent wash by a novel bacterial consortium. *Bioresour. Technol.* 98: 333–339.

Mohanpuria, P., Rana, N.K., and Yadav, S.K. 2007. Cadmium induced oxidative stress influence on glutathione metabolic genes of *Camellia sinensis* (L.) O. Kuntze. *Environmental Toxicology* 22: 368–374.

Mohanty, M., and Kumar, P.H. 2013. Effect of ionic and chelate assisted hexavalent chromium on mung bean seedlings (*Vigna Radiata* l. Wilczek. Var k-851) during seedling growth. *Journal of Stress Physiology and Biochemistry* 9(2): 232–241.

Morais, S., Costa, F.G., and Pereira, M.L. 2012. Heavy metals and human health. In *Environmental health – emerging issues and practice,* Oosthuizen, J. (ed.), InTech Publisher, Croatia. pp. 227–246.

Mukhopadhyay, S., and Maiti, S.K. 2010. Phytoremediation of metal mine waste. *Applied Ecology and Environmental Research* 8(3): 207–222.

Nachiyar, C.V., and Rajkumar, G.S. 2003. Degradation of a tannery and textile dye, Navitan Fast Blue S5R by *Pseudomonas aeruginosa. World Journal of Microbiology and Biotechnology* 19(6): 609–614.

Mirza, N., Mahmood, Q., Maroof Shah, M., Pervez, A., and Sultan, S. 2014. Plants as Useful Vectors to Reduce Environmental Toxic Arsenic Content. The Scientific World Journal Article ID 921581.

Onyancha, D., Mavura, W., Ngila, J., Ongoma, P., and Chacha, J. 2008. Studies of chromium removal from tannery wastewaters by algae biosorbents, Spirogyra condensate and Rhizocolonium hieroglyphicum. *J. Hazard Mater* 158(2–3): 605–614.

Ortiz-Hernández, M.L., Sánchez-Salinas, E., Dantán-González, E., and Castrejón-Godínez, M.L. 2013. Pesticide biodegradation: Mechanisms, genetics and strategies to enhance the process. In *Biodegradation—Life of science,* Chamy, R., and Rosenkranz, F. (eds.), InTech, Croatia. pp. 251–287.

Osugi, M.E., Rajeshwar, K., Ferraz, E.R.A., de Oliveira, D.P., Araujo, A.R., and Zanoni, M.V.W. 2009. Comparison of oxidation efficiency of disperse dyes by chemical and photoelectrocatalytic chlorination and removal of mutagenic activity. *Electrochimica Acta* 54(7): 2086–2093.

Pandey, A., Singh, P., and Iyengar, L. 2007. Bacterial decolorization and degradation of azo dyes. *International Biodeterioration and Biodegradation* 59: 73–84.

Pant, D., and Adholeya, A. 2007. Biological approaches for treatment of distillery wastewater. *Bioresource Technology* 98: 2321–2334.

Patel, Y., Mehta, C., and Gupte, A. 2012. Assessment of biological decolorization and degradation of sulfonated di-azo dye Acid Maroon V by isolated bacterial consortium EDPA. *International Biodeterioration and Biodegradation* 75: 187–193.

Plavsic, M., Cosovic, B., and Lee, C. 2006. Copper complexing properties of melanoidins and marine humic material. *Science of the Total Environment* 366: 310–319.

Preeti, P.P., Tripathi, A.K., and Shikha, G. 2011. Phytoremediation of arsenic using Cassia fistula Linn. seedling. *International Journal of Research and Chemistry of Environment* 1: 24–28.

Preethi, S., Anumary, A.M.A., and Thanikaivelan, P. 2013. Probing horseradish peroxidase catalyzed degradation of azo dye from tannery wastewater. *SpringerPlus* 2: 341.

Rajkumar, M., Ma, Y., and Freitas, H. 2008. Characterization of metal-resistant plant-growth promoting Bacillus weihenstephanensis isolated from serpentine soil in Portugal. *J. Basic Microbiol.* 48: 1–9.

Rameshraja, D., and Suresh, S. 2011. Treatment of tannery wastewater by various oxidation and combined processes. *International Journal of Environmental Research* 5(2): 349–360.

Reddy, A.M., Kumar, S.G., Jyonthsnakumari, G., Thimmanaik, S., and Sudhakar, C. 2005. Lead induced changes in antioxidant metabolism of horse gram (*Macrotyloma uniflorum* (Lam.) Verdc.) and bengal gram (*Cicer arietinum* L.). *Chemosphere* 60: 97–104.

Reshma, S.V., Spandana, S., and Sowmya, M. 2011. *Bioremediation technologies.* World Congress of Biotechnology, Hyderabad, India.

Rice, K.M., Walker, E.M., Jr., Wu, M., Gillette, C., and Blough, E.R. 2014. Environmental mercury and its toxic effects. *Journal of Preventive Medicine and Public Health* 47: 74–83.

Saichek, R.E., and Reddy, K.R. 2005. Electrokinetically enhanced remediation of hydrophobic organic compounds in soil: A review. *Critical Reviews in Environmental Science and Technology* 35: 115–192.

Salmeron-Alcocer, A., Ruiz-Ordaz, N., Juarez-Ramírez, C., and Galındez-Mayer, J. 2007. Continuous biodegradation of single and mixed chlorophenols by a mixed microbial culture constituted by *Burkholderia* sp., *Microbacterium phyllosphaerae*, and *Candida tropicalis*. *Biochemical Engineering Journal* 37: 201–211.

Salido, A.L., Hastly, K.L., Lim, J.M., and Butcher, D.J. 2003. Phytoremediation of arsenic and lead in contaminated soils using Chinese brake ferns (Pteris vittata) and Indian mustard (Brassica juncea). *Int. J. Phytoremediation.* 5(2): 89–103.

Santos, V.L.D., Monteiro, A.D.S., Braga, D.T., and Santoro, M.M. 2009. Phenol degradation by *Aureobasidium pullulans* FF13 isolated from industrial effluents. *Journal of Hazard Material* 161: 1413–1420.

Saratale, R.G., Saratale, G.D., Chang, J.S., and Govindwar, S.P. 2010. Decolorization and biodegradation of reactive dyes and dye wastewater by a developed bacterial consortium. *Biodegradation* 21(6): 999–1015.

Satyawali, Y., and Balakrishnan, M. 2008. Wastewater treatment in molasses-based alcohol distilleries for COD and color removal: A review. *Journal of Environmental Management* 86: 481–499.

Saxena, G., and Bharagava, R.N. 2015. Persistent organic pollutants and bacterial communities present during the treatment of tannery wastewater. In *Environmental waste management,* Chandra, R. (ed.), CRC Press, Taylor and Francis Group, Boca Raton, FL, pp. 217–247.

Saxena, G., and Bharagava, R.N. 2016. Ram Chandra: Advances in biodegradation and bioremediation of industrial waste. *Clean Technology and Environmental Policy* 18: 979–980. doi: http://dx.doi.org/10.1007/s10098-015-1084-9.

Saxena, G., Chandra, R., and Bharagava, R.N. 2016. Environmental pollution, toxicity profile and treatment approaches for tannery wastewater and its chemical pollutants. *Reviews of Environmental Contamination and Toxicology* 240: 31–69. doi: http://dx.doi.org/10.1007/398_2015_5009.

Senthilkumaar, S., Kalaamani, P., and Subburaam, C.V. 2006. Liquid phase adsorption of Crystal violet onto activated carbons derived from male flowers of coconut tree. *Journal of Hazardous Material* 136: 800–808.

Shanker, A.K., Cervantes, C., Loza-Tavera, H., and Avudainayagam, S. 2005. Chromium toxicity in plants. *Environmental International* 31(5): 739–753.

Sharma, B.M., Bharat, G.K., Tayal, S., Nizzetto, L., Cupr, P., and Larssen, T. 2014. Environment and human exposure to persistent organic pollutants (POPs) in India: A systematic review of recent and historical data. *Environment International* 66: 48–64.

Sharma, H. 2011. Metal hyperaccumulation in plants: A review focusing on phytoremediation technology. *Journal of Environmental Science and Technology* 4: 118–138.

Sharma, P., and Dubey, R.S. 2005. Lead toxicity in plants. *Brazilian Journal of Plant Physiology* 17: 35–52.

Sheng, X., Xia, J., Jiang, C., He, L., and Qian, M. 2008. Characterization of heavy metal-resistant endophytic bacteria from rape (Brassica napus) roots and their potential in promoting the growth and lead accumulation of rape. *Environ. Pollut.* 156: 1164–1170.

Singh, N., Kumar, D., and Sahu, A. 2007. Arsenic in the environment: Effects on human health and possible prevention. *Journal of Environmental Biology* 28(2 Suppl): 359–365.

Singh, R., Gautam, N., Mishra, A., and Gupta, R. 2011. Heavy metals and living systems: An overview. *Indian Journal of Pharmacology* 43(3): 246–253.

Singh, S., Singh, B.B., Chandra, R., Patel, D.K., and Rai, V. 2009. Synergistic biodegradation of pentachlorophenol by *Bacillus cereus* (DQ002384), Serratia marcescens (AY927692) and Serratia marcescens (DQ002385). *World J. Microbiol. Biotechnol.* 25: 1821–1828.

Sinha, S., and Mukherjee, S.K. 2008. Cadmium-induced siderophore production by a high Cd-resistant bacterial strain relieved Cd toxicity in plants through root colonization. *Curr. Microbiol.* 56: 55–60.

Sinha, S., Singh, S., and Mallick, S. 2008. Comparative growth response of two varieties of *Vigna radiata* L. (var. PDM 54 and var. NM 1) grown on different tannery sludge applications: Effects of treated wastewater and ground water used for irrigation. *Environmental Geochemistry and Health* 30(22): 407–422.

Sirianuntapiboon, S., Phothilangka, S., and Ohmomo, S. 2004. Decolourization of molasses wastewater by a strain no. BP103 of acetogenic bacteria. *Bioresour. Technol.* 92: 31–39.

Sittig, M. 1981. *Handbook of toxic and hazardous chemicals.* Noyes Publications, Park Ridge, NJ, pp. 525–526.

Srivastava, S., Ahmad, A.H., and Thakur, I.S. 2007. Removal of chromium and pentachlorophenol from tannery wastewaters. *Bioresour. Technol.* 98(5): 1128–1113.

Steinle, P., Stucki, G., Stettler, R., and Hanselmann, K.W. 1998. Aerobic mineralization of 2,6-dichlorophenol by *Ralstonia* sp. strain RK1. *Applied and Environmental Microbiology* 64: 2566–2571.

Sujata and Bharagava, R.N. 2016. Exposure to Crystal Violet, Its Toxic, Genotoxic and Carcinogenic Effects on Environment and Its Degradation and Detoxification for Environmental Safety. *Rev. Environ. Contam. Toxicol.* 237: 71–104.

Sung, R.H., Soydoa, V., and Hiroaki, O. 2000. Biodegradation by mixed microorganism of granular activated carbon loaded with a mixture of phenols. *Biotechnology Letters* 22: 1093–1096.

Tchounwou, P.B., Yedjou, C.G., Patlolla, A.K., and Sutton, D.J. 2012. Heavy metal toxicity and the environment. *EXS* 101: 133–164.

Tewari, C.P., Shukla, S., and Pandey, P. 2011. Biodegradation of pentachlorophenol (PCP) by consortium of Flavobacterium sp. in tannery effluent. *Journal of Environmental Research and Development* 7(2A): 876–882.

Tormoehlen, L.M., Tekulve, K.J., and Agas, K.A. 2014. Hydrocarbon toxicity: A review. *Clinical Toxicology* 52: 479–489.

Tripathi, M., Vikram, S., Jain, R.K., and Garg, S.K. 2011. Isolation and growth characteristics of chromium(VI) and pentachlorophenol tolerant bacterial isolate from treated tannery effluent for its possible use in simultaneous bioremediation. *Indian Journal of Microbiology* 51(1): 61–69.

Trivedi, P., Pandey, A., and Sa, T. 2007. Chromate reducing and plant growth promoting activities of psychrotrophic Rhodococcus erythropolis MtCC 7905. *J. Basic Microbiol.* 47: 513–517.

Trujillo-Tapia, N., Mondragon, C.C., Vasquez-Murrieta, M.S., Cleemput, O.V., and Dendooven, L. 2008. Inorganic N dynamics and N2O production from tannery effluents irrigated soil under different water regimes and fertilizer application rates: A laboratory study. *Applied Soil Ecology* 38(3): 279–288.

USDHHS. 2001. United States Department of Health and Human Services (USDHHS): Public health statement. In: *Toxicological profile for pentachlorophenol*. Prepared by Syracuse Research corporation, Washington, D.C., pp. 1–11.

Vangronsveld, J., Herzig, R., Weyens, N., Jana, B., Adriaensen, K., Ruttens, A., Thewys, T., et al. 2009. Phytoremediation of contaminated soils and groundwater: Lessons from the field. *Environmental Science and Pollution Research* 16(7):765–794.

Vankar, P.S., and Bajpai, D. 2008. Phytoremdiation of chrome-VI of tannery effluent by Trichoderma species. *Desalination* 222(1–3): 255–262.

Varun, M., DSouza, R., Pratas, J., and Paul, M.S. 2012. Metal contamination of soils and plants associated with the glass industry in North Central India: Prospects of phytoremediation. *Environment Science and Pollution Research* 19: 269–281.

Ventura-Lima, J., Mauricio-Reis, B., and Jose, M. 2011. Arsenic toxicity in mammals and aquatic animals: A comparative biochemical approach. *Ecotoxicology and Environmental Safety* 74: 211–218.

Verma, T., and Maurya, A. 2013. Isolation of potential bacteria from tannery effluent capable to simultaneously tolerate hexavalent chromium and pentachlorophenol and its possible use in effluent bioremediation. *International Journal of Engineering and Science* 2: 64–69.

Vivas, A., Biro, B., Ruíz-Lozanoa, J.M., and Azcon, R. 2006. Two bacterial strains isolated from a Zn-polluted soil enhance plant growth and mycorrhizal efficiency under Zn toxicity. *Chemosphere* 52: 1253–1533.

Wang, D., Liu, Y., Lin, Z., Yang, Z., and Hao, C. 2008. Isolation and identification of surfactin producing *Bacillus subtilis* strain and its effect of surfactin on crude oil. *Wei Sheng Wu Xue Bao* 48(3): 304–311.

Wang, M.C., Gong, M., Zang, H.B., Hua, X.M., Yao, J., Pang, Y.J., and Yang, Y.H. 2006. Effect of methamidophos and urea application on microbial communities in soils as determined by microbial biomass and community level physiological profiles. *Journal of Environmental Science and Health Part B* 41: 399–413.

Wang, Y.S., Pan, Z.Y., Lang, J.M., Xu, JM., and Zheng, Y.G. 2007. Bioleaching of chromium from tannery sludge by indigenous, Acidithiobacillus thiooxidans. *J. Hazard Mater* 147(1–2): 319–334.

Wani, A.L., Ara, A., and Usmani, J.A. 2015. Lead toxicity: A review. *Interdisciplinary Toxicology* 8(2): 55–64.

Wani, P.A., Khan, M.S., and Zaidi, A. 2007. Effect of metal tolerant plant growth promoting Bradyrhizobium sp. (vigna) on growth, symbiosis, seed yield and metal uptake by green gram plants. *Chemosphere* 70: 36–45.

Wójcik, M., and Tukiendorf, A. 2004. Phytochelatin synthesis and cadmium localization in wild type of *Arabidopsis thaliana*. *Plant Growth Regulation* 44: 71–80.

Xu, M., Guo, J., and Sun, G. 2007. Biodegradation of textile azo dye by Shewanella decolorationis S12 under microaerophilic conditions. *Appl. Microbiol. Biotechnol.* 76(3): 719–726.

Yadav, S.K. 2010. Heavy metals toxicity in plants: An overview on the role of glutathione and phytochelatins in heavy metal stress tolerance of plants. *South African Journal of Botany* 76: 167–179.

Yang, O., Kim, H.L., Weon, J.-II., and Seo, Y.R. 2015. Endocrine-disrupting chemicals: Review of toxicological mechanisms using molecular pathway analysis. *Journal of Cancer Prevention* 20: 12–24.

Ye, X., Guo, X., Cui, X., Zhang, X, Zhang, H., Wang, M.K., Qiu, L., and Chen, S. 2012. Occurrence and removal of endocrine-disrupting chemicals in wastewater treatment plants in the Three Gorges Reservoir area, Chongqing, China. *Journal of Environmental Monitoring* 14(8): 2204–2211.

Zadeh, B.M., Savaghebi-Firozabadi, G.R., Alikhani, H.A., and Hosseini, H.M. 2008. Effect of sunflower and amaranthus culture and application of inoculants on phytoremediation of the soils contaminated with cadmium. *Americam-Eurasian Journal of Agriculture and Environment Science* 4: 93–103.

Zhang, W.H., and Tyerman, S.D. 1999. Inhibition of water channels by HgCl2 in intact wheat root cells. *Plant Physiology* 120: 849–857.

Zhou, Z.S., Huang, S.Q., Guo, K., Mehta, S.K., Zhang, P.C., and Yang, Z.M. 2007. Metabolic adaptations to mercury-induced oxidative stress in roots of *Medicago sativa* L. *Journal of Inorganic Biochemistry* 10: 1–9.

Zhuang, X., Chen, J., Shim, H., and Bai, Z. 2007. New advances in plant growth-promoting rhizobacteria for bioremediation. *Environment International* 33: 406–413.

# 3 Pesticides Contamination in the Environment

## *Toxicological Effects, and Biodegradation and Bioremediation Mechanisms for Environmental Safety*

*Namasivayam Vasudevan and Odukkathil Greeshma*

**CONTENTS**


3.1 Introduction ... 58
3.2 Use of Pesticide: Worldwide Scenario ... 59
3.3 Pesticide Residues in the Environment ... 62
 3.3.1 Pesticide Residues in Water Resources ... 62
 3.3.2 Pesticide Residues in Soil and Sediment ... 63
 3.3.3 Pesticide Residues in the Air ... 65
 3.3.4 Pesticide Residues in the Biotic System ... 65
 3.3.5 Environmental Fate of Pesticides ... 66
3.4 Toxicological Impact of Pesticides ... 67
 3.4.1 Toxicity in Plants ... 68
 3.4.2 Toxicity in Animals ... 69
 3.4.3 Toxicity in Microbes ... 70
 3.4.4 Toxicity in Humans ... 70
3.5 Biodegradation of Pesticides ... 72
 3.5.1 Microbes Involved in Pesticide Biodegradation ... 72
 3.5.2 Role of Enzymes in Biodegradation of Pesticides ... 72
 3.5.3 Biodegradation Pathway of Pesticide ... 74
  3.5.3.1 Oxidation and Reduction ... 74
  3.5.3.2 Dehalogenation ... 74
  3.5.3.3 Conjugation ... 75
 3.5.4 Biodegradation of Pesticides in Water ... 75
  3.5.4.1 Factors Affecting the Biodegradation of Pesticides ... 76

3.5.5 Biodegradation of Pesticides in Soil ... 78
3.5.5.1 pH and Temperature ... 78
3.5.5.2 Soil Texture ... 79
3.5.5.3 Soil Moisture Content ... 80
3.5.5.4 Adsorption and Bioavailability ... 80
3.6 Bioremediation of Pesticides-Contaminated Soil and Water ... 82
3.6.1 Microbial Catalyst ... 83
3.6.2 *Ex Situ* and *In Situ* Bioremediation Techniques ... 83
3.6.2.1 Bioaugmentation ... 83
3.6.2.2 Biosparging ... 84
3.6.2.3 Biosorption ... 84
3.6.2.4 Composting ... 85
3.6.2.5 Earthworm Assisted Bioremediation ... 85
3.6.2.6 Surfactant Mediated Bioremediation ... 85
3.6.2.7 Organic Amendment Enhanced Bioremediation ... 87
3.6.2.8 Biobed ... 88
3.7 Conclusion ... 90
References ... 90

## 3.1 INTRODUCTION

Pesticides include a wide range of chemical compounds of different chemical formulations intended to prevent, destroy, repel, or mitigate any pest. Chemical formulations of pesticide include an active ingredient and an inert ingredient. An active ingredient is the chemical substance, which averts, wipes out, resists, or alleviates a pest, or is a plant regulator, defoliant, and dessicant or nitrogen stabilizer. Inert ingredients are generally the components that are important for product performance and usability. These include substances like common food commodities, such as edible oils, spices, herbs, and some natural materials like beeswax and cellulose. The inert component of the pesticide formulation acts as a solvent to help in the penetration of active ingredients into the plant parts. It also functions in preventing foam formation, extends product shelf life, improves safety for the user, and prevents photodegradation.

There are different types of pesticides depending on their chemical constituents and on their usage. Based on the chemical constituents, pesticides are classified as organochlorine, organophosphate, carbamate, synthetic pyrethroids, and inorganic pesticides. There is another class of pesticide that is produced from natural materials, called a biopesticide. Based on their usage, pesticides are of different types: insecticides, which are used against the insect pest; nematicides, used against nematodes; fungicides, used against fungi; weedicides, used against weed pests; and so on. An increase in the demand for agro-products and pesticide resistance to pests have resulted in the invention of new formulations of higher toxicity and dosage.

According to the World Health Organization's (WHO) recommended classification of pesticide by hazard, the active ingredients of pesticide are classified as extremely hazardous Class 1A, highly hazardous Class 1B, moderately hazardous Class 2, and slightly hazardous Class 3 active ingredients (WHO 2004). These active ingredients are diluted by a carrier or diluents and are made available in various formulations like dust, granule, emulsifiable concentrate, wettable powder, fumigants, smokes, vapors, and aerosol. Commercially available pesticides are such formulations of active ingredients. Their dosage depends on the nature and extent of pest attack. Formulation and dosage also vary depending on the type of pest, type of crop, and plant parts where the pest had caused damage. Common active ingredients of pesticides now in use are atrazine, chlorpyrifos, cypermethrin, fenvalerate, diazinon, endosulfan, lindane, deltamethrin, ethion, carbofuran, simazine, permethrin, monocrotophos, etc.

## 3.2 USE OF PESTICIDE: WORLDWIDE SCENARIO

About 2 million tonnes of pesticides are consumed globally per year. This accounts to 45% usage by Europe, 25% usage by the United States, and 25% usage by other countries (Figure 3.1).

About 25% of the cultivated land area of the world is dependent on pesticides. Among the different types of pesticides, consumption of herbicides accounts for 47.5%, insecticides 29.5%, fungicides 17.5%, and others 5.5% (Figure 3.2).

Consumption of herbicides is more compared to insecticides and fungicides. Consumption of herbicides is more in the United States and Europe compared to Asian countries like India (Figure 3.3) because herbicide control is mainly done manually by hand. In the 1990s, global consumption of herbicides was less than in 2007 to 2008 when there was an increase. Over the years, there was increase in the sale of pesticides (Figure 3.3). Year 2008 marked a higher sale of pesticides at a global level; until 2003, sales of pesticides were highest in North America and Europe. From 2008 to 2012, there was an increase in sales of pesticide in Asian countries. Sales of pesticide were least in African countries.

Among the various pesticides registered for use, the most commonly used pesticides include dichlorodiphenyltrichloroethane (DDT), lindane, hexachlorocyclohexane (HCH), endosulfan, atrazine, chlorpyrifos, cypermethrin, carbaryl, acephate, carbofuran, Atrazine, and permethrin (Table 3.1).

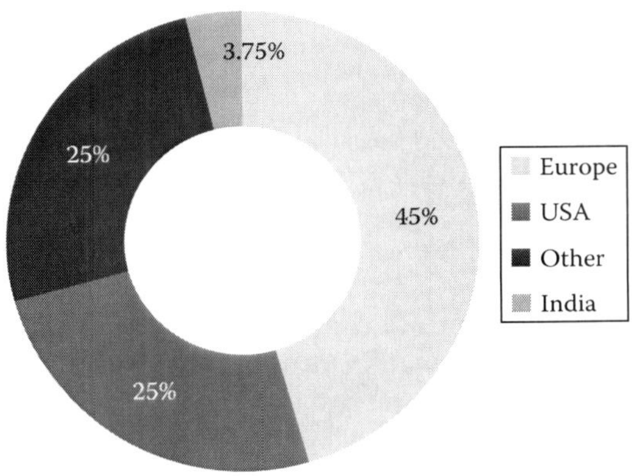


**FIGURE 3.1** Global consumption of pesticides.

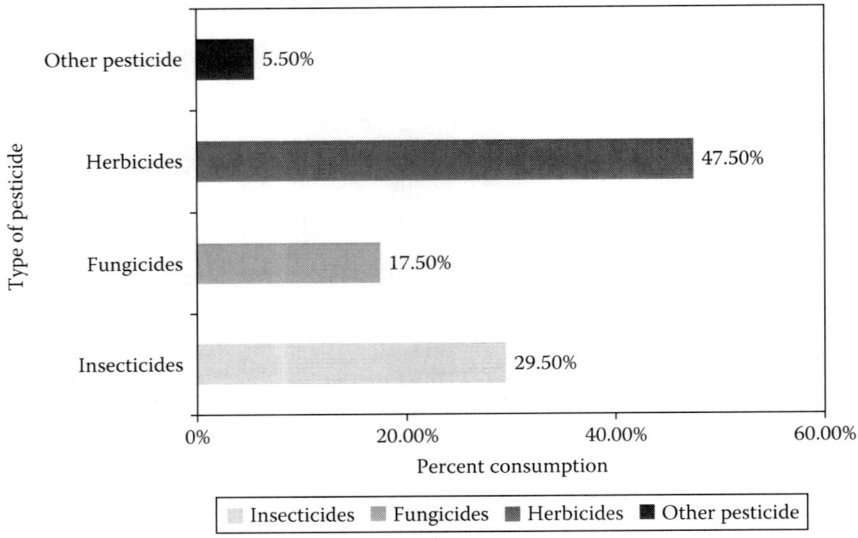


**FIGURE 3.2** Category-wise consumption of pesticides at the global level.

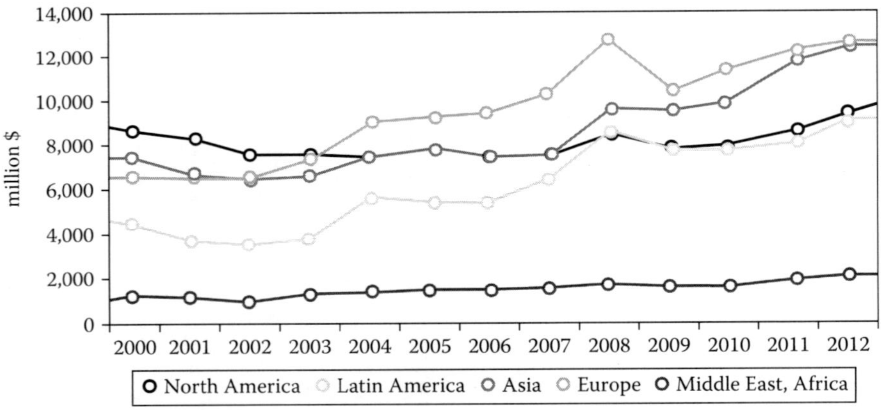


**FIGURE 3.3** Global pesticide sales.

Recently, the usage of organophosphorus insecticides has increased compared with the organochlorine insecticides. Organophosphorus insecticides are mostly used against aphids and mosquitoes causing viral diseases. Organochlorine, Carbamate, and pyrethroids pesticides are used against insect pests like Lepidoptera, Hemiptera, and Diptera. Most of these pesticides are used against insect pests of various crops like rice, wheat, cotton, plantation crops, vegetables, and fruits. All these are stomach and contact insecticides, which show larvicidal, adulticidal, and ovicidal activity.

**TABLE 3.1**
**Various Pesticide Formulations and Their Characteristics**

| Pesticide | Crop | Formulations | Dosage/(ha) | Toxicity Class | Water Solubility (mg/L) | KOC Adsorption Coefficient | Half-life (days) |
|---|---|---|---|---|---|---|---|
| Atrazine | Sugarcane | 50% W.P | 1–4 kg | Moderately toxic | 32 at 20°C | 122 | 110 |
| Chlorpyrifos | All crops | 20% E.C | 0.25–1 kg | Highly toxic | 1.39 at 25°C | 1.6 | 60–120 |
| | Rice | 10% G | 10 kg | | | | |
| | | 1.5 D.P | 25 kg | | | | |
| Cypermethrin | Brinjal, cabbage, rice (mixed with other pesticide), sunflower | 10% E.C | 550 per 800 L of water | Highly toxic | 0.0040 at 25°C | 6.6 | 7–168 |
| Deltamethrin | Rice, cotton | 25% Tablet | 2 g per 500 L of water | Highly toxic | 0.0002 at 25°C | 4.6 | 11–72 |
| | | 1.8% E.C | 625–780 mL per 500 L of water | | | | |
| Endosulfan | Chilly, cotton, jute, potato, cashew | 35% E.C | 5–10 kg | Highly toxic | 0.33 at 25°C | 3.5 | 60–800 |
| | | 4% D.P | 4–5.25 kg | | | | |
| Fenalverate | Cotton | 0.4% D.P | 20–25 kg | Highly toxic | 0.0013 at 25°C | 5.3 | 30–120 |
| Carbofuran | All crops | 3% G | 8–50 kg | Extremely toxic | 320 at 25°C | 22 | 10–200 |
| Diazinon | Soyabean, mango, tea, cauliflower | 20% E.C | 0.18–0.75 kg | Slightly toxic | 40 at 25°C | 2.2 | 30–40 |
| Dichlorvos | Paddy, wheat, cashew, cucurbit | 76% E.C | 470–1253 mL in 1000–2000 L of water | Extremely toxic | 10 g at 25°C | 30 | 50–60 |
| Lindane | Vegetables, paddy, maize | 6.5% W.P | 1.25–2.5 kg | Highly toxic | 7.3–7.8 at 25°C | 1100 | 100–126 |
| | | 6% G 1.3%. D.P | 25 kg | | | | |
| | | | 20 kg | | | | |
| Monocrotophos | Paddy, maize, pea, cotton | 36% S.C | 1250 mL in 500–1000 L of water | Extremely toxic | 154 at 25°C | | 20–30 |
| Permethrin | Cotton | 25% E.C | 400–500 mL in 1000 L of water | Highly toxic | 0.07 at 25°C | | 30–38 |

*Source*: Pesticide Association of India 1999; Pesticide Information Profiles 1996; Atwal, A.S., *Agricultural Pests of India and South East Asia*, New Delhi, India, Kalyani Publishers, 52, 1986; Lorenz, E.S., *Toxicity of Pesticides. Pesticide Safety Fact Sheet*, University Park, Pennsylvania State University, 2006.

*Note:* D.P, dustable powder; E.C, emulsifiable concentrate; G, granule; S.C, soluble concentrate; W.P, wettable powder.

## 3.3 PESTICIDE RESIDUES IN THE ENVIRONMENT

Pesticide residues reach different environments by point source and nonpoint source contamination. The main point source for pesticide contamination is by direct application of the pesticide and from release of wastewater from pesticide industries. Nonpoint source contamination includes agriculture runoff, air drift, and bioaccumulation through food. Due to the development of resistance in insects, there is an additional demand for increased dosage in the use of pesticides. This led to the persistence of pesticide and its metabolites in various environments due to lack of timely degradation. Pesticides, which are ubiquitous, persistent, and bioaccumulative in nature, are organochlorine pesticides like DDT, endosulfan, lindane, etc. (Loganathan and Kannan 1994; Hong et al. 1995; Lee et al. 1997). Residues of many organochlorine pesticides were detected in sediments and soil (Jayashree and Vasudevan 2006), seawater (Bidleman et al. 1995), groundwater, and other water sources (Sivasankaran et al. 2007). Apart from organochlorine pesticides, organophosphates, pyrethroids, and others have also been reported in environmental samples (Beena and Madan 2008).

### 3.3.1 Pesticide Residues in Water Resources

Intensive use of pesticides in agriculture, pesticide industries (point source), atmospheric fallout, agricultural runoff, industrial discharges, etc., had resulted in the occurrence of pesticide residues in various environments (Babu et al. 2005). Some pesticides are used as larvicides in health sanitation purposes and are applied to control the aquatic larval stages of insects of public health significance (e.g., mosquitos for the control of malaria and typhus), and this is one major point source contaminant. These pesticides include temephos, methoprene, pyriproxyfen, permethrin, chlorpyrifos, and DDT. Much surface water and groundwater in different regions around the world is contaminated with pesticides, especially by the organochlorine group of pesticides. Owing to the toxicity of many pesticides in use, the WHO and the U.S. Environment Protection Agency (USEPA) has set standard limits for the levels of pesticide residues in drinking water (Table 3.2).

Isomers of DDT, endosulfan, lindane, chlorpyrifos, cypermethrin, cyhalothrin, atrazine, phorate, parathion, dichlorvos, etc., are some of the pesticide residues detected in drinking and surface water (Jayashree and Vasudevan 2007b). The prevalence of pesticide residues in water indicates its persistence in the environment, for example, DDT, endosulfan, and HCH. Studies over different regions of the world still report the occurrence of DDT residues, which proves its persistence. Pesticide residues were widely detected in Asian Countries like India, China, and Pakistan. Kaushik et al. (2010) reported the presence of DDT residues ranging from 238 to 1000 ng/L in river Ghagger, India; similarly, higher concentrations of about 4000 ng/L of DDT residues were detected in rivers like Yamuna (Agarwal et al. 1986), Ganga (Halder et al. 1989; Sharma et al. 2003), and Krishna (Reddy et al. 1997). Another pesticide, widely reported as HCH, is called lindane. Residues of HCH 24.1 μg/L have been reported from various brands of mineral water, and a high concentration of lindane (up to 0.0042 mg/L) has been found in various brands of soft drinks (CSE 2003), and in many rivers like Ghaggar (Kaushik et al. 2010),

**TABLE 3.2**
**Guideline Limits for Pesticide in Drinking Water**

| Pesticides | Guideline Value (μg/L) |
|---|---|
| Alachlor | 20 |
| Aldicarb | 10 |
| Aldrin and dieldrin | 0.03 |
| Atrazine | 2 |
| Carbofuran | 7 |
| Chlordane | 0.2 |
| Chlorotoluron | 30 |
| Cyanazine | 0.6 |
| 2,4-D (2,4-dichlorophenoxyacetic acid) | 30 |
| 1,2-Dibromo-3-chloropropane | 1 |
| 1,2-Dibromoethane | 0.4 |
| 1,2-Dichloropropane (1,2-DCP) | 40 |
| 1,3-Dichloropropene | 20 |
| Dichlorprop | 100 |
| Dimethoate | 6 |
| Endrin | 0.6 |
| Fenoprop | 9 |
| Isoproturon | 9 |
| Lindane | 2 |
| Mecoprop | 10 |
| Methoxychlor | 20 |
| Metolachlor | 10 |
| Pendimethalin | 20 |
| Simazine | 2 |
| Trifluralin | 20 |
| Chlorpyrifos | 30 |
| DDT and metabolites | 1 |
| Permethrin | 300 |
| Pyriproxyfen | 300 |

*Source*: World Health Organization (WHO), *Report on Pesticide*, Geneva, Switzerland, WHO, 2005.

Ganga (Agnihotri et al. 1994), Gandaki (Srivastava et al. 1996), Krishna and Godavari (Reddy et al. 1997), and Yamuna (CPCB 2000; Kaushik et al. 2008).

## 3.3.2 Pesticide Residues in Soil and Sediment

Soil and sediment serve as a sink for pesticides where most of the pesticide residues get locked up. This mainly depends on the solubility of the pesticides, which influences the sorption of pesticides to organic matter. Binding of pesticide residue to soil is mainly due to adsorption of pesticide to soil particles, which in turn depends on the properties of soil and the pesticide. Soil properties include size, shape, configuration, molecular structure, chemical functions, solubility, polarity, polarizability and charge distribution of interacting species, and the acid-base nature of the pesticide molecule (Bailey and White 1970; Senesi 1992; Pignatello and Xing 1996). A variety of agricultural factors influence the fate and binding of pesticides in soil. They are concentration of pesticide applied, rate and mode of application, repeated application, aging, and use of organic and inorganic soil amendments. The environmental significance of a bound pesticide residue, however, depends not on its nonextractability under laboratory test conditions, but on its bioavailability (Khan 1982; Lock et al. 2002). Generally, bound residues are released by physico-chemical and biochemical processes, but in most of the cases the pesticide residues are more stable to both physico-chemical and biochemical processes. The physico-chemical process is affected by their highly stable chemical structure and less solubility, whereas the biochemical process is affected by their limited availability to plants and microbes (Gevao et al. 2000). Hence, the pesticide residues tend to accumulate in soil without undergoing degradation leading to persistence of pesticide in soil.

The main source of pesticide in soil is agriculture practices, either by direct application to soil or by indirect seepage to soil from plant parts. Pesticide adsorbed on the surface soil will leach down to subsurface soil and thus to various depths of soil. The occurrence of pesticides at different depths of soil depends mainly on sorption and biodegradation, which determine the mobility and longevity of the pesticides in the soil, respectively (Kookana et al. 1998). Sorption of pesticide to soil decreases with a decrease in soil organic matter; hence, as the soil depth increases, the sorption of pesticide decreases with the pattern governed by organic matter content (Clay and Koskinen 2003). The other factor is that biodegradation of pesticide varies with depth along with extent of sorption, because sorption is an indicator of bioavailability (Jensen et al. 2004). Degradation and sorption of pesticides in soils evidently shows three-dimensional variability with highly complex relationships between bioavailability and biodegradation in both subsoil and topsoils (Beulke et al. 2004). Therefore, the persistence or fate of pesticides on the soil surface is dependent on sorption and biodegradation of pesticide along with bioavailability.

The monitoring of pesticide residues at various depths in soil has revealed the downward drift of pesticide residues to various depths of soil. Sonia Rodriguez-Cruz et al. (2006) reported the presence of isoproturon and bentazone residues and variation in their degradation over a period of time at different depths of soil (up to 60 cm). Residues of organochlorines pesticides (endosulfan, aldrin, dicofol, and lindane), organophosphorus insecticides (three pirimiphos, dimethoate, and malathion), pyrethroids (deltamethrin), carbamates (carbaryl and propamocarb), and fungicides (tetradifon and vinclozolin) were detected in surface soil and 10 cm and 20 cm of soil collected from AL-Salhia New-AL-Sharqiyagovernorat, Egypt (Samir et al. 2012). Residues of dichlorvos, diazinon chlorpyrifos, and fenitrothion in soil collected from Borno State, Nigeria, at a soil depth 0–10 cm ranged from 65.87 to 211.03 μg/g; 82.45 to 232.34 μg/g at 11–20 cm, and 104.98 to 267.18 μg/g at 21–30 cm depth (Akan et al. 2013). The above studies revealed that the distribution of pesticides varies with depth. Giri and Rai (2012) reported a homogenous distribution of pesticide residues like dimethoate, methoate, dichlorvos, parathion-methyl, and metalaxyl at different depths of soil collected from soil profiles from Godavari in the Lalitpur district, Nepal, India (10 cm, 30 cm, and 50 cm). High concentrations of HCH residues were detected in soil samples at depths of 20 (0.6 to 3800 mg/Kg) and 50 cm (0.08 to 0.3 mg/Kg) at the Vikuge State Farm Coast Region, Tanzania, East Africa (Kishimba and Mihale 2004). The persistence of pesticide residues at 20–50 cm depth of soil is mainly due to decline in the biodegradation at this depth due to less bioavailability of pesticide and less microbial population (Fomsgaard 1995; Soulas and Lagacherie 2001).

Generally, the sediment is composed of detritus and inorganic and organic particles, and is relatively heterogeneous in terms of its physical, chemical, and biological characteristics (Hakanson 1992). Sediments are usually regarded as the ultimate sink for persistent pollutants discharged into the environment (Gibbs 1973) and are one of the best media for the long-term monitoring of many contaminants (Larsen and Jensen 1989). Pesticides enter into the marine environments through different pathways. Rivers and estuaries transport metal and organic contaminants to coastal marine environments. The input from estuaries are, in particular, important because estuaries are a major interface between the land and the sea (Hong et al. 1995). The applied pesticide in the agriculture land can be transported to surface water bodies through surface runoff, leaching, and vapor phase and ultimately accumulates and settles to the bottom sediments. Hence, bottom sediments represent an integrated measure of particle-bound contaminants that have deposited over a longer period of time (Sarkar et al. 2008).

### 3.3.3 Pesticide Residues in the Air

The major cause of pesticide contamination in air is due to the spray drift during their application and due to postapplication emissions. This accounts for 25%–30% of the total spraying. Postapplication emissions are mainly due to evaporation of pesticides from soil and plant surfaces and due to soil weathering. The soil particles adsorbed with pesticides drift in the air during weathering and can prevail in the air for many days to weeks. These pesticides in air may undergo decomposition before they settle down, or some may get dispersed in air as such and settle down elsewhere. The physical state of pesticides in the air can be in the form of gas or vapor and even dust in the case of pesticide dust formulations. Atmospheric transport of pesticide is one of the main sources for pollution of the environment due to pesticide, even in areas that are located far away from the point of application of pesticides.

### 3.3.4 Pesticide Residues in the Biotic System

The monitoring of pesticide residues in plants, animals, and animal products has great importance; hence, the levels of pesticide residues in flora and fauna are considered as appropriate indicators of environmental contamination by the pesticides. The main source of pesticide residue accumulation in plants is through absorption from soil and water either by direct application of pesticides or by agriculture runoff. Contaminated soil acts as a potential source for the contamination of water, flora and fauna, and finally man due to man's top position as consumer in a majority of food chains (Abrahams 2002). Various studies report the presence of pesticides in flora and fauna via food chain. Among the various fauna studied, pesticide residues are most frequently detected in fishes, birds, and poultry animals like cow, goat, buffaloes, etc. Plants such as *Talinum triangulare*, *Corchorus olitorius*, *Amaranthus caudatus*, *Cratylia argentea*, *Capsicum frutescens*, *Lycopersicum esculentum*, *Raphanus sativus*, *Zea mays*, and *Dioscorea alata* have shown to contain paraquat at 0.01–0.13 ppm (Akinloye et al. 2011). Persistent organochlorines such as DDT and its metabolites, HCH isomers, chlordane (CHL) compounds, hexachlorobenzene (HCB), and polychlorinated biphenyls (PCBs) were detected in whole-body homogenates of resident and migratory birds collected from South India (Tanabe et al. 1998). Residues of DDE at concentration $5.232 \pm 1.3$ ng/g were detected in tilapia fish (*Tilapia zillii*) from Lake Bosomtwi, Ghana; residues of organophosphate insecticides like BHC, heptachlor epoxide were detected in some brackish-water fishes and some shellfishes from Calicut region, Kerala, India. Residues in the flora and fauna around the world reveal the contamination of various ecosystems with pesticide residues and also their persistence in the environment.

Detection and identification of pesticides in the food we eat are a problem of increasing concern. Residues of many pesticides were detected in different food products, mainly fruits and vegetables. Monitoring studies done in India and many developing countries have revealed widespread contamination of fruits, vegetables, cereals, milk and milk products, and animal feed with pesticide residues (Adeniyi and Oladele 1999; Kang et al. 2000, 2002; Shah et al. 2000; Singh 2002; Kole et al. 2002; Battu et al. 2004). Beena et al. (2006) reported the presence of organochlorine pesticides endosulfan and lindane, organophosphate pesticide, and some pyrethroids in fruit samples including guava, grapes, and berries harvested in Hisar, Haryana, India. Residues of organophosphate pesticides like methyl parathion, chlorpyrifos, and malathion ranging from 2.5 to 6 ppb were detected in vegetable samples of North India (Bhanti and Taneja 2007). Most of the studies report the residues within the maximum residue limit (MRL), but continuous dietary intake of such food has resulted in the accumulation of pesticide residues in human beings. Many studies have reported that pesticides can bring chronic toxicological effects in human beings due to this dietary intake of pesticides (Nolan et al. 1984; CDC 2005; Van Audenhaege et al. 2009).

### 3.3.5 Environmental Fate of Pesticides

Most of the pesticide formulations are made in a way that upon environmental exposure they will undergo degradation via physical, chemical, and biological reactions. Increased consumption of the pesticides led to the persistence of pesticide and its metabolites in various environments due to lack of timely degradation. The outcome of their persistency is contamination of water sources, destruction of soil quality, biomagnification, loss of biodiversity, etc. (Hickey and Anderson 1968; Smith et al. 1991; Strachan et al. 1994; Bidleman et al. 1995; Colborn and Thayer 2000). One such class of pesticide, which is ubiquitous, persistent, and bioaccumulative in nature, is organochlorine pesticides like DDT, endosulfan, lindane, etc. The environmental fate of a pesticide depends on the chemical behavior of the pesticide in several distinct environments. It depends on its solubility in water; adsorption to the soil and sediments; rate of input and removal; rate of microbial, chemical, and physical degradation; and physical and chemical factors of the environment (Schnoor 1996; Tinsley 2004). The potential risk caused by the incorporation of pesticide in the environment depends on many factors: toxic properties of pesticide, amount applied, formulation, method and time of application, and especially its mobility and time of treatment and persistence in nature. Excessive and frequent application of pesticides results in contamination of water, soil, food, and air.

## 3.4 TOXICOLOGICAL IMPACT OF PESTICIDES

Toxicity, the ability of a pesticide to produce injury or illness to the living organism, is mostly expressed in terms of its $LC_{50}$ and $LD_{50}$ values of the pesticides. Based on these, pesticides are classified as highly toxic, moderately toxic, and slightly relatively non-toxic, which is labeled with symbols (WHO 2004). Even though pesticides are labeled as slightly toxic or nontoxic, all pesticides can be hazardous to humans, other living beings, and the environment if pesticides are misused. Table 3.3 summarizes the $LD_{50}$ and $LC_{50}$ values for the four toxicity categories and their associated signal words.

Continuous usage of pesticides resulted in many toxicological impacts on living beings due to direct and indirect exposure to pesticides and their residues. This negative impact of pesticides is mainly due to the highly toxic, stable, and less-soluble active ingredients of pesticides. Organochlorine pesticides, which are problematic and serious environmental concerns, include atrazine, DDT, benzene hexa chloride (BHC), lindane, and endosulfan (Pandit et al. 2001).

**TABLE 3.3**
**Range of Toxicity of Pesticides**

| Routes of Exposure | Toxicity Category | | | |
|---|---|---|---|---|
| | I | II | III | IV |
| **Oral $LD_{50}$** | Up to 50 mg/Kg | 50–500 mg/Kg | 500–5000 mg/Kg | >5000 mg/Kg |
| **Inhalation $LC_{50}$** | Up to and including 0.2 mg/L | 0.2–2 mg/L | 2–20 mg/L | >20 mg/L |
| **Dermal $LD_{50}$** | Up to and including 200 mg/Kg | 200–2000 mg/Kg | 2000–20,000 mg/Kg | >20,000 mg/Kg |
| **Eye effects** | Corrosive corneal opacity not reversible within 7 days | Corneal opacity reversible within 7 days; irritation persisting for 7 days | No corneal opacity; irritation reversible within 7 days | No irritation |
| **Skin effects** | Corrosive | Severe irritation at 72 h | Moderate irritation at 72 h | Mild or slight irritation at 72 h |
| **Signal words** | **Danger Poison** | **Warning** | **Caution** | **Caution** |

*Source*: EnviroNews Forum, *Environmental Health Perspectives*, 107, A62, 1999; Odukkathil, G., and Vasudevan, N. *Reviews in Environmental Science and Biotechnology* 12: 421–444, 2013b.

According to the WHO's recommended classification of pesticide by hazard, the active ingredients of pesticide are classified as extremely hazardous Class 1A, highly hazardous Class 1B, moderately hazardous Class 2, and slightly hazardous Class 3 active ingredients (WHO 2004). These active ingredients are diluted by a carrier or diluents and are made available in various formulations like dust, granule, emulsifiable concentrate, wettable powder, fumigants, smokes, vapors, and aerosol. Commercially available pesticides are such formulations of active ingredients. Their dosage depends on the nature and extent of pest attack. Formulation and dosage also vary depending on the type of pest, type of crop, and plant parts where the pest had caused damage. Common active ingredients of pesticides now in use are atrazine, chlorpyrifos, cypermethrin, fenvalerate, diazinon, endosulfan, lindane, deltamethrin, ethion, carbofuran, simazine, permethrin, monocrotophos, etc.

Even though the aim behind the usage of pesticides is to kill or drive away the target pest, their extensive use resulted in several toxicological impacts among the nontarget species including human beings. Death and chronic illness caused by pesticide poisoning numbers about 1 million per year. This toxicological effect is due to operational hazards during manufacture, field application, direct toxic effect, and indirect toxic effect due to toxic residues in food and environment. There are many global incidences regarding the various toxicological impacts of pesticide usage. The WHO has estimated that about 0.2 million people are killed by pesticide, worldwide every year (Bhatnagar 2001). Khamukhaan village in the Philippines was poisoned by the aerial spraying of pesticides on a banana plantation (EnviroNews Forum 1999). Residues of pesticide were detected in food samples because of direct spraying and by transfer through the food chain.

In India, several cases of residues like parathion, endosulfan, DDT, etc., were reported in food samples. Likewise, malathiodofos in Taiwan, methyl parathion in Peru, and carbamate poisoning in the Philippines are some of the incidents recorded earlier (Watts 2005, 2010). There is evidence suggesting that pesticides produce unwanted side effects in the environment and thereby bring risk to all life forms (Lee et al. 1997; Gurug and Tanabe 2001; Watts 2005; Sivasankaran et al. 2007).

Pesticide pollution to the local environment also affects the lives of birds, wildlife, domestic animals, fish, and livestock. The use of unprescribed pesticides in inappropriate doses is not only disturbing soil conditions but is also destroying the healthy pool of biocontrol agents that normally co-exist with vegetation.

### 3.4.1 Toxicity in Plants

Even though pesticides are plant-protecting agents, they can cause negative impacts on plant growth, plant physiology, and plant reproduction. These impacts include effects on plant growth due to inhibition of nitrogen fixation, inhibition of total biomass yield, chlorosis, phytotoxicity due to adsorption of certain nutrient elements like Ca, Mn, Co, and Cu, which is followed by the inhibition of nutrient adsorption. Pesticide affects plant reproduction by killing the pollinating insects, thereby affecting production. Pesticides can also cause direct harmful effects on plants including poor root hair development, shoot yellowing, and reduced plant growth. Khan et al. (2000) reported phytotoxicity of endosulfan at a lower concentration (500 mg/Kg) by facilitating the adsorption of certain nutrient elements like Ca, Mn, Co, and Cu, which is followed by the inhibition of nutrient adsorption. The inhibitory effect of DDT on total biomass and economic yield of oil seed crops such as groundnut (*Arachis hypogaea*), soya bean (*Glycine max*), and Indian mustard (*Brassica juncea*) with a large degree of chlorosis were also reported (Mitra and Raghu 1999). Khandakar et al. (2003) reported an increased availability of nitrogen, phosphorus, and sulfur during the application of pesticides BHC and aldicarb in an eggplant (*Solanum melongena*) cultivating field. Genotoxicity of heptachlor in bean cell and hop plants was reported by Laouedj et al. (1995).

### 3.4.2 Toxicity in Animals

Toxic effects of pesticides on animals can be direct or indirect. Direct impact is caused by direct exposure to pesticides, whereas indirect effect can be mainly through the food chain or a reduction in essential food sources of the animal population, or by feeding upon prey, in which pesticide is bioaccumulated. Pesticide can even affect biodiversity by completely wiping out a particular population. Pesticides affect the growth and reproduction of earthworms. Some pesticides can bioaccumulate, or build up to toxic levels in the bodies of organisms that consume them over time, a phenomenon that especially impacts species high on the food chain. The next group in the kingdom Animalia that is most affected is the target organism of the pesticides, which includes the pest and several beneficial insects. Usage of pesticide has affected several pollinating insects such as honeybees and butterflies. Many of these pollinating groups are even completely wiped out from the environment. Amphibians are also affected by the usage of pesticides. Several species of frogs are extinct due to pesticides, which affect their metamorphoses and is one of the main reasons for their vanishing. Pesticides also affect aquatic organisms such as fish and zooplanktons by entering the aquatic system through agriculture runoff. Repeated exposure to sublethal doses of some pesticides can cause physiological and behavioral changes that reduce fish populations, such as abandonment of nests and broods, decreased immunity to disease, and decreased predator avoidance. Birds are another class of kingdom Animalia that are affected by pesticides. The U.S. Fish and Wildlife Service estimates that 72 million birds are killed by pesticides in the United States each year. Bald eagles are common examples of nontarget organisms that are impacted by pesticide use. Pesticides cause thinning of eggshell, growth abnormalities in embryos, and also death due to direct exposure.

Shanmugam et al. (2000) reported that exposure to lethal concentrations of phosphomidon, carbaryl, and endosulfan for 24–96 hours leads to behavioral changes such as hyperexcitedness, uncoordinated movements, and poor response to external stimuli leading to paralysis and death in *Barytelphusa cunicularis*. The study also revealed that *B. cunicularis* was found to be more sensitive to endosulfan than the other two. Certain pathological changes in fish gills like loss of original shape, curing of secondary gill filaments, necrosis in pillar cell nucleus, and development of vacuoles in secondary gill epithelium were induced by monocrotophos (Yazdandoost and Katdare 2000). Exposure to some pesticides like endosulfan and DDT can lead to immediate and higher mortality of crustaceans and tree frog tadpole *Litoria freycineti* (Broomhal and Shine 2003). Increased use of pesticide in Sub Saharan Africa had resulted in progressive threatening of amphibians and reptilian population. DDT (180 g/ha), endosulfan (200 g/ha), and chlorpyrifos resulted in deaths of snake, lizards, and frogs (Lambert 1997). Pesticide residues in milk can destruct fat globules (Chubiko et al. 1998). Certain herbicides (e.g., 2,4-D and 2,4,5-T) reduced egg production in chicken (8% and 18%, respectively), and DDT resulted in egg breakage and reproduction failure in brown pelican (Rekha et al. 2004).

### 3.4.3 Toxicity in Microbes

Pesticides can affect soil biodiversity, wherein they reduce the growth of many microbes in the soil. Several beneficial microbes that play a key role in nutrient cycles and degradation of organic matter are affected by the usage of pesticide. Many fungicides, bactericides, rodenticides, and herbicides, which are directly sprayed onto soil, also kill other beneficial microbes in the soil. This affects soil fertility. The decrease in viable counts of bacteria and algae due to long-term exposure to DDT was also reported (Megharaj et al. 2002). It also affected the species composition of algae and cyanobacteria. Aldrin, HCH, carbofuran, 2,4-D, basaline, and malchetic at higher doses lead to a reduction in nitrifying bacteria and in the concentration of nitrate and nitrite (Pandey and Rai 1993). The presence of aldrin (50 mg/L) decreased the dentrification activity of dentrifiers, whereas it induced the growth of nitrifying bacteria showing that it is tolerant to aldrin (Lopez et al. 2002). Application of pesticide at 50 μg/g can reduce the population of soil microbes like bacteria and actinomycetes and eliminated certain microbes like fungi and protozoan propagules (Digrak and Ozcelik 1998). Ghadiri et al. (1995) reported that endosulfan applied to the soil reduced the population and activity of soil microbes responsible for the degradation of organochlorines in the soil. A mutagenic effect of endosulfan on bacteria and yeast was reported by Chaudhuri et al. (1999) and ASTDR (2000).

### 3.4.4 Toxicity in Humans

Humans are exposed to pesticides (found in environmental media such as soil, water, air, and food) by different routes of exposure such as inhalation, ingestion, and dermal contact. Exposure to pesticides results in acute and chronic health problems (WHO 2004). Pesticides being used in agricultural tracts are released into the environment and come into human contact directly or indirectly (Bhatnagar 2001). Increasing incidence of cancer, chronic kidney diseases, suppression of the immune system, sterility among males and females, endocrine disorders, and neurological and behavioral disorders, especially among children, has been attributed to chronic pesticide poisoning (Agnihotri 1999). Human health hazards vary with the extent of exposure. Moderate human health hazards from the misapplication of pesticides include mild headaches, flu, skin rashes, blurred vision, and other neurological disorders, while rare but severe human health hazards include paralysis, blindness, and even death (ICAR Report 1967; Ekundayo 2003). Moreover, pesticide residues in food samples have also become a major issue. Some of the chronic and acute toxicological effects of pesticides are chronic liver damage, endocrine and reproductive disorders, immunosuppression, different cancers, inhibition of choline esterases, and Parkinson's and Alzheimer's diseases (Gupta et al. 1984) (Table 3.4).

**TABLE 3.4**
**Toxicological Impact of Pesticides**

| Pesticide | Test Organism | Toxicological Impact | Reference |
|---|---|---|---|
| DDT | Common frog | Metamorphosis | Cooke (1970) |
| | Rat | Rcproduction | Ottoboni (1969) |
| Endosulfan | Snails | Fecundity | Ellis-Tabanor and Hyslop (2005) |
| | Human T-cell leukemic line (external) | Induction of apoptosis | Kannan et al. (2002) |
| Diethylstilbestrol, *o,p*-DDT, Methoxychlor | Mice | Postnatal growth and neurobehavioral development | Paola et al. (2001) |
| Lindane | Mouse | Spermatogenesis | Traina et al. (2003) |
| | Earthworm and | Toxicity to the organism with respect to lindane bioavailability in different soil | Lock et al. (2002) |
| | Spring tail | Life cycle | Brown and Doube (2004) |
| | Copepod | Effect on intercellular calcium level in kidney phagocytes and blood leukocytes | Betoulle et al. (2000) |
| | Rainbow trout | | |
| Chlorpyrifos | Rats | Brain cholinesterase, muscarinic receptor binding, and neurotrophin levels | Betancourt and Carr (2004) |
| | Juvenile Rat | | Carr et al. (2001) |
| | Salmon | Behavioural changes at both pre and post weanling ages Acetylecholinesterase | Sandahl et al. (2005) |
| Parathion, chlorpyrifos, methyl parathion | Rat | Cardiac muscarinic receptor binding | Marcia and Carey (2002) |
| Atrazine | Rat | Pituitary ovarian function | Cooper et al. (2002) |
| | Frog tadpole | Lethal impact | Relyea (2005) |
| Carbofuran | Rat | Induced oxidative stress in mammalian brain | Rai and Sharma (2007); |
| | Amphipods | Energy metabolism and reproductive behavior | Dutra et al. (2009) |
| Cypermethrin and deltamethrin | Ray finned fish, | Lethal toxicity | Marigoudar et al. (2009); |
| | Rohu fish | Biochemical parameters | Saha and Kaviraj (2009) |
| | Freshwater catfish | | |

*Source*: Odukkathil, G., and Vasudevan, N. *Reviews in Environmental Science and Biotechnology* 12: 421–444, 2013b.

## 3.5 BIODEGRADATION OF PESTICIDES

The term *biodegradation* describes the process of the breaking down of organic compounds, usually by microorganisms, into less complex compounds and finally to water and $CO_2$ and oxides or mineral salts of other elements present. The complete breakdown of pesticides into inorganic components is termed *biomineralization*. In this process of biodegradation of pesticides, the microbes utilize pesticide as a carbon source, nitrogen source, any other mineral source, or a final electron acceptor in the respiratory chain. Biotransformation of pesticide occurs by metabolic and co-metabolic processes. This can be via oxidative, reductive, hydrolytic, and conjugation pathways. The biodegradation of pesticides follows different metabolic pathways depending on the nature of the pesticide, environmental conditions, and the nature of microbes. Several microorganisms are reported for their innate capability to degrade pesticides. The following section describes the various microbes that have been reported as pesticide degraders.

### 3.5.1 Microbes Involved in Pesticide Biodegradation

Among the microbes, bacteria and fungi are widely reported for biodegradation of pesticides. Bacteria that can degrade pesticide are *Bacillus* sp., cyanobacteria, *Alcaligenes* sp. *JB1, Terrabacter* sp., *Pseudomonas sp., Agrobacterium radiobacter J14a, Ralstonia* sp. strain M91-3, *Chelatobacter heintzii, Ganoderma* sp., *Serratia marcescens, Klebsiella* sp., *Burkholderia* sp., *Sphingomonas* sp., *Bordetella* sp., *Achromobacter* sp., and *Arthrobacter* sp. (Table 3.5).

Most of the pesticide-degrading bacteria are gram negative. The process of degradation varies among the species of bacteria based on the nature of species and target compounds. Fungi capable of degrading pesticides are *Agrocybe semiorbicularis, Auricularia auricula, Coriolus versicolor, Dichomitus squalens, Flammulina velutipes, Hypholoma fasciculare, Pleurotus ostreatus, Stereum hirsutum*, and *Aspergillus versicolor*. These fungi have shown their ability to degrade various pesticide groups like phenylamide, triazine, phenylurea, dicarboximide, and chlorinated and organophosphorus compounds (Bending et al. 2002). Among the fungi, *Phanerochaete chrysoporium* degrades a wide variety of pesticides. Pesticides like lindane, atrazine, diuron, terbuthylazine, metalaxyl, DDT, $\gamma$-HCH, dieldrin, aldrin, heptachlor, CHL, mirex, etc., have been degraded to different extents by white-rot fungi (Kennedy et al. 1990; Hickey et al. 1994; Mougin et al. 1994; Singh and Kuhad 1999; Bending et al. 2002; Quintero et al. 2007).

### 3.5.2 Role of Enzymes in Biodegradation of Pesticides

Enzymes, called the biological catalyst, play a key role in the biodegradation of pesticide. The biodegradation of pesticide includes a stepwise oxidation, reduction, dechlorination, dehydration, and hydrolysis reactions, which are catalyzed by several enzymes. This involves (1) oxidative enzymes, which catalyze oxidative transformation (cytochrome p450, peroxidases, and polyphenol oxidases); (2) hydrolytic enzymes (hydrolases), which catalyze hydrolytic transformation, which cleaves bonds of the substrate by adding hydrogen or hydroxyl from water molecules; (3) reductive

**TABLE 3.5**
**Bacteria Capable for the Degradation of Pesticides**

| Sl. No | Bacteria | Pesticide |
|---|---|---|
| 1. | *Bacillus* sp. | DDT, endosulfan |
| 2. | Cyanobacteria | DDT |
| 3. | *Alcaligenes* sp. JB1 | DDT |
| 4. | *Terrabacter* sp. strain | DDT |
| 5. | *Pseudomonas* sp., *Pseudomonas aeruginosa* | DDT, endosulfan, chlorpyrifos |
| 6. | *Agrobacterium radiobacter* strain J14a and *Ralstonia* sp. strain M91-3 | Atrazine |
| 7. | *Chelatobacter heintzii* Citl | Atrazine |
| 8. | *Achromobacter xylosoxidans* | Endosulfan |
| 9. | *Bordetella petrii* | Endosulfan |
| 10. | Algae | Endosulfan |
| 11. | *Ganoderma* sp. | Lindane |
| 12. | *Serratia marcescens* | Chlorpyrifos |
| 13. | *Bacillus cereus* | Chlorpyrifos |
| 14. | *Klebsiella* sp. | Chlorpyrifos |
| 15. | *Burkholderia* sp. | Fenitrothion |
| 16. | *Sphingomonas* sp. | 2,4-Dichlorophenoxy acetic acid |

*Source:* Odukkathil, G., and Vasudevan, N. *Reviews in Environmental Science and Biotechnology* 12: 421–444, 2013b.

enzymes (nitroreductase), which catalyze reductive transformation by which removal of anion occurs by reduction; (4) dehydrohalogenase enzyme, which catalyzes reductive dehalogenation in which an organohalide acts as a terminal electron acceptor for ATP production; and (5) conjugating enzymes, which catalyze conjugation reactions by which an exogenous or endogenous natural compound is added to pesticide facilitating mineralizazation, which occurs by using existing enzymes. This includes xyloxylation, alkylation, acylation, and nitrosylation. This type of biotransformation occurs in fungal biodegradation of pesticides. Different enzymes will be involved in the degradation of different pesticides as the bacteria capable of biodegrading pesticide produce different enzymes. For example, enzymes involved in the oxidative biodegradation of endosulfan are endosulfan monooxygenase I, endosulfan monooxygenase II, endosulfate monooxygenase, sulfatase, and endosulfan hemisulfate sulfatase. In a microbial catalyst like *Stenotrophomonas maltophilia, Rhodococcus erythropolis, Arthrobacter* sp. KW, *Klebsiella oxytoca*, and *Mycobacterium* sp. ESD, biodegradation of endosulfan follows oxidative biodegradation (Sutherland et al. 2002; Weir et al. 2006). In addition, there are some ligninolytic enzymes that are also involved in pesticide degradation (Chandra et al. 2008; Chandra and Chowdhary 2015). In bacteria like *Pseudomonas aeruginosa* and *Burkholderia cepacia*, biodegradation of endosulfan occurs by hydrolysis, which involves major enzymes like endosulfan diol hydrolyase, endosulfan diol dehydrogenase, endosulfan hydroxyl ether dehydrogenase, and endosulfan lactone lactonase.

### 3.5.3 Biodegradation Pathway of Pesticide

The pathway of biodegradation of pesticide varies with the target compounds, microbes involved, and environmental conditions. One of the environmental conditions which alter the biodegradation pathways are presence or absence of oxygen, the media in which the target compound is present. Sometimes pH and temperature can also alter the pathway of biodegradation of pesticides. Availability of oxygen is one of the major factors that alter the metabolic pathway. Under aerobic condition most of the pesticide degradation follows oxidative pathway whereas, under anaerobic condition reductive degradation occurs. As mentioned above, different microbes produce different enzymes, hence, based on this the biodegradation pathway also varies. The steps involved in the biodegradation of pesticides are oxidation, dehalogenation, reduction, and conjugation.

#### 3.5.3.1 Oxidation and Reduction

During the process of oxidation, pesticides accept oxygen or remove hydrogen or remove electrons. During this process, one substrate get oxidized and other gets reduced. Hence, it is called as redox reaction and is catalysed by a class of enzymes called oxidoreductases. Oxidoreductases are of five types oxidases, oxygenases, aerobic dehydrogenases, anaerobic dehydrogenases, and hyperperoxidases. Monooxygenase is the most commonly reported enzyme in the biodegradation of pesticides. For example, in the biodegradation process of endosulfan most of the metabolites are formed by oxidative biodegradative pathway. These include enzymes like endosulfan mono oxygenase I, endosulfan monoxygenase II, endosulfate monoxygenase, sulfatase and endosulfan hemisulfate sulfatase (Figure 3.4). Conversion of DDNU (2,2-bis (4-chlorophenyl)ethylene) a metabolite of DDT to DDOH (2,2-bis (4-chlorophenyl) ethanol and DDA (bis (4 chlorophenol acetic acid) occurs by oxidation reaction.

#### 3.5.3.2 Dehalogenation

During the process of dehalogenation, halogens present in the pesticide are removed and is catalysed by dehalogenase. This occurs by hydrolytic displacement, elimination, reductive displacement and oxidative displacement. Hydrolytic displacement is reported mainly for aliphatic compounds and their occurrence is comparatively less in the case of aromatic compounds. Elimination reaction involves removal of HCl and formation of cycloalkenes. Such types of reactions are reported in the degradation $\gamma$ Hexachlorocyclohexane by *Pseudomonas paucimobilis*. Formation of DDE from DDT is also an elimination reaction. Reductive displacement is one of the predominant reactions for chlorine displacement from organochlorine compounds under oxygen deficient condition. During this reaction a halogen atom is removed from the chlorinated compound with simultaneous addition of electrons. Dechlorination of DDT by Proteus vulgaris occurs by reductive displacement. Conversion of DDT to DDD is a reductive dechlorination process which involves substitution of aliphatic chlorine for a hydrogen atom in the presence of transition metals and metal complexes

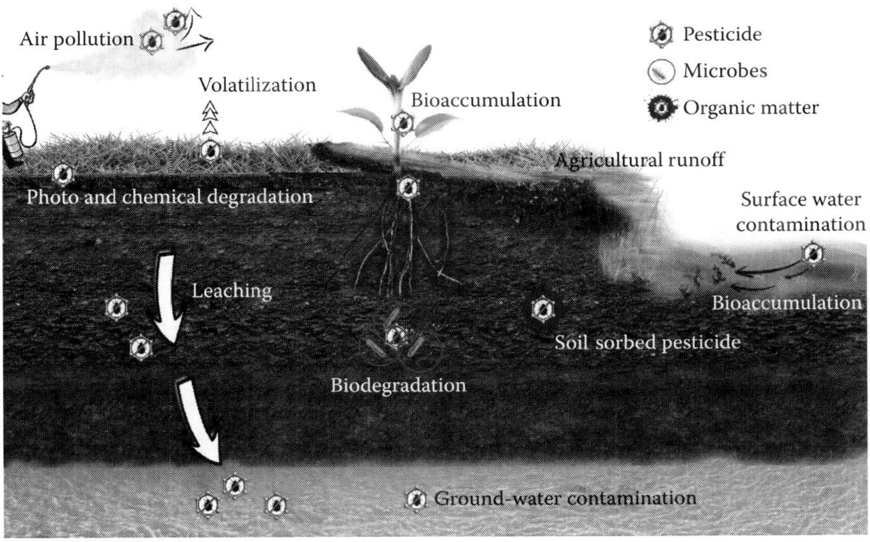


**FIGURE 3.4** Fate of pesticides in the environment.

as reducing agent. Oxidative displacement is mediated by dioxygenase enzymes and during this reaction a halogen atom is removed from the chlorinated compound with simultaneous removal of electron.

#### 3.5.3.3 Conjugation

During conjugation of pesticide exogenous or endogenous natural compound is added to pesticide facilitating mineralization, which occurs by using existing enzymes. This includes xyloxylation, alkylation, acylation and nitrosylation. This type of biotransformation occurs in fungal biodegradation of pesticides. During conjugation, the pesticide is converted to more complex substrates or organism combines pesticide with cell metabolites. These types of reactions are carried out by those organisms catalysing the addition of aminoacid, organic acid or methyl crown to the substrate. Degradation of sodium dimethyldithiocarbamate is one such example for conjugation. During this process the above mentioned pesticide is combined with amino acid in the cell and thereby pesticides get converted to inactive form.

### 3.5.4 Biodegradation of Pesticides in Water

Biodegradation of pesticides in water mainly depends on the availability of a specific microorganism that is capable of degrading pesticide. Specific microorganisms capable of degrading pesticide in water can be isolated from pesticide-contaminated soil, water, and wastewater by enrichment techniques providing pesticide as the carbon source. Several factors affect the degradation of pesticide in water. These include pH, temperature, availability of nutrients, availability of oxygen, solubility of the pesticide, salinity, toxicity, and in some cases availability of other carbon sources and electron acceptors. Several studies have been carried out on the biodegradation of pesticide in liquid media using specific microbes capable of degrading pesticides. The effects of the above-mentioned factors on biodegradation of pesticide have also been studied. The following sections discuss those factors affecting the biodegradation of pesticide in water.

### 3.5.4.1 Factors Affecting the Biodegradation of Pesticides

#### *3.5.4.1.1 pH and Temperature*

The growth of microbes in water is mainly controlled by the pH and temperature of the water. The optimum pH and temperature required for growth of most microbes are 7 and 37°C. This pH and temperature range can also vary based on the nature of the microbes. Hence, for each organism there is an optimum pH and temperature for its maximum growth and utilization of substrate. Optimum pH and temperature required for the degradation of pesticides may vary with the nature of the pesticide and also with the microbial catalyst. Hence, optimization of pH and temperature is necessary for the removal of pesticide from water. Most of the studies on biodegradation of pesticides reports a pH range of 7–8 and temperature of 37ºC as optimum, but some studies report a lower and higher range of pH and temperature as optimum.

#### *3.5.4.1.2 Availability of Nutrients*

Microbes need specific nutrients for their growth. These include carbon, nitrogen, phosphorus, magnesium, calcium, potassium, sodium, and some trace elements. Availability of these nutrients in the water is necessary for the biodegradation of pesticides in water. In the case of pesticide-degrading microbes, a mineral salt medium with simple composition with essential nutrients like carbon, nitrogen, phosphorus, magnesium, calcium, potassium, and sodium can serve as a good media for microbial growth. Pesticides are the main carbon source and only carbon source in the media. In the biodegradation of pesticides like endosulfan, lindane, atrazine, chlorpyrifos, and cypermethrin, no additional carbon source is needed. In the case of DDT, since the biodegradation follows a co-metabolic pathway, the presence of an additional carbon source is necessary (Figure 3.5).

#### *3.5.4.1.3 Availability of Oxygen*

Oxygen is one of the major factors affecting the biodegradation pathway of pesticides in water. The availability of oxygen is necessary for a certain group of microbes without which they cannot survive, whereas for some other groups the presence of oxygen is not necessary where the degradation follows an anaerobic degradation pathway. There is also another group that can survive in both aerobic and anaerobic conditions. Hence, based on the microbial catalyst used, proper oxygen should be supplied for the biodegradation of pesticide in water. In the case of anaerobes, an anaerobic condition should be maintained. Degradation of DDT by *E. coli* and *E. aerogenes*, Endosulfan by *Staphylococcus* sp., *Bacillus circulans*-I and –I, Lindane by *Desulfococcus, Desulfobacter*, *Citrobacter* and *Dehalobacter*, Cypermethrin by *Rhodobacter sphaeroides*, are few examples for anaerobic degradation.

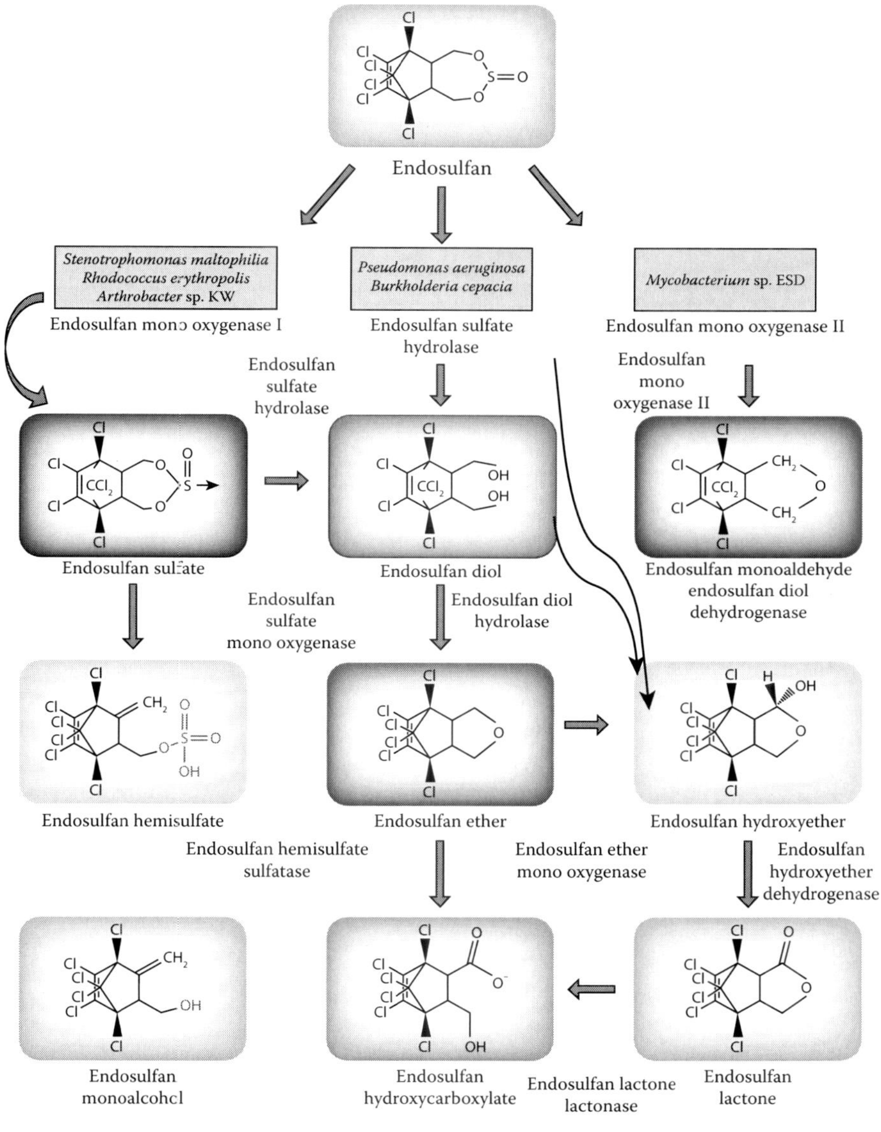


**FIGURE 3.5** Biodegradation pathway of endosulfan. (From Klema, V., Pathway of endosulfan degradation, University of Minnesota, BioC/MicE5309, and Edward LaBelle, University of Minnesota, 2011. With permission.)

#### *3.5.4.1.4 Solubility of Pesticides*

The extent and rate of biodegradation of pesticide in water depend on its solubility in water. Different pesticides have different solubilities in water, which limit their biodegradation in water. For example, pesticides like Endosulfan, DDT, cypermethrin, lindane, deltamethrin, fenvalerate, etc., have less water solubility; hence their biodegradation in water requires more time, and the growth of bacteria during the initial period will be very slow. In such cases, a microbial catalyst capable of solubilizing the pesticide by producing some extracellular substance like biosurfactant will be more suitable and offers promising prospects. For example, biodegradation of endosulfan can be enhanced by using biosurfactant-Producing bacterium *Bordetella petrii* and *Achromobacter xylosoxidan*s (Odukkathil and Vasudevan 2015).

### 3.5.5 Biodegradation of Pesticides in Soil

Soil is a heterogeneous system in which the distribution of pesticides will not be uniform and the environmental conditions will limit the biodegradation of pesticides in soil depending on the environmental conditions as well as the nature of the pesticides. Most of the pesticides undergo partial degradation in soil leading to the formation and accumulation of metabolites, which are sometimes more toxic and less soluble than the parent compound. They can also inhibit the microbial population in the soil and further reduce the degradation resulting in the partial degradation of the pesticide. For example, degradation of endosulfan in soil by fungi and bacteria follows such a pathway via oxidation and hydrolysis leading to the formation of toxic endosulfan sulfate and less toxic endosulfan diol (Kullman and Matsumura 1996; Sutherland et al. 2002; Weir et al. 2006). The best example for partial degradation is DDT, which undergoes degradation to form metabolites like DDD and DDE that are toxic and more persistent than the parent compound (Foght et al. 2001). Introducing a microbial population, which can degrade the pesticides completely, and optimizing the environmental condition can enhance the degradation of pesticides and their metabolites in the soil. Environmental factors that affect the biodegradation of pesticide in soil are pH, temperature, moisture content, adsorption, and bioavailability.

#### 3.5.5.1 pH and Temperature

The growth of microorganisms in soil is also controlled by the pH of the soil. Variation of pH in water will be comparatively more, but in a stable system like soil, the variation in pH will be comparatively less. Optimization of pH and temperature prevailing in the soil is a difficult task. This again depends on the soil type. Since most microbial species can survive only within a certain pH range and pH affects the availability of nutrients, optimization of pH is a vital step in the biodegradation of pesticide in soil. Soil temperature is another factor that controls the activity and survival of microbes as well as the rate of organic matter decomposition (Sims and Bass 1982). The degradation rate of pesticides in soil increases with a rise in temperature (USEPA 1982). A temperature rise at the rate of 10°C increases the growth of microorganisms (Thibault and Elliott 1979) and also increases desorption of pesticides in soil (Mohit Kumar et al. 2008). Studies in this field also support the need for optimization of these factors for the successful biodegradation of pesticide in soil.

The optimum pH and temperature required for the degradation of a pesticide may vary with the nature of pesticide as well as microbes involved in it. For example, the highest degradation of HCH isomers (α and γ) was observed at an initial pH of 9 in soil slurry and an effective degradation of both the isomers in soil at an optimum temperature of 30°C (Sididque et al. 2003). In the case of endosulfan, the accelerated biodegradation by mixed bacterial culture was achieved by optimization of the pH and temperature in soil to 8 and 30°C, respectively (Kumar and Philip 2006). In the case of chlorpyrifos, fenamiphos, and ethoprofos, a pH of 5.7 is found to be optimum (Karpouzas and Walker 2000; Vidali, 2000; Singh et al. 2004). Soil pH is also important in maintaining enhanced degradation of pesticide for a prolonged period. The enhanced degradation of chlorpyrifos in a bioslurry phase reactor was observed at an optimized pH and temperature of 7 and 30°C, respectively (Brajesh et al. 2006). The enhanced degradation of endosulfan at pH 8 was observed in optimized environmental conditions (Venkata Mohan et al. 2007). Low pH also decreases activity of bacteria and/or the enzyme involved in pesticide transformation (Roberts et al. 1998; Cycon et al. 2009). Degradation of diazinon was inhibited in acidic and alkaline pH (Abo-Amer and Ale 2011).

The production and emulsification stability of biosurfactant produced by bacteria and thus, bioavailability of pesticides in soils is affected by the soil pH and temperature. Extreme pH and temperature have a negative impact on the production and emulsification stability of biosurfactant. Of these two parameters, pH has the greatest effect on the surface-active properties of the biosurfactant. Sidal and Yilmaz (2012) reported that an initial pH value of 7 was found to be most suitable for *Pseudomonas* sp. to produce rhamnolipid, and the highest biosurfactant concentration (600 mg/L) was obtained at 30°C. Surface-active properties of biosurfactant produced by *Lactobacillus pentosus* also were affected by decrease in pH and temperature, with a maximum stability and reduction of surface tension at pH 8 (Bello et al. 2012). Optimal production of biosurfactant by *Pseudomonas aeruginosa* and *Bacillus subtilis* was observed at pH 7 and 37°C (Makkar and Cameotra 2000; Al-Araji et al. 2004). Microbial degradation of most of the hydrophobic organic compounds like pesticides in soil is mostly aided by biosurfactant production by which soil-bound pesticides are made bioavailable.

#### 3.5.5.2 Soil Texture

Soil texture directly affects even distribution of the nutrients, oxygen, microbes, water, and contaminants in the contaminated soil. Soil is a heterogeneous mixture of several inorganic and organic compounds of varying compositions and surface activities. These compounds may bind with pesticides and reduce the bioavailability (Torrents and Jayasundera 1997). Hence, bioavailability of pesticides varies with the composition of soil. Texture affects a range of soil environmental factors, including redox potential and moisture content, which affect the rate of biodegradation of pesticide in soil. The extent of adsorption and rate of desorption of pesticides affect the bioavailability of adsorbed pesticides. Studies on these aspects suggest that the sorption and desorption characteristics vary among various soil types and pesticides.

The adsorption of endosulfan isomers on the soil follows the order of clay soil, composted soil, red soil, and sandy soil, whereas its degradation follows the order of sandy soil, red soil, composted soil, and clayey soil (Mathava Kumar and Ligy Philip 2006). This difference in degradation of endosulfan in various soils may be attributed to the difference in bioavailability of the pesticide with the soil type. Since clay soil has high organic content, the adsorption of pesticide is high. This was also observed in the case of nitroaromatic pesticide "dinoseb" (Martin and Mermoid 1998). Soil texture had 10% effect on biodegradation of endosulfan on clay loam soil. Similarly, a difference in degradation rate of another pesticide Fipronil in different soil was also reported (Kumar et al. 2012).

#### 3.5.5.3 Soil Moisture Content

Soil moisture content has considerable effect on the degradation of pesticides. All soil microorganisms require moisture for cell growth and function. Diffusion of water and soluble nutrients in and out of the microbial cells depends on the availability of water. In supersaturated soil, excess water produces some undesirable effects like reduction in oxygen content, which in turn affects the aerobes. This anoxic condition prevailing in the environment leads to anaerobic respiration, which follows a pathway with less energy and slows the rate of biodegradation. Soil moisture content is also essential for availability of pesticides and improvement and proliferation of microorganisms (Daniel and Timothy 1991). In soil, pesticides in water-soluble forms are available to microbes, and only those fractions can be degraded by microbes. Hence, the maintenance of moisture content of soil is important in the biodegradation of pesticide in soil.

The optimum moisture content required for the biodegradation of pesticide in soil is 20%, but this can also vary with the pesticide and nature of the microbes. One such variation reported is in the case of degradation of chlorpyrifos and fenamiphos. The degradation of chlorpyrifos and fenamiphos was slow at 20% moisture content, whereas at 40% and above the degradation rate was high (Singh et al. 2006). Kumar and Philip (2006) reported that the degradation of HCH isomers was optimal at 15% moisture content and it decreased with moisture content greater than 30%. According to Scholz et al. (1998), a decrease in the degradation rate of pesticide in soil at moisture content greater than 30% may be due to agglomeration of soil under high water content, resulting in limited oxygen supply and thereby reducing microbial activity. Arshad et al. (2008) studied the effect of the soil-to-water ratio ranging from 10 to 40 w/v% on the degradation of endosulfan isomers by *P. aeruginosa* in loam soil slurry. Several studies have reported that biodegradation of pesticide decreased with a decrease in soil moisture content less than 15%. Hence, maintaining optimum soil moisture content is essential for the biodegradation of pesticide in soil.

#### 3.5.5.4 Adsorption and Bioavailability

Adsorption is the process of binding any substance onto the surface of an adsorbent. In the context of biodegradation of pesticide in soil, the adsorption refers to the process of binding of pesticide to the soil particles, especially soil organic matter. The pesticides that are adsorbed are called *bound pesticide residues*. According to Gevao et al. 2000, bound pesticide residues can be defined as the fraction of pesticides that cannot be readily extracted from the soil without modifying the chemical structure of the original pesticide or its metabolites. Adsorption of pesticides to soil depends on the properties of soil and the compound, which include size, shape, configuration, molecular structure, chemical functions, solubility, polarity, polarizability and charge distribution of interacting species, and the acid-base nature of the pesticide molecule (Bailey and White, 1970; Senesi 1992; Pignatello and Xing 1996). Adsorption and fate of pesticide in soil are influenced by a variety of agricultural factors like concentration of pesticide applied, rate and mode of application, repeated application, aging, and use of organic and inorganic soil amendments. The environmental significance of a bound pesticide residue depends not on its nonextractability under laboratory test conditions, but on its bioavailability (Khan 1982; Calderbank 1989). These bound pesticides can be bioavailable or nonbioavailable. The bioavailable bound pesticide residue is that fraction of a pesticide in soils that can be taken up by plants and/or soil-inhabiting microbes/other animals, while the nonavailable fraction cannot (Figure 3.6).

In nature, bound residues are released by physico-chemical and biochemical process, but in most of the cases the pesticide residues are more stable to both physico-chemical and biochemical process. The highly stable chemical structure of a pesticide and its less solubility affect the physico-chemical process, whereas their limited availability to plants and microbes affects the biochemical process. Thus, the pesticide residues start accumulating in soil without undergoing degradation, leading to the persistence of pesticide in soil. One option to release pesticide is by physico-chemical treatment or by enhancing the bioavailability of pesticide, thereby increasing the biochemical degradation process.

Bioavailability of pesticide is a major constraint for biodegradation pesticide in soil. The bioavailability in the context of biodegradation of pesticide can be defined as the amount of pesticide that can be readily taken up by microbes

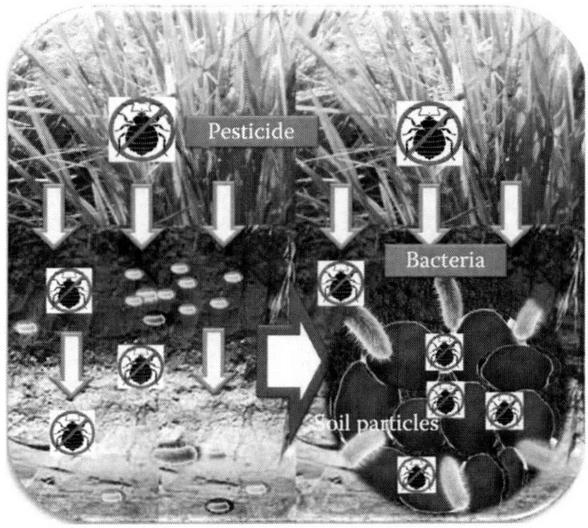


**FIGURE 3.6** Soil-adsorbed pesticide and its nonbioavailability.

(Maier and Valdes 2000). It affects the biodegradation of pesticide as follows: (1) At low pesticide concentration, microbes fail to produce some energy, which induces the catabolic gene systems involved in biodegradation. (2) At low pesticide concentration in an environment with less nutrients, microbial cells may degrade the pesticide, but low levels of nutrients reduce their growth rate, which eventually leads to a decrease in the uptake of pesticide by the microbes. Sometimes, the bioavailable pesticides are high to induce biodegradation in a growing stage of microbes, thereby leading to optimization of the biodegradation rate. (3) Moreover, most of the biodegradation processes are enzyme-catalyzed reactions. So, according to Michaelis–Menten kinetics, in a system where the contaminant concentration and microbial growth are low, the rate of enzymatic reaction will be less. This, in turn, slows the biodegradation process. In soil, the reduced bioavailability of pesticide for a prolonged period of time is due to the (1) chemical oxidation reaction, which incorporates them into soil organic matter; (2) slow diffusion into very small pores and absorption into organic matter; and (3) formation of semirigid films around nonaqueous phase liquids (NAPLs) with a high resistance toward NAPL–water mass transfer. In the case of pesticides, NAPLs exist in superfund sites like hazardous waste dumpsites or manufacturing unit areas, and they exist in the form of dense NAPLs (DNAPLs) (Bosma et al. 1997). Residues of aldrin; β endosulfan; chlorpyrifos; diazinon; malathion; parathion; phorate; and 1,2,4-trichloroethane exist in DNAPLs in the superfund site of Casmalia Resources. A DNAPL hotspot of organophosphorus pesticides was reported in dunes at North Sea Western Coast of Jutland, Denmark (Asker et al. 2004). Reduced bioavailability of pesticide is also due to (4) covalent bond formation between the humic substances and contaminants, (5) the insoluble nature of contaminants, (6) the sorption reaction of contaminants to solid natural surfaces, and (7) the toxicity of the compounds or its intermediate.

Bioavailability is a function of water solubility and adsorption/desorption equilibria, which are controlled by chemical structure and soil organic matter content (Rao and Davidson 1998). Studies by Ogram et al. (1985) strongly suggest that only material dissolved in soil solution is available for metabolism. Sorbed material may be metabolized, but only after desorption has occurred. The bioavailability is driven by the mass transfer of contaminant from soil phase to solution phase (Kumar et al. 2006). Hence, increased microbial conversion capacities do not necessarily lead to higher biotransformation rates when mass transfer of contaminant from the soil phase to the solution phase is the limiting factor (Sahu et al. 1993). In most of the bioremediation studies, it was observed that there was considerable reduction in the rate of degradation of pesticides with decrease in moisture content and vice versa because of reduced bioavailability.

## 3.6 BIOREMEDIATION OF PESTICIDES-CONTAMINATED SOIL AND WATER

Bioremediation is an eco-friendly technique that uses the innate capability of plants or microbes to remove or detoxify the contaminant from the environment. Several groups of microbes and plants are reported to be efficient biological agents for remediating pesticide-contaminated sites. Remediation of a site can be carried out within the site or can be transported away from the site and treated elsewhere. This bioremediation is of two types: *in situ* or *ex situ*. Successful bioremediation needs a proper microbe and proper environmental conditions. These microbes have the capability of degrading pesticides biochemically. This method is less expensive and features minimal site disruption, elimination of waste permanently, elimination of long-term liability, and greater public acceptance with regulatory encouragement, and it can be coupled with other physical or chemical treatment methods. Disadvantages include limited degradation of persistent pesticide due to less bioavailability, production of toxic metabolites other than the parent pesticide (e.g., formation of endosulfate during endosulfan degradation in soil), competition for survival between enriched microbial catalyst and the native microbes, and environmental conditions prevailing in the site. Recent developments in bioremediation overcome these limitations to a certain extent.

### 3.6.1 Microbial Catalyst

Microbial catalysts capable of biodegrading the pesticide can be enriched from the pesticide-contaminated soil or water by enrichment techniques providing the pesticide as the carbon source. The source of pesticide-contaminated soil can be soil in the pesticide industry or pesticide industry waste, agriculture soil/runoff exposed to pesticide, or soil from any other contaminated site and sediments. Essential requirements for the microbial catalyst are energy source, carbon source, nutrients, and optimum environmental conditions like pH, temperature, and moisture content. Microbial catalyst, which was already isolated and proved to be an efficient catalyst in degrading pesticide, was discussed in Section 3.5.1.

### 3.6.2 *Ex Situ* and *In Situ* Bioremediation Techniques

*Ex situ* bioremediation technique which involves excavation of pesticide contaminated soil and treatment elsewhere away from the site whereas *in situ* bioremediation involves onsite treatment of pesticide contaminated soil or water. *In situ* bioremediation is of two types: intrinsic and engineered in situ bioremediation. Conversion of toxic pesticides into less toxic form using naturally occurring microbial population is called intrinsic bioremediation. This technique is a very slow process as it is dependent on variety of factors such as naturally occurring microbes, limited electron acceptor, nutrients and environmental conditions. In such cases where site conditions are not suitable, construction of some engineered structures are required to supply nutrients and to stimulate microbes. This technique is called as engineered intrinsic bioremediation. *Ex situ* and in situ bioremediation techniques applied in pesticide bioremediation includes approaches like composting, bioaugumentation, biosparging, biosorption and biobed treatment.

#### 3.6.2.1 Bioaugmentation

Bioaugumentation refers to the technique of accelerating the biodegradation of pesticide in soil by inoculating pesticide degrading microbes enriched from pesticide contaminated soil and water. Microorganisms capable of degrading pesticides have been isolated from various contaminated sites and are demonstrated for their capability to remediate soil and water. This can be carried out both on site and offsite. Bioaugumentation technique can be carried out in three different ways, one method is by using single isolated microbe capable of degrading pesticide, second method is by using microbial consortium capable of degrading pesticide and third method is by using genetically modified microbes. These microbes capable of degrading pesticides are isolated from pesticide contaminated soil; water and wastewater by enrichment technique or by isolating the microbes on a solid media and then growing it on pesticides. Greeshma and Vasudevan (2016) proposed that bioaugumentation of biosurfactant producing pesticide degrading bacterial consortium has enhanced the bioremediation of endosulfan contaminated agricultural surface and subsurface soil with complete removal of endosulfan in 10 to 15 days. Pimmeta et al. 2013 reported that bioaugumentation of Burkholderia cepacia PCL3 in carbofuran contaminated soil showed a shortest carbofuran $t_{1/2}$ of 1.6 d. Several successful studies on bioremediation of DDT, endosulfan, lindane, chlorpyrifos etc in laboratory scale or field scale have also been reported. In the case of genetically modified microbes, specific genes contributing for the pesticide degradation or for thriving in extreme environmental condition or for biosurfactant/enzyme production etc are incorporated into the microbes to produce a recombinant microbe. For example a recombinant Rhodococcus TE1 produced by introducing a gene encoding s-triazine hydrolase from Rhodococcus corallines is reported to degrade atrazine (Shao et al. 1995). However introduction genetically modified microbes into natural environment is restricted due to ethical issues and risk of these organisms getting dispersed into the environment thereby affecting the natural microbial communities.

#### 3.6.2.2 Biosparging

Biosparging is an *in-situ* remediation technique that uses indigenous microorganisms to biodegrade pesticide in the saturated zone. This involves the introduction of oxygen into permeable soil to increase the activity of aerobic microorganisms. Biosparging also introduces oxygen to the saturated zone, which will increase the rate of biodegradation. This technique has not been used commonly in pesticide-contaminated sites.

#### 3.6.2.3 Biosorption

Biosorption is the process of binding of pollutants to the biomass by physico-chemical mechanisms such as adsorption or ion exchange which does not involve metabolic energy or transport. This involves a number of physicochemical processes which are independent of metabolism like adsorption, absorption, surface complexation, ion exchange and precipitation. It's an instanteaneous process and consist of passive process that take place in the cellwall of both living and dead cells. Generally bio-sorption technique is applied for the removal of heavy metals from water/wastewater and also for removal dye from textile industry wastewater. Mostly fungal, algal and bacterial biomass is used as sorbent. There are also other biological biosorbents like plant products/parts, animal biomass like shells/hair and other waste materials. Studies related to biosorption of pesticides has also been carried out for removal of pesticide from water. Deepika and Anil (2011) proposed that the endosulfan can be removed from water using biosorbent prepared from fungal species *Aspergillus nidulans* sorption capacity of 43.55 mg/g$^{-1}$. Important fungal biosorbents that adsorbs atrazine include *Aspergillus, Penicillium and Rhizopus*. Pesticides are also removed by algae. Pesticide reported to be removed by using algae includes aldrin, dieldrin, endrin, lindane, carbaryl and chlordane etc. (Kleknes and Kosaric, 1992). Algae capable of bosorption include *Chlamydomonas sp., Chlorlla sp., Scenedesmus sp., Selenastrum capricornulum* etc.

### 3.6.2.4 Composting

Composting is another ex situ bioremediation technique in which the contaminated soil or waste is treated biologically with aid of soil microbes or earthworms. The harzardous compounds like pesticide and organic matter serve as the substrate for the microbes. The pesticide present in the substrate will degraded biochemically with help of enzymes. Composting can enhance the process of bioremediation of pesticide as they inhabit wide group of microbes. Approaches like adding compost to the pesticide contaminated soil to enhance the bioremediation process is also in practice. Degradation of carabaryl in sewage wood chip composting mixture at 1.3 ppm and 2.2 ppm concentration is an experimental proof of biodegradation during composting. Likewise, addition of manure and vermi compost also enhanced degradation of atrazine and metamitron (Mohsen et al. 2005). Enhanced degradation of endosulfan with a half-life of degradation 1.01 to 1.29 days in soil amended with vermi compost and farmyard manure was also reported by Mukherjee (2012).

### 3.6.2.5 Earthworm Assisted Bioremediation

Earthworm helps in bioremediation by providing the condition which meets the limitation of bioremediation process like sorption, aeration, temperature, moisture content. Earthworm accelerates the microbial degradation of pesticide by making burrow, producing casts, its gut interaction and other biochemical physical interaction. It has the potential to release the soil adsorbed pesticide and thereby make it bioavailable. A number of studies have investigated the use of earthworms in the bioremediation of pesticides. Enhanced mineralization of atrazine was observed by using earthworm as it altered the size and diversity of microbial community (Binet et al. 2006). In another study, use of species *L.terrestris* increased mineralization of atrazine and bioavailability was observed. Earthworm gut microflora was also found to enhance degradation of endosulfan, HCH. Biological, chemical, physical action makes earthworm a suitable mediator for bioremediation by enhancing the bioavailability of contaminants. Gevao et al. 2000 proposed that earthworm enhances the release of bound pesticide residues such as isoproturon, C-dicamba and C-atrazine, prevented the formation of bound pesticide residues and subsequently enhanced the biodegradation of pesticides in soil.

### 3.6.2.6 Surfactant Mediated Bioremediation

Surfactants are surface-active agents capable of lowering of surface and interfacial tensions of liquids. They are molecules having both polar and nonpolar end which makes them amphipathic in nature. Surfactants aggregate to form micelles in solution. The micelle formation occurs when the lipophyllic part of the surfactants that is unable to form hydrogen bonding with the aqueous phase causes an increase in free energy of the system. Hence, to decrease the free energy of the system, the hydrocarbon tail must get isolated from the water, adsorbs to the organic matrix, form micelles/vesicles where the hydrocarbon moiety of the surfactant become situated towards the centre with the hydrophilic part in contact with water (Haigh 1996). The presence of surfactant can lower the interfacial tension between immiscible fluids enabling them to be miscible through the creation of additional surfaces. Thus a single interface consisting of an immiscible and miscible constituent is transformed into smaller interface of the two constituents. Micelle formation allows the partitioning of hydrophobic structures into the central hydrophobic pseudophase core enabling solubility. This can lead to increased dispersion of a compound in solution above its water solubility limit (Rouse et al. 1994). In surfactant mediated bioremediation, the surfactant can improve the solubilisation of pollutants from soil, which in turn improves their bioavailability (Li and Chen 2009).

Both synthetic surfactant and biosurfactant are used for enhancing the bioavailability of organic contaminants. Synthetic surfactants like Tween 80, Afonic 1412-7, Triton X-100 and Brij 30 were experimented widely on their capability to enhance the bioavailability of the hydrophobic organic contaminants. Effect of rhamnolipid and Triton X-100 surfactants on the degradation of trifluralin and atrazine by Streptomyces and on the biodegradation of coumaphos by a bacterial consortium was assessed and the results suggest that the surfactants have a good potential to increase the water solubility of pesticides, there by enhances degradation (Juan et al. 2001). A higher concentration of surfactants is needed for degradation of pesticide (Juan et al. 2001). Enhanced degradation of endosulfan in soil up to 94% was observed with the addition of Tween 80 and at pH 8.5 (Jerald 1996). In another study on bioremediation of chlorinated pesticide using anerobic sludge, addition of tween 80 resulted in decrease in recovery of pesticide HCH, methoxychlor, DDT residues due to desorption of these pesticides and thereby increased degradation (Juan et al. 2001).

Besides synthetic surfactants, biosurfactant were also widely used in the bioremediation of contaminated soil. Biosurfactants are more ecologically acceptable than synthetic surfactants in the bioremediation of contaminated soil and subsurface environments. Diverse range of bacteria capable of producing biosurfactants have been known such as *Pseudomonas aeruginosa* (mono- and di-rhamnolipid biosurfactants), *Corynebacterium, Nocardia* and *Rhodococcus*, spp. (phospholipids, trehalosedimycolates/dicorynomycolates, and glycolipids), *Bacillus subtilis* (surfactin), *Bacillus licheniformis* (lipopeptide), *Arthrobacter paraffineus* (trehalose and sucrose lipids (Lang 2002). Many pesticides have limited solubility and as such, often exist as a second organic phase in the soil environment which is a major limiting factor for biodegradation. These hydrophobic pesticides remain bound to soil and remain unavailable for biodegradation due to interfacial tension between aqueous and organic phases. Many indigenous microbes in the soil may produce biosurfactants under these limited contaminant availability conditions. Presence of biosurfactant induces micellization or pseudo solubilization of pesticide so that a hydrophilic microbe may interact with a hydrophilic micelle in which hydrophobic substrate (pesticide) is attached. This pseudo solublized pesticide containing micelle coalesces with microbial membranes and deliver a contaminant molecule directly to the outer membrane of a microbial cell (Miller and Bartha 1989).

Biosurfactant producing bacteria were also isolated from pesticide contaminated sites and were proved for their enhancing capability for solubility and bioavailability (Jennings and Tanner 2000). Awasthi et al. (1999) reported the enhanced biodegradation of endosulfan using biosurfactant produced by MTCC 14 *Bacillus subtilis* in soil. Greeshma and Vasudevan (2013) isolated an *Arthrobacter sp* capable of producing biosurfactant on endosulfan isomers. Biosurfactant producing bacteria enhanced the biodegradation of endosulfan isomers and its major metabolite endosulfate by enhancing the bioavailability of endosulfan and endosulfate (Greeshma and Vasudevan 2013). Biosurfactant capable of emulsifying pesticides were produced using *Bacillus subtilis* to disperse Fenthion, an organophosphrous pesticide (Veenanadig et al. 2000) using packed column. Biosurfactant which enhances the solubility of hexachlorobiphenyl was produced from *Pseudomonas aeruginosa* (Berg et al. 1990). Biosurfactant can enhance the biodegradation of pesticide when it is added externally to the contaminated soil. Enhanced bioavailability of HCH isomers in soil was observed in the presence of sorpholipid, rhamnolipid and trehalose containing lipid surfactants (Manickam et al. 2012). Even though biosurfactants are more ecologically acceptable than synthetic surfactants in the bioremediation of contaminated soil and subsurface environments, their high cost of production limits their large scale utilization.

#### 3.6.2.7 Organic Amendment Enhanced Bioremediation

The application of organic waste in agricultural soils, is a practice that has been carried out for centuries, due to its fertilizer properties and contribution to the physicochemical and biological properties of the soil (Balesdenta et al. 2000; Cooperband 2002; Palma et al. 2002; Vorkamp et al. 2002; Said-Pullicino et al. 2004). Recently, an increasing interest has been focused on assessing the consequences of exogenous sources of organic carbon on the behaviour of pesticides in soils (Albarran et al. 2004; Blacksaw et al. 2005) because environmental contaminants depend on their retention and degradation in soil results in an increase in microbial activity due to availability of simple organic nutrients like sugar and aminoacids (Cox et al. 2001). Moreover, the addition of organic amendments increases the dissolved organic matter in soil which affects the pesticide movement (Cox et al. 2001). Import of organic carbon to soils in the form of compost, sludge, and crop residues has become a very popular practice owning to its low costs and recycling of nutrients (Sigua et al. 2005; Majumdar and Singh 2007). Dissolved organic matter (DOM) also interacts with organic pollutants and exerts a great effect on mobilization and other behaviours of pesticides in soils (Thom et al. 1997; Esposito et al. 1998; Song et al. 2008).

Organic amendments can accelerate or increase biodegradation by stimulation of microorganisms due to structural changes in the porosity induced by higher organic content (Worrall et al. 2001). Addition of organic waste to soil contributes to enhancement of active humified components, such as humic acid and fulvic acid (Plaza et al. 2003), which exert an important role in geochemical processes as sources of nutrients for plants and microorganisms, in acid–base buffering capacity of soils, and promoting a good soil structure, thereby improving aeration and moisture retention (Graber et al. 2001; Marschner et al. 2003). On agronomic aspect, addition of organic waste enhances biological activity and fertility (Namkoong et al. 2002; Zagal et al. 2002;

Dungan et al. 2003; Bohme et al. 2005; Wanner et al. 2005) and on environmental aspects, the addition plays an important role on the fate of xenobiotic compounds such as heavy metals, aromatic hydrocarbons, and pesticides (Fogarty and Tuovinen, 1991; Buyuksonmez et al. 2000; Alexander, 2003; Harlieb et al. 2003). Moreover, addition of organic amendments favours the growth of fungi in the soil. The enzymes secreted by the fungi non-selectively act on pesticide and helps in the initial transformation of pesticide. This partially degraded pesticide thus becomes more amenable to bacterial mineralization. Addition of organic waste also increases the bacterial population in the soil (Chiu et al. 1998; Purnomo et al. 2010). The addition of organic amendments frequently modifies the rate and pathway of pesticide degradation in soils, according to the nature of the organic amendment and their effect on microbial community (Alvey and Crowely 1995). Most of the lignocellulosic organic materials are prone to fungal attack and during this degradation process extracellular enzymes are released which inturn serve as catalysts for bacterial degradation of pesticides. Addition of organic amendments also enhances the cometabollic biotransformation of pesticides by increasing the microbial activity (Hance, 1973).

Dissolved organic matter have characteristics similar to the surface active agents like surfactants which lowers the surface tension and increase the solubility of compounds making it available to the soil microbes (Li et al. 2005). Ling and Hong (2011) reported higher and rapid degradation of flouroxypar in soil amended with straw and sludge. Hence, in order to stimulate the microbial degradation, amendments like spent mushroom compost, farmyard manure, biogas slurry, poultry litter, coir pith compost, vermi compost, charcoal, leaf compost and wheat bran can be added during bioremediation process. Addition of bioprocessed materials farmyard manure biogas slurry increased the degradation of atrazine (Neeru et al. 2008). Addition of manure, vermi compost also enhanced degradation of atrazine and metamitron (Mohsen et al. 2005). Addition of seaweed as an amendment and sodium source in soil also showed enhanced bioavailability (Kantachote et al. 2004). Addition of organic nutrients enhanced aerobic degradation of alachlor in subsurface soils and in one aquifer sample, indicating that nutrient availability limits biodegradation (Pothuluri and Moorman 1990). Mukherjee (2012) reported enhanced degradation of endosulfan with a half-life of degradation 1.01 to 1.29 days in soil amended with vermi compost and farmyard manure. Organic amendments can have reverse effect on degradation. In an herbicide degradation study amendments (e.g. compost, and sawdust) stimulated general microbial populations and activity without concurrent increase in herbicide degradation. This action has probably resulted from the fact that specific microbial populations responsible for degrading the contaminants were not stimulated (Moorman et al. 2001). Sometimes addition of auxiliary carbon source/organic amendments will decrease the degradation of pesticides in soil since the auxiliary carbon source being easily available than the toxic substrate (Supriya and Dileep, 2009).

#### 3.6.2.8 Biobed

One of the significant contributors for deterioration of natural water sources are pesticides. Biobed treatment system is one of the onsite treatment techniques to minimize this non-point source contamination due to pesticides. This system was first implemented in Sweden in 1993. Over the years it has developed into an efficient treatment system in Latin America, Chile and several modified biobed systems came into existence. Many researches have been carried out in this area especially on the effect of operating conditions, components of biobed, stabilization time of biomixture, incorporation of lignocellulosic residues and biomixture-pesticide-microbe's interaction (Tortella et al. 2012).

A biobed is a modified pit consisting of biobed mixture comprising of a lignocellulosic substrate, organic amendment(s), native soil and biocover (Figure 3.7). This composition harbours millions of microbes of which a few are capable of degrading pesticides. The lignocellulosic component of biobed stimulates the growth of lignolytic microorganisms and production of extracellular lignolytic enzymes such as peroxidases. The organic amendment part of the biobed enhances the sorption of pesticides and controls moisture of the biobed. Soil also enhances the sorption capacity of the biobed and also harbours pesticide degrading microbes. The biocover also enhances the degradation of pesticide in the upper part of the biomixture. It also enhances the root zone microbial activity. The efficiency of a biobed system depends on the absorption capacity of the biobed mixture (Castillo et al. 2008). Generally, a biobed system designed for treating hazardous compounds like pesticides are lined biobeds in order to prevent seepage. The minimum required dimension of a biobed for treating the agriculture wastewater (1000 L) is: depth-1m and surface area 1m2. Several modified biobed treatment systems are now in existence and still studies are being carried out in improvising the design of biobed and absorption capacity of the biobed mixture for efficient removal of pesticides from agriculture runoff and agriculture wastewater. Recently, Cooper et al. (2016) assessed the efficiency of a three staged biobed system in treating the wastewater from an estate of area 20 $Km^2$. The study on this onsite biobed system revealed that the system is capable of removing pesticides by an average of 91.6%. Biobed system is also proved for its efficiency in removing multiple pesticides contaminated water.

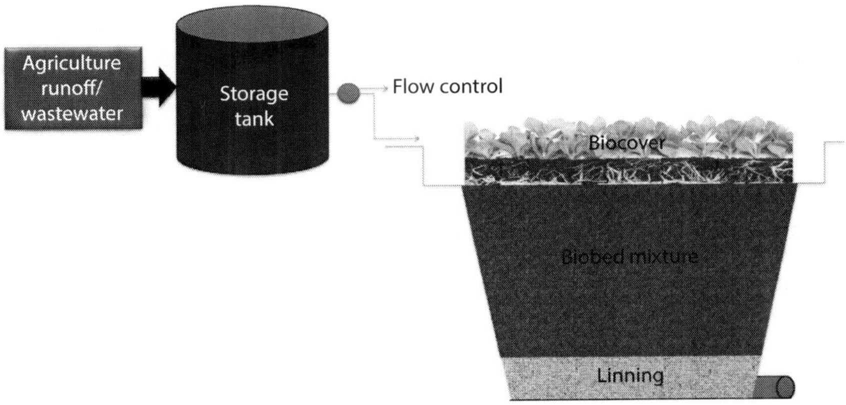


**FIGURE 3.7** Schematic diagram for Biobed Treatment System.

## 3.7 CONCLUSION

As long as human dependence on crop-protecting agents like pesticides continues, problems related to pesticide usage like acute and chronic toxicity, contamination of water and soil, and bioaccumulation of pesticide at different trophic levels of the food chain will prevail in the world. The only precautionary measure that can be adopted for prevention of such harmful effects of pesticide is more restricted and scientific use of pesticides. Proper training should be given for proper handling of these toxic compounds in the field. A study over different regions of the world concludes that pesticide residues prevail in the environment, and there is an immediate need for removal of pesticides from the environment. Bioremediation is one of the effective and environmentally safe options for removal of pesticide from the environment.

## REFERENCES

Abo-Amer., and Aly, E. 2011. Biodegradation of diazinon by *Serratiamarcescens*DI101 and its use in bioremediation of contaminated soil environment. *Journal of Microbiology and Biotechnology* 21(1): 71–80.

Abrahams, P.W. 2002. Soils: Their implications to human health. *Science of the Total Environment* 291: 1–32.

Adeniyi, A., and Oladele, O. 1999. Residues of organochlorine pesticides in fruits, vegetables and tubers from Nigerian markets. *Science of the Total Environment* 231(2–3): 227–233.

Agarwal, H.C., Mittal, P.K., Menon, K.B., and Pillai, M.K.K. 1986. DDT residues in the river Yamuna in Delhi, India'. *Water, Air, and Soil Pollution* 28: 89–104.

Agnihotri, N.P. 1999. Pesticide safety and monitoring, All India coordinated research project on pesticides residues. *Indian Council of Agricultural Research*, New Delhi, India.

Agnihotri, N.P., Gajbhiye, V.T., Kumar, M., and Mohapatra, S.P. 1994. Organochlorine insecticides in Ganga river water near Farrukhabad, India. *Environmental Monitoring and Assessment* 30: 105–112.

Aislabie, J.M., Richards, N.K., and Boul, H.L. 1997. Microbial degradation of DDT and its residues—A review. *New Zealand Journal of Agricultural Research* 40: 269–282.

Akan, J.C., Jafiya, L., Mohammed, Z., and Abdulrahman, F.I. 2013. Organophosphorus pesticide residues in vegetables and soil samples from Alau Dam and Gongulong agricultural areas, Borno State, Nigeria. *International Journal of Environmental Monitoring and Analysis* 1(2): 58–64.

Akinloye, O.A., Adamson, I., Ademuyiwa, O., and Arowolo, T.A. 2011. Occurrence of paraquat residues in some Nigerian crops, vegetables and fruits. *Journal of Environmental Chemistry and Ecotoxicology* 3(7): 195–198.

Al-Araji, Y., and Issa, L. 2004. Biosurfactant production by *Pseudomonas aeruginosa*181. Ph. D. thesis, Universiti Putra Malaysia, Selangor, Malaysia.

Albarran, A., Celis, R., Hermosin, M., Lopez-Pineiro A., and Cornejo, J. 2004. Behavior of simazine in soil amended with the finalresidue of the olive-oil extraction process. *Chemos* 54: 717–724.

Alexander, R. 2003. *Landscape architect specifications for compost utilization*. CWC/PNWER. The US Composting Council, Washington, 91.

Alvey, S., and Crowley, D.E. 1995. Influence of organic amendments on biodegradation of atrazine as a nitrogen source. *Journal of Environmental Quality* 24, 1156–1162.

Arshad, M., Hussain, S., and Saleem, M. 2008. Optimization of environmental parameters for biodegradation of alpha and beta endosulfan in soil slurry by *Pseudomonas aeruginosa*. *Journal Applied Microbiology* 104: 1364–5072.

Asker, G., Vibeke, F.L., and Kim, D. 2004. *A DNAPL hotspot of Organophosphorus pesticides*. Høfde 42 Harboøre Tange, County of. Ringkøbing, Denmark.

ASTDR. 2000. *Toxicological profile for endosulfan*. US Department of Health and Human Services; Agency for Toxic Substances and Disease Registry, Atlanta, GA.

Atwal, A.S. 1986. *Agricultural pest of India and South East Asia*. Kalyani Publishers, New Delhi, p. 52.

Awasthi, N., Kumar, A., Makkar, R., and Cameotra, S.S. 1999. Biodegradation of soil-applied endosulfan in the presence of a biosurfactant. *Journal of Environmental Science and Health Part B* 34: 793–803.

Bailey, G.W, and White, J.L. 1970. Factors influencing the adsorption, desorption, and movement of pesticides in soils. *Residue Reviews* 32: 29–92.

Balesdenta, J., Chenub, C., and Balabane, M. 2000. Relationship of soil organic matter dynamics to physical protection and tillage. *Soil and Tillage Research* 53: 215–230.

Battu, R.S., Singh, B., and Kang, B.K. 2004. Contamination of liquid milk and butter with pesticide residues in the Ludhiana district of Punjab state, India. *Ecotoxicology and Environmental Safety* 59: 324–331.

Bello, X.V., Devesa-Rey, R., Cruz, J.M., and Moldes, A.B. 2012. Study of the synergistic effects of salinity, pH, and temperature on the surface-active properties of biosurfactants produced by Lactobacillus pentosus. *Journal of Agricultural and Food Chemistry* 60: 1258−1265.

Bending, G., Friloux, M., and Walker, A. 2002. Degradation of contrasting pesticides by white rot fungi and its relationship with ligninolytic potential. *FEMS Microbiology Letters* 212: 59–63.

Berg, G., Seech, A.F., Lee, H., and Trevors, J.T. 1990. Identification and characterization of soil bacterium with emulsifying activity. *Journal of Environmental Science and Health* 7: 753–764.

Betancourt, A.M., and Carr, R.L. 2004. The effect of chlorpyrifos and chlorpyrifos-oxon on brain cholinesterase, muscarinic receptor binding, and neurotrophin levels in rats following early postnatal exposure. *Toxicological Sciences* 77: 63–71.

Betoulle, S., Duchiron, C., and Deschaux, P. 2000. Lindane differently modulates intracellular calcium levels in two populations of rainbow trout (*Oncorhynchus mykiss*) immune cells: Head kidney phagocytes and peripheral blood leucocytes. *Toxicology* 145(2–3): 14, 203–215.

Beulke, S., Brown, C.D., Dubus, I.G., Fryer, C.J., and Walker, A. 2004. Evaluation of probabilistic modelling approaches against data on leaching of isoproturon through undisturbed lysimeters. *Ecological Modelling* 179: 131–144.

Bhanti, M., and Taneja, A. 2007. Contamination of vegetables of different seasons with organophosphorous pesticides and related health risk assessment. *Chemosphere* 69: 63–68.

Bhatnagar, V.K. 2001. Pesticides pollution: Trends and perspectives. *ICMR Bulletin* 31: 87–88.

Bidleman, T.F., Falconer, R.L., and Walla, M.D. 1995. Toxaphene and other organochlorine compounds in air and water at Resolute Bay, NWT, Canada. *Science of the Total Environment* 161: 55–63.

Blacksaw, R., Molnar, L., and Larney, F. 2005. Fertilizer manure and compost effects on weed growth and competition with winter wheat in western Canada. *Crop Protection* 24: 971–980.

Bohme, L., Langer, U., and Bohme, F. 2005. Microbial biomass, enzyme activities and microbial community structure in two European long-term field experiments. *Agriculture Ecosystems Environment* 109: 141–152.

Bosma, T.M.P., Middeldorp, P.J.M., Schraa, G., and Zehnder, A.J.B. 1997. Mass transfer limitation of biotransformation: Quantifying bioavailability. *Environmental Science and Technology* 31: 248–252.

Broomhal, S., and Shine, R. 2003. Effect of the insecticide endosulfan and presence of congeneric tadpoles on Australian Tree frog (*Litoria freycineti*) Tadpoles. *Archives of Environmental Contamination and Toxicology* 45: 221–226.

Brown, G.B., and Doube, B.M. 2004. Earthworm ecology, In *Earthworm Ecology*. Edwards C.A, (ed.), 2nd edn. CRC Press, Boca Raton, FL, 213–239.

Buyuksonmez, F., Rynk, R., Hess, T., and Bechinski, E. 2000. Occurrence, degradation and fate of pesticides during composting. Part II. Occurrence and fate de pesticides in compost and composting systems. *Compost Science Utilization* 8: 61–81.

Calderbank, A., 1989. The occurrence and significance of bound pesticide residues in soil. *Environmental Contamination and Toxicology* 108: 71–103.

Carr, R.L., Chambers, H.W., Guarisco, J.A., Richardson, J.R., Tang, J, and Chambers, J.E. 2001. Effects of repeated oral postnatal exposure to chlorpyrifos on open-field behavior in juvenile rats. *Toxicological Science* 59: 260–267.

Castillo, M.D., Torstensoon, L., and Stenstrom, J. 2008. Biobeds for Environmental Protection from Pesticide Uses- A Review. *J. Agric. Food Chem.* 56: 6206–6219.

CDC. 2005. *Third national report on human exposure to environmental chemicals.* U.S. Department of Health and Human Services, Centers for Disease Control and Prevention, Atlanta, GA. pp. 349–377.

Central Pollution Control Board (CPCB). 2000. Water quality status of Yamuna river. ADSORBS/32/1999-2000, Annexure III, pp. 107–115.

Chandra, R., Bharagava, R.N., and Rai, V. 2008. Melanoidins as major colourant in sugarcane molasses based distillery effluent and its degradation. *Bioresource Technology* 99: 4648–46601.

Chandra, R., and Chowdhary, P. 2015. Properties of bacterial laccases and their application in bioremediation of industrial wastes. *Environmental Science Processes and Impacts* 17: 326–342.

Chaudhuri, K., Selvaraj, S., and Pal, A.K. 1999. Studies on the genotoxicology of endosulfan in bacterial system. *Mutation Research* 439: 63–67.

Chiu, S.W., Ching, M.L., Fong, K.L., and Moore, D. 1998. Spent oyster mushroom substrate performs better than many mushroom mycelia in removing the biocide pentachlorophenol. *Mycological Research* 102: 1553–1562.

Chubiko, M.I., Smol-Sike, G.M., and Basova, G.M. 1998. The effect of pesticides on the quality of milk products. *Gigiena and Sanitariia* 2: 23–25.

Clay, S.A., and Koskinen, W.C. 2003. Effect of variability of soil properties as a function of depth on pesticide sorption-desorption. Terrestrial field dissipation studies: Purpose, design and interpretation. *ACS Symposium Series* 842: 102–116.

Colborn, T., and Thayer, K. 2000. Aquatic ecosystem: Harbingers of endocrine disruption. *Ecological Applications* 10(40): 949–957.

Cooke, A.S. 1970. The effect of pp′-DDT on tadpoles of the common frog (*Rana temporaria*). *Environmental Pollution* 57–71.

Cooperband, L. 2002. *Building soil organic matter with organic amendments*. Report by Center for Integrated Agricultural Systems, University of Wisconsin–Madison, Madison, WI, pp. 1–16.

Cooper, R.L., Stoker, T.E., Tyrey, L., Goldman, J.M., and McElroy, W.K. 2002. Atrazine disrupts the hypothalamic control of pituitary-ovarian function. *Toxicological Science* 53: 297–307.

Cox, L., Cecchi, A., Celis, R., Hermos, M., Koskinen, W., and Cornejo, J. 2001. Effect of exogenous carbon on movement of simazine and 2, 4-D in soils. *Soil Science Society of American Journal* 65: 1688–1695.

CSE Report: analysis of pesticide residues in soft drinks. 2006. Centre for Science and Environment, New Delhi, India.

Cycon, M., Wojcik, M., and Piotrowska-Seget, Z. 2009. Biodegradation of the organophosphorus insecticide diazinon by *Serratia sp. and Pseudomonas sp.* and their use in bioremediation of contaminated soil. *Chemotherapy* 76: 494–501.

Daniel, R.S., and Timothy, B.P. 1991. Effect of moisture on sorption and biodegradation of carbofuran in soil. *Journal of Agricultural and Food Chemistry* 39: 2063–2060.

Deepika Dave, and Anil Kumar Dikshit. 2011. Evaluation of Kinetic Parameters of Endosulfan Biosorbed on Biosorbent Prepared from *Aspergillusnidulans*. *Journal of Microbiology Research* 1(1): 8–16.

Demand Pattern of Pesticide Use (Tech. Grade) for Agriculture. 1999. Pesticide Association of India, Mumbai, India.

Digrak, M., and Ozcelik, S. 1998. Effect of some pesticides on soil microorganisms. *Bulletin of Environmental Contamination and Toxicology* 60: 916–922.

Dungan, R., Ibekwe, A., and Yates, S. 2003. Effects of propargyl bromide and 1.3- dichloropropene on microbial communities in an organically amended soil *FEMS Microbiology Ecology* 43: 75–87.

Dutra, B.K., Fernandes, F.A., Lauffer, A.L., and Oliveira, G.T. 2009. Carbofuran-induced alterations in the energy metabolism and reproductive behaviors of *Hyalella castroi* (Crustacea, Amphipoda). *Comparative Biochemistry and Physiology-Part C: Toxicology and Pharmacology* 149(4): 640–646.

Ekundayo, E.O. 2003. Effect of common pesticides used in the Niger Delta basin of southern Nigeria on soil microbial populations. *Environmental Monitoring and Assessment* 89(1): 35–41.

Ellis-Tabanor, M., and Hyslop, E. 2005. Effect of sub lethal concentrations of endosulfan on growth and fecundity of two species of snails. *Bulletin of Environmental Contamination and Toxicology* 74: 1173–1178.

Espositom, E., Paulillo, S.M., and Manfio, G.P. 1998. Biodegradation of the herbicide diuron in soil by indigenous actinomycetes. *Chemosphere* 37: 541–548.

Extension Toxicology Network Pesticide Information Profiles. 1996. National Pesticide Information Centre, Orgeon State University, Corvallis, OR.

Fogarty, A., and Tuovinen, O. 1991. Microbiological degradation of pesticides in yard waste composting. *Microbiology Reviews* 55: 225–233.

Foght, J., April, T., Briggar, K., and Aislabie, J. 2001. Bioremediation of DDT contaminated soils, A review. *Bioremediation Journal* 5(3): 225–246.

Fomsgaard, I.S. 1995. Degradation of pesticides in subsurface soils, unsaturated zone—A review of methods and results. *International Journal of Environmental Analytical Chemistry* 58: 231–245.

Gevao, B., Semple, K.T., and Jones, K.C. 2000. Bound pesticide residues in soils: A review. *Environmental Pollution* 108: 3–14.

Ghadiri, H., Rose, C.W., and Connell, D.W. 1995. Degradation of organochlorine pesticides in soils under controlled environment and outdoor conditions. *Journal of Environmental Management* 43(2): 141–151.

Gibbs, R.J. 1973. Mechanisms of trace metal transport in rivers. *Science* 180: 71–2.

Giri, K., and Rai, J.P.N. 2012. Biodegradation of endosulfan isomers in broth culture and soil microcosm by *Pseudomonas fluorescens* isolated from soil. *International Journal of Environmental Studies* 69(5): 729–742.

Goswami, S., and Singh, D.K. 2009. Biodegradation of α and β endosulfan in broth medium and soil microcosm by bacterial strain *Bordetella sp.* B9. *Biodegradation* 20: 199–207.

Graber, E., Dror, I., Bercovich, F., and Rosner, M. 2001. Enhanced transport of pesticides in a field trial with treated sewage sludge. *Chemosphere* 44: 805–811.

Gupta, S.K., Jani, J.P., Saiyed, H.N., and Kashyap, S.K. 1984. Health hazards in pesticide formulators exposed to a combination of pesticides. *Indian Journal of Medical Research* 79: 666–672.

Gurug, K.S., and Tanabe, S. 2001. Contamination by persistent organochlorines and butyltin compounds in the west coast of Sri Lanka. *Marine Pollution Bulletin* 42(3): 179–186.

Haigh, S.D. 1996. A review of the interaction of surfactants with organic contaminants in soil. *Sci. Total Environ.* 185: 161–170.

Hakanson, L. 1992. Sediment variability. In *Sediment toxicity assessment,* Burton, G.A., Jr. (ed.),. Lewis Publishers, Boca Raton, FL, pp. 19–36.

Halder, P., Raha, P., Bhattacharya, P., Chowdhury, A., and Aditya Chaudhary, N. 1989. Studies on the residues of DDT and endosulfan occurring in Ganga water. *Indian Journal of Environmental Health* 31: 156–161.

Hance, R.J. 1973. The effect of nutrients on decomposition of herbicide atrazine and linuron incubated with soil. *Pesticide Science* 4: 817–822.

Hartlieb, N., Ertunc, T., Schaeffer, A., and Klein, W. 2003. Mineralization, metabolism and formation of non-extractable residues of $_{14}$C-labelled organic contaminants during pilot-scale composting of municipal biowaste. *Environmental Pollution* 126: 83–91.

Hickey, J.J., and Anderson, D.W. 1968. Chlorinated hydrocarbons and eggshell changes in raptorial and fish-eating birds. *Science* 162: 271–273.

Hickey, W.J., Fuster, D.J., and Lamar, R.T. 1994. Transformation of atrazine in soil by *Phanerochaete chrysosporium. Soil Biology and Biochemistry* 26: 1665È71.

Hong, H., Xu, L., Zhang, L., Chen, J.C., Wong, Y.S., and Wan, T.S.M. 1995. Environmental fate and chemistry of organic pollutants in the sediment of Xiamen and Victoria harbors. *Marine Pollution Bulletin* 31: 22–36.

Howard, M.D., and Pope, C.N. 2002. In vitro effects of chlorpyrifos, parathion, methyl parathion and their oxons on cardiac muscarinic receptor binding in neonatal and adult rats. *Toxicology* 170: 1–10.

ICAR. 1967. *Report of the special committee on harmful effects of pesticides*. ICAR, New Delhi, pp. 78.

Jayashree, R., and Vasudevan, N. 2006. Residues of organochlorine pesticides in agricultural soils of Thiruvallur district. *J. Food Agric. Environ.* 4(1): 313–316.

Jayashree, R., and Vasudevan, N. 2007a. Effect of tween 80 added to the soil on the degradation of endosulfan by *Pseudomonas aeruginosa. International Journal of Environmental Science and Technology* 4(2): 203–10.

Jayashree, R., and Vasudevan, N. 2007b. Organochlorine pesticide residues in ground water Thiruvallur district, India. *Environmental Monitoring and Assessment* 128: 209–215.

Jennings, E.M., and Tanner, R.S. 2000. Biosurfactant-producing bacteria found in contaminated and uncontaminated soils. *Proceedings of the Conference on Hazardous Waste Research*. Environmental Challenges and Solutions to Resource Development, Production, and Use Holiday Inn, Denver, CO, May 23–25.

Jensen, P.H., Hansen, H.C.B., Rasmussen, J., and Jacobsen, O.S. 2004. Sorption-controlled degradation kinetics of MCPA in soil. *Environmental Science & Technology* 38: 6662–6668.

Kang, B.K., Singh, B., Chahal, K.K., and Battu, R.S. 2000. Insecticide residues in market samples of cucumber and radish. *Pestology XXIV*: 4: 57–59.

Kang, B.K., Singh, B., Chahal, K.K., and Battu, R.S. 2002. Contamination of feed concentrate and green fodder with pesticide residues. *Pesticide Residues Journal* 14: 308–312.

Kannan, K.S., Holcombe, R.F., Jain, S.K., Alvarez-Hernandez, X., Chervenak, R., Wolf, R.E., and Glass, J. 2002. Evidence for the induction of apoptosis by endosulfan in a human T-cell leukemic line. *Molecular and Cellular Biochemistry* 205: 53–66.

Kantachote, D., Naidu, R., Williams, B., McClure, N., Megharaj, M., and Singleton, I. 2004. Bioremediation of DDT-contaminated soil: Enhancement by seaweed addition. *Journal of Chemical Technology and Biotechnology* 79(6): 632–638.

Karpouzas, D.G., and Walker, A. 2000. Factors influencing the ability of *Pseudomonas putida* epI to degrade ethoprophos in soil. *Soil Biology and Biochemistry* 32, 1753–1762.

Kaushik, C.P., Sharma, H.R., Jain, S., Dawra, J., and Kaushik, A. 2008. Pesticide residues in river Yamuna and its canals in Haryana and Delhi, India. *Environmental Monitoring and Assessment* 144: 329–340.

Kaushik, A., Sharma, H.R., Jain, S., Dawra, S., and Kaushik, C.P. 2010. Pesticide pollution of river Ghaggar in Haryana. India. *Environmental Monitoring and Assessment* 160: 61–69.

Kennedy, D.W., Aust, S.D., and Bumpus, J.A. 1990. Comparative biodegradation of alkyl halide insecticides by the white rot fungus *Phanerochaete chrysosporium* (BKM-F-1767). *Applied Environmental Microbiology* 56: 2347È53.

Khan, S.U. 1982. Bound pesticide residues in soil and plants. *Residue Review* 84: 1–25.

Khan, S., Khan, J.A., and Jabin, S. 2000. Effect of endosulfan on the seed germination, growth and nutrients uptake of fenugreek plant. *Journal of Industrial Pollution Control* 16: 225–230.

Khandakar, U.R., Shrikhande, W.T., and Shinde, D.A. 2003. Nutrient availability to egg plant (*Solanum melongena* L.) as influenced by pesticide application. *Journal of Indian Society of Soil Science* 42: 568–571.

Killer environment. 1999. *Environmental Health Perspectives* 107: A62.

Kishimba, M.A., and Mihale, M.J. 2004. Levels of pesticide residues and metabolites in soil at Vikuge farm, Kibaha district, Tanzania—A classic case of soil contamination by obsolete pesticides. *Tanzania Journal of Science* 30(2): 77–86.

Klema, V. 2011. Pathway of endosulfan degradation University of Minnesota, BioC/MicE5309, and Edward LaBelle, University of Minnesota. http://umbbd.ethz.ch/end/end_map.html

Kole, R.K., Banerjee, H. and Bhattacharyya, A. 2002. Monitoring of pesticide residues in farmgate vegetable samples in West Bengal. *Pesticide Research Journal* 14: 77–82.

Kookana, R.S., Baskaran, S., and Naidu, R. 1998. Pesticide fate and behaviour in Australian soils in relation to contamination and management of soil and water: A review. *Australian Journal of Soil Research* 36: 715–764.

Kullman, S.W., and Matsumura, F. 1996. Metabolic pathway utilized by *Phanerochetechrysosporium* for degradation of the cyclodiene pesticide endosulfan. *Applied & Environmental Microbiology* 62: 593–600.

Kumar, M., Gupta, S.K., Garg, S.K., and Kumar, A. 2006. Biodegradation of hexachlorocyclohexane-isomers in contaminated soils. *Soil Biology and Biochemistry* 38(8): 2318–2327.

Kumar, M, Lakshmi, C.V., and Khanna, S. 2008. Biodegradation and bioremediation of endosulfan contaminated soil. *Bioresearch Technology* 9(8): 3116–3122.

Kumar, M., and Philip, L. 2006. Bioremediation of endosulfan contaminated soil and water optimization of operating conditions in laboratory scale reactors. *Journal of Hazardous Material* 136: 354–364.

Kumar, R., Singh, B., and Gupta, V.K. 2012. Biodegradation of Fipronil by *Paracoccus sp.* in different types of soil. *Bulletin of Environmental Contamination Toxicology* 88: 781–787.

Kumari, B., Madan, V.K., and Kathpal, T.S. 2006. Monitoring of pesticide residues in fruits. *Environmental Monitoring and Assessment* 123(1–3): 407–412.

Kumari, B., Madan, V.K., and Kathpal, T.S. 2008. Status of contamination of soil and water in Haryana, India. *Environmental Monitoring and Assessment* 136: 239–244.

Lambert, M.R.K. 1997. Effect of pesticides on amphibians and reptiles in sub-Saharan Africa. *Reviews of Environmental and Contamination Toxicology* 150: 31–73.

Lang, S. 2002. Biological amphiphiles (microbial biosurfactants). *Current Opinion in Colloid and Interface Science* 7: 12–20.

Laouedj, A., Schenk, C., Pfohl-Leskowicz, A., Keith, G., Schonta, D., Guillemaut, P., and Rether, B. 1995. Detection of DNA adducts in declining hop plants grown on fields formerly treated with hepatochlor a persistent insecticide. *Environmental Pollution* 90: 409–414.

Larsen, B., and Jensen, A. 1989. Evaluation of the sensitivity of sediment stations in pollution monitoring. *Marine Pollution Bulletin* 20: 556–60.

Lee, J.S., Tanabe, S., Takemoto, N., and Kubodera, T. 1997. Organochlorine residues in deep-sea organisms from Suruga Bay, Japan. *Marine Pollution Bulletin* 34: 250–258.

Li, J.L., and Chen, B.H. 2009. Surfactant-mediated biodegradation of polycyclic aroamatic hydrocarbons. *Materials* 27: 6–94.

Li, K., Xing, B., and Torello, W. 2005. Effects of organic fertilizer derived dissolved organic matter on pesticides sorption and leaching. *Environ. Pollut.* 134: 187–194.

Ling, T., and Hong, Y. 2011. Fluroxypyr biodegradation in soils by multiple factors. *Environ. Monit. Asses.* 175(1–4): 227–238.

Lock, K., De Schamphelaere, K.A., and Janssen, C.R. 2002. The effect of lindane on terrestrial invertebrates. *Archives of Environmental Contamination and Toxicology* 42: 217–221.

Loganathan, B.G., and Kannan, K. 1994. Global organochlorine contamination trends: An overview. *Ambio* 23: 87–191.

Lopez, L., Pozo, M.A., Gomez, A., Calvo, L., and Lopea, J.G. 2002. Studies on the effects of the insecticide aldrin on aquatic microbial populations. *International Biodeterioration Biodegradation* 50: 83–87.

Lorenz, E.S. 2006. Toxicity of pesticides. Pesticide safety fact sheet, Pennsylvania State University.

Maier, R.M., and Valdes, J.J. 2000. Bioavailability and its importance to bioremediation. In *Bioremediation*, Valdes, J.J. (ed.). Academic Press, Dordrecht, The Netherlands, pp. 59–78.

Majumdar, K., and Singh, N. 2007. Effect of soil amendments on sorption and mobility of metribuzin in soils. *Chemosphere* 66: 630–637.

Makkar, R.S., and Cameotra, S.S. 2000. Effects of various nutritional supplements on biosurfactant production by a strain of Bacillus subtilis at 45°C. *Journal of Surfactants and Detergents* 5: 11–17.

Manickam, N., Bajaj, A., Harvinder, S.S., and Shanker, R. 2012. Surfactant mediated enhanced biodegradation of hexachlorocyclohexane (HCH) isomers by *Sphingomonas* sp. *NM05 Biodegradation* 23(5): 673–682.

Marcia, D.H., and Carey, N.P. 2002, In vitro effects of chlorpyrifos, parathion, methyl parathion and their oxons on cardiac muscarinic receptor binding in neonatal and adult rats, *Toxicology* 170(1–2): 151–210.

Marigoudar, S.R., Nazeer Ahmed R., and David, M. 2009. Impact of cypermethrin on behavioural responses in the freshwater teleost, *Labeo rohita* (Hamilton). *World Journal of Zoology* 94(1): 19–23.

Marschner, P., Kandeler, E., and Marschner, B. 2003. Structure and function of the soil microbial community in a long-term fertilizer experiment. *Soil Biology and Biochemistry* 35: 453–461.

Martin, J.M., and Mermoid, A. 1998. Sorption and degradation of four nitroaromatic herbicides in mono and multi-solute saturated/unsaturated soil batch systems. *Journal of Contaminant Hydrology* 33: 187–210.

Mata–Sandoval, J.C., Karns, J., and Torrents, A. 2001. Influence of rhamnolipids and triton X-100 on the biodegradation of three pesticides in aqueous phase and soil slurries. *Journal of Agricultural and Food Chemistry* 49: 3296–3303.

Megharaj, M., Kantachote, D., Singleton, I., and Naidu, R. 2002. Effects of long-term contamination of DDT on soil microflora with special reference to soil algae and algal transformation of DDT. *Environmental Pollution* 109: 35–42.

Miller, R.M., and Bartha, R. 1989. Evidence from liposome encapsulation for transport-limited microbial-metabolism of solid alkanes. *Applied Environmental Microbiology* 55: 269–274.

Mitra, J., and Raghu, K. 1999. Effect of DDT residues in soil on productivity of oil seed crops. *Tropical Agricultural Research and Extension* 2: 33–39.

Mohsen, F., Gholam, H.H., and Alireza, K. 2005. Organic amendments to enhance atrazine and metamitron degradation in two contaminated soils with contrasting textures. *Soil and Sediment Control* 14(4): 345–355.

Moorman, T., Cowan, J., Arthur, E., and Coats, J. 2001. Organic amendments to enhance herbicide biodegradation in contaminated soils. *Biology and Fertility of Soils* 33: 541–545.

Mougin, C., Laugero, C., Asther, M., Dubroca, J., Frasse, P., and Asther, M. 1994. Biotransformation of the herbicide atrazine by the white rot fungus *Phanerochaete chrysosporium*. *Applied Environmental Microbiology* 60: 705–708.

Mukherjee, I. 2012. Influence of organic amendments on the degradation of endosulfan. *Bulletin of Environmental Contamination and Toxicology*. 89: 334–339.

Namkoong, W., Hwang, E., Park, J., and Choi, J. 2002. Bioremediation of diesel contaminated soil with composting. *Environmental Pollution* 119: 23–31.

Neeru, K., Asha, G., Santosh, S., Ramesh, K.M., and Anushree, M. 2008. Biodegradation of herbicide (atrazine) in contaminated soil using various bioprocessed materials. *Bioresource Technology* 99: 4642–4647.

Nolan, R.J., Rick, D.L., Freshour, N.L., and Saunders, J.H. 1984. Chlorpyrifos: Pharmacokinetics in human volunteers. *Toxicology and Applied Pharmacology* 73: 8–15.

Odukkathil, G., and Vasudevan, N. 2013a. Enhanced biodegradation of endosulfan and its major metabolite endosulfate by a biosurfactant producing bacterium. *Journal of Environmental Science and Health, Part B* 48: 462–469.

Odukkathil, G., and Vasudevan, N. 2013b. Toxicity and bioremediation of pesticides in agricultural soil. *Reviews in Environmental Science and Biotechnology* 12: 421–444.

Odukkathil, G., and Vasudevan, N. 2015. Biodegradation of endosulfan isomers and its metabolite endosulfate by two biosurfactant producing bacterial strains of *Bordetella petrii*. *Journal of Environmental Science and Health, Part B* 50(2): 81–89.

Odukkathil, G., and Vasudevan, N. 2016. Residues of endosulfan in surface and subsurface agricultural soil and its bioremediation. *Journal of Environmental Management* 165: 72–80.

Ogram, A.V., Jessup, R.E., Ou, L.T., and Rao, P.S.C. 1985. Effects of sorption on biological degradation rates of (2,4-dichlorophenoxy)-acetic acid in soils. *Applied Environmental Microbiology* 35(7), 582–587.

Ottoboni, A. 1969. Effect of DDT on reproduction in the rat. *Toxicology and Applied Pharmacology* 14: 74–81.

Palma, G., Alvear, M., and Salazar, I. 2002. Utilization of waste cells in the preparation for release formulations control of simazine and trifluralin. *Bulletin of Chilean Society of Chemistry* 47: 175–180.

Pandey, R.K., and Rai, S.N. 1993. Effect of varying doses of pesticides on nitrifying bacteria and nitrogen transformation. *Indian Journal of Agricultural Chemistry* 26: 123–132.

Pandit, G.G., Mohan Rao, A.M., Jha, S.K., Krishnamoorthy, T.M., Kale, S.P., and Raghu, K. 2001. Monitoring of organochlorine pesticide residues in the Indian marine environment. *Chemosphere* 44: 301–305.

Paola, P., Parmigiani, S., and vom Saal, F.S. 2001. Effects of prenatal exposure to low doses of diethylstilbestrol, o,p'-DDT, and methoxychlor on postnatal growth and neurobehavioral development in male and female mice. *Hormones and Behavior* 40(2): 252–65.

Pignatello, J.J., and Xing, B. 1996. Mechanisms of slow sorption of organic chemicals to natural particles. *Environmental Science and Technology* 30: 1–11.

Plaza, C., Polo, A., Brunetti, G., Garcia-Gil, J., and Orazio, V. 2003. Soil fulvic acid properties as a means to assess the use of pigamendment. *Soil Tillage Res.* 74: 179–190.

Pothuluri, J.V., and Moorman, T.B. 1990. Aerobic and anaerobic degradation of alachlor in samples from a surface-to-groundwater profile. *Journal of Environmental Quality* 19: 525–530.

Purnomo, A.S., Mori, T., Kamei, I., Nishii, T., and Kondo, R. 2010. Application of mushroom waste medium from *Pleurotus ostreatus* for bioremediation of DDT-contaminated soil. *International Biodeterioration and Biodegradation* 64: 397–402.

Quintero, J.C, Lu Chau, T., Moreira, M.T, Feijoo, G., and Lema, J.M. 2007. Bioremediation of HCH present in soil by the white-rot fungus *Bjerkandera adusta* in a slurry batch bioreactor. *International Biodeterioration and Biodegradation* 60: 319–326.

Rai, D.K., and Sharma, B. 2007. Carbofuran-induced oxidative stress in mammalian brain. *Molecular Biotechnology* 37(1), 66–71.

Rajendran, R.B, Venugopalan, V.K., and Ramesh, R. 2005. Pesticide residues in air from coastal environment, south India. *Chemosphere* 39 1699–1706.

Rao, P.SC., and Davidson, J.M. 1998. Estimation of pesticide retention andtransformation parameters required in nonpoint source pollution models, In *Environmental Impact of nonpoint source pollution*. Overcash, M.R., Davidson, J.M. (eds.), Ann Airbor Science Publishers, Ann Airbor, MI, pp. 23–68.

Richard et al. 2016. Assessing the effectiveness of a three-stage on-farm biobed in treating pesticide contaminated wastewater. *Journal of Environmental Management* 181: 874–882.

RB Associates, Inc. 1982. *Handbook of remedial action at waste disposal sites, EPA Report* 1982 No. 625/6-82-006, Cincinnati, OH.

Reddy, D.J., Rao, B.N., Reddy, K.N., Sultan, M.A., and Babu, T.R. 1997. Monitoring of pesticide residues in river, tank and canal water. *Pesticide Residue Journal* 9(1): 97–100.

Rekha, S.N., Naik, R., and Prasad, R. 2004. Pesticide residue in organic and conventional food-risk analysis. *Chemical Health and Safety* 13: 12–19.

Relyea, R.A. 2005. The lethal impact of roundup on aquatic and terrestrial amphibians. *Ecological Applications* 15(4): 1118–1124.

Roberts, T.R, Hutson, D.H., and Jewess, P.J. 1998. Metabolic pathways of agrochemicals: Insecticides and fungicides. *Part 2, Royal Society of Chemistry, Great Britain* 1134–1137.

Rodriguez-Cruz, M.S., Julie, E.J., Gary, D., and Bending, W. 2006. Field-scale study of the variability in pesticide biodegradation with soil depth and its relationship with soil characteristics. *Soil Biology and Biochemistry* 38: 2910–2918.

Rouse, J.D., Sabatini, D.A., Suflita, J.M., and Harwell, J.H. 1994. Influence of surfactants on microbial degradation of organic compounds. *Critical Reviews in Environmental Science and Technology* 24: 325–370.

Saha, S., and Kavira, A. 2009. Effects of cypermethrin on some biochemical parameters and its amelioration through dietary supplementation of ascorbic acid in freshwater catfish *Heteropneustes fossilis*. *Chemosphere* 74(9): 1254–1259.

Sahu, S.K., Patnaik, K.K., Bhuyan, S., and Sethunathan, N. 1993. Degradation of soil-applied isomers of hexachlorocyclohexane by a *Pseudomonas* sp. *Soil Biology and Biochemistry* 25: 87–391.

Said-Pullicino, D., Gigliotti, G., and Vella, A. 2004. Environmental fate of triasulfuron in soils amended with municipal waste compost. *Journal of Environmental Quality* 33, 1743–1751.

Samir, I.G., Zidan, Z.H., Hassan, M.S., Wafai, Z.A.M., and Selim, M.T. 2012. Monitoring of pesticide residues in strawberry and soil from different farming systems in Egypt. *American-Eurasian Journal of Agricultural & Environmental Science* 12(2): 177–187.

Sandahl, J.F., Baldwin, D.H., Jenkins, J.J., and Nathaniel, L.S. 2005. Comparative thresholds for acetyl cholinesterase inhibition and behavioral impairment in Coho salmon exposed to chlorpyrifos. *Environmental Toxicology and Chemistry* 24(1): 136–145.

Sarkar, S.K., Bhattacharya, B.D., Bhattacharya, A., Chatterjee, M., Alam, A., Satpathy, K.K., and Jonathan, M.P. 2008. Occurrence, distribution and possible sources of organochlorine pesticide residues in tropical coastal environment of India: An overview. *Environment International* 34: 1062–1071.

Schnoor, J.L. 1996. *Environmental modelling; fate and transport of pollutants in water, and soil*. Wiley, Hoboken, NJ.

Scholz, J.M., Schwedes, J., and Deckwer, W.D. 1998. Use of a rotary drum reactor with grinding beads for microbial soil remediation. *Chemical Engineering and Technology* 21: 479–483.

Senesi, N. 1992. Binding mechanisms of pesticides to soil humic substances. *Science of the Total Environment* 123(124): 6376.

Shah, P.G., Raj, M.F., Patel, B.A., Patel, B.K., Diwan, K.D., Patel, J.A., and Talati, J.G. 2000. Pesticidal contamination status in farmgate vegetables in Gujarat. *Pesticide Research Journal* 12: 195–199.

Shanmugam, M., Venkateswarlu, M., and Naveed, A. 2000. Effect of pesticides on the freshwater crab, *Barytelphusa cunicularis* (Westwood). *Journal of Ecotoxicology and Environmental Monitoring* 10: 273–279.

Sharma, H.R., Trivedi, R.C., Akolkar, P., and Gupta, A. 2003. Micropollutants levels in macroinvertebrates collected from drinking water sources of Delhi, India. *The International Journal of Environmental Studies* 60(2): 99–110.

Sidal, U., and Yilmaz, E.S. 2012. Production of Rhamnolipid (A Biosurfactant) using free and immobilized cells of *Pseudomonas* sp. *Kafkas Univ Vet Fak Derg* 18(2): 285–289.

Sididque, T., Okeke, B.C., Arshad, M., and Frankenberger, W.T. 2003. Biodegradation kinetics of endosulfan by *Fusarium ventricosum* and a *Pandoraea* sp. *Journal of Agricultural and Food Chemistry* 51(27): 8015–8019.

Sigua, G.C., Adjei, M., and Rechcigl, J. 2005. Cumulative and residual effects of repeated sewage sludge applications: Forage productivity and soil quality implications in south Florida, USA. *Environmental Science and Pollution Research International* 12, 80–88.

Sims, R., and Bass, J. 1982. *Review of in-place treatment techniques for contaminated surface soil*. Vol 1: Technical evaluation. EPA, Report No. EPA- 540/2-84-003a, Cincinnati, OH.

Singh, B., 2002. Pesticidal contamination of the environment in Punjab. *Indian Journal of Ecology* 29: 189–198.

Singh, B.K., and Kuhad, R.C. 1999. Biodegradation of linden (hexachlorocyclohexa) by the white rot fungus *Trametes hirsutus*. *Letters in Applied Microbiology* 28: 238–241.

Singh, B.K., Walker, A., and Wright, D.J. 2004. Biodegradation of chlorpyrifos by *Enterobacter strain B-14* and its use in bioremediation of contaminated soils. *Applied Environmental Microbiology* 70: 4855–4863.

Singh, B.K., Walker, A., and Wright, D.J. 2006. Bioremediation potential of fenamiphos and chlorpyrifos degrading isolates: Influence of different environmental conditions. *Soil Biology and Biochemistry* 38: 2682–2693.

Sivasankaran, M.A., Reddy, S.S., Govindaradjan, S., and Ramesh, R. 2007. Organochlorine residuals in groundwater of Pondicherry region. *Journal of Environmental Science and Engineering* 49(1): 7–12.

Smith A.G., Hayes, W.J., and Laws, E.R. 1991. Chlorinated hydrocarbons insecticides. In *Handbook of Pesticide Toxicology volume 2, Classes of pesticides*, (Eds). Academic Press, Inc, San Diego, CA, pp. 731–915.

Song, N.H., Chen, L., and Yang, H. 2008. Effect of dissolved organic matter on mobility and activation of chlorotuluron in soil and wheat. *Geoderma* 146: 344–352.

Soulas, G., and Lagacherie, B. 2001. Modelling of microbial degradation of pesticides in soils. *Biology and Fertility of Soils* 33: 551–557.

Srivastava, VK., Singh, BB., Kumar, U., Gupta, RP., and Shukla, KC. 1996. Organochlorine pesticide residues analysis in river water and fishes. *Journal of Environmental Research* 6(4): 163–170.

Strachan, W.M.J., Erikso, G., Kylin, H., and Jensen, S. 1994. Organochlorine compounds in pine needles- methods and trends. *Environmental. Toxicology and Chemistry* 13(3), 5403–5409.

Sutherland, T.D., Horne, I., Harcourt, R.L., Russell, R.J., and Oakeshott, J.G. 2002. Isolation and characterization of a Mycobacterium strain that metabolizes the insecticide endosulfan. *Journal of Applied Microbiology* 93: 380–389.

Tanabe, S., Senthilkumar, K., Kannan, K., and Subramanian, A.N. 1998. Accumulation features of polychlorinated biphenyls and organochlorine pesticides in resident and migratory birds from South India. *Archives of Environmental Contamination and Toxicology* 34: 387–397.

The WHO recommended classification of pesticide by hazard and guidelines to classification 2004. WHO, Geneva, ISBN: 9241546638.

Thibault, G.T., and Elliott, N.W. 1979. *Accelerating the biological cleanup of hazardous materials spills*. In conference proc. on:oil and haz. mat. spills: prevention- control-cleanup-recovery disposal.

Thom, E., Ottow, J.C.G., and Benckiser, G. 1997. Degradation of the fungicide difenoconazolein in a silt loam soil as affected by pretreatment and organic amendment. *Environmental Pollution* 96: 409–414.

Tinsley, I.J. 2004. *Chemical concepts in pollutant behavior*. 2nd edn. Wiley, Hoboken, NJ.

Torrents, A., and Jayasundera, S. 1997. The sorption of nonionic pesticides onto clays and the influence of natural organic carbon. *Chemosphere* 35(7): 1549–1565.

Traina, M.E., Rescia, M., Urbani, E., Mantovani, A., Macri, C., Ricciardi, C., Stazi, A.V., et al. 2003. Long-lasting effects of lindane on mouse spermatogenesis induced by in utero exposure. *Reproductive Toxicology* 17: 25–35.

Van Audenhaege, M., Heraud, F., Menard, C., Bouyrie, J., Morois, S., Calamassi-Tran, G., Lesterle, S., Volatier, J.L., and Leblanc, J.C. 2009. Impact of food consumption habits on the pesticide dietary intake: Comparison between a French vegetarian and the general population. *Food Additives & Contaminants. Part A, Chemistry, Analysis, Control, Exposure and Risk Assessment* 26(10): 1372–1388.

Veenanadig, N.K., Gowthama, M.K., and Karanth, N.G.K. 2000. Scale up studies for the production of biosurfactant in packed column bioreactor. *Bioprocess Engineering* 22: 95–99.

Venkata Mohan S, Sirisha, K., Chandrasekhara Rao, V., Sarma, P.N., and Jayarama Reddy, S. 2007 Bioslurry phase remediation of chlorpyrifos contaminated soil: Process evaluation and optimization by Taguchi design of experimental (DOE) methodology. *Ecotoxicology and Environmental Safety* 68: 252–262.

Vidali, M. 2000. Bioremediation an overview. *Pure and Applied Chemistry* 73: 1163–1172.

Vorkamp, K., Kellner, E., Taube, E., Moller, K., and Herrmann, R. 2002. Fate of methidathion residues in biological waste during anaerobic digestion. *Chemosphere* 48: 287–297.

Wanner, U., Fuhr, F., and Burauel, P. 2005. Influence of the amendment of corn straw on the degradation behaviour of the fungicide dithianon in soil. *Environmental Pollution* 133: 63–70.

Watts, M. 2005. *Pesticide: Sowing poison, growing hunger, reaping sorrow,* vol. 2. Pesticide Action Network Asia and Pacific Policy Research and Analysis, Penang, Malaysia, pp. 1–27.

Watts, M. 2010. *Pesticides: Sowing Poison, Growing Hunger, Reaping Sorrow*, 2nd Edition, Pesticide Action Network (PAN) Asia and the Pacific. Penang, Malaysia.

Weir, K.M., Sutherland, T.D., Horne, I., Russell, R.J., and Oakeshott, J.G. 2006. A single moonoxygenase is involved in the metabolism of the organochloridesendosulfan and endosulphate in an Arthrobactersp. *Applied Environmental Microbiology* 72: 3524–3530.

World Health Organization (WHO). 2005. Report on Pesticide, Geneva, Switzerland.

Worrall, F., Fernandez-Perez, M., Johnson, A., Flores-Cesperedes, F., and Gonzalez-Pradas, E. 2001. Limitations on the role of incorporated organic matter in reducing pesticide leaching. *Journal of Contaminant Hydrology* 49: 241–262.

Yazdandoost, M.Y., and Katdare, M.S. 2000. Water quality index of major rivers in Pune. *Journal of Industrial Pollution Control* 16: 231–238.

Zagal, E. Rodryguez, N., Vidal, I., and Quezada, L. 2002. Microbial activity in a volcanic ash soil under different agricultural management. *Agricultural Technology* 62: 297–309.

# 4 Uranium Radionuclide Contamination in the Environment

## *Ecotoxicological Effects, Health Hazards, and Bioremediation*

*Meenakshi Nandal and Mansi Rastogi*

## CONTENTS


4.1 Introduction ..... 104
4.2 General Properties of Uranium ..... 105
4.3 Sources of Uranium Contamination in Environments ..... 105
 4.3.1 Natural Sources of Uranium Contamination ..... 105
 4.3.2 Anthropogenic Sources of Uranium Contamination ..... 107
4.4 Toxicological Effects of Uranium Contamination in Environments and Health Hazards ..... 107
 4.4.1 Chemical Toxicity ..... 107
 4.4.2 Radiological Toxicity ..... 108
 4.4.3 Effects of Uranium Toxicity on Human and Animal Health ..... 109
4.5 Bioremediation of Uranium-Contaminated Environments ..... 109
 4.5.1 Principles of Bioremediation ..... 110
 4.5.2 Types of Bioremediation ..... 111
  4.5.2.1 *In Situ* Bioremediation ..... 111
  4.5.2.2 *Ex Situ* Bioremediation ..... 112
 4.5.3 Role of Microorganisms in Biotransformation of Metals, Metalloids, and Radionuclides ..... 112
 4.5.4 Microbial Processes and Mechanism of Radionuclides Bioremediation ..... 113
 4.5.5 Various Mechanisms of Uranium Bioremediation ..... 114
  4.5.5.1 Biosorption ..... 114
  4.5.5.2 Bioprecipitation ..... 115
  4.5.5.3 Bioreduction ..... 115
  4.5.5.4 Bioreactive Barrier ..... 117

4.5.5.5 Chelation ........ 117
4.5.5.6 Injection Wells ........ 118
4.5.6 Advantages and Disadvantages of Different Bioremediation Approaches ........ 119
4.6 Enzymatic Reduction of U(VI) ........ 119
4.7 Constructed Wetlands ........ 120
4.8 Conclusion ........ 121
References ........ 122

## 4.1 INTRODUCTION

Bioremediation is an eco-sustainable technology used to clean up contaminated soil and water by using biological agents. However, certain conventional techniques like high-temperature incineration, chemical methods, and various higher decomposition techniques are practiced for cleaning up contaminated sites. But, these processes do not lead to the final destruction of pollutants. Therefore, to overcome these problems, environmentalists are looking toward microbes such as bacterial/fungi/algae/yeast/cyanobacteria, etc., green plants, or their enzymes to remediate environmental pollutants from contaminated sites. Microbes have the ability to thrive in "adverse conditions" of high acidity/alkalinity/toxicity and temperature. Under favorable conditions, these degrade/transform toxic environmental pollutants into simpler and harmless ones. Some bacteria can also ingest the most toxic "cyanide" from contaminated soils/water. After the use of "super bugs" in the cleaning up of oil spills, there have been several successful stories of microbial techniques in the cleaning up of the contaminated sites. Plants involved in bioremediation processes are also adapted to thrive in very harsh environmental conditions and absorb, tolerate, transfer, assimilate, degrade, and stabilize toxic pollutants such as heavy metals, radionuclides, solvents, crude oil, pesticides, explosives, chlorinated compounds, and polyaromatic hydrocarbons from the contaminated sites (Chandra et al. 2015).

Organic pollutants may ideally be degraded into simpler molecule-like carbon dioxide and water. More than 400 plant species ranging from herbs to trees have now been identified as hyperaccumulators of toxic metals. They can accumulate 100–500 times higher levels of metals in their aboveground parts (Sinha et al. 2009). However, microbes also play an important role in the bio-geological cycle of metals and radionuclides. Microbial activity can affect the stability and mobility of radionuclides; thus, it becomes important to understand the mechanism of interaction between the microbes and radionuclides to remediate the contaminated environments. In this chapter, we briefly describe the general properties of uranium (U), its sources of contamination in the environment, toxic effects in the environment as well as on human and animal health, and various bioremediation approaches used to clean up U-contaminated environments. However, bioremediation is considered a safer, cleaner, cost-effective, and environmentally friendly technology that is publicly accepted.

## 4.2 GENERAL PROPERTIES OF URANIUM

Pure U is a silvery white, ductile, and malleable heavy metal that melts at 1,132.3°C and boils at about 3,818°C. Its density is about 19.05 g·cm$^{-1}$ (Kleykamp 1991). However, elemental U hardly exists in nature because U metal is readily subjected to surface oxidation (Whitehead et al. 1971; Beddow et al. 2006). Consequently, it ignites spontaneously on contact with air at ambient temperature and form U oxides with six oxidation states in nature, U(IV) and U(VI) are the most common oxidation states in nature because U(III) gets easily oxidized into U(IV) under most redox conditions, while U(V) is readily disproportionate to U(IV) and U(VI) in nature (Lovley et al. 1993; Benders et al. 2000; Markich 2002).

U forms more than 234 mineral species and accounts for only 5% of all known minerals (Barthelmy 2008). The most common forms of U oxide are Triuranium octoxide ($U_3O_8$) and uraninite ($UO_2$), which have low solubility in water but are relatively stable over a wide range of environmental conditions. Mostly, the $U_3O_8$ exists as $U_2O_5$ and $UO_3$ in nature (Wersin et al. 1994). Generally, U forms many compounds that tend to have yellowish or greenish color. In respect to this chemical property, U has been historically used as a colorant for glass and ceramic (i.e., producing orange-red to lemon yellow hues) and for tinting and shading in early photography.

U is the heaviest naturally occurring element with atomic number 92. Despite its high atomic number, it is a ubiquitous element with an average concentration of 3 mg kg$^{-1}$ in the Earth's crust (IAEA 1998). The concentrations of U in geological materials are highest in continental-type rocks. U is important not only because it is the heaviest element occurring in nature in weighable amounts and nuclear fuel but also due to its chemical and radioactive implications to human health and environments.

## 4.3 SOURCES OF URANIUM CONTAMINATION IN ENVIRONMENTS

### 4.3.1 Natural Sources of Uranium Contamination

In nature, U isotopes are distributed as 99.284% of $^{238}U$, 0.72% of $^{235}U$, and a very small amount (~0.0054%) of $^{234}U$ shown in Figure 4.1. Thus, U with its isotope distribution is called natural U. U is one of the most mobile radioactive metals, which gets easily washed down through soil together with percolating water to both surface and underlying groundwater. Thus, moving water acts as a major source of U exposure, although dust in the air may also act as a source of U exposure.

In nature, U does not occur in elemental form. Its geochemical properties, especially the chemical form, strongly influence its environmental transport. U exists predominantly in a hexavalent oxidation state U(VI), which forms readily soluble species over a range of pH and redox potential (Eh) conditions in aquatic environments. Furthermore, U(VI) ions are exclusively available as species of uranyl ion ($UO_2^{2+}$), which sometimes behave like divalent metal ions (Mkandawire et al. 2005). A small amount of U is also found almost everywhere in soil, rock, and water.

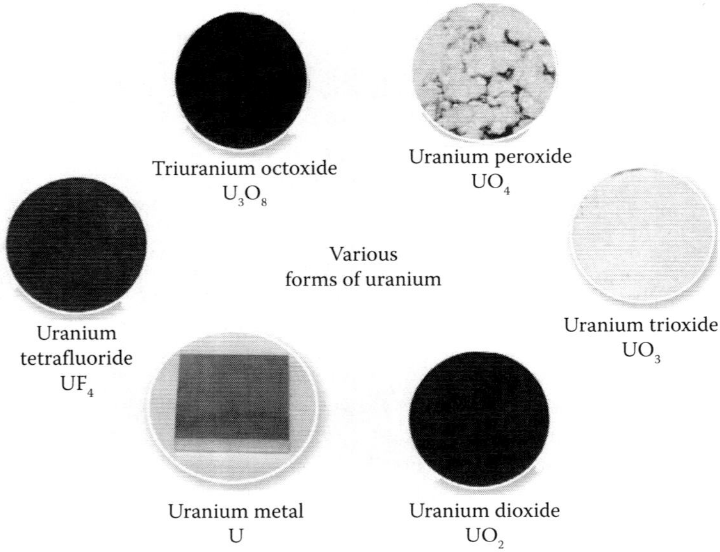


**FIGURE 4.1** Distribution of uranium in nature in various forms.

As a result, it is normal to find dissolved U at very low concentrations in most natural waters, but concern arises only when the concentration of U increases above to microgram per liter (Kalin et al. 2005). Thus, trace contamination occurs naturally, but most of the alarming U contaminations are anthropogenic, associated with activities of the nuclear fuel cycle, phosphate fertilizer production process, and improper disposal of U mine tailings (Chen et al. 2005; Jerden and Sinha 2006). The cycle for nuclear fuel includes the mining, milling, and processing of U ores or U products.

In humid air, U corrodes rapidly and forms yellow uranyl (VI) compounds, where the linear ½ O ¼ U ¼ $O_2$ C unit forms the characteristic structural unit. In the presence of phosphate (V), silicate, arsenate (V), and vanadate (V), the insoluble orthophosphate, orthosilicate, orthoarsenate, and orthovanadate compounds are formed. If these compounds are not present in significant quantities in water, the chemistry of hexavalent U is governed by its interactions with water and carbonate.

Uranium has sixteen radioactive isotopes, but only three isotopes—$^{238}U$, $^{235}U$, and $^{234}U$—occur naturally (Jerden and Sinha 2006). The U isotopes emit mostly alpha and less beta particles accompanied with the little emission of gamma radiation especially during the decay of $^{235}U$ (Cherdyntsev et al. 1982; Ivanovich and Harmon 1992). All the emissions are ionizing radiation, strong enough to break chemical bonds, thereby possessing the ability to damage or destroy the living cells (ATSDR 1999; World Health Organization [WHO] 2001, 2004). In terms of the extent of radioactivity, $^{235}U$, $^{238}U$, and $^{234}U$ contribute 2.2%, 48.6%, and 49.2%, respectively. The total activity level of U depends on the isotopic composition ratio. The isotopes $^{235}U$ and $^{234}U$ pose a greater radiological toxicity risk than $^{238}U$ because they have much shorter half-lives.

### 4.3.2 Anthropogenic Sources of Uranium Contamination

The $^{235}U$ has high fission capacity; thus, it is used as fuel in nuclear reactors. However, a nuclear reactor requires higher than 0.72% content of $^{235}U$ to function. Thus, the natural U is centrifuged to enrich the content of $^{235}U$ from 0.72% to about 1.5%–3% (IAEA 1998, 2007). Then, $^{234}U$ and $^{235}U$ isotopes are extracted to produce high radioactive enriched U, which is primarily used as fuel in nuclear reactors. Consequently, the U that remains after the removal of the enriched fraction is called depleted U and it contains at least three times less $^{234}U$ and $^{235}U$ by mass. The depleted U is weakly radioactive and its radiation dose is only about 60% of natural U with the same mass (WHO 2001; Meinrath et al. 2003). Natural and depleted U differ only in their radioactivity, but their chemical properties are the same.

The large-scale extraction of U from its ore in industries (i.e., open-pit and underground mining) is archived by the following two main steps: (1) crush the ore-bearing rocks into a fine powder; then add to the pulverized ore huge amounts of water, leaching agents, and thickener in order to oxidize and dissolve the U minerals. The U minerals available as $UO_2$ are oxidized to $UO_3$ followed by the formation of either a uranyl sulfate $UO_2\ (SO_4)_{34}$ or a uranyl carbonate predominantly $UO_2(CO_3)_{34}$ in acid leaching or carbonate leaching, respectively. Obviously, there is no crushing required in the *in situ* leaching (ISL) and borehole mining procedure. After leaching, U is concentrated and purified using either ion exchange or solvent extraction processes.

The solvent extraction is the most applied procedure, which uses tertiary amines ($R_3N$) in kerosene diluents (El-Nadi et al. 2003). The amines react with sulfuric acid and remove impurities from the loaded solvents. Cations are removed by using sulfuric acid, while anions are removed with gaseous ammonia (Juznic et al. 1989; Levy and Kearney 1999). The solvents are stripped with ammonium sulfate solution. Then, exposure to gaseous ammonia to neutralize the solution leads to the precipitation of ammonium diuranate. The diuranate is subsequently dewatered and dried to produce hydrated U peroxide ($UO_4{\cdot}2H_2O$). Last, the centrifugation and precipitation of $UO_4{\cdot}2H_2O$ yields the yellowcake ($U_3O_8$) containing 75% U oxides (Moyes et al. 2000; Abdelouas 2006). A number of uranyl hydroxides, oxyhydrates, and uranates are known. The affinity of U(VI) to oxygen is extreme, and there are very few oxygen-free U(VI) compounds. U(V) has been used in laboratory experiments, but it is not stable enough to allow detailed investigations, and it does not play an observable role in nature. U(IV) is hydrolyzed even at very low pH. Freshly prepared U(VI) hydroxides have much higher solubility as compared to uraninite minerals (Meinrath 1998).

## 4.4 TOXICOLOGICAL EFFECTS OF URANIUM CONTAMINATION IN ENVIRONMENTS AND HEALTH HAZARDS

### 4.4.1 Chemical Toxicity

Uranium toxicity has been reported to cause kidney malfunction, respiratory disorders, DNA damage, mutagenicity, cancer, and neurological disorders (Yazzie et al. 2003; Mitrovic et al. 2005; Stearns et al. 2005). Insoluble forms of U exhibit the least chemical toxicity risk because these are readily excreted via the digestive tract, while the soluble forms of U interact with biological ligands and lead to severe toxicological effects. Once U reaches the bloodstream, kidneys filter the U compounds and excrete it out of the body. U ions, specifically uranyl cations, are well reported to act as classic nephrotoxins (Dang et al. 1995; Russell et al. 1996; Periyakaruppan et al. 2007).

### 4.4.2 Radiological Toxicity

The radiological toxicity of U is primarily manifest in bone sarcomas, leukemia, lung cancer, and a general increase in cancer risk, whereas combined chemical and radiological effects are responsible for the nephrotoxicity (Nagarkatti et al. 1996; Veiga et al. 1998; Belles et al. 2007). Radiation hazards from U occur mainly through the injection of insoluble U compounds and inhalation of U containing aerosols, because all U isotopes mainly emit alpha particles that have little penetrating ability (Lloyd et al. 1996). However, the communities near abandoned U mines are exposed to radiation from U decay products, which are sometimes stronger and more dangerous than the U radiation. The development of cancer occurs years after the exposure takes place. The probability of developing an ionizing radiation-induced cancer rises with increasing U intakes as shown in Figure 4.2.

A high radiation dose may cause damage to lung tissues, which may lead to lung cancer. Risks for other radiation-induced cancers including leukemia are lower than for lung cancer. Generally, the WHO and the International Labour Organization (ILO) recommend that the exposure of U radiation should not exceed a dose of more than 1 mSv in a year (WHO 2001; Simpson et al. 2004). In special circumstances, an effective dose of up to 5 mSv in a single year is permitted if the average dose over five consecutive years does not exceed 1 mSv/year, whereas an equivalent dose to the skin should not exceed 50 mSv in a year.

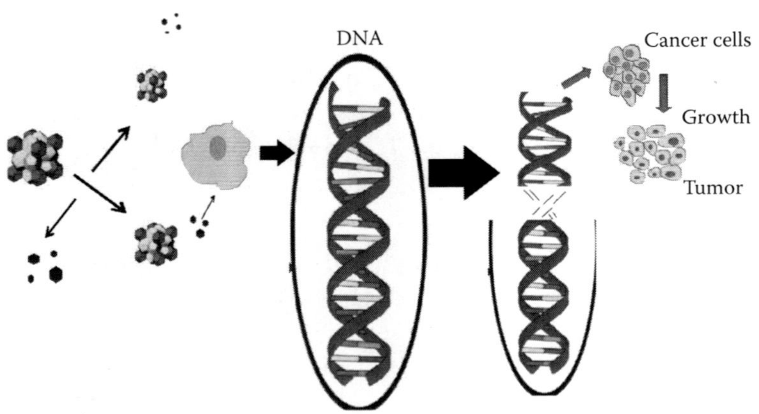


**FIGURE 4.2** Mutagenic and carcinogenic effects of uranium radiations on living cells.

### 4.4.3 Effects of Uranium Toxicity on Human and Animal Health

There are several pathways through, which humans, animals and the entire environment can be exposed to U toxicity hazards. These include the inhalation of aerosol particles, dermal exposures, and ingestion through contaminated water and foodstuffs. Inhalation, ingestion, and external radiation are the most common in human exposure risks. Although the absorption of some soluble compounds through skin is also possible, such dermal exposures generally are not significant. The inhalation of insoluble compounds such as U dioxide ($UO_2$) and triuranium octoxide ($U_3O_8$) leads to deposition in lungs that can remain for long periods. In such conditions, the inhaling aerosols with short-lived U progeny may lead to radiation toxicity as well as chemical toxicity. For ingested U, insoluble compounds are poorly absorbed from the gastrointestinal tract and are only retained in the body for a short time and this generally causes a low toxicity, but the ingested soluble U compounds enter the bloodstream and eventually reach the kidney and other internal organs and cause chemical toxicity.

About 98% of U entering the body via ingestion is not absorbed (Hoffmann et al. 1993; Kusiak et al. 1993; Simon and Garnier-Laplace 2005). The typical gut absorption rates for U in food and water are about 2% for soluble and about 0.2% for insoluble U compounds (Veiga et al. 1998). The fraction of U absorbed into the blood is generally greater following inhalation than following ingestion of the same chemical form. The fraction will also depend on the particle size distribution. The distribution of U in the human body is estimated at 66% in skeleton, 16% in liver, 8% in kidneys, and 10% in other tissues (Russell et al. 1996).

## 4.5 BIOREMEDIATION OF URANIUM-CONTAMINATED ENVIRONMENTS

Unlike most of the organic compounds, U and all other metals do not degrade into nontoxic small components. Biological activity can (1) only transform the toxic metals into nontoxic species or (2) immobilize the metals from the pathway. Figure 4.3 demonstrates the interactions of U with biotic components and summarizes the mechanisms involved in metal bioremediation. The uptake, sorption, and precipitation by purified biopolymers derived from microbial cells provide alternative methods and additive processes for the conventional physicochemical bioremediation processes (Malekzadeh et al. 2002). Some plants and algal species can sequestrate U and other toxic metals into their biomass, while other species exude bioligands that trap U into biometallic colloids. A few bacterial species are capable of reducing the mobile U (VI) to immobile U (IV) species (Smeaton et al. 2008). The algae–microbial biofilms have the ability to sequestrate U and other metals (Hsieh et al. 1994; Allan et al. 2002; Kalin et al. 2005).

The bioreduction, sequestration, and trapping of U in biocolloids can nucleate the biomineralization processes of U. For instance, U can be immobilized through adsorption onto cell surfaces, extracellular polysaccharides, and low molecular weight organic acids including biogenic colloids (Malekzadeh et al. 2002; Lloyd and Renshaw 2005). Furthermore, U can also be internalized into the cells through metabolically active procedures. In cells, U may be sequestrated as inert compounds in cell

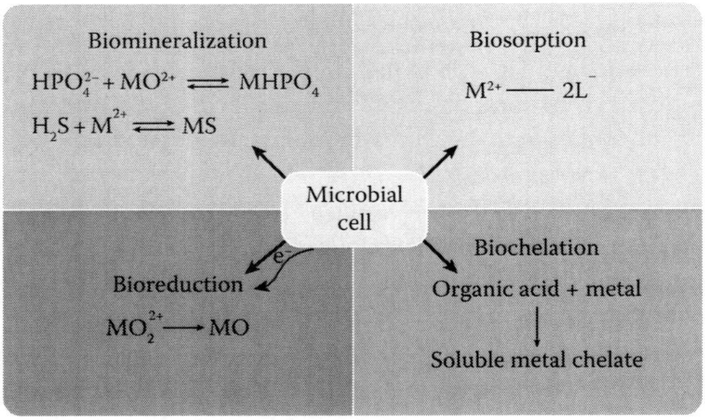


**FIGURE 4.3** Mechanisms of microbial interactions with toxic metals during the bioremediation process.

organelles such as vacuoles, cell membrane, and cell wall. Some species excrete the U compounds back into the water as precipitates or organometallic colloids. Under conducive conditions, the interaction may result in U biomineralization. For instance, some marine and freshwater algae are able to incorporate the U into aragonite ($CaCO_3$) mineral structures.

### 4.5.1 Principles of Bioremediation

Bioremediation is defined as a process whereby the organic pollutants are biologically degraded under controlled conditions (*in situ*) to an innocuous state, or to a level below the concentration limits established by the regulatory authorities (Mueller et al. 1996). By definition, bioremediation is the use of living organisms, primarily microorganisms, to degrade the toxic environmental pollutants into harmless products. It uses naturally occurring bacteria and fungi or plants to degrade or detoxify the substances hazardous to human health and/or the environment. The microorganisms may be indigenous to a contaminated area or may be isolated from elsewhere and brought to the contaminated sites (Kumar et al. 2011). Microorganisms obtain their food from the medium in which they are serving/growing and if a contaminant becomes present, then it can act as an additional food source.

The contaminants serve two useful purposes for the microbes. First, the contaminants act as a source of carbon and nitrogen for the microbes; second, the microbes obtain energy by breaking the chemical bonds and transferring the electrons away from the contaminants, known as oxidation–reduction reactions. The contaminants that lose electrons (electron donor) are oxidized, whereas the chemicals that receive electrons (electron acceptor) are reduced. The energy gained from electron transfer is used along with the carbon and some electrons to produce more cells. Microbes generally use oxygen as an electron acceptor, but sometimes nitrate, sulfate, iron, and $CO_2$ are used for the same purpose. Mostly, the bioremediation process of environmental pollutants occurs under aerobic conditions, but anaerobic conditions also permit microorganisms to degrade the toxic environmental pollutants (Colberg and young 1995). Bioremediation involves manipulation of environmental parameters to allow the microbial cells to grow and degrade the environmental pollutants at a faster rate, but this process also has certain limitations. Some contaminants such as chlorinated organic or highly aromatic hydrocarbons are resistant to microbial attack or are degraded slowly or not at all. Hence, it is not easy to predict the rates of cleanup for a bioremediation exercise, and there are no rules to predict whether a contaminant can be degraded or not.

## 4.5.2 Types of Bioremediation

Based on the removal and transportation of environmental wastes for treatment, bioremediation is basically of two types: *in situ* and *ex situ*.

### 4.5.2.1 *In Situ* Bioremediation

*In situ* bioremediation involves the supply of oxygen and nutrients by circulating aqueous solutions through contaminated soils to stimulate the naturally occurring bacteria to degrade the environmental contaminants. It can be used for the decontamination of soil and groundwater. It is a superior method of cleaning contaminated environments because it is cheaper and uses harmless microorganisms to degrade environmental pollutants. Types of *in situ* bioremediation include intrinsic and engineered.

#### *4.5.2.1.1 Intrinsic Bioremediation*

This approach deals with the stimulation of indigenous or naturally occurring microbial populations by feeding them with nutrients and oxygen to increase their metabolic activity as shown in Figure 4.4.

#### *4.5.2.1.2 Engineered* In Situ *Bioremediation*

This involves the introduction of certain genetically engineered microorganisms to the contaminated sites to speed up the growth of microbes and ultimately the biodegradation

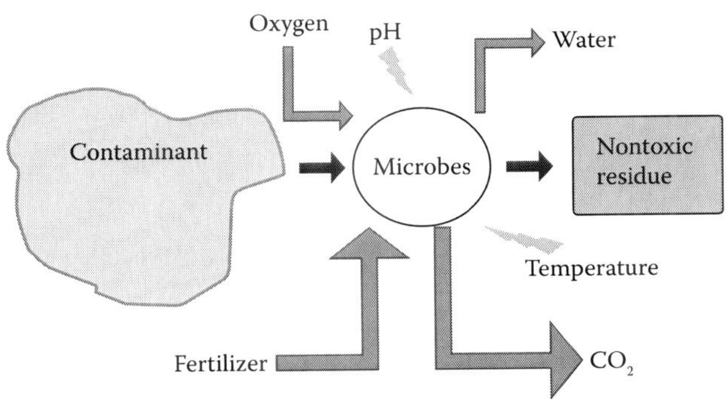


**FIGURE 4.4** *In situ* biodegradation and bioremediation of environmental pollutants from contaminated environments.

and bioremediation of environmental pollutants by enhancing physicochemical conditions like oxygenation, electron acceptors, and nutrients availability.

#### 4.5.2.2 *Ex Situ* Bioremediation

This process involves the excavation/removal and transportation of contaminants from original site to treatment site. Depending on the state of the contaminant to be removed, *ex situ* bioremediation is classified as solid-phase system treatment including land treatment and soil piles and slurry-phase systems including solid liquid suspensions in bioreactors.

##### *4.5.2.2.1 Solid-Phase Treatment*

This includes the treatment of organic wastes such as leaves, animal manures and agricultural wastes, domestic and industrial wastes, sewage sludge, and municipal solid wastes. Solid-phase soil treatment processes also include landfarming, soil biopiles and composting, etc.

##### *4.5.2.2.2 Slurry-Phase Bioremediation*

In this process, the treatment of contaminated soil and water is done in a bioreactor under controlled conditions for microbial growth along with nutrient and oxygen supplements. It includes the treatment of contaminants through bioreactors, bioventing, biosparging, and bioaugmentation.

*4.5.2.2.2.1 Bioreactors* Bioremediation in reactors involves the processing of contaminated solid material (e.g., soil, sediment, sludge, or wastewaters) through an engineered containment system.

*4.5.2.2.2.2 Bioventing* This involves the supply of air and nutrients through the wells to contaminated soil to stimulate the growth of indigenous bacteria and finally the bioremediation process.

*4.5.2.2.2.3 Biosparging* Biosparging involves the injection of air under pressure below the water table to increase groundwater oxygen concentration. It enhances the rate of biological degradation of contaminants by naturally occurring bacteria.

*4.5.2.2.2.4 Bioaugmentation* This involves the addition of microorganisms indigenous or exogenous to the contaminated sites to enhance the bioremediation process.

### 4.5.3 Role of Microorganisms in Biotransformation of Metals, Metalloids, and Radionuclides

Currently, a wide range of microorganisms (bacteria, archaebacteria, yeasts, fungi, and algae) has been widely used in the process of environmental cleanup, and these are known as bioremediators. Due to the wide versatility and adaptability, microbes can be used to degrade or remediate a wide range of toxic environmental pollutants from a contaminated site. Among bacteria, *Bacillus* (Tunali et al. 2006), *Pseudomonas*

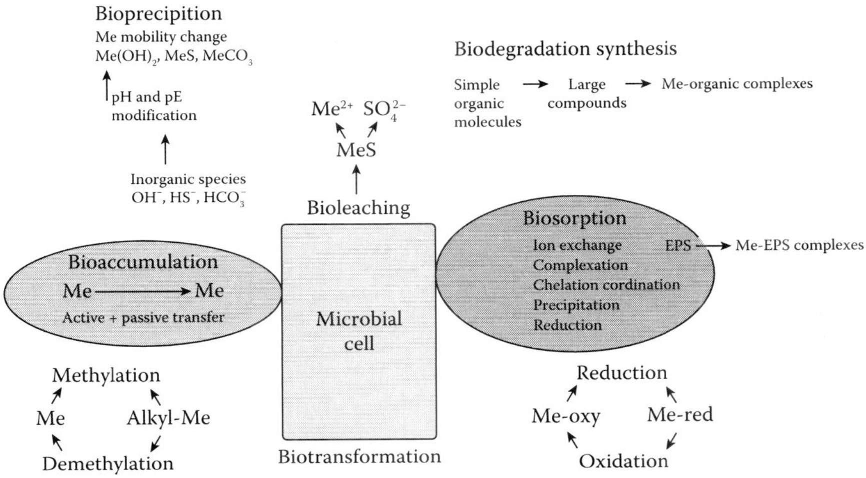


**FIGURE 4.5** Various bioremediation approaches of uranium radionuclides.

(Uslu and Tanyol 2006), and *Streptomyces* (Soltan 2001) have been reported as potent metal biosorbents as shown in Figure 4.5.

The toxicity and mobility of metals depend primarily on their speciation, which is significantly influenced by the soil pH, redox conditions, as well as surface chemistry. All of these are environmental factors that can be optimized by manipulating the microbial activities to reduce the risks posed by heavy metals in aquifer environments. Microorganisms cannot convert metals to different elements, but they can modify the microenvironment around the microbial cell and can catalyze the oxidation, reduction, methylation, and dealkylation reactions that affect the solubility and mobility of toxic metals (Hatzikioseyian 2010). In addition, the microbial cells offer a number of possible physicochemical mechanisms of interaction (e.g., complexation, coordination, chelation, ion exchange, adsorption, and micro precipitation) with soluble metals, radionuclides, and metalloids resulting in the immobilization of metals, metalloids, and radionuclides (White and Gadd 1993; Ehrlich 2001; Gadd 1993, 2004; Remoudaki et al. 2007; Singh and Tripathi 2007; Violante et al. 2008).

### 4.5.4 Microbial Processes and Mechanism of Radionuclides Bioremediation

The microbial processes and mechanism of radionuclides are largely dependent on the active metabolizing capabilities of microorganisms through biologically encoded changes in the oxidation state of bioremediation. The radionuclides can be solubilized by direct and indirect enzymatic reduction through oxidation–reduction, change in pH and Eh (activity of electrons), biosorption by mass, biodegradation of radionuclide-organic complexes, or biosorption by biomass (Holker et al. 2002; Law et al. 2010; Hegazy and Emam 2011) as shown in Figure 4.6. Microbial activity during the

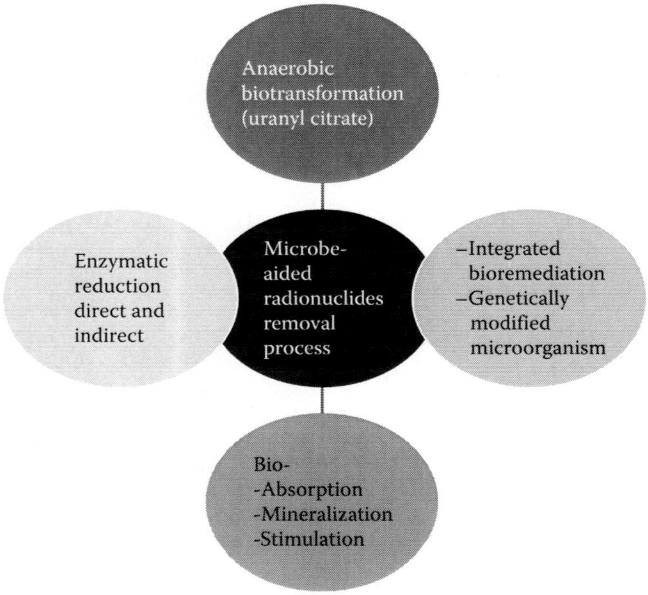


**FIGURE 4.6** Direct and indirect approaches for uranium bioremediation.

biotransformation of radionuclides is greatly influenced by electron donors, acceptors, nutrients, as well as environmental factors.

However, the microbial remediation of toxic metals mainly occurs in two ways: (1) by direct reduction due to the activity of bacterial enzymes such as metal reductases and (2) by indirect reduction due to the biologically produced hydrogen sulfide ($H_2S$) by sulfate-reducing bacteria to reduce and precipitate the metals.

However, the interaction between bacteria and uranium is the use of microbial "bioleaching" mechanisms to extract the uranium from low-grade ores. Preventing the uncontrolled dispersion and transport of radionuclides in groundwater is the overarching remediation goal at many nuclear sites. Stimulating the bacterial interactions to fix the aqueous uranium into insoluble minerals *in situ* may provide a relatively inexpensive and nonintrusive solution to remediate radionuclide contamination (Prakash et al. 2013).

### 4.5.5 Various Mechanisms of Uranium Bioremediation

#### 4.5.5.1 Biosorption

Biosorption of radionuclides to cell surface and polymer substances is a promising technology used for the remediation of contaminated water resources. U biosorption involves the passive deposition of metal ions into the biosorbents through various sorption processes such as ion exchange, complexation, chelation, and microprecipitation (Benders et al. 2000; Bruins et al. 2000; Uslu and Tanyol 2006). In this process, the water contaminated with uranium is brought into contact with either living or dead biomass that possesses abandoned functional groups such as carboxyl, hydroxyl, amine, and phosphate groups within the surface layer. The charged group within the cell surface layer is able to interact with ions or charged molecules present in the uranium-contaminated water. As a result, metal cations become electrostatically attracted and bound to the cell surface layer (Uslu and Tanyol 2006).

The polymers secreted by many metabolizing microbes can also immobilize the toxic metals. The effectiveness of this process is highly affected by the pH of solution and saturation of the biosorbents when metal interactive sites are occupied (Mkandawire et al. 2005). The cell wall structure and characteristics of the U ions determine the capacity and efficiency of U bioremediation from the contaminated water by using sorption techniques. The capacity of U adsorption onto the cell surface largely depends on the (1) number of available ligand binding sites, (2) U speciation, and (3) the physicochemical properties of media. The chemical composition and cell size are reasonably constant for a given species, implying that the number of ligands group on a surface is also fixed. Furthermore, the complexation is also limited to a single layer (Ariff et al. 1999; Prikryl et al. 2001; Kalin et al. 2005). Thus, the cell surface carries a constant concentration of U with respect to the cell composition, even if the concentration of U ions is more than that of the affinity sites (Kalin et al. 2005). The number of available surface binding sites can only increase with cell growth; thus, the rate at which the U ions are removed from the contaminated water by the biosorption procedure is a function of growth rate as shown in Figure 4.7.

The technical application potential of uranium biosorption is substantially dependent on the recovery of uranium, which has been sequestered by the microbial biomass as well as the potential for reusing the regenerated biomass in multiple biosorption–desorption cycles. The recovery of adsorbed uranium can be achieved by the use of an appropriate elution solution capable of effectively stripping the adsorbed uranium from the exhausted biomass and bringing it back to a solution. The elution must be completed without damaging the microbial biomass structure.

#### 4.5.5.2 Bioprecipitation

Bioprecipitation is also known as biocrystallization or biomineralization in which the toxic metals and radionuclides can be precipitated with microbially generated ligands such as phosphate, sulfide, or oxalate (Chen et al. 2005). The process can take place either internal or external of the cell. In this process, metals precipitate with microbially generated ligands such as sulfide or phosphate, or as carbonates or hydroxides in response to localized alkaline conditions at the cell surface. Biomineralization is often more efficient than chemical precipitation in dilute solutions because the ligands are concentrated near the cell surface, which provide nucleation foci for precipitation (Lloyd and Renshaw 2005). For example, nickel can be removed from solution via intercalation into hydrogen metal or through microbially enhanced chemisorptions of heavy metals (MECHEM).

#### 4.5.5.3 Bioreduction

The bioreduction strategy is based on the injection of physiological electron donors such as acetate, lactate, ethanol, or glucose to stimulate U(VI) reduction by a microbial community native to the contaminated aquifers (Allan et al. 2002). Microorganisms

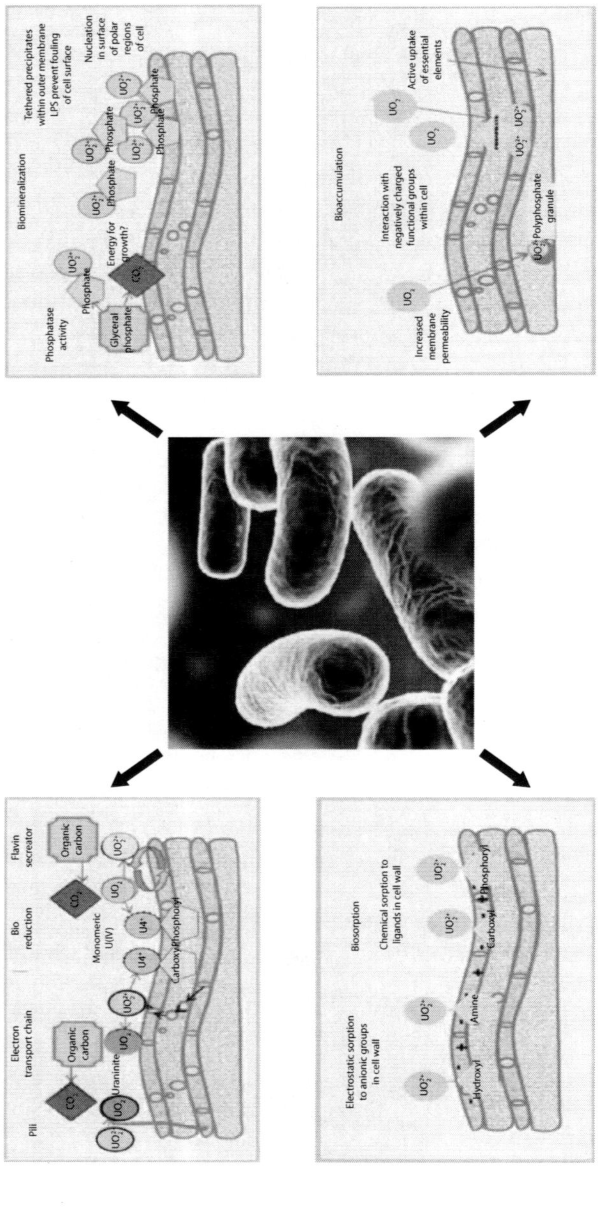


**FIGURE 4.7** Microbe–uranium interaction mechanisms: *bioreduction*, *biomineralization*, *biosorption*, and *bioaccumulation*.

are known to have evolved the biochemical pathway for the degradation or transformation of toxic compounds from their immediate environment either for survival or to derive energy by using the toxic compounds as electron donors or acceptors.

#### 4.5.5.4 Bioreactive Barrier

It is a bioremediation strategy targeting leaching and groundwater. This technique consists of a permeable zone, which cleans up the contaminated plume through the immobilization or transformation of the toxic pollutants into the nontoxic forms. In a subsurface system, the flow rate of water is intercepted by a perpendicular "wall" of reactive materials that can degrade, precipitate, sorb, or exchange the contaminants, which can reach the innocuous or legal concentration down the gradient barrier (El-Nadi et al. 2003). The strategies capitalize on the process of bioaugmentation, where microorganisms are introduced in the form of a barrier to clean the leaching water. Figure 4.8 illustrates how the barriers are constructed so that the contaminated water passes through the barriers and is treated through the process of U reduction using either the Fe-dissimilating bacteria or sulfur-reducing bacteria (SRB). To stimulate the bioremediation processes, a source of SRB such as cow manure or organic matter is added to the passive treatment systems.

#### 4.5.5.5 Chelation

Chelators are the organic molecules that trap or encapsulate the highly reactive and toxic trace metal cations and prevent them from being involved in unwanted chemical reactions. Sometimes the binding between metals and chelators may lead to precipitation (i.e., form insoluble compounds). Thus, most plants and microorganisms use chelators to aid the uptake and transportation of essential metal nutrients. Due to unique binding properties, chelators are often used to stabilize toxic metals, which makes them critical compounds in bioremediation (Kalin et al. 2005). Chelating agents

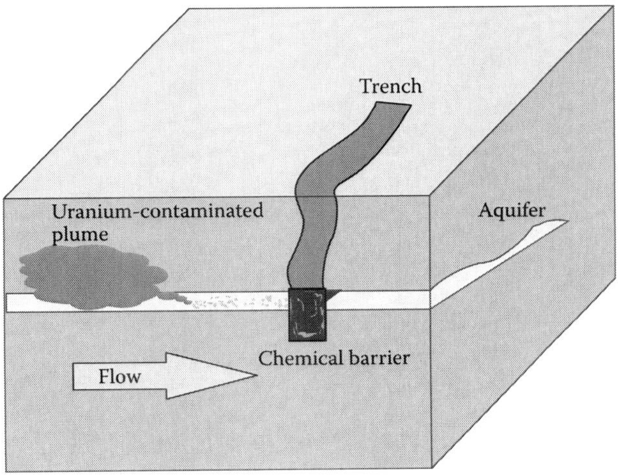


**FIGURE 4.8** Chemical bioreactive barrier to control groundwater contamination.

produced by microorganisms are mostly siderophores, which enhance the dissolution of Fe and increase its mobility and bioavailability. Siderophores have low molecular weight and function primarily as Fe transport compounds. There are some similarities in the biological behavior of ferric and uranyl ions; thus, chelators can bind U in a manner similar to those produced for Fe (Kalin et al. 2005).

Plants first release special chelators known as phytosiderophores that bind to metals in the rhizosphere and allow them to be absorbed by the plant roots. The functioning of chelators is stereoselective depending on the recognition of metal ion coordination geometry. Nonetheless, the benefits of chelating agents on U remediation are enormous. The low molecular weight organic acids like citric acids and oxalic acids exuded by most of the aquatic plants facilitate the dissolution of U and uptake by macrophytes and microorganisms. Other chelating agents include metallothioneins and phytochelatins, which facilitate both the uptake and transportation of metals in organisms.

#### 4.5.5.6 Injection Wells

Well injection is another promising bioremediation strategy applicable in U abandoned mines. It is one of the major procedures for U bioremediation in groundwater, especially in abandoned mines, leaching contaminated water, and waste heaps. Mostly, the Fe and sulfur-reducing microorganisms are responsible for the bioremediation of U-contaminated groundwater through the bioreduction of U as shown in Figure 4.9.

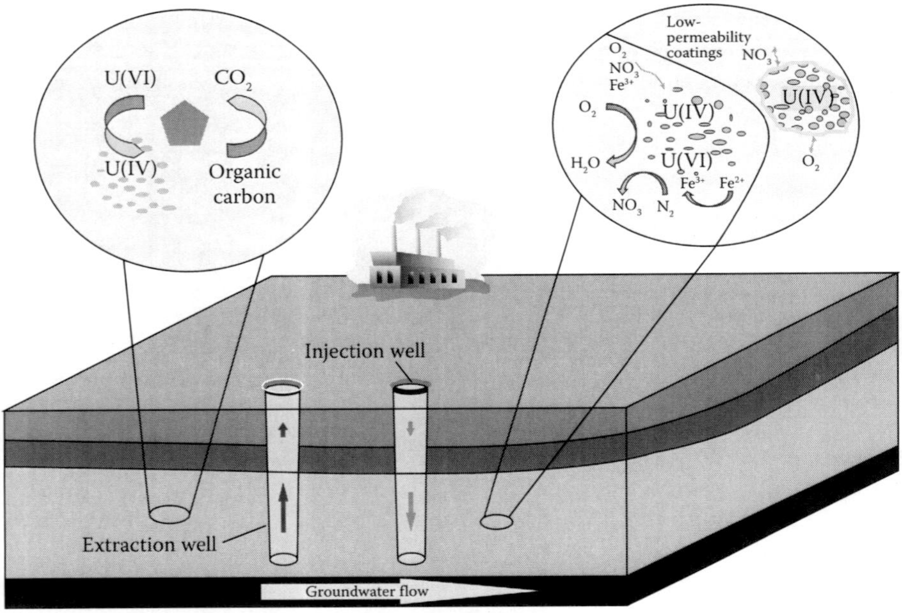


**FIGURE 4.9** Injection well showing U bioremediation.

Figure 4.9 shows a conceptual illustration of uranium bioremediation processes, where (1) Indigenous microorganisms present in soils, sediments, and groundwater contaminated by nuclear energy and weapons production activities are stimulated through the introduction of organic carbon compounds via the injection wells, and the selected organisms may couple the oxidation of organic carbon to the reduction of aqueous uranium as U(VI) gets converted from a soluble to an insoluble form as U(IV). (2) Reduced U(IV) may be re-oxidized to U(VI) following the cessation of organic carbon injection accompanied by subsequent delivery of oxidants such as $O_2$, $NO_3$, and $Fe^{3+}$. The presence of diffusional barriers (e.g., biomass or low permeability sediments) or preferential reductants (e.g., FeS) can suppress re-oxidation and maintain the stability of immobilized U(IV) (Benders et al. 2000).

In most cases, the underground environments are rich with microorganisms, but the environmental conditions are usually not conducive for their growth and reduction of U. The underground environments usually lack carbon sources and contain high nitrates. Thus, the biostimulation of microbial activities is necessary for the U bioremediation to occur. A common method used to deliver carbon sources is injection wells. For such constructions, a series of wells are installed to control the flow of groundwater and allow the injection of solutes such as ethanol, required to create a geochemical environment conducive for microbial growth and subsequent U(VI) reduction. For instance, the ethanol is added to stimulate the *in situ* denitrification of residual nitrate and reduction of U(VI). The addition of acetate to groundwater promotes the growth and dissimilatory activity of Fe(III) reducing microorganisms resulting in the removal of soluble U(VI) from the contaminated groundwater (Yazzie et al. 2003; Stearns et al. 2005).

### 4.5.6 Advantages and Disadvantages of Different Bioremediation Approaches

Bioremediation is considered to be a very safe and helpful technology as it relies on microbes that occur naturally in soil and pose no threats on the environment or the people living in that area. It has certain advantages and disadvantages that have been summarized in Table 4.1.

## 4.6 ENZYMATIC REDUCTION OF U(VI)

The members of genera *Shewanella, Desulfovibrio, Clostridium, Geobacter, Thermus* and *Pyrobaculum,* and *Desulfosporosinus* have displayed U(VI) reduction activity (Lovley et al. 1993; Kalin et al. 2005; Lloyd and Renshaw 2005). The two approaches (i.e., biochemical and genetic) are responsible for identifying the U(VI) reductases in dissimilatory metal reducing bacteria [DMRB] (Lovley et al. 1993). The pathway of enzymatic reduction of U(VI) involves C-type cytochromes as a reductase in the presence of an electron donor, such as hydrogen gas. In $H_2$ utilizing bacteria, the electrons are transferred by the tetraheme cytochromes C3, while the triheme periplasmic cytochrome C7 plays an important role in the transfer of electrons from acetate to U(VI). In addition, the periplasmic and outer membrane cytochromes play an important role in the transfer of electrons from acetate to U(VI) (Kalin et al. 2005; Lloyd and Renshaw 2005).

**TABLE 4.1**
**Merits and Demerits of Different Bioremediation Strategies**

| Techniques | Examples | Advantages | Limitations |
|---|---|---|---|
| *In situ* | Bioremediation<br>Bioventing<br>Biostimulation<br>Bioaugmentation | Cost-efficient<br>Noninvasive<br>Passive relatively<br>Naturally attenuates<br>Treatment of soil and water | Environmental constraints<br>Require longer duration<br>Monitoring laborious |
| *Ex situ* | Landfarming<br>Composting<br>Biopiles/biocells | Economic<br>Environment friendly<br>Onsite active | Longer treatment time<br>Need to control abiotic loss<br>Bioavailability and mass transfer problems |
| Bioreactors | Slurry/aqueous reactors | Faster degradation<br>Optimized environmental parameters<br>Mass transfer enhancement<br>Inoculants and surfactants supportive | Requires excavation<br>Expensive<br>Requires maintenance |

However, the reduction of U(VI) to U(IV) requires two electrons, but microorganisms can transfer only one electron. The one-electron reduction process initially generates U(V), but due to the instability of U(V) complexes, it disproportionates to form U(IV). The enzymatic reduction of U can sometimes be hindered by the presence of some naturally occurring chemicals. In case of *Shewanella* and *Geobacter* species, an electron transfer chain containing C-type cytochromes is thought to pass through the periplasm and terminate at the outer membrane facilitating electron transfer to the extracellular solid phase. U(IV) is reported to precipitate in the periplasm and outside of both gram-negative and gram-positive cells suggesting that U(VI) complexes do not generally have access to intracellular enzyme. Thus, it indicates that the best candidates for the reductases would be the electron carrier proteins or enzymes exposed to the outside of the cytoplasmic membrane, within the periplasm, and/or in the outer membrane (Parihar et al. 2013).

## 4.7 CONSTRUCTED WETLANDS

The constructed wetland treatment approach is an environmentally friendly and cost-effective approach that is applicable for the treatment of industrial wastewaters containing inorganic and organic pollutants using aquatic plants (Saxena and Bharagava 2015; Saxena et al. 2016). A general scheme of bioremediation of U is presented in Figure 4.10. In this treatment approach, the contaminated groundwater is pumped into the constructed wetland for passive treatment. The wetland treatment strategy relies on using several basic physical, chemical, and biological processes to remove or immobilize the U or other organic and inorganic pollutants from the aquatic system.

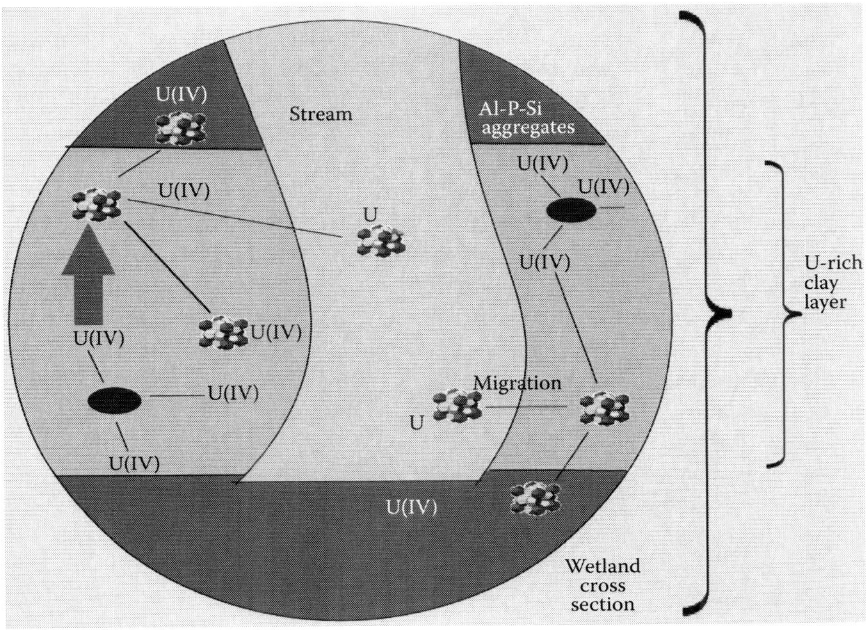


**FIGURE 4.10** Constructed wetland showing the U bioremediation.

In a constructed wetland, the three main compartments (i.e., substrate [sediment], hydrology, and vegetation) play an important role in the remediation of U or other pollutants. The substrate harbors consortia of detritus microbes, which are responsible for uranyl reduction and transformation, which may cause U biomineralization. The expansive rhizosphere of autotrophic wetland plant species provides an enriched culture zone for microbes involved in U species biotransformation. The physical processes in constructed wetlands through which U is removed from contaminated waters include settling and sedimentation following the adsorption on particulate matter. To some extent, mats of floating plants serve as sediment traps, while rooted and floating plants regulate the flow rate of water by providing the resistance. More than 50% of U is easily adsorbed onto the particulate matter in the constructed wetlands. In sediments, U is adsorbed onto the particles by either cation exchange or chemisorptions. The U speciation changes with time as the sediment condition changes.

## 4.8 CONCLUSION

The bioremediation approach provides an environmentally friendly means for the cleaning up and reduction of pollution load from contaminated environments. Without any doubt, this technique is certainly in process of paving a way to greener pastures, but with certain optimistic conjecture, its merits are defaced. Regardless of this aspect, bioremediation is considered as a good option for the treatment of contaminated soils and groundwater. Microbes play a vital role in this process either naturally present or externally inoculated within the contaminated vessel. The microbial transformation processes of radionuclides, heavy metals, or other contaminants are quite complex and not easy to understand. This requires a vigorous study of the molecular mechanisms that occur during the microbial transformations of radionuclides. As microbes show a wide range of mechanisms, there are still few mechanisms that are not known; thus, bioremediation is still considered as a developing technology. For sustainable development and environmental management, it will play a significant role by restoring degraded environments. So, for better exploitation of this technique in sustainable development and environmental management, knowledge of these microbial interactions, their genetics, and their responses to natural environments and pollutants, along with field trials conducted for new cost-effective bioremediation techniques are necessary.

However, considering the uranium contamination in the environment, the largest quantities of wastes are generated from the mining and milling of U minerals containing toxic radionuclides. The major threats of U contamination on human beings are the chemitoxical and radiological effects caused due to the improper storage of U. Thus, there is strong demand for the proper management of U-contaminated environments. Currently, the technologies available for management of U-contaminated environments include both physicochemical and biological methods. The main advantages of biological methods are their low cost and high efficiency. By applying these biological techniques to stabilize the U in subsurface soils appears promising in the near future. Applications like the use of microorganisms to remediate the contaminated soils and water by means of bioreduction of U(VI) to U(IV) with different techniques like biotransformation, chelation, constructed wetlands, bioreactive barriers, and injection wells are being practiced effectively. The inoculation of efficient microbes and supply of better conditions could lead to wider applications and better outcomes of this technique in the sustainable development and management of contaminated environments.

## REFERENCES

Abdelouas, A. 2006. Uranium mill tailings: Geochemistry, mineralogy, and environmental impact. *Elements* 2(6): 335–341.

Allan, V.J.M., Callow, M.E., Macaskie, L.E., and Paterson-Beedle, M. 2002. Effect of nutrient limitation on biofilm formation and phosphatase activity of a *Citrobacter* sp. *Microbiology* 148: 277–288.

Ariff, A.B., Mel, M., Hasan, M.A., and Karim, M.I.A. 1999. The kinetics and mechanism of lead(II) biosorption by powderized *Rhizopus oligosporus*. *World Journal Microbiology and Biotechnology* 15: 291–298.

ATSDR. 1999. *Toxicological profile for uranium*. Agency for Toxic Substances and Disease Registry. U.S. Department of Health and Human Services, Atlanta, GA.

Barthelmy, D. 2008. *Mineralogy database*. http://webmineral.com (accessed July 18, 2015).

Beddow, H., Black, S., and Read, D. 2006. Naturally occurring radioactive material (NORM) from a former phosphoric acid processing plant. *Journal of Environmental Radioactivity* 86(3): 289–312.

Bellés, M., Linares, V., Albina, M.L., Sirvent, J., Sánchez, D.J., and Domingo, J.L. 2007. Melatonin reduces uranium-induced nephrotoxicity in rats. *Journal of Pineal Research* 43(1): 87–95.

Benders, J., Duff, M.C., Phillips, P., and Hill, M. 2000. Bioremediation and bioreduction of dissolved U(VI) by microbial mat consortium supported on silica gel particles. *Environmental Science and Technology* 34: 3235–3241.

Bruins, M.R., Kapil, S., and Oehme, F.W. 2000. Microbial resistance to metals in the environment. *Ecotoxicology and Environmental Safety* 45(3): 198–207.

Chandra, R., Saxena, G., and Kumar, V. 2015. Phytoremediation of environmental pollutants: An eco-sustainable green technology to environmental management. In *Advances in biodegradation and bioremediation of industrial waste,* Chandra, R. (ed.), CRC Press, Boca Raton, FL, pp. 1–30.

Chen, B., Roos, P., Borggaard, O.K., Zhu, Y.G., and Jakobsen, I. 2005. Mycorrhiza and root hairs in barley enhance acquisition of phosphorus and uranium from phosphate rock but mycorrhiza decreases root to shoot uranium transfer. *New Phytologists* 165(2): 591–598.

Cherdyntsev, V.V., Ivanovich, M., and Harmon, R.S. 1982. *Uranium series disequilibrium: Applications to environmental problems.* Clarendon Press, Oxford.

Dang, H.S., Pullat, V.R., and Sharma, R.C. 1995. Distribution of uranium in human organs of an urban Indian population and its relationship with clearance half-lives. *Health Physics* 68(3): 328–331.

Ehrlich, H.L. 2001. *Geomicrobiology*. 4th Ed. Marcel Dekker. Inc, New York.

El-Nadi, Y.A., El-Hefny, N.E., and Daoud, J.A. 2003. Mechanism of extraction of hexavalent uranium from alkaline medium by aliquat-336/ kerosene solution. *Journal of Nuclear and Radiochemical Sciences* 4(2): 19–22.

Gadd, G.M. 1993. Microbial formation and transformation of organometallic and organometalloid compounds. *FEMS Microbiology Reviews* 11: 297–316.

Gadd, G.M. 2004. Microbial influence on metal mobility and application for bioremediation. *Geoderma* 122: 109–119.

Hatzikioseyian, A. 2010. Principles of bioremediation processes. In *Trends in bioremediation and phytoremediation*, Grażyna Płaza. (ed.), Research Signpost, Trivandrum, Kerala, India, pp. 23–54.

Hegazy, A.K., and Emam, M.H. 2011. Accumulation and soil-to-plant transfer of radionuclides in the Nile Delta coastal black sand habitats. *International Journal of Phytoremediation* 13: 140–155.

Hoffmann, W., Kranefeld, A., and Schmitz-Feuerhake, I. 1993. Radium-226-contaminated drinking water: Hypothesis on an exposure pathway in a population with elevated childhood leukemia. *Environmental Health Perspectives* 101(3): 113–115.

Holker, U., Schmiers, H., Grosse, S., Winkelhofer, M., Polsakiewicz, M., and Ludwig, S. 2002. Solubilization of low-rank coal by *Trichoderma atroviride*: Evidence for the involvement of hydrolytic and oxidative enzymes by using 14C-labelled lignite. *Journal of Industrial Microbiology and Biotechnology* 28: 207–212.

Hsieh, K.M., Murgel, G.A., Lion, L.W., and Shuler, M.L. 1994. Interactions of microbial biofilms with toxic trace metals: Observation and modeling of cell growth, attachment, and production of extracellular polymer. *Biotechnology and Bioengineering* 44(2): 219–231.

IAEA. 1998. *Guidebook on good practice in the management of uranium mining and mill operations and the preparation for their closure*. International Atomic Energy Agency, Vienna, p. 81.

IAEA. 2007. *Management of reprocessed uranium current status and future prospects*. Nuclear fuel cycle and materials section, International Atomic Energy Agency, Vienna, 108 p.

Ivanovich, M., and Harmon, R.S. 1992. *Uranium-series disequilibrium: Applications to earth, marine, and environmental science*. Clarendon Press, Oxford.

Jerden, J.J.L., and Sinha, A.K. 2006. Geochemical coupling of uranium and phosphorous in soils overlying an unmined uranium deposit: Coles Hill, Virginia. *Journal of Geochemical Exploration* 91(1–3): 56–70.

Juznic, K., Koponen, M.H., and Juznic, A. 1989. Transfer of 210Pb and 210Po to the environment during mining and processing of uranium ores. *Journal of Radioanalytical and Nuclear Chemistry* 135(4): 293–298.

Kalin, M., Wheeler, W.N., and Meinrath, G. 2005. The removal of uranium from mining wastewater using algal/microbial biomass. *Journal of Environmental Radioactivity* 78(2): 151–177.

Kleykamp, H. 1991. Thermodynamics of the uranium-platinum metals systems. *Pure and Applied Chemistry* 63(10): 1401–1408.

Kumar, A., Bisht, B.S., Joshi, V.D., and Dhewa, T. 2011. Review on bioremediation of polluted environment: A management tool. *International Journal of Environment Science* 1(6): 1079–1093.

Kusiak, R.A., Ritchie, A.C., Muller, J., and Springer, J. 1993. Mortality from lung cancer in Ontario uranium miners. *British Journal of Industrial Medicine* 50(10): 920–928.

Law, G.T., Geissler, A., Lloyd, J.R., Livens, F.R., Boothman, C., and Begg, J.D. 2010. Geomicrobiological redox cycling of the transuranic element neptunium. *Environmental Science and Technology* 44: 8924–8929.

Levy, D.B., and Kearney, W.F. 1999. Irrigation of native rangeland using treated wastewater from in situ uranium processing. *Journal of Environmental Quality* 28(1): 208–217.

Lloyd, J.R., and Renshaw, J.C. 2005. Bioremediation of radioactive waste: Radionuclide–microbe interactions in laboratory and field-scale studies. *Current Opinion in Biotechnology* 16(3): 254–260.

Lloyd, R.D., Polig, E., Taylor, G.N., Bruenger, F.W., and Miller, S.C. 1996. Uranium skeletal dosimetry and distribution in young adult beagles: A guide for calculating uranium skeletal doses in humans. *Health Physics* 70(3): 396–401.

Lovley, D.R., Widman, P.K., Woodward, J.C., and Phillips, E.J.P. 1993. Reduction of uranium by cytochrome C3 of *Desulfovibrio vulgaris*. *Applied and Environmental Microbiology* 59(11): 3572–3576.

Malekzadeh, F., Latifi, A.M., Shahamat, M., Levin, M., and Colwell, R.R. 2002. Effects of selected physical and chemical parameters on uranium uptake by the bacterium *Chryseomonas* MGF-48. *World Journal of Microbiology and Biotechnology* 18(7): 599–602.

Markich, S.J. 2002. Uranium speciation and bioavailability in aquatic systems: An overview. *Scientific World Journal* 2: 707–729.

Meinrath, G. 1998. Aquatic chemistry of uranium: A review focusing on aspects of environmental chemistry. *Freiberg On-line Geoscience Freiberg* 1: 1–92.

Meinrath, A., Schneider, P., and Meinrath, G. 2003. Uranium ore and depleted uranium in the environment, with reference to uranium in the biosphere from Erzgebirge/Sachsen, Germany. *Journal of Environmental Radioactivity* 64: 175–193.

Mitrovic, S.M., Allis, O., Furey, A., and James, K.J. 2005. Bioaccumulation and harmful effects of microcystin-LR in the aquatic plants *Lemna minor* and *Wolffia arrhiza* and the filamentous alga *Cladophora fracta*. *Ecotoxicology and Environmental Safety* 61(3): 345–352.

Mkandawire, M., Lemna, G., Shaker, L., Kalin, M.A., Wheeler, W.N., and Meinrath G. 2005. Fate and effect of uranium and arsenic in surface waters of abandoned uranium mines: Studies with the removal of uranium from mining wastewater using algal/microbial biomass. *Journal of Environmental Radioactivity* 78(2): 151–177.

Moyes, L.N., Parkman, R.H., Charnock, J.M., Vaughan, D.J., Livens, F.R., Hughes, C.R., and Braithwaite, A. 2000. Uranium uptake from aqueous solution by interaction with geothite, lepidocrocite, muscovite and macknawite: An X-ray absorption spectroscopy study. *Environmental Science and Technology* 34: 1062–1068.

Mueller, J.G., Cerniglia, C.E., and Pritchard, P.H. 1996. Bioremediation of environments contaminated by polycyclic aromatic hydrocarbons. In *Bioremediation: Principles and applications.* Ronald, L., and Don L. Crawford (eds.,). Cambridge University Press, Cambridge, pp. 125–194.

Nagarkatti, M., Nagarkatti, P.S., and Brooks, A. 1996. Effect of radon on the immune system: Alterations in the cellularity and functions of T-cells in lymphoid organs of mouse. *Journal of Toxicology and Environmental Health* 47(6): 535–552.

Parihar, L., Jasmeen, K.J., and Vikramjit, S. 2013. Bioremediation of uranium in contaminated water samples of Bathinda, Punjab by *Desulfovibrio* genus. *Journal of Soil Science and Environmental Management* 4(1): 1–5.

Periyakaruppan, A., Kumar, F., Sarkar, S., Sharma, C., and Ramesh, G. 2007. Uranium induces oxidative stress in lung epithelial cells. *Archives of Toxicology* 81(6): 389–395.

Prakash, D., Gabani, P., Chandel, A.K., Ronen, Z., and Singh, O.V. 2013. Bioremediation: A genuine technology to remediate radionuclides from the environment. *Microbiology and Biotechnology* 6(4): 349–360.

Prikryl, J.D., Jain, A., Turner, D.R., and Pabalan, R.T. 2001. Uranium(VI) sorption behavior on silicate mineral mixtures. *Journal of Contaminant Hydrology* 47(2–4): 241–253.

Remoudaki, E., Hatzikioseyian, A., and Tsezos, M. 2007. A study of the operating parameters of a sulphate-reducing fixed bed reactor for the treatment of metal-bearing waste water. *Advanced Material Research* 20–21: 230–234.

Russell, J.J., Kathren, R.L., and Dietert, S.E. 1996. A histological kidney study of uranium and non-uranium workers. *Health Physics* 70(4): 466–472.

Saxena, G., and Bharagava, R.N. 2015. Persistent organic pollutants and bacterial communities present during the treatment of tannery wastewater. In *Environmental waste management,* Chandra, R. (ed.), CRC Press, Boca Raton, FL, pp. 217–247.

Saxena, G., Chandra, R., and Bharagava, R.N. 2016. Environmental pollution, toxicity profile and treatment approaches for tannery wastewater and its chemical pollutants. *Reviews of Environmental Contamination and Toxicology* 240: 31–69. doi: http://dx.doi.org/10.1007/398_2015_5009.

Simon, O., and Garnier-Laplace, J. 2005. Laboratory and field assessment of uranium trophic transfer efficiency in the crayfish *Orconectes limosus* fed the bivalve *C. fluminea*. *Aquatic Toxicology* 74(4): 372–383.

Simpson, S.L., Angel, B.M., and Jolley, D.F. 2004. Metal equilibration in laboratory-contaminated (spiked) sediments used for the development of whole-sediment toxicity tests. *Chemosphere* 54(5): 597–609.

Singh, S.N., and Tripathi, R.D. 2007. Nitrate pollution and its remediation. In *Environmental Bioremediation Technologies,* Shree N. Singh and Rudra D. Tripathi (eds.), Springer, NY, pp. 353–389.

Sinha, R.K., Valani, D., Sinha, S., Singh, S., and Herat, S. 2009. Study of micro-remediation, phytoremediation & vermi-remediation biotechnologies for contaminated lands & soil by the environmentally adapted microbes, plants & earthworms. In *Solid waste management and environmental remediation,* Faerber, T., and Herzog, J. (eds.), Nova Science Publishers, Inc., NY, pp. 1–72.

Smeaton, C.M., Weisener, C.G., Burns, P.C., Fryer, B.J., and Fowle, D.A. 2008. Bacterially enhanced dissolution of meta-autunite. *American Mineralogists* 93(11–12): 1858–1864.

Soltan, E.M. 2001. Isolation and characterization of antibiotic and heavy metal resistant *Pseudomonas aeruginosa* from different polluted waters in Sohag District, Egypt. *Microbial Biotechnology* 11: 50–55.

Stearns, D.M., Yazzie, M., Bradley, A.S., Coryell, V.H., Shelley, J.T., Ashby, A., Asplund, C.S., and Lantz, R.C. 2005. Uranyl acetate induces hprt mutations and uranium-DNA adducts in *Chinese hamster* ovary EM9 cells. *Mutagenesis* 20(6): 417–423.

Tunali, S., Cabuk, A., and Akar, T. 2006. Removal of lead and copper ions from aqueous solutions by bacterial strain isolated from soil. *Chemical Engineering Journal* 115: 203–211.

Uslu, G., and Tanyol, M. 2006. Equilibrium and thermodynamic parameters of single and binary mixture biosorption of lead and copper ions onto *Pseudomonas putida*: Effect of temperature. *Journal of Hazardous Material* 135: 87–93.

Veiga, L.H.S., Amaral, E.C.S., and Fernandes, H.M. 1998. Human health risk screening of radioactive and non-radioactive contaminants due to uranium industry operation. *Journal of Environmental Radioactivity* 39(1): 69–85.

Violante, A., Huang, P.M., and Gadd, G.M. 2008. *Bio-physico-chemical processes of heavy metals and metalloids in soil environments*. Wiley/John Wiley & Sons Inc., NJ.

Wersin, P., Hochella, M.F., Persson, P., Redden, G., Leckie, J.O., and Harris, D.W. 1994. Interaction between aqueous uranium (VI) and sulfide minerals: Spectroscopic evidence for sorption and reduction. *Geochimica Cosmochimics Acta* 58(13): 2829–2843.

White, C., and Gadd, G.M. 1993. Microbial treatment of metal pollution: A working biotechnology. *Trends in Biotechnology* 11: 353–359.

Whitehead, N.E., Brooks, R.R., and Peterson, P.J. 1971. The nature of uranium occurrence in the leaves of *Coprosma australis* (A. Rich) Robinson. *Australian Journal of Biological Science* 24: 67–73.

WHO. 2001. *Depleted uranium sources, exposure and health effects*. Department of Protection of the Human Environment, World Health Organization, Geneva.

WHO. 2004. *Uranium in drinking water: Background document for development of WHO guidelines for drinking-water quality*. World Health Organization, Geneva, p. 23.

Yazzie, M., Gamble, S.L., Civitello, E.R., and Stearns, D.M. 2003. Uranyl acetate causes DNA single strand breaks in-vitro in the presence of ascorbate (vitamin C). *Chemical Research and Toxicology* 16(4): 524–530.

# 5 Dyes Contamination in the Environment

## *Ecotoxicological Effects, Health Hazards, and Biodegradation and Bioremediation Mechanisms for Environmental Cleanup*

*Uma Shankar Prasad Uday, Nibedita Mahata, Soumya Sasmal, Tarun Kanti Bandyopadhyay, Abhijit Mondal, and Biswanath Bhunia*

**CONTENTS**

5.1 Introduction .......... 128
5.2 Dyes: Nature, Characteristics, and Classification .......... 129
5.2.1 Acid Dyes .......... 129
5.2.2 Basic Dyes .......... 129
5.2.3 Direct Dyes .......... 129
5.2.4 Disperse Dyes .......... 129
5.2.5 Mordant Dyes .......... 130
5.2.6 Sulfur Dyes .......... 130
5.2.7 Reactive Dyes .......... 130
5.2.8 Solvent Dyes .......... 130
5.2.9 Vat Dyes .......... 130
5.3 Transport and Fate of Dyes in the Environment .......... 130
5.4 Sources of Dyes Contamination in Environments .......... 131
5.5 Ecotoxicological Effects and Health Hazards from Dyes .......... 131
5.6 Approaches for the Treatment of Dye-Containing Wastewaters .......... 132
5.6.1 Physicochemical Treatment Approaches .......... 133
5.6.1.1 Equalization and Homogenization .......... 134
5.6.1.2 Flotation .......... 135
5.6.1.3 Coagulation–flocculation .......... 137

5.6.1.4 Adsorption ..... 141
5.6.1.5 Advanced Oxidation Process ..... 146
5.6.2 Biological Treatment Approaches ..... 155
5.6.2.1 Biodegradation of Dyes ..... 156
5.6.2.2 Enzymatic Degradation of Dyes ..... 160
5.7 Biodegradation of Dyes by Genetically Engineered Microbes ..... 160
5.8 Conclusion ..... 161
Acknowledgments ..... 161
References ..... 161

## 5.1 INTRODUCTION

Increasing populations along with progressive adoption of an industrial-based lifestyle have inevitably led to an increased anthropogenic impact on the biosphere, which leads to environmental pollution. Environmental pollution control has been a concern across the globe. The most concerning and major source of environmental pollution is wastewater pollution. Wastewater pollution causes dreadful effects on public water supplies. Skin irritation, intestinal tract infection, and in extreme cases kidney failure and cancer are main health-associated problems (Gupta 2009). Industry is a major source of wastewater pollution due to industrial expansion to meet the needs of an increasing population, especially in developed countries like India. Wastewater pollution comes from different industrial effluents, such as textile, printing, pharmaceuticals, etc. (Uday et al. 2016). As with other industries, dye industries were set up during rapid industrialization and urbanization.

The industries that commonly use dyes are printing, paper, textile, leather, tanning, pharmaceutical, and other allied industries (Chakraborty et al. 2011). Recently, industries have increased significantly the use of synthetic complex organic dyes as coloring materials. It has been observed that approximately 12% of synthetic textile dyes are lost during manufacturing and processing operations. However, 20% of these dyes enter the environment through effluents, which creates an environmental threat (Chakraborty et al. 2012). Quality of water is altered by the presence of dye materials; therefore, removal of this kind of pollutant is of prime importance. It has been noticed that dyes are very difficult to treat with municipal waste treatment operations because of their complicated chemical structures (Chowdhury et al. 2013). It is obvious that a very small quantity of dye causes high visibility and undesirability, which makes it esthetically unpleasant (Crini 2006).

Some scientists have focused on developing various remediation purposes to deal with the dye wastewaters problem. Conventional primary and secondary systems are not suitable for treating effluents containing dyes that are nonbiodegradable (Sanroman et al. 2004). Some investigations have focused on the development of a treatment process for dye wastewater, such as chemical and biological decolorization methods such as coagulation/flocculation treatment, biodegradation processes, oxidation methods, membrane filtration, and adsorption (Anjaneyulu et al., 2005; Rai et al., 2005; Crini et al., 2007; Gimbert et al., 2008). Many physical and chemical treatment processes are being used for color removal. Photocatalytic and sono-chemical degradation, micellar-enhanced ultrafiltration, cation exchange membranes, electrochemical degradation, adsorption/precipitation processes, integrated chemical–biological degradation, integrated iron(III) photo-assisted biological treatment, solar photo-Fenton and biological processes, and Fenton-biological treatment schemes are commonly used (Robinson et al. 2001a). The main disadvantages of these processes include incomplete removal, expensive plant requirements, high capital and operating costs, labor intensive processes, etc. A remediation process is expensive and not easily processed; therefore, it can be manipulated using several techniques, which can offer an economical solution to the problem of environmental pollution in the near future. In this chapter, we try to foster an in-depth understanding of the degradation process by trying to cover all types of chemical and biochemical routes through describing processes, degradation abilities, and factors affecting this process. In addition, we address recent progress on remediation mechanisms for environmental cleanup.

## 5.2 DYES: NATURE, CHARACTERISTICS, AND CLASSIFICATION

A dye molecule consists of two key components such as chromophore and auxochrome. Chromophores are responsible for producing color, and auxochromes give additional support to the chromophores. An auxochrome also decreases the solubility of dye and gives enhanced affinity toward the fibers. Dyes can be classified in many ways due to high structural diversity (Gupta 2009). However, some properties of dyes can be broadly classified based on their usage (Lazar 2005; Gupta 2009).

### 5.2.1 Acid Dyes

These dyes are sodium salts of a sulfonic, carboxylic, or phenol organic acid. They are freely soluble in water and posses an affinity for amphoteric fibers. Therefore, these dyes are commonly used for nylon, wool, paper, leather, food, cosmetics, etc. Most synthetic food colors fall in this category.

### 5.2.2 Basic Dyes

They produce colored cations in solution; therefore, these are also called *cationic dyes*. These types of dyes are used in paper, modified nylons and polyesters, medicines, etc.

### 5.2.3 Direct Dyes

They are water-soluble anionic dyes and are also called *biological stains*. The dyeing process is normally carried out in a neutral or slightly alkaline condition at the boiling point along with sodium chloride, sodium sulfate, or sodium carbonate (dyeing of cotton and rayon, paper, leather, etc., are more useful by them).

### 5.2.4 Disperse Dyes

These are frequently called *insoluble nonionic dyes*. Polyester, nylon, cellulose, acrylic fibers, etc., are suitable for dyeing with disperse dye.

### 5.2.5 Mordant Dyes

They improve the fastness of the dye against water, light, and perspiration. Different mordants can change the final color significantly; therefore, the choice of mordant is very important. They are mostly applied for the dyeing of wool.

### 5.2.6 Sulfur Dyes

They contain a relatively small group of dyes. These have intermediate structures. These are more useful from an economic point of view because of low cost and good wash-fastness properties. Cotton and rayon are more targeted for their use.

### 5.2.7 Reactive Dyes

They can directly react with the fiber substrate as they have a chromophore attached to a substituent. These create a covalent bond, which attaches reactive dyes to natural fibers. These are used for the dyeing of cotton, wool, and nylon.

### 5.2.8 Solvent Dyes

These are insoluble in water since they are generally nonpolar or a little polar. These are used for dyeing plastics, gasoline, lubricants, oils, etc.

### 5.2.9 Vat Dyes

These are water-insoluble dyes. They are not capable of dyeing fibers directly but are manipulated to increase affinity toward textile fiber through a reduction in alkaline liquor, which will produce the water-soluble alkali metal salt of the dye. These dyes are mainly used for the dyeing of cotton, wool, and nylon.

## 5.3 TRANSPORT AND FATE OF DYES IN THE ENVIRONMENT

During the coloration process, a small proportion of colorants (which by their nature are highly visible substances) may be released into open waters due to incomplete exhaustion of the colorants onto textile fibers. Thus, even a minor release may cause an obvious appearance of color in the body of discharged wastewater. It is also noted that a typical limit of visibility of colored discharge in a river varies between 0.1 and 1 g/L, depending on color, illumination, and degree of clarity of the water. In pure water, a reactive dye solution with a concentration of as little as 0.005 mg/L may be visible, especially if the dye is red or violet.

Reactive dyes have been known to produce large volumes of colored effluent compared to other dyes, due to their typically low rates of fixation on the fiber. The fixation rate for reactive dyes in particular is low due to the tendency of the dyes to react with water during the dyeing process. This reaction results in a hydrolyzed dye that has a comparatively low affinity for the fiber. The use of antimicrobial agents in natural fibers such as cotton and wool can often result in the resistance of effluents to biological degradation. Those factors therefore hinder removal of dyes from textile wastewater.

## 5.4 SOURCES OF DYES CONTAMINATION IN ENVIRONMENTS

The toxicity of synthesized dyes mainly depends on structures of precursors used for synthesis thereof. Synthetic dyes are extensively used in textile dyeing, paper printing and photography, pharmaceuticals, food, paint, cosmetics, and petroleum products industries (Zollinger 2003). According to their chemical structure, dyes may be classified into several groups such as azo, anthraquinone, benzodifuranone, quinophthalones, and others (Uday et al. 2016). Of these, the azo compounds, characterized by the presence of one or more azo groups, are the most used (Stolz 2001). Presently, there are at least 3,000 azo dyes in use, representing about 60% of the total dyes applied throughout the world (Hunger 1994).

Most of the commercially used dyes, especially food dyes used for coloring bakery goods, sweets, cereals, and drugs, should be monitored due to their toxic effects (Inomata et al. 2006). Many dyes like Amaranth entered the market and then they were banned because they were discovered to be carcinogenic. Despite their allergenic properties and the potential to cause hyperactivity and histological changes in the thyroid, some dyes like Allura Red and Erythrosine are still permitted as food colorants in the United States (Scotter and Castle 2004).

## 5.5 ECOTOXICOLOGICAL EFFECTS AND HEALTH HAZARDS FROM DYES

Azo dyes are mainly used in textile industries for coloring wool, cotton, polyester, and various other substrates (Hunger 1994). Environmental pollution with these compounds causes, apart from visual pollution, changes in the biological cycles mainly affecting processes of photosynthesis (Kunz et al. 2002b). Furthermore, nonionic azo dyes are considered potentially toxic not only because of the dye itself but also because azo compounds can generate carcinogenic/mutagenic products, such as aromatic amines due to the metabolism of intestinal microflora and/or mammalian azo reduction and chemical reduction (Chequer et al. 2009).

Thousands of different types of dyes are mainly produced from chemical synthesis. The first synthetic dye "Mauveine," synthesized by William Henry Perkin, laid a new foundation and announced the arrival of these xenobiotics. A serious threat of pollution and deterioration of our valuable water resources came after a few years of the industrial revolution. Most of the untreated textile effluent dyes decrease sunlight penetration in the depths of water bodies resulting in reduction of the rate of photosynthesis of algae, ultimately hampering water quality and causing a wide array of adverse effects on the aquatic ecosystem (Saratale et al. 2011) and other aquatic vegetation.

Textile industries are the major contributors to dyes-based water pollution. Yahagi et al. (1975) earlier reported that azo dyes and their derivatives have some mutagenic and carcinogenic properties. It was also reported that the complex structure of aminoazobenzene dyes and their various derivatives may cause mutagenesis, which is a major cause of cancer (Garg et al. 2002). According to the International Agency for Research on Cancer (IARC 1982), benzidine-like dyes are extremely potent carcinogens to many mammals and human beings.

Erythrosine, a xanthene dye, causes allergy, cancer, damage to DNA, and neurotoxic and xenoestrogenic effects on a variety of animals including human beings (Mittal et al. 2006). Similarly, Disperse Orange 1 and Disperse Blue 291 are responsible for genotoxic and mutagenic effects like increased micronuclei in human hepatoma derived HepG2 cells (Tsuboy et al. 2007). It was also reported that malachite green and its metabolite leucomalachite green persist in two human cell lines, namely epithelium resembling Caco-2 and HepG2, causing significant decreases in cell viability, total protein content, and colony-forming ability (Stammati et al. 2005). Reactive dyes are famous as most efficient colors among all but a survey of dye manufacturing industry workers revealed that these dyes are responsible for enhancing the rate of bladder cancer (Rehn 1895). Furthermore, in Spain and Algeria, spraying of a textile paint chemical was reported to cause a pulmonary disease called "Ardystil syndrome" in many spray workers leading to deaths.

In 2007, Oliveira et al. reported some mutagenic properties in dyes like Disperse Orange 37, Disperse Blue 373, Disperse Violet 93, and some components of an unknown black color and their products to contain aromatic amine rings. Commonly used malachite green has some adverse effects and is to be responsible for a 20-times increase in egg mortality rate in largemouth bass *Micropterus salmoides* (Wright 1976). In addition to this malachite green was found to induce mitochondrial damages, nuclear alterations, liver focal necrosis, and sinusoidal congestions in rainbow trout fishes (Gerundo et al. 1991). Furthermore, malachite green also decreases serum calcium and protein levels leading to increased blood cholesterol levels in freshwater catfish *Heteropneustes fossilis* (Srivastava et al. 1995).

It was also observed that various dyes and textile effluents have very strong genotoxic effects on the root cells of *Allium cepa* (Phugare et al. 2011; Watharkar and Jadhav 2014). Disperse Orange 1 was found to impart mutagenic effects on *Salmonella typhimurium* strains viz. TA98, YG1041, TA100, and YG1042 (Chequer et al. 2009). Similarly, mutagenicity evaluation of Disperse Blue 291 was also reported using *Salmonella* assay (Aragão de Umbuzeiro et al. 2005). Congo Red and Reactive Orange 16 azo dyes were reported to cause inhibition of algal growth and affected the natural bacterial luminescence (Novotný et al. 2006). However, textile wastewaters containing a mixture of dyes and have high biological oxygen demand (BOD), chemical oxygen demand (COD), total suspended solid (TSS), total dissolved solid (TDS), and alarming toxicities (Casieri et al. 2008).

## 5.6 APPROACHES FOR THE TREATMENT OF DYE-CONTAINING WASTEWATERS

Wastewater treatment is a mixture of unit processes, some physical, others chemical or biological in their action. A conventional treatment process is composed of a series of individual unit processes, with the output (or effluent) of one process becoming the input (influent) of the next process. The first stage is usually made up of physical processes. The basic chemical oxidation process with air and oxygen also occurs in nature, but it is no longer sufficient for highly polluted wastewater. Therefore, there is a significant need to develop a wastewater treatment process that can remove the pollutants effectively by a simple method. The conventional methods used in the textile industry for color/dye removal from effluents include physicochemical methods, which have been widely used in sewage treatment plants.

### 5.6.1 Physicochemical Treatment Approaches

The technologies often used in the dye decolorization process of textile effluent are physical and chemical approaches (Forgacs et al. 2004; dos Santos et al. 2007) like chemical precipitation (Tunay et al. 1996), adsorption (Al-Degs et al. 2000), oxidation (Arslan et al. 1999a), ozonation, and Fenton's oxidation (Solozhenko et al. 1995; Lin and Lo 1997; Kang and Hwang 2000) have been investigated to treat dye-containing wastewater. Some examples of physical and chemical approaches are summarized in Table 5.1.

These physical and chemical treatment techniques are well known for the treatment of industrial wastewaters, but are less effective, very costly, and have limited applicability due to the low biodegradability of dyes (Seesuriyachan et al. 2007; Ayed et al. 2011). But, such treatment technologies generally involve complex procedures and

**TABLE 5.1**
**Advantages and Disadvantages of Physical and Chemical Treatment Technologies**

| Physical and Chemical Treatment Methods | Advantages | Disadvantages |
|---|---|---|
| Ozonation | Applied in gaseous state, no toxic metabolites complete chemical oxygen demand | Short half-life, stability affected by auxiliary chemicals expensive |
| Oxidation | Rapid and proficient process | Chemical required, high energy cost, byproducts production |
| Coagulant/flocculants | Simple, economically feasible | High amount sludge production, high amount chemicals required for pH adjustment |
| Adsorption on activated carbon | Effective, high capacity | Regeneration is expensive, loss of absorbent, ineffective against disperse and vat dye |
| Fenton reagents | Effective decolorization for soluble and insoluble dyes | Sludge generation |
| Activated carbon | Good elimination of a variety of dyes | Very costly |
| Electrokinetic coagulation | Economically feasible | Produce high sludge |

are economically impracticable. In the past several decades, many techniques have been developed to find an economic and efficient way to treat the colored wastewater from the textile industry, including biochemical, physicochemical, combined treatment processes, and other approaches (Forgacs et al. 2004). Fenton's oxidation is one of the oldest advanced oxidation processes (AOPs), which is used successfully, as it is comparatively cheap and uses easy to handle reagents. Fenton's reagent, a mixture of hydrogen peroxide and ferrous iron, is effective for color and COD removal of dye effluent. Several investigators have demonstrated that AOPs are effective for complete color removal and partial degradation of organic matter (Arslan and Balcioglu 1999; Kang and Hwang 2000).

#### 5.6.1.1 Equalization and Homogenization

Pretreatment of textile dyeing wastewater is generally required to ensure treatment effects and stable operation, because amount of water pollution and quantity fluctuations are complex components. In general, the regulating tank is set to treat the wastewater. In the tank, the wastewater is usually mixed with air or mechanical mixing equipment to prevent lint, cottonseed shell, and the slurry from settling to the bottom of the tank. The hydraulic retention time is generally about 8 hours. Flow control (equalization) and/or homogenization of contaminated flows are the two objectives of the equalization tank.

Flow equalization is used to minimize the variability of water and wastewater flow rates and composition. Each unit operation in a treatment train is designed for specific wastewater characteristics. When all unit operations are carried out at uniform flow conditions, there are the possibilities to improve efficiency and control. If there exists a wide variation in flow composition over time, the treatment efficiency of the overall process performance may degrade severely. The influent water with varying flow composition enters this basin first before it is allowed to go through the rest of the treatment process. Many processes use equalization basins to accumulate and consolidate smaller volumes of wastewater such that full-scale batch reactors can be operated. Other processes incorporate equalization basins in continuous treatment systems to equalize the waste flow so that the effluent at the downstream end can be discharged at a uniform rate.

Many investigators have ascribed various benefits to the use of flow equalization in wastewater treatment systems. Some of the most important benefits are listed as follows (Lovell 1973; Wesley and Eckenfelder 1989; Tchobanoglous and Burton 1991; Alley 2007):

1. Equalization improves sedimentation efficiency by improving hydraulic detention time.
2. Efficiency of a biological process can be increased because of uniform flow characteristics and minimization of the impact of shock loads and toxins during operation.
3. Manual and automated control of flow-rate-dependent operations, such as chemical feeding, disinfection, and sludge pumping, are simplified.
4. Treatability of the wastewater is improved and some BOD reduction and odor removal is provided, if aeration is used for mixing in the equalization basin.
5. A point of return for recycling concentrated waste streams is provided, thereby mitigating shock loads to primary settlers or aeration basin.

Sometimes, it is thought that equalization tanks also serve the purpose of dilution. The U.S. Environmental Protection Agency's (USEPA's) viewpoint is that dilution is mixing of waste that is more concentrated with greater volumes of less concentrated waste such that the resulting wastewater does not need further treatment. Equalization basins in a treatment system can be located in-line or off-line. In in-line equalization, 100% incoming raw wastewater directly enters into the equalization basin, which is then pumped directly to other treatment units (e.g., primary treatment units). However, the equalization basin does not directly receive the incoming wastewater for sideline or off-line equalization.

#### 5.6.1.2 Flotation

Flotation is a rather unique separation process that produces a large number of microbubbles in order to form the three-phase substances of water, gas, and solid. Dissolved air may be added under pressure to cause the formation of tiny bubbles, which will attach to particles. The microbubble adheres to the tiny fibers under the effect of interfacial tension, buoyancy of bubble rising, hydrostatic pressure, and a variety of other forces. The mixtures float to the surface due to the low density so that the oil particles are separated from the water, which helps to remove the fibers in wastewater effectively. It originated in minerals processing (known as froth flotation). The first apparatus of froth flotation was recorded, in 1905, in Broken Hill of Australia where air bubbles were produced by intense stirring of the pulp (Hines and Vincent 1962). Quite early, applications of this method were reported in water and wastewater treatment (Hansen et al. 1943).

According to the technique used for the generation of gas bubbles (usually air), which are the transport medium of this separation process, two broad categories of flotation are generally well known:

1. Dispersed-air flotation, where electrolytic and froth flotation belong; the latter is applied mainly in the mineral processing field, with rather large air bubbles being generated.
2. Dissolved-air flotation—in which vacuum flotation is also classified—finds applications mainly in water and wastewater treatment; smaller bubbles of diameter are generated.

Many parameters affect the flotation process; the most important are reagent concentration, pH value of the solution, ionic strength, gas flow rate, flotation time, and feed concentration (Zouboulis et al. 1990).

##### *5.6.1.2.1 Flotation methods for the removal of dyes*

The flotation technique can remove a large fraction of pigments and dyes to a lower extent (Carre et al. 2000). The application of flotation to dyes and paint effluents is quite old, but still active (McCormick et al. 1964; Cheng and Huang 1988). The removal of color from wastewater can also be carried out by a flotation technique (Kabil and Ghazy 1994; Basu and Malpani 2001). The air flotation technique was usually applied for dye effluent after the particles had been destabilized. Due to the low density of color flocs formed after the coagulation process, flotation was preferred as an alternative to sedimentation (Mavros et al. 1994), where the hydrophobic dye flocs would remain at the bubble surface and rise to the top of the liquid phase (Gregory 2005). Recent work on the applicability of air flotation in the remediation of dye wastewater has been conducted (Dafnopatidou et al. 2007; Dafnopatidou and Lazaridis 2008) showing the removal of dyes by flotation was closely related to the characteristics of wastewater such as pH and background electrolyte and surfactant concentration (Dafnopatidou et al. 2007). Dafnopatidou and Lazaridis (2008) further observed that with the assistance of chemical treatment, dissolved-air flotation could manage to remove more than 97% of dyeing mill effluent. Although, due to the exceptional foaming, surfactant-rich dye wastewater could not be satisfactorily decolorized, and thus, in textile industries, this technique is not currently used for effluent decolorization.

*5.6.1.2.1.1 Ion Flotation* The method generally involves the removal of surface-inactive ions from aqueous solutions by adding the proper surfactant, so that a scum appears on the surface of the solution. After the passage of gas bubbles through the solution, it consists of solids and surfactant as a chemical constituent.

*5.6.1.2.1.2 Precipitate Flotation* Precipitate flotation involves initially the formation of some "kind" of precipitate of the ionic species from dilute aqueous dispersions and its subsequent removal by transfer on air bubbles to the surface of the solution. It is an evolution of the ion flotation method.

*5.6.1.2.1.3 Adsorbing Colloid Flotation* This method is the removal of a solute by adsorption on or coprecipitation with a carrier floc, which is then floated, usually with the help of a suitable surfactant (Clarke and Wilson 1983). For the case of As(V) removal (Stalidis et al. 1988), the presence of surfactants was not necessary for efficient flotation. But As(III) was also removed (Stalidis et al. 1986) using the same method.

*5.6.1.2.1.4 Dissolved-air Flotation* This bubble generation technique has been used for the selective separation of copper by ion flotation, in the presence of As(V) (Stalidis et al. 1988). A statistical approach to the (dissolved-air) flotation of copper and zinc, precipitated as sulfides, by cationic surfactants at a pH of about 2 (Stalidis et al. 1988).

*5.6.1.2.1.5 Electrolytic Flotation* This technique has been investigated for the removal of metals from their dilute solutions; chromium (VI) ions were effectively removed (Zouboulis and Matis 1989). Dissolved-air and dispersed-air flotation techniques were also tested and the results compared (Zouboulis et al. 1992). It should be pointed out that toxic inorganic materials, even in trace amounts, have to be removed to a high degree from industrial discharges. This is a promising field for the application of flotation as a pretreatment method. Further, other conventional methods are not as effective; for example, sedimentation and flocculation remove less than 80% of them.

#### 5.6.1.3 Coagulation–flocculation

Coagulation–flocculation sedimentation is one of the most used and important methods, especially in the conventional treatment process, due to its simplicity and cost-effectiveness. The efficiency of coagulation–flocculation strongly affects the overall wastewater treatment performance. The whole treatment process of coagulation-flocculation can be divided into two distinct procedures that should be applied consecutively. The first is coagulation, the process whereby destabilization of a given colloidal suspension or solution is taking place. The function of coagulation is to overcome the factors that promote the stability of a given system, which can be achieved with the use of appropriate chemicals, usually aluminum or iron salts, the so-called coagulant agents. The second subprocess is flocculation, which refers to the induction of destabilized particles in order to come together, to make contact, and thereby, to form large agglomerates, which can be separated easier usually through gravity settling (Bratby 2006).

Particles smaller than $10^{-5}$ mm may be referred to as colloids, and particles smaller than $10^{-6}$ mm as solutions. With decreasing size, the time required for settling increases, up to several years for certain solution ingredients. The only way for settling and the subsequent separation is to come closer due to the very small size, to make contacts, and to form larger particles, which can be settled easier. However, this procedure is hindered due to the homonymous negative charge of the material carries. The electrostatic repulsive forces constrain the particles from approaching each other, and the suspension is characterized as stable; therefore, a long time is required for settling. In order to accelerate the settling time, destabilization is required, denoting the importance of coagulation. Destabilization can be achieved with one or a combination of two or more of the following mechanisms, after the addition of a coagulant agent (Duan and Gregory 2003; Crittenden et al. 2005):

1. Compression of the electrical double layer
2. Adsorption and charge neutralization
3. Adsorption and interparticle bridging
4. Enmeshment in precipitate (by using of excess coagulant dose “sweep flocculation”)

After destabilization, flocculation promotes the aggregation and flocs formation, usually after the addition of an appropriate flocculent agent. Two general types of flocculation can be identified: microflocculation (or perikinetic flocculation), in which particle aggregation is brought about by the thermal motion of fluid molecules (Brownian motion) and macroflocculation (orthokinetic flocculation), in which particle aggregation is brought about by inducing velocity gradients and mixing in the suspension (Crittenden et al. 2005).

Coagulation–flocculation is used widely during water or wastewater treatment. It is an integral treatment step in surface or underground waters treatment, intended for human consumption. Typical applications are the removal/separation of colloids and suspended particles, of natural organic matter, or of metal ions. In wastewater treatment, additional applications include the removal of toxic metals, anions (i.e., phosphates), color, odor, etc.

##### *5.6.1.3.1 Brief Overview on Coagulation and Flocculation Process: Destabilization Mechanisms*

Generally, coagulation and flocculation in water/wastewater treatment involve the addition of chemicals to alter the physical state of dissolved and suspended solids to facilitate removal via subsequent sedimentation process (Alexander et al. 2012). Coagulation is generally defined as destabilization of suspension, giving rise to aggregations. Flocculation describes the process in which the destabilized particles are induced to make contact for formation of larger aggregates (Gregory 2005; Bratby 2006).

In water, the behavior of colloidal particles is strongly influenced by their electrokinetic charge. For hydrophilic colloids, these primary charges are mainly contributed by polar groups such as carboxylic and amine groups (Sincero and Sincero 2002). Due to the size of colloidal particles (about 0.01 to 1 μm), attractive body forces between particles are less than the repelling forces of the electrical charges (Tchobanoglous et al. 2003), and thus, these particles tend to remain discrete and dispersed in suspension. In order to remove colloidal particles by settling, it is practical only if they settle rapidly in the order of several hundreds of millimeters per hour (Sincero and Sincero 2002).

In a solution, the primary charge of the colloids attracts ions of the opposite charges, also known as counter-ions. These ions, held by electrostatic and van der Waals forces, will form a compact layer (Stern layer) around the primary (Tchobanoglous et al. 2003). The counter-ions attached to the surface will in turn attract their own counter-ions (the co-ions of the primary charge), forming the diffuse layer. Unlike the charges in the Stern layer, only a part of the diffused layer will move along with the colloid by shearing at the shear plane (Sincero and Sincero 2002). The potential at this surface of shear is often called the zeta potential. The measured zeta potential gives a good approximation of the surface charge of the colloidal particle (Pavon-Silva et al. 2009).

The stability of colloidal particles can be explained in a quantitative manner by using the DLVO theory (named after Derjaguin, Landau, Verwery, and Overbeek). This theory involves the estimations of attraction energy (van der Waals forces) and the energy of repulsion (overlapping of electrical double layers) in terms of distance between particles (Bratby 2006). The net interaction energy, also known as the energy barrier, is the difference between the repulsion and attraction forces. Approaching particles would have to overcome this large energy barrier to come into contact. Under normal conditions, as the barrier height is usually much larger than the average thermal energy of particles, it is almost impossible for colliding particles to surmount the barrier, and the suspension will remain stable (Gregory 2005). In the case when the kinetic energy of the particles is large enough to overcome the energy barrier, the particles will coalesce (Bratby 2006).

Four mechanisms that can bring on particle destabilization are double-layer compression, charge neutralization, colloid entrapment, and intraparticle bridging (Sincero and Sincero 2002). The double-layer compression mechanism involves a reduction in the double layer around the colloidal particle by a change in ionic strength induced from the addition of an indifferent electrolyte, which resulted in destabilization of the colloid (Nharingo et al. 2015). Under stable conditions in which the concentration of counter-ions is low, colloid particles are unable to get close to each other because of their thick electric double layer (Sincero and Sincero 2002). However, as the concentration of counter-ions increases via the addition of salts, the diffuse layer becomes thinner, and particles can approach closer before experiencing repulsion.

In industry, double-layer compression is not a feasible coagulation mechanism to be used for wastewater treatment due to the massive amount of salt needed (Tchobanoglous et al. 2003). It is unlikely that a sufficient increase in ionic strength would give a practical destabilization method. On the other hand, absorption of counter-ions on the surface of the particles is promising. This destabilization method is called charge neutralization and is often achieved by adsorption of mononuclear and polynuclear metal hydrolysis species or polyelectrolytes on the surfaces of the particles (Tchobanoglous et al. 2003).

Hydrolyzing coagulants are found to be effective in neutralizing the negative surface charge of many types of particles, including bacteria and clays (Duan and Gregory 2003). The most commonly used metal coagulants for this purpose can be categorized into aluminum coagulants (alum, aluminum chloride, and sodium aluminate) and iron coagulants (ferric chloride, ferric sulfate, ferrous sulfate, and ferric chloride sulfate) (Bratby 2006).

Furthermore, continuous studies on the aquatic chemistry and behavior of simple aluminum salt leads to the development of highly efficient pre-polymerized inorganic coagulants (Bratby 2006; Zouboulis and Tzoupanos 2009). Pre-polymerized coagulants such as polyaluminum chloride (PAC), polyaluminum sulfate (PAS), and polyaluminum chloro-sulfate (PACS) have been used extensively worldwide during the last two decades (Zouboulis and Tzoupanos 2009). These coagulants are able to function efficiently over a wide range of pH, temperature, and colloid concentration ranges as compared to the conventional coagulants (Bratby 2006; Zouboulis and Tzoupanos 2009). Due to the precise nature of charge-neutralizing species, charge reversal that can lead to restabilization will occur at high dosages (Gregory 2005).

For a practical water treatment operation, it is favorable to add metal coagulants at a dosage higher than the solubility of the amorphous hydroxide precipitates (Gregory 2005). In this case, particles may provide the nucleation site for precipitation to occur, and this phenomenon leads to entrapment of the particle as the precipitate forms. The settling hydrous precipitates then enmeshes with other colloidal particles in water and aids its removal (Sincero and Sincero 2002; Parsons and Jefferson 2006). This method often leads to faster aggregation as compared to charge neutralization by giving larger and stronger flocs (Gregory 2005). One of the major advantages of colloid entrapment is that it is the least dependent on the nature of the impure particles to be removed, whether the particle consists of bacteria, clays, oxides, or others. However, production of a large amount of sludge remains a problem for this mechanism (Gregory 2005).

The next mechanism of coagulation is interparticle bridging, usually achieved by an addition of water-soluble polymers (cationic, anionic, or nonionic). This mechanism requires the polymer chain to be adsorbed on the particle surface, either by chemical bonding or by mere physical attachment (Sincero and Sincero 2002). Only several segments of the polymer chain are attached, while the bulk of the chain extends to the surrounding solution for contact and adherence with other particles (Teh et al. 2014).

Bridged particles then intertwine with other bridged particles during mixing and produce flocs (Tchobanoglous et al. 2003). Although the flocs are fast settling and have the ability to withstand high shear, they may not readily re-form when the flocs are broken. Linear and high molecular weight polymer is found to be a very effective coagulant in this mechanism as compared to the branched or cross-linked structures with the same molecular weight (Gregory 2005).

Supplementary to destabilization, flocculation is required as the next step to induce aggregation and settling of large agglomerates from destabilized particles. The overall process of flocculation is divided into two stages known as (1) perikinetic (microscale) and (2) orthokinetic (macroscale) flocculation. The first stage of flocculation occurs immediately after destabilization and throughout random thermal agitation of fluid molecules known as Brownian motion. In this case, the rate of flocculation is dependent on the temperature and concentration of particles as presented in the theory of Smoluchowski (Kitchener 1972). Normally, perikinetic flocculation occurs automatically. The floc formed during perikinetic flocculation has poor settling characteristics as the process is limited by floc size (Bratby 2006) and only capable of removing minute particles.

Unlike perikinetic flocculation, orthokinetic flocculation in the second stage induces contact of particles through bulk fluid motion (gentle motion of fluid). In general, bulk movement can be prompted by mechanical agitation to induce velocity gradients in the liquid for improved contact between particles. Orthokinetic action can be applied using the conventional jar test, under controlled shear in a Couette apparatus or capillary flow (Gregory and Ives 1978). The rate of flocculation based on orthokinetic action is dependent on the particle nature (i.e., size and concentration) and velocity shear gradient of the fluid (Scholz 2015). Based on a general approximation, orthokinetic floc formation is predominant in a system with particles larger than 1 μm in diameter and velocity gradients higher than 5 seconds (Sincero and Sincero 2002).

Both stages of flocculation are important depending on the nature of the particles. The chances of collision are proportional to the concentration of the particles; hence, perikinetic flocculation is more rapid in concentrated suspensions (Scholz 2015). On the whole orthogenesis, flocculation assists as a conditioning step in the later stages of flocculation to form macroflocs for effective settling of flocs.

##### *5.6.1.3.2 Coagulation–flocculation Process on Dye/Textile Wastewater*

The textile industry presents a global pollution problem owing to the dumping or accidental discharge of dye wastewater into waterways, impacting the quality and esthetics of water resources (Chan et al. 2011). Wastewater from textile dyeing facilities is difficult to treat because of its highly fluctuating composition and color variability, depending on the dyestuff, fabric, and concentration of added fixing compounds (Singh and Arora 2011). Dyes are extensively used in many industries, especially in textile manufacturing as it was estimated that more than 100,000 commercial dyes are used. The coagulation–flocculation process has been successfully employed for color removal of water-insoluble sulfur dispersed dyes as well as water-soluble dyes.

A study was conducted to compare the effectiveness of chemical coagulation with electrocoagulation using aluminum electrodes for the decolorization of Disperse Red Dye (Merzouk et al. 2011). It was shown in their study that the use of electrocoagulation as an alternative process should not be ignored due to its lower equipment cost and high robustness to pH change as compared to chemical coagulation (Merzouk et al. 2011).

In recent years, interest has been shifted toward investigating the performance of natural coagulants in single dye wastewater treatment as compared to inorganic ones. One of the many promising natural coagulants/flocculants is chitosan. The wide application of chitosan is due to its biodegradability, nontoxic nature, and unique physicochemical properties because of the presence of primary amino groups and high nitrogen content. Moreover, chitosan possesses several intrinsic characteristics such as high cationic charge density and long polymer chains that make it an effective coagulant/flocculant for wastewater treatment purposes. It was found that effective Acid Blue 92 removal (~99%) could be achieved by using low dosages of chitosan (100 mg/L) in coagulation–flocculation treatment (Szygua et al. 2009). Additionally, dye recovery was possible, and the chitosan used in the treatment process could be recycled.

The potential use of *Moringa oleifera* seed extract as a coagulant to remove several dyes in wastewater has also been attempted. Its use in water and wastewater treatment relies on the presence of water-soluble protein in its seed—a natural cationic polyelectrolyte that causes coagulation (Bhatia et al. 2007). Beltran-Heredia et al. (2009) and Beltran-Heredia and Sanchez-Martin (2008) reported that a coagulation–flocculation process using *M. oleifera* on Chicago Sky Blue (azo dye) and Carmine Indigo (indigoid dye) dye wastewater yielded color removal of 99% and 40%, respectively.

The prospect of using snail shell as a coagulant aid together with alum in the treatment of aniline blue (Acid Blue 93) was also investigated (Oladoja et al. 2011). Preliminary study showed that no dye removal was achieved by using alum alone and only considerable reduction of dye was observed when snail shell was used alone. Modification of natural organic resources to produce cationic polymeric flocculants for industrial applications has also been attempted. For instance, a lignin-based cationic polyelectrolyte (L-DAF) was successfully synthesized from Mannich reaction product of dimethylamine, acetone, and formaldehyde grafted onto hydroxymethylated (Fang et al. 2010). It was found that this new cationic flocculant enabled the removal of anionic azo dye (Acid Black, Reactive Red, Direct Red) more than 85%.

Many researchers have advised the reuse of sludge generated from wastewater treatment plants as a potential coagulating agent, as it could significantly reduce the costs of handling and disposing sludge. In particular, waterworks sludge (ferric chloride sludge) was found to be a suitable low-cost material in the coagulation–flocculation process (Moghaddam et al. 2010).

#### 5.6.1.4 Adsorption

Adsorption is an important unit operation in a number of natural and industrial systems, such as fundamental biological studies, separation and purification processes, recovery of chemical compounds, and waste treatment processes. It can be replaced with other separation processes and contribute efficiently to the removal of contaminants from aqueous solution (Ranjan Jena et al. 2004). Adsorption is a physicochemical wastewater treatment process in which dissolved molecules are attached to the surface of an adsorbent by physical–chemical forces. Since adsorption is sometimes reversible, the regeneration of the adsorbent with consequential economy of operation is also possible. Rather than attempting to remove the dye and degrade it at the same time, adsorption offers the alternative of splitting up these two steps once the dye has been adsorbed onto the substrate. Adsorption, as a treatment process, has garnered considerable attention in this respect, since a well-designed adsorption system can produce an effluent with virtually no dyestuffs present (McKay et al. 2000).

Recently, adsorption techniques have been considered due to their efficiency in the removal of pollutants. Adsorption produces a high-quality product and is a process that is economically feasible. Decolonization may be a result of two mechanisms—adsorption and ion exchange—and is influenced by many physicochemical factors, such as dye/adsorbent interaction, adsorbent surface area, particle size, temperature, pH, and contact time (Robinson et al. 2001b). The dye adsorption process can be represented by four consecutive steps (Al-Godah 2000; Baup et al. 2000; Sanghi and Bhattacharya 2002):

1. Diffusion/convection of dye molecules through the bulk of solution
2. Diffusion of dye molecules through a diffusional boundary layer (film diffusion)
3. Diffusion of dye molecules from the surface into the interior of the adsorbent materials
4. Adsorption of dye molecules on the surface of the materials through molecular interactions

The concentration of the dye and agitation may affect Step 2. Step 3 is usually considered as the rate-determining stage, which certainly should affect the adsorption of dyes on the substrate. Step 4 is dependent on the nature of the dye molecules, such as anionic and cationic structures. It is important to note that Step 3 could involve two different phenomena:

1. Porous diffusion (adsorbate first diffuses in the liquid filling the pores and then is adsorbed)
2. Surface diffusion (adsorbate is first adsorbed then diffuses from one site to another)

The most widely used dye adsorbent and most readily available from commercial sources is activated carbon (Thomas and Yeh 1995). However, several adsorbents studied to determine their ability to adsorb dyes from aqueous effluents are given in Table 5.2.

#### *5.6.1.4.1 Appropriate Isotherms for Dye Adsorption*

Equilibrium data, commonly known as adsorption isotherms, are the basic requirements for the design of adsorption systems. Obtaining equilibrium data for a specific adsorbate-adsorbent system can be performed experimentally, with a time-consuming procedure that is incompatible with the growing needs for adsorption systems design. The physical chemistry involved with the adsorption process may be complex, and no single theory of adsorption could adequately present all experimental data satisfactorily. Although the

**TABLE 5.2**
**Some Adsorbents Used for Dyes Removal**

| Adsorbate | Adsorbent |
|---|---|
| Acid dyes | Cationized sawdust (Naouab et al. 2001), compost sorbent (Tsui et al. 2003), maize waste (Davila et al. 2005), orange peel (Sivaraj et al. 2001), peat (Ramakrishna and Viraraghavan 1996), polysaccharides (Delval et al. 2002), red mud (Namasivayam et al. 2001), shale oil ash (Al-Godah 2000), bottom ash and de-oiled soya (Gupta et al. 2006) |
| Cationic dyes | Bagass fly ash (Indra et al. 2005), blast furnace sludge (Bhatnagar and Jain 2005), clinoptilolite and amberlite (Julide et al. 2006), compost sorbent (Tsui et al. 2003), dried activated sludge (Osman et al. 2004), fly ash (Basava Rao and Mohan Rao 2006), giant duckweed (Waranusantigul et al. 2003), linseed cake (Liversidge et al. 1997), maize waste (Davila et al. 2005), natural zeolite (Wang et al. 2006; Wang and Ariyanto 2007), peat (Allen et al. 2004), shale oil ash (Al-Godah 2000), silkworm pupa (Noroozi et al. 2007) |
| Direct dyes | Chitosan (Bhavani and Dutta 1999), coal-based sorbent (Mohan et al. 2002), compost sorbent (Tsui et al. 2003), fly ash (Basava Rao and Mohan Rao 2006) |
| Disperse dyes | Polysaccharides (Delval et al. 2002) |
| Reactive dyes | Chitin (Mckay et al. 1984), eggshell membrane (Koumanova et al. 2002), eucalyptus bark (Morais et al. 1999), maize waste (Davila et al. 2005), polysaccharides (Delval et al. 2002), activated carbon (Al-Degs et al. 2007), sugar beet pulp (Aksu and Isoglu 2007) |

assumptions on which they were based were seen in later years to be not entirely valid, some of the oldest theories of adsorption are still the most useful due to their simplicity.

However, most industries discharge effluents containing several components; hence adsorption systems design must be based on multicomponent effluents, making multicomponent equilibrium data a necessity. Compared with single-solute isotherms, multicomponent isotherms require a tedious procedure to be obtained experimentally; therefore, many attempts were carried out to obtain the latter from single-solute equilibrium data. However, multicomponent systems are complicated due to the interaction and competition effects involved. For single-solute systems, Langmuir and Freundlich isotherms have been the most common in addition to the BET (Brunauer et al. 1938) multilayer model (Mckay and Al Duri 1989; Khan et al. 1997).

Competitive interactions between adsorbate have been recognized as a factor that affects adsorption. The magnitude of these competitive interactions needs to be determined in order to better predict and model contaminant migration (Gutierrez and Fuentes 1993). Multicomponent equilibrium studies have passed through many stages since the beginning of this century when none of the multicomponent equilibrium studies dealt with competitive adsorption and a few were only concerned with the selectivity of adsorption processes (Mckay and Al Duri 1989). The complexity of multicomponent adsorption is even severe when the adsorbent exhibits a microporous structure. This is mainly because of an inherent adsorption heterogeneity known for this type of adsorbent (Nguyen and Do 2001).

#### 5.6.1.4.2 Multicomponent Adsorption

Several methodologies have been developed. The earliest attempt (Butler and Ockrent 1930) extended the Langmuir isotherm to account for competitive adsorption, and this was later modified to include a separate term for adsorption without competition (Jain and Snoeyink 1973). McKay and Al Duri (1989) applied the competitive Langmuir isotherm to basic dyes on carbon. Prediction of the equilibrium adsorption of mixtures from single-solute isotherms is important both in theory of adsorption and for practical applications. The ideal adsorbed solution theory (IAST) of Myers and Prausnitz (1965) can be used to predict mixture adsorption equilibrium from corresponding single-solute isotherms. However, it is not suitable for mixture systems that show non-ideal behaviors. These theories include the real adsorbed solution theory (Krishna 2001), the vacancy solution theory (Čermáková et al. 2008), the heterogeneous extended Langmuir model, and the heterogeneous IAST (Bai et al. 2003). For its simplicity, the extended Langmuir model remains the most widely used isotherm. Among the mentioned multicomponent isotherm equations, only some of them have been applied to model the dye systems. Table 5.3 summarizes the theories applied for the adsorption of multicomponent systems of dyes.

At low concentrations, competitive interactions are not expected to play a significant role in the adsorption process. However, in the case of a highly concentrated accidental release, a full understanding of the competitive interactions among adsorbates would be necessary to predict contaminant migration. The study of competitive interactions also serves the purpose of defining the concentrations at which the effect due to competitive interactions starts to be significant in adsorption models (Gutierrez and Fuentes 1993).

Allen et al. (1988) in their results indicated that the presence of two or more dyes in solution can and will have an effect on the amount of dyes adsorbed. The amount of any one dye adsorbed was reduced in the presence of a second or third dye. The extent of this effect varied with individual dyes. Factors that could be influencing mixture adsorption will be interactions between dyes in solutions; the effect of each dye in contributing to the surface charge on the adsorbent; and competition between the different dyes for the available surface. It is evident from the equilibrium studies that equilibrium adsorption capacities decreased in multicomponent systems as compared to single dye systems. The different dye ions will experience different physical and electrical forces according to their structure, molecular size, and functional groups. Hence, competition to take available sites cannot be a factor alone (Allen et al. 1988).

For the adsorption of dyes, it is known that molecular characteristics of dyes and physicochemical properties of adsorbents can affect the adsorption mechanism. It is found that different molecular weights and also the molecular sizes of dyes play a major role in the physiosorption system; the smaller molecule will occupy the pores in the structure of adsorbent in advance. However, for the systems where the predominant adsorption mechanism is chemisorption, the role of molecular size will not play a major role. For example, on comparing the structures of two different dyes, one can find that the availability of the ionic charges may be diverse due to their special positions in the structures affecting the adsorption efficiency.

## TABLE 5.3
## Multicomponent Adsorption Studies of Dyes Systems

| Multicomponent System | Adsorbent | Type of Models Applied | Reference |
|---|---|---|---|
| C.I. Basic Blue 41, C.I. Basic Red 18 | Activated carbon, Silkworm pupa | Langmuir, IAST | Noroozi et al. 2008 |
| Remazol reactive yellow, Remazol reactive black, Remazol reactive red | Activated carbon | Langmuir, Redlich-Peterson, Freundlich | Al-Degs et al. 2007 |
| Malachite green, Pb ions | Natural zeolite | Freundlich, Langmuir | Wang and Ariyanto 2007 |
| Gemazol Turquoise Blue-G reactive dye, copper(II) ions | Dried sugar beet pulp | Langmuir | Aksu and Isoglu 2007 |
| Metanil yellow, C.I. Reactive Blue 15 | Cross-linked chitosan | Langmuir | Chiou and Chuang 2006 |
| C.I. Acid Blue 80, C.I. Acid Red 114, C.I. Acid Yellow 117 | Activated carbon | IAST, Langmuir | Mckay et al. 2004a; McKay et al. 2000; McKay et al. 2004b; Porter et al. 1999 |
| C.I. Basic Blue 3, C.I. Basic Yellow 21, C.I. Basic Red 22 | Peat | Freundlich, Langmuir, Redlich-Peterson | Allen et al. 1988; Allen et al. 2004 |
| C.I. Basic Red 46, C.I. Basic Yellow 28 | Bentonite | Freundlich, Langmuir | Turabik 2008 |
| C.I. Acid Red 18, C.I. Acid Yellow 23 | Bentonite and modified bentonite | Langmuir | Qiao et al. 2009 |
| C.I. Acid Blue 80, C.I. Acid Blue 324, C.I. Acid Green 25, C.I. Acid Green 27, C.I. Acid Orange 7, C.I. Acid Orange 8, C.I. Acid Orange 10, C.I. Acid Red 1 | Husk of mango seed | Freundlich, Langmuir | Dávila-Jiménez et al. 2009 |
| C.I. Acid Blue 25, C.I. Acid Yellow 117 | Bamboo-derived active carbon | Langmuir, Freundlich Redlich-Peterson, Sips | Chan et al. 2009 |
| C.I. Acid Blue 25, C.I. Acid Yellow 117 | Tyre demineralized activated | correlative extended Freundlich isotherm | Chan et al. 2012 |
| Malachite green, methylene blue | Rarasaponin-bentonite | Extended Langmuir model | Kurniawan et al. 2012 |
| Remazol brilliant blue R, C. I. Reactive Orange 16 | Polysaccharides composite | Sheindorf-Rehbun-Sheintuch equation, extended Langmuir model | Janaki et al. 2012 |

Furthermore, the dye molecule could provide conditions for a chemical reaction between the adsorbent and adsorbate. As a result, the chemical properties and functional groups of dye and adsorbent can alter the process of adsorption and play the main role in adsorption of dyes (Noroozi et al. 2008). The disadvantage of adsorption technique is that the pollutants may only transfer to the adsorbent, which needs to be regenerated regularly, resulting in additional costs.

#### 5.6.1.5 Advanced Oxidation Process

Advanced oxidation processes (AOPs) for wastewater treatment have received a great deal of attention in recent years. AOPs generate the highly reactive hydroxyl radical ($^{\bullet}OH$) to degrade the recalcitrant chemicals present in wastewater (Muruganandham et al. 2007; Wu et al. 2008; Wu et al. 2011). These OH radicals attack the most organic molecules rapidly and nonselectively. The versatility of AOPs is also enhanced by the fact that they offer various alternative methods of hydroxyl radical production, thus allowing better compliance with specific treatment requirements. The eco-friendly end product is the special feature of these AOPs, which are more efficient as they are capable of mineralizing a wide range of organic pollutants. Interestingly, AOPs can make use of solar energy rather than artificial light sources. The latter rely on high electrical power, which is costly and hazardous.

AOPs such as ozonation ($O_3$), ozone combined with hydrogen peroxide ($O_3/H_2O_2$) and UV irradiation ($O_3$/UV) or both ($O_3/H_2O_2$/UV), ozone combined with catalysts ($O_3$/catalysts), UV/$H_2O_2$, Fenton and photo-Fenton processes ($Fe^{2+}/H_2O_2$ and $Fe^{2+}/H_2O_2$/UV), and the ultrasonic process and photocatalysis have been successfully used for wastewater treatment (Mahamuni and Adewuyi 2010; Matilainen and Sillanpää 2010; Muruganandham et al. 2010, 2012; Vilhunen et al. 2010).

##### *5.6.1.5.1 Principles of AOPs*

AOPs have common principles in terms of the participation of hydroxyl radicals that are assumed to be operative during the reaction. Although it is claimed that there are other species involved, the active species responsible for the destruction of contaminants in most cases seems to be the hydroxyl radical ($OH^{\bullet}$), which is unstable and quite reactive. Due to the instability of the $OH^{\bullet}$ radical, it must be generated continuously *in situ* through chemical or photochemical reactions described in the literature (Hao et al. 2000). Hydroxyl radicals produced in either way described above may attack organic molecules by abstracting a hydrogen atom from the molecule (Clarke and Knowles 1982). Carey (1992) described a common pathway for the degradation of organics by the hydroxyl radicals as follows:

$$OH^{\bullet} + RH \rightarrow H_2O + R^{\bullet}$$
$$R^{\bullet} + H_2O_2 \rightarrow ROH + OH^{\bullet}$$
$$R^{\bullet} + O_2 \rightarrow ROO^{\bullet}$$
$$ROO^{\bullet} + RH \rightarrow ROOH + R^{\bullet}$$

##### *5.6.1.5.2 Positives Aspects and Limitations*

Several authors have pointed out the advantages and disadvantages of different AOP systems, such as UV/$H_2O_2$, UV/$TiO_2$, photo-Fenton ozone-based processes.

In general, the literature has highlighted the following positive aspects of AOPs (Parsons 2004; Teixeira and Jardim 2004; Pignatello et al. 2006).

1. Unlike conventional technologies, which use strongly oxidant species, under certain conditions AOPs can provide the complete mineralization of pollutants.
2. They are used for the destruction of refractory compounds resistant to other treatments, such as biological processes.
3. They allow the conversion of recalcitrant compounds and refractory contaminants submitted to biodegradation systems.
4. They can be used in combination with other processes for pre- or post-treatment.
5. They possess strong oxidizing power with high reaction rates.
6. They are ideal to decrease the concentration of compounds formed by alternative pretreatment, such as disinfection.
7. The formation of by-products can be minimized if optimized amounts of reactants are employed.
8. In many cases, AOPs consume less energy compared to thermal destruction processes (incineration) of liquid wastewaters.
9. They enable *in situ* treatment.

However, like any other form of treatment, AOPs cannot be indiscriminately applied to arbitrary residue processing.

There are certain conditions that limit their implementation as highlighted by previous works (Domènech et al. 2001; Morais 2005):

1. Not all processes are readily scaled-up to commercial or industrial needs.
2. Costs can be high, mostly due to the consumption of electric energy by radiating sources in the case of photo-oxidative degradation processes.
3. Restrictions are present under conditions of high pollutant concentrations.
4. In some cases, strict control of oxidant concentration and pH correction are required for the operation of the process as for post-processing or disposal of treated effluent (e.g., Fenton and photo-Fenton).
5. Process limitations are also related to pH changes, such as particle aggregation and modification of surface properties of catalysts used in heterogeneous systems, and the distribution of iron complexes in Fenton systems (and their UV-visible radiation properties as well), or even iron precipitation.

#### *5.6.1.5.3 Chemical Fundamentals*

*5.6.1.5.3.1 Hydroxyl Radical: Generation and Interferences* AOPs exhibit an elevated capacity to produce hydroxyl radicals ($^{\bullet}OH$) (Kusic et al. 2006; Poulopoulos et al. 2006). Due to its high standard reduction potential (Equation 5.1), this radical oxidizes a wide variety of organic compounds to $CO_2$, $H_2$, and inorganic ions from any heteroatoms:

$$\begin{gathered} {}^{\bullet}OH + e^- + H^+ \rightarrow H_2O \\ k = 7 \times 10^9 \text{ L mol s}^{-1} \quad E^o = 2.8 \text{ V} \end{gathered} \tag{5.1}$$

This is a great advantage in relation to treatments that simply transfer pollutants from one phase to another, like filtration, adsorption, precipitation, coagulation, flocculation, sedimentation, flotation, membrane separation, centrifugation, inverse osmosis, distillation, extraction, and evaporation (Teixeira and Jardim 2004). In most of these operations, the final disposal of passive solid phases remains an unresolved problem (Kunz et al. 2002a). Hydroxyl radicals are obtained from strong oxidants such as $H_2O_2$ and $O_3$, combined or not with UV radiation, iron II or III salts, semiconductor photo catalysts (such as $TiO_2$), or from the photolysis of water with vacuum UV radiation (Nogueira and Jardim 1998; Catorceno et al. 2010; Loures et al. 2013). After fluorine, hydroxyl radical (OH) is the strongest oxidant (Dorfman and Adams 1973). Table 5.4 presents the oxidation potentials of some species.

According to Kiwi and co-workers (2000), the efficiency of AOPs is strongly influenced by the presence of high concentrations of hydroxyl radical scavengers, such as carbonate, bicarbonate, and ionic chloride species, reducing the process efficiency. Essam and co-workers (Kiwi et al. 2000) investigated the AOP oxidation of treated wastewater containing chloride ions, noting that there is a marked reduction in the oxidation of organic load as a function of $Cl^-$ concentration.

In the Fenton and photo-Fenton systems, chloride ions can form complexes with iron ($FeCl^{2+}$), which effectively compete with the $^\bullet OH$ radical. In fact, in the typical pH range of the photo-Fenton reaction, the complexes $Fe(OH)^{2+}$ and $FeCl^{2+}$ coexist and can undergo photochemical reactions as shown by Equations 5.2 and 5.3:

$$Fe(OH)^{2+} + h\upsilon \rightarrow Fe^{2+} + HO^\bullet \tag{5.2}$$

$$FeCl^{2+} + h\upsilon \rightarrow Fe^{2+} + Cl^\bullet \tag{5.3}$$

**TABLE 5.4**
**Standard Reduction Potential of Selected Oxidant Species**

| Species | E° (V, 25°C)[a] |
|---|---|
| Fluorine ($F_2$) | 3.03 |
| Hydroxyl radical (•OH) | 2.80 |
| Atomic oxygen ($O_2$) | 2.42 |
| Ozone ($O_3$) | 2.07 |
| Hydrogen peroxide ($H_2O_2$) | 1.78 |
| Perhydroxyl radical ($HO_2$) | 1.70 |
| Chlorine dioxide | 1.57 |
| Hypochlorous acid (HClO) | 1.49 |
| Chlorine ($Cl_2$) | 1.36 |
| Bromine ($Br_2$) | 1.09 |
| Iodine ($I_2$) | 0.54 |

[a] Potential refers to the standard hydrogen electrode (SHE)
*Source*: Loures, C.C.A., et al., *Int. Rev. Chem. Eng.*, 5(2), 102–120, 2013. With permission.

The $FeCl^{2+}$ complex generates Fe(II) and $Cl^{\bullet}$ radicals from photon absorption, at wavelength 347 nm. At that same wavelength, the complex $Fe(OH)^{2+}$ generates hydroxyl radicals less efficiently with quantum yield. The species $Cl^{\bullet}$ combines with chloride ions in solution, giving the radical anion $Cl_2^{\bullet-}$, with standard reduction potential equal to 2.10 V (SHE). This species can generate undesirable chlorinated oxidation products, thus competing with hydroxyl radicals. Moreover, as reported by Kiwi and co-workers (2000), the same hydroxyl radical reacts with chloride ions in solution generating $HOCl^{\bullet}$ species, that, in an acidic medium, generate $Cl^{\bullet}$ radicals (Equations 5.3 through 5.6):

$$^{\bullet}OH + Cl^{-} \rightleftharpoons HOCl^{\bullet-} \xrightarrow{H+} [H_2OCl^{\bullet-}]^{\pm} \quad (5.4)$$

$$\downarrow$$

$$H_2O + Cl^{\bullet} \quad (5.5)$$

$$\downarrow$$

$$Cl^{\bullet} + Cl^{-} \rightarrow Cl_2^{\bullet-} \quad (5.6)$$

$$\downarrow$$

$$FeCl^{2+} + h\upsilon \rightarrow Fe^{2+} + Cl^{\bullet} \quad (5.3)$$

Depending on the structure of the organic contaminant, different reactions may occur involving the hydroxyl radical, such as hydrogen atom abstraction, electrophilic addition to substances containing unsaturated and aromatic rings, electron transfer, and radical–radical reactions. Ramos and co-workers (2012) used molecular modeling techniques to study the reactions of hydroxyl radicals during phenol oxidation.

*5.6.1.5.3.2 Abstraction of Hydrogen Atoms* Hydroxyl radicals are capable of oxidizing organic compounds by hydrogen abstraction, producing organic radicals (Equation 5.7). Subsequent addition of molecular oxygen forms peroxyl radicals (Equation 5.8); reactive intermediates initiate chain reactions leading to thermal degradation to $CO_2$, water, and inorganic salts. The reaction by hydrogen abstraction usually occurs with aliphatic hydrocarbons (Legrini et al. 1993):

$$HR + {}^{\bullet}OH \rightarrow R^{\bullet} + H_2O \quad (5.7)$$

$$R^{\bullet} + O_2 \rightarrow RO_2^{\bullet} \quad (5.8)$$

*5.6.1.5.3.3 Electrophilic Addition of Hydroxyl Radical* The electrophilic addition of hydroxyl radical to organic compounds containing π bonds results in the formation of organic radicals (Equation 5.9). This usually occurs with hydrocarbons containing unsaturated bonds or aromatic systems (Legrini et al. 1993):

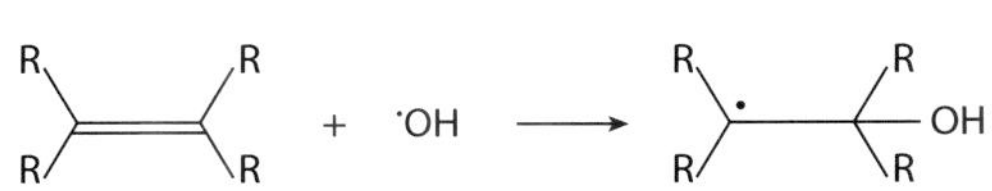


(5.9)

Fast dechlorination of chlorophenols is also promoted by the electrophilic generating chloride ions (Equation 5.10) (Legrini et al. 1993):

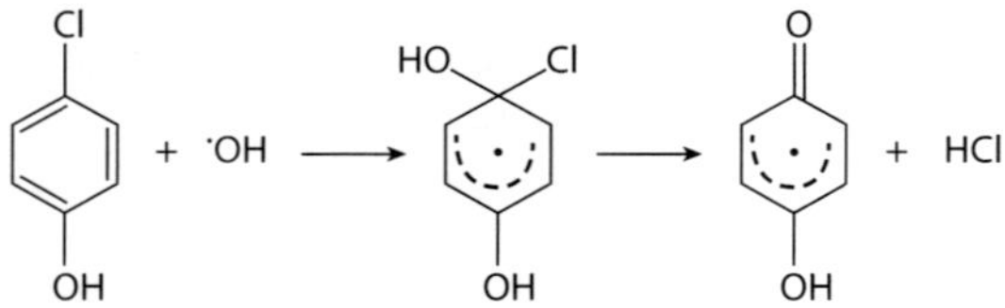

(5.10)

*5.6.1.5.3.4 Electron Transfers* Electron transfer reactions occur when electrophilic addition and hydrogen abstraction are deprived, as in the case of chlorinated hydrocarbons (Equation 5.11) (Legrini et al. 1993).

$$RX + {}^{\bullet}OH \rightarrow RX^{\bullet+} + OH^- \quad (5.11)$$

*5.6.1.5.3.5 Radical–Radical Reactions* Other radical reactions can also occur (Equations 5.12 and 5.13). However, such reactions are undesirable from the point of view of the oxidation of organic compounds due to $^{\bullet}OH$ consuming and subsequent impairment of the photodegradation process efficiency:

$$\begin{aligned} &2OH^{\bullet} \rightarrow H_2O_2 \\ &k = 5.3 \times 10^9 \ L \ mol^{-1} \ s^{-1} \end{aligned} \quad (5.12)$$

$$\begin{aligned} &H_2O_2 + {}^{\bullet}OH \rightarrow HO_2{}^{\bullet} + H_2O \\ &k = 2.7 \times 10^7 \ L \ mol^{-1} \ s^{-1} \end{aligned} \quad (5.13)$$

The predominance of one or another reaction depends on several factors, among which the concentration of organic substrate, as well as its recalcitrant character.

### *5.6.1.5.4 Systems Used in AOPs*

The systems used in AOPs are divided into two major groups, depending on the number of phases involved: homogeneous processes and heterogeneous processes.

These different processes have been used alone or in combination, resulting in different treatment systems. In each group, a source of radiation, such as ultraviolet (UV) or ultraviolet-visible (UV-Vis), may or may not be used. Table 5.5 represents the examples of treatment systems based on AOPs (Loures et al. 2013).

*5.6.1.5.4.1 Homogeneous Systems and Radical Generators* Homogeneous processes occur in a single phase, and use ozone ($O_3$), hydrogen peroxide, and UV ($UV/H_2O_2$) or Fenton reagent (mixture of $H_2O_2$ with Fe[II] salt) to generate hydroxyl radicals. Some combinations of technologies like photo-Fenton and UV/ozone are also included in this group.

5.6.1.5.4.1.1 Ozone Ozone is a powerful oxidizing agent for water and wastewater. Once dissolved in water, ozone reacts with a great number of organic compounds in two different ways: by direct oxidation as molecular ozone or by indirect reaction through formation of secondary oxidants like hydroxyl radical (Baig and Liechti 2001).

**TABLE 5.5**
**Systems Used in the Generation of Hydroxyl Radical**

| System | Irradiated | Nonirradiated |
|---|---|---|
| Homogeneous system | $O_3$/UV | $O_3$/ $H_2O_2$ |
| | $H_2O_2$/UV | $O_3/OH^-$ |
| Heterogeneous systems | $H_2O_2/Fe^{2+}$/UV | $H_2O_2/Fe^{2+}$ |
| | Sc/$O_3$/UV | Electro-Fenton |
| | Sc/$H_2O_2$/UV | — |
| | Sc/UV | — |

*Source*: Loures, C.C.A., et al., *Int. Rev. Chem. Eng.*, 5(2), 102–120, 2013. With permission.
*Note:* Sc, semiconductor (e.g., $TiO_2$, ZnO).

Ozone application can be generalized into a powerful disinfection and a strong oxidant to remove color and odor, eliminating trace toxic synthetic organic compounds and assisting in coagulation (Am Water Works Res et al. 1991). Results presented by a few researchers revealed that ozone decolorize all dyes, except nonsoluble disperse and vat dyes, which react slowly and take a longer time (Namboodri et al. 1994; Marmagne and Coste 1996; Rajeswari 2000). Furthermore, it has been documented that color removal using ozonation from textile wastewater is dependent on dye concentration (Lin and Lin 1993; Sevimli and Sarikaya 2002; Konsowa 2003). A higher initial dye concentration of textile wastewater causes more ozone consumption. Increasing ozone concentration enhances mass transfer that causes an increase in ozone concentration in the liquid phase, which increases color removal.

According to Ahmet et al. (Baban et al. 2003) and Arslan and Isil (Alaton and Balcioğlu 2002), 40–60 minutes of ozonation of biotreated textile wastewater yielded 99% of decolorization efficiency. High color removal can be achieved on wastewater, which contains high initial dye concentration and low initial COD. Alkaline pH and high temperature were also found as favorable conditions for high total organic carbon (TOC) and COD removals. In spite of having high color removal efficiency, limited COD and TOC removal efficiencies were obtained. This could be explained by incomplete oxidation of organic materials and production of small organic molecular fragments along with destruction of dyestuff that was not completely oxidized.

5.6.1.5.4.1.2 Combined Use of $H_2O_2$/UV Oxidization of the textile wastewater with hydrogen peroxide alone has been found ineffective at both acid and alkali values (Tunay et al. 1996), while under UV irradiation, $H_2O_2$ are photolyzed to form two hydroxyl radicals ($2OH^{\bullet}$) that react with organic contaminants (Crittenden et al. 1999). Galindo and Kalt (1999) documented that the $H_2O_2$/UV process was more effective in an acid medium in terms of discoloration. Nonefficient color removal at alkaline pH is contributed to the fact that hydrogen peroxide undergoes decomposition leading to dioxygen and water, rather than producing hydroxyl radicals under UV radiation. Therefore, the instantaneous concentration in $HO^{\bullet}$ is lower than expected. Furthermore, the $H_2O_2$/UV process is more sensitive to the scavenging effect of carbonate at higher pH values. Temperature did not have a significant effect on discoloration (Galindo and Kalt 1999). Others documented that removal of textile dye by $H_2O_2$/UV increased as doses of effective hydrogen peroxide increased up to a "critical" value (Arslan et al. 1999b; Galindo and Kalt 1999; Ince 1999). High concentrations of $H_2O_2$ acts as a radical scavenger, while low concentration of $H_2O_2$ does not generate enough hydroxyl radicals ($OH^{\bullet}$) that are consumed by dye, and this leads to a slow rate of oxidation. Arslan and colleagues (Alaton and Balcioğlu 2002; Arslan and Balcioğlu 2001) also documented that the treatment of textile wastewater was not effective via $H_2O_2$/UV oxidation unless a preliminary ozonation period was introduced to produce sufficient $OH^{\bullet}$ to observe a significant color and COD reduction.

Complete destruction of reactive dyes and azo dye has been recorded in 30–90 minutes by previous researchers (Cisneros et al. 2002; Georgiou et al. 2002; Neamtu et al. 2002; Kurbus et al. 2003), while Perkowski and Kos (2003) documented more than 99% color reduction of dye house wastewater after 2 hours. Kang et al. (1999) documented that the longer the reaction time, the more advantageous it is for color removal, where color removal at 10 minutes was only 9% of removal at 120 minutes. Galindo and Kalt (1999) documented the inverse relationship between the initial dye concentration and the efficiency of UV/$H_2O_2$ treatment. An increase in the dye concentration induces a rise of the internal optical density. Dyes react; consequently, the solution becomes more and more impermeable to UV radiation. Therefore, the rate of photolysis of hydrogen peroxide directly depends on the incident intensity. Consequently, when initial dye concentration rises, the production of hydroxyl radicals decreases.

5.6.1.5.4.1.3 Combined Use of $O_3$/$H_2O_2$ The addition of both hydrogen peroxide and ozone to wastewater accelerates the decomposition of ozone and enhances production of the hydroxyl radical. At acidic pH, $H_2O_2$ reacts only very slowly with $O_3$, whereas at pH values above 5 a strong acceleration of $O_3$ decomposition by $H_2O_2$ has been observed (Staehelin and Hoigne 1982). At higher pH, even a very small concentration of $H_2O_2$ will be dissociated into $HO_2^-$ ions that can initiate the ozone decomposition more effectively than the $OH^-$ ion (Staehelin and Hoigne 1982; Glaze and Kang 1989). Arslan et al. (1999b) documented that $H_2O_2$/$O_3$ treatment of synthetic dye highly depended on the pH of the effluent. Hydrogen peroxide in alkaline medium reacts with sodium hydroxide, and as a result lower concentrations of hydrogen peroxide are available for the formation of hydroxyl radicals.

5.6.1.5.4.1.4 Combined Use of $O_3$/$H_2O_2$/UV The addition of $H_2O_2$ to the $O_3$/UV process accelerates the decomposition of ozone, which results in an increased rate of $OH^{\bullet}$ generation (Al-Kdasi et al. 2004). Among all AOPs, for dye wastewater and acetate, the polyester fiber dyeing process effluent combination of $H_2O_2$/$O_3$/UV appeared to be the most efficient in terms of decoloration (Azbar et al. 2004; Perkowski and Kos 2003). Many workers have reported that COD removal efficiency of raw textile wastewater increased from 18% to 27% by using sequential ozonation and $H_2O_2$/UV, and in the case of biotreated textile effluent, a preliminary ozonation step increased COD removal of the $H_2O_2$/UV treatment system from 15% to 62% (Arslan and Balcioğlu 2001; Alaton and Balcioğlu 2002). Azbar et al. (2004) documented 99% COD removal from acetate and polyester fiber dying process effluent in batch mode operation using a combination of $O_3/H_2O_2$/UV in 90 minutes.

5.6.1.5.4.1.5 Catalytic Ozonation Catalytic ozonation may be regarded as a homogeneous process, which is based on activation of ozone by metal ions present in aqueous solution, or as a heterogeneous process in the presence of supported metal oxides or metal oxides. Catalytic ozonization corresponds to an effective method for removal of various organic compounds in water and in aqueous effluents. However, this method has been used mainly at the laboratory scale (Kasprzyk-Hordern et al. 2003). The proposed catalysts for homogenous catalytic ozonization process are transition metals such as $Fe^{2+}$, $Mn^{2+}$, $Ni^{2+}$, $Co^{2+}$, $Cd^{2+}$, $Cu^{2+}$, $Ag^{+}$, $Cr^{3+}$, $Zn^{2+}$. The nature of the transition metal applied not only determines the rate of reaction, but also the selectivity and consumption of ozone (Kasprzyk-Hordern et al. 2003). Catalytic ozonization shows to be effective in the removal of several organic compounds present in water and aqueous effluents (Kasprzyk-Hordern et al. 2003). It is important to note that the homogeneous catalytic ozonation must be associated to a technique of removing the metal ion from treated effluents. The toxic character of some metals may limit the application of this process. In spite of these limitations, catalytic ozonation is a promising technology for application in the decontamination of aqueous effluents.

5.6.1.5.4.1.6 Fenton Reaction The Fenton process has its roots in the finding reported in 1894 that ferrous ion strongly elevated the oxidation of tartaric acid by hydrogen peroxide (Herney-Ramirez et al. 2010). In the Fenton process, hydrogen peroxide is added to wastewater in the presence of ferrous salts, generating species that are strongly oxidative with respect to organic compounds. $^{\bullet}OH$ is traditionally regarded as the key oxidizing species in Fenton processes. The Fenton process mechanism is quite complex and is described in detail with equations in the literature (Gallard and De Laat 2000; Pera-Titus et al. 2004). Fenton's oxidation process is composed of four stages including pH adjustment, oxidation reaction, neutralization and coagulation, and precipitation. The organic substances are removed at two stages of oxidation and coagulation (Kang and Hwang 2000; Li et al. 2010). $^{\bullet}OH$ radicals are responsible for oxidation, and coagulation is ascribed to the formation of ferric hydroxo complexes (Li et al. 2010; Tang 2016). The relative importance of oxidation and coagulation depends primarily on the $H_2O_2/Fe^{2+}$ ratio. Chemical coagulation predominates at a lower $H_2O_2/Fe^{2+}$ ratio, whereas chemical oxidation is dominant at higher $H_2O_2/Fe^{2+}$ ratios (Neyens and Baeyens 2003; Li et al. 2010).

Many chemicals are refractory to Fenton oxidation, however, such as acetic acid, acetone, carbon tetrachloride, methylene chloride, oxalic acid, maleic acid, malonic acid, n-paraffins, and trichloroethane (Bautista et al. 2008). It has been demonstrated that these compounds are resistant under the usual mild operating conditions of Fenton oxidation (Zazo et al. 2005, 2006; Molina et al. 2006). In addition to these basic studies, the process has been applied to industrial wastewaters (such as chemical, pharmaceutical, textile, paper pulp, cosmetic, and cork processing wastewaters), sludge, and contaminated soils (Kanmani and Muthuselvi 2005), resulting in significant reductions of toxicity, improvement of biodegradability, and color and odor removal (Bautista et al. 2008).

5.6.1.5.4.1.7 Photo-Fenton Processes The photo-Fenton process, as its name suggests, is rather similar to the Fenton one, but also employs radiation (Malato et al. 2002; Gogate and Pandit 2004; Pera-Titus et al. 2004; Herney-Ramirez et al. 2010). The photo-Fenton reaction is also well known in the literature (Kiwi et al. 1993; Pera-Titus et al. 2004) as an efficient and inexpensive method of wastewater and soil treatment (Sauleda and Brillas 2001; Pera-Titus et al. 2004). The photo-Fenton process is known to be capable of improving the efficiency of dark Fenton or Fenton-like reagents by means of the interaction of radiation (UV or UV-Vis) with Fenton's reagent (Tokumura et al. 2008). This technique has been suggested as feasible and promising for removing pollutants from natural and industrial waters and increasing the biodegradability of chlorophenols when used as a pretreatment method to decrease water toxicity (Pera-Titus et al. 2004). Some of its most innovative applications include oxalate as a ligand of iron ions (Aplin et al. 2000; Pera-Titus et al. 2004).

The effectiveness of photo-Fenton processes is attributed to the photolysis of Fe (III) cations in acidic media yielding Fe(II) cations, in conjunction with reaction between Fe(II) and $^{\bullet}OH$ to yield hydroxyl radicals (Fenton's reaction). In this process, the photolytic decomposition of Fe $(OH)^{2+}$ is accelerated, providing an additional source of highly oxidative hydroxyl radicals compared to the "simple" Fenton process (Herney-Ramirez et al. 2010). The photo-Fenton process produces more hydroxyl radicals than the conventional Fenton method (Fe[II] with hydrogen peroxide) or photolysis, thus promoting organic pollutant degradation rates. This process consists of a combination of Fenton reagents ($Fe^{2+}/H_2O_2$) and light energy (Muaoz et al. 2005; Rodríguez et al. 2005) and thus of two reactions (Tokumura et al. 2008). The first reaction is a reaction of $Fe^{2+}$ with $H_2O_2$, which generates the powerful reactive species $^{\bullet}OH$ radicals and oxidizes $Fe^{2+}$ to $Fe^{3+}$. In other words, the hydroxyl radical generation in Fenton processes is due to the iron-catalyzed decomposition of $H_2O_2$. The first photo-Fenton reaction causes the formation of hydroxyl radicals. The second reaction of the photo-Fenton process is a reaction of $Fe^{3+}$ with water, which occurs when light is used at a wavelength from 300 to 650 nm. This generates $^{\bullet}OH$ radicals and reduces $Fe^{3+}$ to $Fe^{2+}$. These two oxidation–reduction reactions occur repeatedly and completely mineralize organic pollutants to $CO_2$ and $H_2O$ (Tokumura et al. 2008). The oxidation power of the photo-Fenton process is attributed to the generation of OH radicals. Without irradiation, a Fenton-like reaction occurred instead of a photo-Fenton reaction. The Fenton-like reaction is a reaction of $Fe^{3+}$ with $H_2O_2$, which causes the reduction of $Fe^{3+}$ to $Fe^{2+}$. Appropriate implementation of the photo-Fenton treatment depends mainly on the operating variables—$H_2O_2$/COD molar ratio, $H_2O_2/Fe^{2+}$ molar ratio, and irradiation time.

*5.6.1.5.4.2 Heterogeneous Systems* Heterogeneous processes use semiconductor catalysts (Tobaldi et al. 2008). As examples of heterogeneous systems, UV/titanium dioxide ($TiO_2$) and UV/zinc oxide (ZnO) are included. Once the semiconductor is exposed to UV radiation, it is promoted to an electronically excited state, characterized by the generation of electron-hole pairs, due to the promotion of electrons from the valence band to the conduction band by photon absorption.

$TiO_2$ is the most commonly used semiconductor in photodegradation of organic compounds. Several characteristics contribute to this: the possibility of activation by sunlight, its insolubility in water, chemical stability over a wide pH range, the possibility of immobilization on surfaces, low cost, and its nontoxic character (Chen et al. 1999; Wang and Hong 1999). In addition to $TiO_2$, other semiconductors such as CdS, ZnO, $WO_3$, ZnS, and $Fe_2O_3$ are mentioned in literature (Wilcoxon et al. 1999; Yeber et al. 1999). There are many studies comparing the effectiveness of $TiO_2$ and ZnO as photocatalysts (Liu et al. 1998; Silva et al. 2010). Such studies have shown various aspects related to the use of these semiconductors: the photodegradation mechanism of ZnO is similar to $TiO_2$ (Daneshvar et al. 2004); both are efficient semiconductors, inexpensive, and nontoxic; and efficiency is increased when nanometric particles are used. The major advantage of ZnO is due to its higher absorption band in comparison with $TiO_2$ (Behnajady et al. 2006). The metal oxide semiconductors have shown good photocatalytic activity toward the degradation of harmful organics into less harmful molecules under light illumination such as $TiO_2$ and ZnO. The synthesized ZnO/$TiO_2$ has been extensively characterized in terms of morphology as well as structural, crystalline, and photocatalytic properties.

### 5.6.2 Biological Treatment Approaches

Dyes fall into different classes with respect to their chemical and structural diversity and spectrum of usage. Few dye classes never have an equivalent, which occurs in the natural environment, such as xanthenes, phenothiazines, and sulfur-containing similar dyes. Textile dyes are wide-ranging and heterogeneous types of molecules used in numerous technological fields (i.e., textile and paper industry, food processing, and hair coloring) (Forgacs et al. 2004). Many chemical classes of synthetic dyes are regularly exploited on an industrial large scale such as azo dyes, triarylmethanes, anthraquinones, and phenothiazine are among the most important and prevalent (Al-Rashed and Al-Gaid 2012). It is proven that the physicochemical method of industrial effluent treatment does not eliminate the dyes effectively and completely. Microbial degradation and decolorization of the azo dyes has gained more consideration because of its environmentally friendly and inexpensive nature. Microbes are able to decolorize the dyes by the processes of both anaerobic and aerobic metabolisms.

Physical as well as chemical treatment methods of degradation of dyes such as precipitation, adsorption, coagulation, flocculation, flotation, electrochemical destruction, and mineralization and decolorization processes have some serious disadvantages such as cost, time, and release of residues into the environment (Gogate and Pandit 2004). All these useful techniques are applied by industries to diminish the toxicity level but not to neutralize the level of toxicity (Maier et al. 2004).

Microorganisms can be employed to wholly degrade those azo dyes by means of secreting enzymes, such as azo reductase, laccase, peroxidase, and hydrogenase over these dyes, which leads to reduction of those dyes. The reduced forms of such azo dyes are further mineralized into simpler compounds and are used as their energy source (Stolz 2001). Based on the survey of literature, the microbial decolorization of the azo dyes is more effective under combined effects of anaerobic and aerobic conditions. Bioaccumulation and biosorption are the two principal technologies in the biological process for posttreatment of dye bearing industrial effluents. They have good potential to replace traditional methods for the treatment of dyes present in industry effluents. The biological process of remediation can be carried out *in situ* at the contaminated site, and this process is usually environmentally benign (i.e., zero secondary pollution) and cost effective. These are the principal advantages of biological technologies for the treatment of dye industry effluents. Hence, research attention has been focused seriously on biological methods for the treatment of effluents (Vijayaraghavan et al. 2008). The disadvantage of the above-mentioned degradation process is that it goes from low degradation efficiency or even zero degradation for a few days and has practical difficulty in a continuous process. The difference between bioaccumulation and biosorption of dyes was also reported in a review (Pearce et al. 2003).

Bioaccumulation can be defined as a phenomenon of uptake of toxicants by living cells; biosorption can be defined as the passive uptake of toxicants by inactive or dead biological materials. The significant advantage of biosorption over the bioaccumulation process is the use of living microorganisms is not necessary for the continuous treatment of toxic effluents. To address this problem, biomass materials, which are adaptable to environmental conditions and toxicant concentrations, can be used.

Wang et al. (2009) reported the increasing trend of biological processes, such as biodegradation, bioaccumulation, and biosorption for degradation of different dyes due to their cost effectiveness, ability to produce less sludge, and environmental neutrality. Fungi and algae have also been reported to play an important role in dye decolorization. Apart from that they also reported that adsorption rather than degradation shows a foremost role during the decolorization process by fungi and algae (Wang et al. 2009). Various biological methods of dye degradation are illustrated in Figure 5.1.

### 5.6.2.1 Biodegradation of Dyes

#### *5.6.2.1.1 Biodegradation of Dyes by Bacteria*

Reactive dyes are generally colored compounds that comprise one or two functional groups and are able to form covalent bonds through the active sites in fiber molecules. A carbon or phosphorous atom of the dye molecule may bind to hydroxyl groups in cellulose, thiol, amino, and hydroxyl groups in wool, or amino groups in polyamides. Most fiber-reactive azo dyes are applied for dyeing of cellulosic materials, such as cotton; therefore, these dyes are the major source of waste in textile effluents. Microorganisms can break down dyes for their self-growth and/or energy requirements for sustainability. In a few cases, metabolic pathways, which organisms follow for their own natural growth and development, may also be used for breaking down pollutant molecules. During this process, microorganisms do not get any benefit, but researchers have considered advantages of this phenomenon and use it for the bioremediation process. Complete degradation of any biological compound finally yields water and either carbon dioxide or methane. Partial degradation will yield products that may or may not be less toxic than its native pollutant. Over the past few decades,

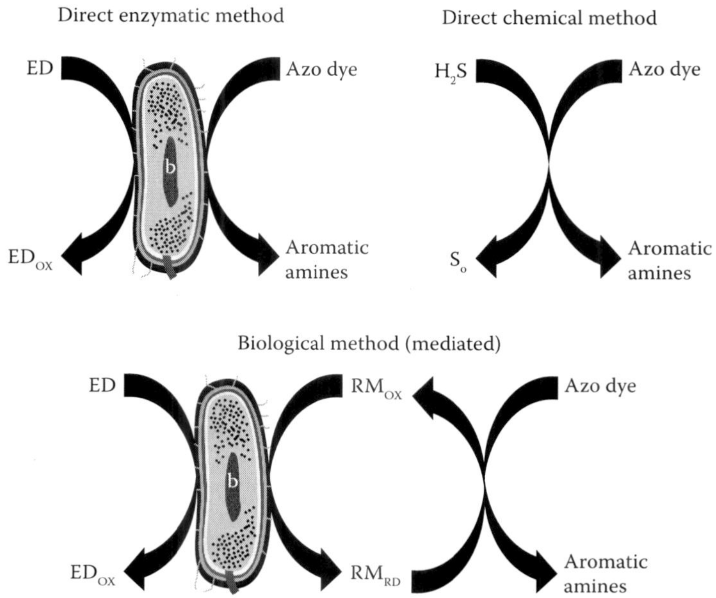


**FIGURE 5.1** Various biological methods of degradation and decolorization of dyes.

many researchers reported that microorganisms are capable of dye decolorization at the laboratory scale, but there are few reports available on their exploitation in treatment processes.

Efforts to isolate bacterial culture accomplished of degrading azo dyes began in the 1970s with reports of strain *Bacillus subtilis* (Banat et al. 1996). Bacterial isolates from sludge and soil samples from various origins containing *Bacillus* sp., *Alcaligenes* sp., and *Aeromonas* sp. were reported to have very high dye decolorization capability (Sharma et al. 2004). Cyanobacteria (*Phormidium ceylanicum* and *Gloeocapsa pleurocapsoides*) reported decolorized Acid Red 97 and FF sky Blue dye more than 80% after 26 days (Parikh and Madamwar 2005). The decolorization of azo red dyes and direct yellow by bacterial and *Actinomycetes* was observed (Moawad et al. 2003). Other reports suggested that *Pseudomonas* sp. (Kothari et al. 2005), *Escherichia coli* and sulfate-reducing bacteria (Yang et al. 2000) are also efficient dye decolorizers. The potentiality of these treatments depends mostly on the endurance and adaptability of microorganisms during the treatment process environment.

##### *5.6.2.1.2 Biodegradation of Dyes by Fungi*

White-rot basidiomycetes belong to the group of fungi known for depolymerizing lignin with their own extracellular and nonspecific ligninolytic enzymes. In 1980s, this fact stimulated research on the ability of ligninolytic fungi to degrade organic pollutants (Gao et al. 2010). It was well known that *Phanerochaete chrysosporium* is accomplished of biodegrading various kind of biopollutants, and it soon became a model white-rot fungus in which maximum research was conducted. The enzymes produced with this fungus are lignin peroxidase (LiP) and manganese peroxidase (MnP) (Faraco et al. 2009). In the next decade, a few new species of white-rot fungi like *Pleurotus ostreatus* and *Trametes versicolor* (Pazarlioglu et al. 2010) were thoroughly studied for degradation of dyes. A more intense study with *Irpex lacteus* (Novotný et al. 2011), *Bjerkandera adusta* (Eichlerová et al. 2007), and *Alternaria alternata* (Chakraborty et al. 2013) initiated in the last decade, and interest about the decolorization capability of *Ceriporiopsis subvermispora* (Tanaka et al. 2009) and *Dichomitus squalens* (Pavko and Novotný 2008) has increased in the last few years. It is well known that most of the white-rot fungi produce at least two of the three highly nonspecific enzymes like LiP, MnP, and Lac, which enable the generation of free radicals when conducting a variety of reactions (Velic et al. 2001).

Nonphenolic compounds are oxidized by lignin peroxidase and manganese peroxidase, and laccase oxidizes phenolic compounds (McMullan et al. 2001). In general, acidic conditions (pH 4–5) favor the activity of these enzymes, which is a severe limitation since textile dyes usually have high pH values (Kokol et al. 2007). However, the extracellular nature of these enzymes means that the problem of substrate diffusion into the cell is not encountered in fungi (Kaushik and Malik 2009). High-density nitrogen sources (25–60 mM) tend to impede the decolorization, while in general glucose seems the best carbon source supplement. There are numerous studies depicting that the multifarious ligninolytic enzyme system has connection to dye degradation. The white-rot fungus *Phlebia radiata* was used to decolorize 12 synthetic dyes—amaranth, chromotrope 2B and 2R, tropaeolin, metanil yellow, congo red, amido black, chlorazole black E, trypan blue, Chicago blue, procion red HE-7B, and procion red M8B—six of which were decolorized completely (Chengalroyen and Dabbs 2013). Manganese peroxidase and laccase of *Ischnoderma resinosum* reported up to 80% to 90% decolorization of four reactive dyes within 24 hours (Kokol et al. 2007). Promising studies include the use of purified enzymes to mineralize dyes. The laccase isolated from *Pycnoporus cinnabarinus* degraded 70% of Chicago sky blue within 1 hour (Kaushik and Malik 2009).

##### *5.6.2.1.3 Biodegradation of Dyes by Yeast*

The degradation of azo dyes by several yeast species has recently begun, whereas enzymatic decolorization has been reported since the 1990s. In 1992, Kakuta and co-workers used an immobilized yeast strain, *Candida curvata,* for the pretreatment of dye wastewaters (Kakuta et al. 1992). Another study conducted by Kim et al. (1995) showed that *Geotrichum candidum* decolorized different kinds of acidic and reactive dispersive dyes. Martins et al. (1999) demonstrated that liquid aerated cultures (batch mode) of the yeast *Candida zeylanoides* decolorized numerous azo dyes derived from 2-naphtholaminoazobenzene and *p*-*N*,*N*-dimethylaminoazobenzene. Many other researchers also established that cultures of *Candida tropicalis, Debaryomyces polymorphus,* and *Issatchenkia occidentalis* also decolorized several other azo dyes (Yang et al. 2003; Ramalho et al. 2004).

Similar activities have also been reported of several other ascomycetous and yeasts such as *Pseudozyma rugulosa* (Yu and Wen 2005), *Candida oleophila* (Lucas et al. 2006), *Saccharomyces cerevisiae* (Jadhav et al. 2007), *Galactomyces geotrichum* (Jadhav et al. 2008), *Candida albicans* (Vitor and Corso 2008), *Yarrowia lipolytica*

(Aracagok and Cihangir 2013), and *Issatchenkia orientalis* (Jafari et al. 2014) that are able to decolorize rather complex azo dyes such as Reactive Black 5 (Aracagok and Cihangir 2013; Jafari et al. 2014), Direct Violet 51 (Vitor and Corso 2008), Reactive Brilliant Red K-2BP (Yang et al. 2003; Yu and Wen 2005), Reactive Yellow 84, Reactive Red 141 (Martorell et al. 2012), and Reactive Blue 171 (Saratale et al. 2009). The efficiency of dye decolorization largely depends on the structure and complexity of dye to be decolorized. Relatively small structural alterations in dye such as the different positions of identical substituents, can affect the whole decolorization process (Verma and Madamwar 2005). Only limited reports include *basidiomycetous* yeasts in azo dye degradation, including *Trichosporon multisporum, Trichosporon akiyoshidainum* (Pajot et al. 2008), *Trichosporon beigelii* (Saratale et al. 2009), and *Trichosporon porosum* (Martorell et al. 2012). However, various yeast species reported capable for the degradation and decolorization of azo dyes are given in Table 5.6.

##### *5.6.2.1.4 Biodegradation of Dyes by Algae*

The degradation of azo dyes by different algae species was assessed, and it was discovered that certain algae species could only degrade a good number of azo dyes to some extent. The reduction rate seems to be connected to the molecular structure of the dyes and the species of algae used. The azo reductase enzyme of algae is accountable for degrading azo dyes into aromatic amine by breaking the azo linkage. After that, the aromatic amine is then subjected to further metabolism by different algae. Many algae species exhibited their efficiency in degradation and decolorizing of azo dyes in wastewater effluents. Some species of *Chlorella pyrenoidosa, Oscillatoria,* and *Chlorella vulgaris* degraded azo dyes and decolorized wastewater, which contains dye. Few species of algae could utilize aniline, a degradation product of azo dye breakdown. Also reported is the capability of seven microalgae, which were species dependent belonging to green and blue-green algae, to remove purine from solution either by biotransformation or bioaccumulation.

**TABLE 5.6**
**Various Yeast Species Reported Capable of Azo Dyes Degradation**

| Yeast Strain | Azo Dye | Reference |
|---|---|---|
| *Candida zeylanoides* | Several azo dyes derived from 2-naphtholaminoazobenzene and *p-N,N*-dimethylaminoazobenzene | Martins et al. 1999 |
| *Debaryomyces polymorphous* | Reactive Black 5<br>Reactive Red M_3BE | Yang et al. 2003 |
| *Candida tropicalis* | Procion Scharlach H_E3G | Yang et al. 2003 |
| *Issatchenkia occidentalis* | Methyl Orange | Ramalho et al. 2004 |
| *Pseudozyma rugulosa Y_48* | Reactive Brilliant Red K_2BP | Yu and Wen 2005 |

#### 5.6.2.2 Enzymatic Degradation of Dyes

Dye remediation can be possible by using enzymes. In the case study of enzymatic remediation of azo dyes, azo reductases and laccases enzymes seem to be the most promising enzymes. Laccases enzymes have proven to decolorize a wide spectrum of industrial dyes (Reyes et al. 1999). Low molecular weight compounds like 2,2′-azino-bis-(3-ethylbenzothiazoline-6-sulfonic acid) (ABTS) may also be essential to arbitrate the actual electron transfer steps of laccases (Wong and Yu 1999). In the presence of redox mediators, decolorization actions of degradable dyes could be significantly enhanced.

Similarly, in the case study of peroxidases, the addition of veratryl alcohol was shown to positively influence the decolorization of anthraquinone and azo dyes catalyzed by lignin peroxidase. However, this kind of effect may either be attributed to the protection of the enzyme of being inactivated by hydrogen peroxide or to the completion of the oxidation–reduction cycle of the lignin peroxidase rather than to just redox-mediation. Polyphenol oxidase (PPO) (EC 1.14.18.1) is a tetramer enzyme that comprises four atoms of copper per molecule, and binding sites for two the aromatic compounds and oxygen. The enzyme catalyzes hydroxylation of monophenols to *O*-diphenol. They can further catalyze the oxidation of *O*-diphenols to *O*-quinones (Mukherjee et al. 2013).

## 5.7 BIODEGRADATION OF DYES BY GENETICALLY ENGINEERED MICROBES

Engineered bacteria can be classified into two categories. They are special engineered bacteria and generalized engineered bacteria. Special engineered bacteria are prepared by separating certain target genes with various degrading properties. Therefore, they show various degradative effects within it through gene operations that can degrade a variety of new types. However, generalized engineered bacteria are considered as a mixed flora, which has high degradation efficiency. They can be separated, screened, and identified from the natural environment, a polluted environment, and treatment systems, and can degrade various organics efficiently (Ma et al. 2008).

Researchers have been genetically engineering bacteria to deal with environmental issues. The main challenges for the engineered strategy include the availability of tools, which can be modified by recombinant DNA technology, and the application of these tools so that the desired degradation can be achieved under industrial conditions. In this strategy, it is necessary to identify genes that are more suitable for industrial applications (Beg et al. 2001). The main challenge for recombinant DNA technology is to improve the reaction characteristics of genetically engineered organisms by introducing genes. It has been observed that the robustness of engineered bacteria is often required for industrial applications. There are several reports where genetically engineered microorganisms were used for dye removal. It has been found that genetically engineered bacteria are useful for removal of contaminant. Yujie et al. (2005) constructed genetically engineered bacteria *pGEX-AZR/E. coli JM-109* and evaluated the ability to degrade azo dyes. They reported that genetically engineered bacteria possess high efficiency in various azo dyes treatments. It also reported that degradation efficiency was higher, especially the small molecular. Gou et al. (2012) experimented that the activated sludge Fosmid library can be used to screen the functional genes. In addition, it is more useful for development of new genes, which are required for building the genetically engineered bacteria.

## 5.8 CONCLUSION

Water pollution continues to be a great concern due to increasing population and industrial expansion, especially from the developed countries, and the problem of removing pollutants from wastewater. The effluents generated from domestic and industrial activities constitute the major sources of the natural water pollution load of dye. This pollution load is a great obligation in terms of wastewater management, which not only increases treatment cost considerably but also introduces a wide range of chemical pollutants and microbial contaminants to water sources. The prevention of pollution of water sources and protection of public health by safeguarding water supplies against the spread of diseases are the two fundamental reasons for treating wastewater. Currently there are several methods, chemical, physical, and biological, that can be utilized for the effective degradation and decolorization of dyes. The chemical and physical methods tend to be complex and costly. Biological treatment, however, is becoming less costly with advances in biotechnology and can be just as effective. This chapter examines both the chemical and biological destruction processes that transform dye into less toxic compounds, and discusses the feasibility of implementing biological treatment as an alternative to the current standard.

## ACKNOWLEDGMENTS

This research is supported by the NIT Agartala, Ministry of Human Resource and Development, Government of India.

## REFERENCES

Aksu, Z., and Isoglu, I.A. 2007. Use of dried sugar beet pulp for binary biosorption of Gemazol Turquoise Blue-G reactive dye and copper(II) ions: Equilibrium modeling. *Chemical Engineering Journal* 127: 177–188.

Alaton, Ã.A., and Balcioğlu, I.A. 2002. The effect of pre-ozonation on the H2O2/UV-C treatment of raw and biologically pre-treated textile industry wastewater. *Water Science and Technology* 45(12): 297–304.

Al-Degs, Y., Khraisheh, M.A.M., Allen, S.J., and Ahmad, M.N. 2000. Effect of carbon surface chemistry on the removal of reactive dyes from textile effluent. *Water Research* 34(3): 927–935.

Al-Degs, Y., Khraisheh, M.A.M., Allen, S.J., Ahmad, M.N., and Walker, G.M. 2007. Competitive adsorption of reactive dyes from solution: Equilibrium isotherm studies in single and multisolute systems. *Chemical Engineering Journal* 128: 163–167.

Alexander, J.T., Hai, F.I., and Al-aboud, T.M. 2012. Chemical coagulation-based processes for trace organic contaminant removal: Current state and future potential. *Journal of Environmental Management* 111: 195–207.

Al-Godah, Z. 2000. Adsorption of dyes using shale oil ash. *Water Research* 34: 4295–4303.

Al-Kdasi, A., Idris, A., Saed, K., and Guan, C.T. 2004. Treatment of textile wastewater by advanced oxidation processes—A review. *Global NEST: The International Journal* 6(3): 222–230.

Allen, J.S., McKay, G., and Khader, K.Y.H. 1988. Multi-component sorption isotherms of basic dyes onto peat. *Environmental Pollution* 52: 39–53.

Allen, J.S., Mckay, G., and Porter, J.F. 2004. Adsorption isotherm models for basic dye adsorption by peat in single and binary component systems. *Journal of Colloid and Interface Science* 280: 322–333.

Alley, E.R. 2007. *Water quality control handbook*. McGraw-Hill, New York.

Al-Rashed, S.M., and Al-Gaid, A.A. 2012. Kinetic and thermodynamic studies on the adsorption behavior of Rhodamine B dye on Duolite C-20 resin. *Journal of Saudi Chemical Society* 16(2): 209–215.

Am Water Works Res, F., Langlais, B., Reckhow, D.A., and Brink, D.R. 1991. *Ozone in water treatment: Application and engineering*. CRC press, Boca Raton, FL.

Anjaneyulu, Y., Chary, N.S., and Raj, D.S.S. 2005. Decolourization of industrial effluents—available methods and emerging technologies—A review. *Reviews in Environmental Science and Bio/Technology* 4(4): 245–273.

Aplin, R., Feitz, A.J., and Waite, T.D. 2000. Effect of Fe (III)-ligand properties on effectiveness of modified photo-Fenton processes. *Water Science and Technology: A Journal of the International Association on Water Pollution Research* 44(5): 23–30.

Aracagok, Y.D., and Cihangir, N. 2013. Decolorization of reactive black 5 by Yarrowia lipolytica NBRC 1658. *American Journal of Microbiological Research* 1(2): 16–20.

Arslan, I., and Balcioğlu, I.A. 1999. Degradation of commercial reactive dyestuffs by heterogenous and homogenous advanced oxidation processes: A comparative study. *Dyes and Pigments* 43(2): 95–108.

Arslan, I., and Balcioglu, I.A. 2001. Advanced oxidation of raw and biotreated textile industry wastewater with O3, H2O2/UV-C and their sequential application. *Journal of Chemical Technology and Biotechnology* 76(1): 53–60.

Arslan, I., Balcioğlu, I.A., and Tuhkanen, T. 1999a. Oxidative treatment of simulated dyehouse effluent by UV and near-UV light assisted Fenton's reagent. *Chemosphere* 39(15): 2767–2783.

Arslan, I., Balcioglu, I.A., and Tuhkanen, T. 1999. Advanced oxidation of synthetic dyehouse effluent by O3, H2O2/O3 and H2O2/UV processes. *Environmental Technology* 20(9): 921–931.

Ayed, L., Achour, S., and Bakhrouf, A. 2011. Application of the mixture design to decolourise effluent textile wastewater using continuous stirred bed reactor. *Water SA* 37(1): 21–26.

Azbar, N., Yonar, T., and Kestioglu, K. 2004. Comparison of various advanced oxidation processes and chemical treatment methods for COD and color removal from a polyester and acetate fiber dyeing effluent. *Chemosphere* 55(1): 35–43.

Baban, A., Yediler, A., Lienert, D., Kemerdere, N., and Kettrup, A. 2003. Ozonation of high strength segregated effluents from a woollen textile dyeing and finishing plant. *Dyes and Pigments* 58(2): 93–98.

Bai, R., Deng, J., and Yang, R.T. 2003. Improved multisite langmuir model for mixture adsorption using multiregion adsorption theory. *Langmuir* 19: 2776–2781.

Baig, S., and Liechti, P.A. 2001. Ozone treatment for biorefractory COD removal. *Water Science and Technology* 43(2): 197–204.

Banat, I.M., Nigam, P., Singh, D., and Marchant, R. 1996. Microbial decolorization of textile-dyecontaining effluents: A review. *Bioresource Technology* 58(3): 217–227.

Basava Rao, V.V., and Mohan Rao, S.R. 2006. Adsorption studies on treatment of textile dyeing industrial effluent by flyash. *Chemical Engineering Journal* 116: 77–84.

Basu, S., and Malpani, P.R. 2001. Removal of methyl orange and methylene blue dye from water using colloidal gas aphron—Effect of processes parameters. *Separation Science and Technology* 36(13): 2997–3013.

Baup, S., Jaffre, C., Wolbert, D., and Laplanche, A. 2000. Adsorption of pesticides onto granular activated carbon: Determination of surface diffusivities using simple batch experiments. *Adsorption* 6: 219–228.

Bautista, P., Mohedano, A.F., Casas, J.A., Zazo, J.A., and Rodriguez, J.J. 2008. An overview of the application of Fenton oxidation to industrial wastewaters treatment. *Journal of Chemical Technology and Biotechnology* 83(10): 1323–1338.

Beg, Q.K., Kapoor, M., Mahajan, L., and Hoondal, G.S. 2001. Microbial xylanases and their industrial applications: A review. *Applied Microbiology and Biotechnology* 56(3–4): 326–338.

Behnajady, M.A., Modirshahla, N., and Hamzavi, R. 2006. Kinetic study on photocatalytic degradation of CI Acid Yellow 23 by ZnO photocatalyst. *Journal of Hazardous Materials* 133(1): 226–232.

Beltrán-Heredia, J., and Martin, J.S. 2008. Azo dye removal by Moringa oleifera seed extract coagulation. *Coloration Technology* 124(5): 310–317.

Beltran-Heredia, J., Martin, J.S., and Delgado-Regalado, A. 2009. Removal of carmine indigo dye with *Moringa oleifera* seed extract. *Industrial & Engineering Chemistry Research* 48(14): 6512–6520.

Bhatia, S., Othman, Z., and Ahmad, A.L. 2007. Coagulation-flocculation process for POME treatment using *Moringa oleifera* seeds extract: Optimization studies. *Chemical Engineering Journal* 133(1): 205–212.

Bhatnagar, A., and Jain, A.K. 2005. A comparative adsorption study with different industrial wastes as adsorbents for the removal of cationic dyes from water. *Journal of Colloid and Interface Science* 281: 49–55.

Bhavani, K.D., and Dutta, P.K. 1999. Physico-chemical adsorption properties on chitosan for dyehouse effluent. *American Dyestuff Reporter*, April, pp. 53–58.

Bratby, J. 2006. *Coagulation and flocculation in water and wastewater treatment*. IWA Publishing, UK.

Brunauer, S., Emmett, P.H., and Teller, E. 1938. Adsorption of gases in multi-molecular layers. *Journal of American Chemical Society* 60: 309–316.

Butler, J.A.V., and Ockrent, C. 1930. Studies in electrocapillarity. III. *Journal of Physical Chemistry* 34: 2841–2859.

Carey, J.H. 1992. Introduction to advanced oxidation processes(AOP) for destruction of organics in wastewater. *Water Quality Research Journal of Canada* 27(1): 1–21.

Carre, B., Magnin, L., Galland, G., and Vernac, Y. 2000. Deinking difficulties related to ink formulation, printing process, and type of paper. *Tappi Journal* 83(6): 60.

Casieri, L., Varese, G., Anastasi, A., Prigione, V., Svobodova, K., Marchisio, V.F., and Novotný, Č. 2008. Decolorization and detoxication of reactive industrial dyes by immobilized fungi *Trametes pubescens* and *Pleurotus ostreatus*. *Folia Microbiologica* 53(1): 44–52.

Catorceno, L.L.C., Nogueira, K.R.B., and Teixeira, A. 2010. Treatment of aqueous effluents containing phenol by the O3, O3-UV, and O3-H2O2 processes: Experimental study and neural network modeling. *Separation Science and Technology* 45(11): 1521–1528.

Čermáková, J.Ř., Marković, A., Uchytil, P., and Seidel-Morgenstern, A. 2008. Single component and competitive adsorption of propane, carbon dioxide and butane on Vycor glass. *Chemical Engineering Science* 63: 1586–1601.

Chakraborty, S., Basak, B., Dutta, S., Bhunia, B., and Dey, A. 2013. Decolorization and biodegradation of congo red dye by a novel white rot fungus Alternaria alternata CMERI F6. *Bioresource Technology* 147: 662–666.

Chakraborty, S., Chowdhury, S., and Saha, P.D. 2011. Adsorption of crystal violet from aqueous solution onto NaOH-modified rice husk. *Carbohydrate Polymers* 86(4): 1533–1541.

Chakraborty, S., Chowdhury, S., and Saha, P.D. 2012. Biosorption of hazardous textile dyes from aqueous solutions by hen feathers: Batch and column studies. *Korean Journal of Chemical Engineering* 29(11): 1567–1576.

Chan, L.S., Cheung, W.H., Allen, S.J., and McKay, G. 2009. Separation of acid-dyes mixture by bamboo derived active carbon. *Separation and Purification Technology* 67: 166–172.

Chan, O.S., Cheung, W.H., and McKay, G. 2012. Single and multicomponent acid dye adsorption equilibrium studies on tyre demineralised activated carbon. *Chemical Engineering Journal* 191: 162–170.

Chan, S.H.S., Wu, T.Y., Juan, J.C., and Teh, C.Y. 2011. Recent developments of metal oxide semiconductors as photocatalysts in advanced oxidation processes (AOPs) for treatment of dye waste-water. *Journal of Chemical Technology and Biotechnology* 86(9): 1130–1158.

Chen, J., Ollis, D.F., Rulkens, W.H., and Bruning, H. 1999. Photocatalyzed oxidation of alcohols and organochlorides in the presence of native TiO 2 and metallized TiO 2 suspensions. Part (I): Photocatalytic activity and pH influence. *Water Research* 33(3): 661–668.

Cheng, M.-H., and Huang, S.-D. 1988. Solvent sublation and adsorbing colloid flotation of direct red. *Journal of Colloid and Interface Science* 126(1): 346–354.

Chengalroyen, M., and Dabbs, E. 2013. The microbial degradation of azo dyes: Minireview. *World Journal of Microbiology and Biotechnology* 29(3): 389–399.

Chequer, F.M.D., Angeli, J.P.F., Ferraz, E.R.A., Tsuboy, M.S., Marcarini, J.C., Mantovani, M.S., and de Oliveira, D.P. 2009. The azo dyes Disperse Red 1 and Disperse Orange 1 increase the micronuclei frequencies in human lymphocytes and in HepG2 cells. *Mutation Research/Genetic Toxicology and Environmental Mutagenesis* 676(1): 83–86.

Chiou, M., and Chuang, G. 2006. Competitive adsorption of dye metanil yellow and RB15 in acid solutions on chemically cross-linked chitosan beads. *Chemosphere* 62: 731–740.

Chowdhury, S., Chakraborty, S., and Das, P. 2013. Adsorption of crystal violet from aqueous solution by citric acid modified rice straw: Equilibrium, kinetics, and thermodynamics. *Separation Science and Technology* 48(9): 1339–1348.

Cisneros, R.L., Espinoza, A.G., and Litter, M.I. 2002. Photodegradation of an azo dye of the textile industry. *Chemosphere* 48(4): 393–399.

Clarke, N., and Knowles, G. 1982. High purity water using H2O2 and UV radiation. *Effluent and Water Treatment Journal* 23: 335–341.

Clarke, A.N., and Wilson, D.J. 1983. Foam flotation. In *Theory and application*. Dekker, M., (ed.), NY.

Crini, G. 2006. Non-conventional low-cost adsorbents for dye removal: A review. *Bioresource Technology* 97(9): 1061–1085.

Crini, G., Badot, P., Morin-Crini, N., and Torri, G. 2007. *Wastewater treatment processes: A recent review of the available methods*. Press Universitaires de Franche-Comte (PUFC), pp. 16–62.

Crittenden, J.C., Hu, S., Hand, D.W., and Green, S.A. 1999. A kinetic model for H2O2/UV process in a completely mixed batch reactor. *Water Research* 33(10): 2315–2328.

Crittenden, J.C., Trussel, R.R., Hand, D.W., Howe, K.J., and Tchobanoglous, G. 2005. *Coagulation, mixing and flocculation. Water Treatment: Principles and Design*, 2nd edition, Wiley, Hoboken, NJ 664–691.

Dafnopatidou, E.K., Gallios, G.P., Tsatsaroni, E.G., and Lazaridis, N.K. 2007. Reactive dyestuffs removal from aqueous solutions by flotation, possibility of water reuse, and dyestuff degradation. *Industrial & Engineering Chemistry Research* 46(7): 2125–2132.

Dafnopatidou, E.K., and Lazaridis, N.K. 2008. Dyes removal from simulated and industrial textile effluents by dissolved-air and dispersed-air flotation techniques. *Industrial & Engineering Chemistry Research* 47(15): 5594–5601.

Daneshvar, N., Salari, D., and Khataee, A.R. 2004. Photocatalytic degradation of azo dye acid red 14 in water on ZnO as an alternative catalyst to TiO2. *Journal of Photochemistry and Photobiology A: Chemistry* 162(2): 317–322.

Dávila-Jiménez, M.M., Elizalde-González, M.P., and Hernández-Montoya, V. 2009. Performance of mango seed adsorbents in the adsorption of anthraquinone and azo acid dyes in single and binary aqueous solutions. *Bioresource Technology* 100: 6199–6206.

Dávila-Jiménez, M.M., Elizalde-González, M.P., and Peláez-Cid, A.A. 2005. Adsorption interaction between natural adsorbents and textile dyes in aqueous solution. *Colloids and Surfaces A: Physicochemical and Engineering Aspects* 254: 107–114.

de Aragão Umbuzeiro, G., Freeman, H., Warren, S.H., Kummrow, F., and Claxton, L.D. 2005. Mutagenicity evaluation of the commercial product CI Disperse Blue 291 using different protocols of the Salmonella assay. *Food and Chemical Toxicology* 43(1): 49–56.

Delval, F., Crini, G., Morin, N., Vebrel, J., Bertini, S., and Torri, G. 2002. The sorption of several types of dye on crosslinked polysaccharides derivatives. *Dyes and Pigments* 53: 79–92.

DomÃ¨nech, X., Jardim, W.F., and Litter, M.I. 2001. Procesos avanzados de oxidación para la eliminación de contaminantes. *Eliminiación de Contaminantes por FotocatÃ¡lisis HeterogÃ©nea, cap*, 1.

Dorfman, L.M., and Adams, G.E. 1973. Reactivity of the hydroxyl radical in aqueous solutions. (No. NSRDS-NBS-46). National Standard Reference Data System, USA.

dos Santos, A.B., Cervantes, F.J., and van Lier, J.B. 2007. Review paper on current technologies for decolourisation of textile wastewaters: Perspectives for anaerobic biotechnology. *Bioresource Technology* 98(12): 2369–2385.

Duan, J., and Gregory, J. 2003. Coagulation by hydrolysing metal salts. *Advances in Colloid and Interface Science* 100: 475–502.

Eichlerová, I., Homolka, L., Benada, O., Kofroňová, O., Hubálek, T., and Nerud, F. 2007. Decolorization of Orange G and Remazol Brilliant Blue R by the white rot fungus Dichomitus squalens: Toxicological evaluation and morphological study. *Chemosphere* 69(5): 795–802.

Fang, R., Cheng, X., and Xu, X. 2010. Synthesis of lignin-base cationic flocculant and its application in removing anionic azo-dyes from simulated wastewater. *Bioresource Technology* 101(19): 7323–7329.

Faraco, V., Pezzella, C., Miele, A., Giardina, P., and Sannia, G. 2009. Bio-remediation of colored industrial wastewaters by the white-rot fungi *Phanerochaete chrysosporium* and *Pleurotus ostreatus* and their enzymes. *Biodegradation* 20(2): 209–220.

Forgacs, E., Cserhati, T., and Oros, G. 2004. Removal of synthetic dyes from wastewaters: A review. *Environment International* 30(7): 953–971.

Galindo, C., and Kalt, A. 1999. UV/H2O2 oxidation of monoazo dyes in aqueous media: A kinetic study. *Dyes and Pigments* 40(1): 27–35.

Gallard, H., and De Laat, J. 2000. Kinetic modelling of Fe (III)/H2O2 oxidation reactions in dilute aqueous solution using atrazine as a model organic compound. *Water Research* 34(12): 3107–3116.

Gao, D., Du, L., Yang, J., Wu, W.-M., and Liang, H. 2010. A critical review of the application of white rot fungus to environmental pollution control. *Critical Reviews in Biotechnology* 30(1): 70–77.

Garg, A., Bhat, K.L., and Bock, C.W. 2002. Mutagenicity of aminoazobenzene dyes and related structures: A QSAR/QPAR investigation. *Dyes and Pigments* 55(1): 35–52.

Georgiou, D., Melidis, P., Aivasidis, A., and Gimouhopoulos, K. 2002. Degradation of azo-reactive dyes by ultraviolet radiation in the presence of hydrogen peroxide. *Dyes and Pigments* 52(2): 69–78.

Gerundo, N., Alderman, D., Clifton-Hadley, R., and Feist, S. 1991. Pathological effects of repeated doses of malachite green: A preliminary study. *Journal of Fish Diseases* 14(5): 521–532.

Gimbert, F., Morin-Crini, N., Renault, F., Badot, P.-M., and Crini, G. 2008. Adsorption isotherm models for dye removal by cationized starch-based material in a single component system: Error analysis. *Journal of Hazardous Materials* 157(1): 34–46.

Glaze, W.H., and Kang, J.W. 1989. Advanced oxidation processes. Test of a kinetic model for the oxidation of organic compounds with ozone and hydrogen peroxide in a semibatch reactor. *Industrial & engineering chemistry research* 28(11): 1580–1587.

Gogate, P.R., and Pandit, A.B. 2004. A review of imperative technologies for wastewater treatment I: Oxidation technologies at ambient conditions. *Advances in Environmental Research* 8(3): 501–551.

Gou, M., Qu, Y.-Y., Zhou, J.-T., Xu, B.-W., and Cao, X.-Y. 2012. Construction of metagenomic fosmid library from activated sludge. *Journal of South China University of Technology* 40(1): 120–124.

Gregory, J. 2005. *Particles in water: Properties and processes*. CRC Press, Boca Raton, FL.

Gregory, J., and Ives, K.J. 1978. *The scientific basis of flocculation*. Sijthoff and Noorhoff International Publishers, BV, Alpen van der Rijn, 101.

Gupta, V. 2009. Application of low-cost adsorbents for dye removal—A review. *Journal of Environmental Management* 90(8): 2313–2342.

Gupta, V.K., Mittal, A., Krishnan, L., and Mittal, J. 2006. Adsorption treatment and recovery of the hazardous dye, Brilliant Blue FCF, over bottom ash and de-oiled soya. *Journal of Colloid and Interface Science* 293: 16–26.

Gutierrez, M., and Fuentes, H.R. 1993. Modeling adsorption in multicomponent systems using a Freundlich-type isotherm. *Journal of Contaminant Hydrology* 14: 247–260.

Hansen, C.A., Gotaas, H.B., and Eliassen, R. 1943. Sewage treatment by flotation [with Discussion]. *Sewage Works Journal* 15: 242–254.

Hao, O.J., Kim, H., and Chiang, P.-C. 2000. Decolorization of wastewater. *Critical Reviews in Environmental Science and Technology* 30(4): 449–505.

Herney-Ramirez, J., Vicente, M.A., and Madeira, L.M. 2010. Heterogeneous photo-Fenton oxidation with pillared clay-based catalysts for wastewater treatment: A review. *Applied Catalysis B: Environmental* 98(1): 10–26.

Hines, P.R., and Vincent, J.D. 1962. The early days of froth flotation. In *Froth flotation,* Fuerstenau, D. W., (ed.), 50th Anniversary Volume. AIME, New York, pp. 11–38.

Hunger, K. 1994. On the toxicology and metabolism of azo dyes. *CHIMIA International Journal for Chemistry* 48(11): 520–522.

Ince, N.H. 1999. Critical effect of hydrogen peroxide in photochemical dye degradation. *Water Research*, 33(4): 1080–1084.

Indra, D.M., Vimal, C.S., Nitin, K.A., and Indra, M.M. 2005. Adsorptive removal of malachite green dye from aqueous solution by bagasse fly ash and activated carbon-kinetic study and equilibrium isotherm analyses. *Colloids and Surfaces A: Physicochemical Engineering Aspects* 264: 17–28.

Inomata, N., Osuna, H., Fujita, H., Ogawa, T., and Ikezawa, Z. 2006. Multiple chemical sensitivities following intolerance to azo dye in sweets in a 5-year-old girl. *Allergology International* 55(2): 203–205.

International Agency for Research on Cancer. 1982. IARC Monographs on the evaluation of the carcinogenic risk of chemicals to humans, chemicals, industrial processes and industries associated with cancer in humans. *IARC*, Lyon, France.

Jadhav, J., Parshetti, G., Kalme, S., and Govindwar, S. 2007. Decolourization of azo dye methyl red by Saccharomyces cerevisiae MTCC 463. *Chemosphere* 68(2): 394–400.

Jadhav, S., Kalme, S., and Govindwar, S. 2008. Biodegradation of methyl red by *Galactomyces geotrichum* MTCC 1360. *International Biodeterioration & Biodegradation* 62(2): 135–142.

Jafari, N., Soudi, M.R., and Kasra-Kermanshahi, R. 2014. Biodecolorization of textile azo dyes by isolated yeast from activated sludge: Issatchenkia orientalis JKS6. *Annals of Microbiology* 64(2): 475–482.

Jain, J.S., and Snoeyink, V.L. 1973. Adsorption from bisolute systems on active carbon. *Journal of Water Pollution Control Federation* 45: 2463–2479.

Janaki, V., Vijayaraghavan, K., Ramasamy, A.K., Lee, K., Oh, B.T., and Kamala-Kannan, S. 2012. Competitive adsorption of Reactive Orange 16 and Reactive Brilliant Blue R on polyaniline/bacterial extracellular polysaccharides composite—A novel eco-friendly polymer. *Journal of Hazardous Materials* 241–242: 110–117. doi: http://dx.doi.org/10.1016/j.jhazmat.2012.09.019.

Julide, Y., Turkan, K., Gulsen, D., and Timur, D. 2006. Adsorption of Basic Yellow 28 from aqueous solutions with clinoptilolite and amberlite. *Journal of Colloid and Interface Science* 294: 255–264.

Kabil, M.A., and Ghazy, S.E. 1994. Separation of some dyes from aqueous solutions by flotation. *Separation Science and Technology* 29(18): 2533–2539.

Kakuta, T., Tateno, Y., Koizumi, T., Kodama, K., Yoshizawa, K., and Nojiro, K. 1992. Azo dye wastewater treatment with immobilized yeast. *Journal of Fermentation and Bioengineering* 74(4): 254.

Kang, S.-F., Liao, C.-H., and Hung, H.-P. 1999. Peroxidation treatment of dye manufacturing wastewater in the presence of ultraviolet light and ferrous ions. *Journal of Hazardous Materials* 65(3): 317–333.

Kang, Y.W., and Hwang, K.-Y. 2000. Effects of reaction conditions on the oxidation efficiency in the Fenton process. *Water Research* 34(10): 2786–2790.

Kanmani, S., and Muthuselvi, S. 2005. Soil remediation by Fenton/photo Fenton oxidation processes. *Journal of the Institute of Public Health Engineers* 2: 5–9.

Kasprzyk-Hordern, B., Ziółek, M., and Nawrocki, J. 2003. Catalytic ozonation and methods of enhancing molecular ozone reactions in water treatment. *Applied Catalysis B: Environmental* 46(4): 639–669.

Kaushik, P., and Malik, A. 2009. Fungal dye decolourization: Recent advances and future potential. *Environment International* 35(1): 127–141.

Khan, A.R., AL-Badri, T.A., and AL-Haddad, A. 1997. Adsorption of phenol based organic pollutants on activated carbon from multi-component dilute aqueous solutions. *Water Research* 31: 2102–2112.

Kim, S.J., Ishikawa, K., Hirai, M., and Shoda, M. 1995. Characteristics of a newly isolated fungus, Geotrichum candidum Dec 1, which decolorizes various dyes. *Journal of Fermentation and Bioengineering* 79(6): 601–607.

Kitchener, J.A. 1972. Principles of action of polymeric flocculants. *British Polymer Journal* 4(3): 217–229.

Kiwi, J., Lopez, A., and Nadtochenko, V. 2000. Mechanism and kinetics of the OH-radical intervention during Fenton oxidation in the presence of a significant amount of radical scavenger (Cl-). *Environmental Science & Technology* 34(11): 2162–2168.

Kiwi, J., Pulgarin, C., Peringer, P., and Grätzel, M. 1993. Beneficial effects of homogeneous photo-Fenton pretreatment upon the biodegradation of anthraquinone sulfonate in waste water treatment. *Applied Catalysis B: Environmental* 3(1): 85–99.

Kokol, V., Doliška, A., Eichlerová, I., Baldrian, P., and Nerud, F. 2007. Decolorization of textile dyes by whole cultures of *Ischnoderma resinosum* and by purified laccase and Mn-peroxidase. *Enzyme and Microbial Technology* 40(7): 1673–1677.

Konsowa, A.H. 2003. Decolorization of wastewater containing direct dye by ozonation in a batch bubble column reactor. *Desalination* 158(1): 233–240.

Kothari, S., Vyas, R., Ameta, R., and Punjabi, P.B. 2005. Photoreduction of congo red by ascorbic acid and EDTA over cadmium sulphide as photocatalyst. *Indian Journal of Chemistry Section A* 44(11): 2266.

Koumanova, B., Peeva, P., Allen, S.J., Galagher, K.A., and Healy, M.G. 2002. Biosorption from aqueous solutions by eggshell membranes and *Rhizopus oryzae*: Equilibrium and kinetic studies. *Journal of Chemical Technology and Biotechnology* 77: 539–545.

Krishna, R. 2001. Diffusion of binary mixtures across zeolite membranes: Entropy effects on permeation selectivity. *International Communications in Heat and Mass Transfer* 28: 337–346.

Kunz, A., Peralta-Zamora, P., de Moraes, S.G., and Durán, N. 2002. Novas tendências no tratamento de efluentes têxteis. *Química Nova* 25(1): 78–82.

Kurbus, T., Le Marechal, A.M., and VoncÌ†ina, D.B. 2003. Comparison of H2O2/UV, H2O2/O3 and H2O2/Fe2+ processes for the decolorisation of vinylsulphone reactive dyes. *Dyes and Pigments* 58(3): 245–252.

Kurniawan, A., Sutiono, H., Indraswati, N., and Ismadji, S. 2012. Removal of basic dyes in binary system by adsorption using rarasaponin–bentonite: Revisited of extended Langmuir model. *Chemical Engineering Journal* 189–190: 264–274.

Kusic, H., Koprivanac, N., and Srsan, L. 2006. Azo dye degradation using Fenton type processes assisted by UV irradiation: A kinetic study. *Journal of Photochemistry and Photobiology A: Chemistry* 181(2): 195–202.

Lazar, T. 2005. Industrial dyes—Chemistry, properties, applications. Wiley Online Library, USA.

Legrini, O., Oliveros, E., and Braun, A.M. 1993. Photochemical processes for water treatment. *Chemical Reviews* 93(2): 671–698.

Li, W., Zhou, Q., and Hua, T. 2010. Removal of organic matter from landfill leachate by advanced oxidation processes: A review. *International Journal of Chemical Engineering* 2010: 1–10.

Lin, S.H., and Lin, C.M. 1993. Treatment of textile waste effluents by ozonation and chemical coagulation. *Water Research* 27(12): 1743–1748.

Lin, S.H., and Lo, C.C. 1997. Fenton process for treatment of desizing wastewater. *Water Research* 31(8): 2050–2056.

Liu, B.-J., Torimoto, T., and Yoneyama, H. 1998. Photocatalytic reduction of CO 2 using surface-modified CdS photocatalysts in organic solvents. *Journal of Photochemistry and Photobiology A: Chemistry* 113(1): 93–97.

Liversidge, R.M., Lloyd, G.J., Wase, D.A.J., and Forster, C.F. 1997. Removal of Basic Blue 41 from aqueous solution by linseed cake. *Process Biochemistry* 32: 473–477.

Loures, C.C.A., Alcântara, M.A.K., Izário Filho, H.J., Teixeira, A., Silva, F.T., Paiva, T.C.B., and Samanamud, G.R.L. 2013. Advanced oxidative degradation processes: Fundamentals and applications. *International Review Chemical Engineering* 5(2): 102–120.

Lovell, H.L. 1973. An appraisal of neutralization processes to treat coal mine drainage. *National Service Center for Environmental Publications (NSCEP)*, USA.

Lucas, M.S., Amaral, C., Sampaio, A., Peres, J.A., and Dias, A.A. 2006. Biodegradation of the diazo dye Reactive Black 5 by a wild isolate of *Candida oleophila*. *Enzyme and Microbial Technology* 39(1): 51–55.

Ma, F., Guo, J., Zhao, L., and Shan, D. 2008. The construction and application of engineering bacteria for bioaugmented treatment of petrochemical wastewater. *Acta Scientiae Circumstantiae* 28: 885–891.

Mahamuni, N.N., and Adewuyi, Y.G. 2010. Advanced oxidation processes (AOPs) involving ultrasound for waste water treatment: A review with emphasis on cost estimation. *Ultrasonics Sonochemistry* 17(6): 990–1003.

Maier, J., Kandelbauer, A., Erlacher, A., Cavaco-Paulo, A., and Gübitz, G.M. 2004. A new alkali-thermostable azoreductase from *Bacillus* sp. strain SF. *Applied and Environmental Microbiology* 70(2): 837–844.

Malato, S., Blanco, J., Vidal, A., and Richter, C. 2002. Photocatalysis with solar energy at a pilot-plant scale: An overview. *Applied Catalysis B: Environmental* 37(1): 1–15.

Marmagne, O., and Coste, C. 1996. Color removal from textile plant effluents. *American Dyestuff Reporter* 85(4): 6.

Martins, M., Cardoso, M., Queiroz, M., Ramalho, M., and Campus, A. 1999. Biodegradation of azo dyes by the yeast *Candida zeylanoides* in batch aerated cultures. *Chemosphere* 38(11): 2455–2460.

Martorell, M.M., Pajot, H.F., and de Figueroa, L.I. 2012. Dye-decolourizing yeasts isolated from Las Yungas rainforest. Dye assimilation and removal used as selection criteria. *International Biodeterioration & Biodegradation* 66(1): 25–32.

Matilainen, A., and Sillanpää, M. 2010. Removal of natural organic matter from drinking water by advanced oxidation processes. *Chemosphere* 80(4): 351–365.

Mavros, P., Daniilidou, A.C., Lazaridis, N.K., and Stergiou, L. 1994. Colour removal from aqueous solutions. Part I. Flotation. *Environmental Technology* 15(7): 601–616.

McCormick, J., Berger, H.F., and Coates, J. 1964. *Color removal and BOD reduction in kraft effluents by foam separation*. Technical Bulletin 177. National Council for Stream Improvement, USA.

Mckay, G., and AL Duri, B. 1989. Prediction of multicomponent adsorption equilibrium data using empirical correlations. *Chemical Engineering Journal* 41: 9–23.

Mckay, G., Blair, H.S., and Gardner, J.R. 1984. The adsorption of dyes onto chitin in fixed bed columns and batch adsorbers. *Journal of Applied Polymer Science* 29: 1499–1514.

McKay, G., Choy, K.K.H., and Porter, J.F. 2000. Langmuir isotherm models applied to the multicomponent sorption of acid dyes from effluent onto activated carbon. *Journal of Chemical and Engineering Data* 45: 575–584.

Mckay, G., Choy, K.K.H., and Porter, J.F. 2004a. Intraparticle diffusion in single and multicomponent acid dye adsorption from wastewater onto carbon. *Chemical Engineering Journal* 103: 133–145.

McKay, G., Choy, K.K.H., and Porter, J.F. 2004b. Single and equilibrium studies for the adsorption of acidic dyes on carbon from effluents. *Langmuir* 20: 9646–9656.

McMullan, G., Meehan, C., Conneely, A., Kirby, N., Robinson, T., Nigam, P., Banat, I., Marchant, R., and Smyth, W. 2001. Microbial decolourisation and degradation of textile dyes. *Applied Microbiology and Biotechnology* 56(1–2): 81–87.

Merzouk, B., Gourich, B., Madani, K., Vial, C., and Sekki, A. 2011. Removal of a disperse red dye from synthetic wastewater by chemical coagulation and continuous electrocoagulation. A comparative study. *Desalination* 272(1): 246–253.

Mittal, A., Mittal, J., Kurup, L., and Singh, A. 2006. Process development for the removal and recovery of hazardous dye erythrosine from wastewater by waste materials—Bottom ash and de-oiled soya as adsorbents. *Journal of Hazardous Materials* 138(1): 95–105.

Moawad, H., El–Rahim, W.M.A., and Khalafallah, M. 2003. Evaluation of biotoxicity of textile dyes using two bioassays. *Journal of Basic Microbiology* 43(3): 218–229.

Moghaddam, S.S., Moghaddam, M.R.A., and Arami, M. 2010. Coagulation/flocculation process for dye removal using sludge from water treatment plant: Optimization through response surface methodology. *Journal of Hazardous Materials* 175(1): 651–657.

Mohan, S.V., Rao, N.C., and Kartikeyan, J. 2002. Adsorptive removal of direct azo dye from aqueous phase onto coal based sorbents: A kinetic and mechanistic study. *Journal of Hazardous Materials* B90: 189–204.

Molina, C.B., Casas, J.A., Zazo, J.A., and Rodriguez, J.J. 2006. A comparison of Al-Fe and Zr-Fe pillared clays for catalytic wet peroxide oxidation. *Chemical Engineering Journal* 118(1): 29–35.

Morais, J.L.d. 2005. *Study of the potential of advanced, isolated and integrated oxidative processes with traditional biological processes for the treatment of slurry*. PhD Thesis: Federal University of Paraná, Curitiba, Brazil.

Morais, L.C., Freitas, O.M., Goncalves, E.P., Vasconcelos, L.T., and Gonzalez, C.G. 1999. Reactive dyes removal from wastewater by adsorption on eucalyptus bark: Variables that define the process. *Water Research* 33: 979–988.

Munoz, I., Rieradevall, J., Torrades, F., Peral, J., and Domènech, X. 2005. Environmental assessment of different solar driven advanced oxidation processes. *Solar Energy* 79(4): 369–375.

Mukherjee, S., Basak, B., Bhunia, B., Dey, A., and Mondal, B. 2013. Potential use of polyphenol oxidases (PPO) in the bioremediation of phenolic contaminants containing industrial wastewater. *Reviews in Environmental Science and Bio/Technology* 12(1): 61–73.

Muruganandham, M., Amutha, R., Lee, G.-J., Hsieh, S.-H., Wu, J.J., and Sillanpää, M. 2012. Facile fabrication of tunable Bi2O3 self-assembly and its visible light photocatalytic activity. *The Journal of Physical Chemistry C* 116(23): 12906–12915.

Muruganandham, M., Amutha, R., Repo, E., Sillanpää, M., Kusumoto, Y., and Abdulla-Al-Mamun, M.D. 2010. Controlled mesoporous self-assembly of ZnS microsphere for photocatalytic degradation of Methyl Orange dye. *Journal of Photochemistry and Photobiology A: Chemistry* 216(2): 133–141.

Muruganandham, M., Chen, S.H., and Wu, J.J. 2007. Mineralization of N-methyl-2-pyrolidone by advanced oxidation processes. *Separation and Purification Technology* 55(3): 360–367.

Myers, A.L., and Prausnitz, J.M. 1965. Thermodynamics of mixed-gas adsorption. *AIChE Journal* 11: 121–127.

Namasivayam, C., Yamuna, R.T., and Arasi, D.J.S.E. 2001. Removal of acid violet from wastewater by adsorption on waste red mud. *Environmental Geology* 41: 269–273.

Namboodri, C.G., Perkins, W.S., and Walsh, W.K. 1994. Decolorizing dyes with chlorine and ozone: Part II. *American Dyestuff Reporter* 83(4): 17–27.

Naouab, M., Guthier, R., Guthier, H., and Rammah, M. 2001. Cationized sawdust as ion exchanger for anionic residual dyes. *Journal of Applied Polymer Science* 82: 31–37.

Neamtu, M., Siminiceanu, I., Yediler, A., and Kettrup, A. 2002. Kinetics of decolorization and mineralization of reactive azo dyes in aqueous solution by the UV/H2O2 oxidation. *Dyes and Pigments* 53(2): 93–99.

Neyens, E., and Baeyens, J. 2003. A review of classic Fenton's peroxidation as an advanced oxidation technique. *Journal of Hazardous Materials* 98(1): 33–50.

Nguyen, C., and Do, D.D. 2001. Multicomponent supercritical adsorption in microporous activated carbon materials. *Langmuir* 17: 1552–1557.

Nharingo, T., Zivurawa, M.T., and Guyo, U. 2015. Exploring the use of cactus Opuntia ficus indica in the biocoagulation—Flocculation of Pb (II) ions from wastewaters. *International Journal of Environmental Science and Technology* 12(12): 3791–3802.

Nogueira, R.F.P., and Jardim, W.F. 1998. A fotocatãlise heterogÃªnea e sua aplicaÃ§Ã£o ambiental. *QuÃ-mica nova* 21(1): 69–72.

Noroozi, B., Sorial, G.A., Bahrami, H., and Arami, M. 2007. Equilibrium and kinetic adsorption study of a cationic dye by a natural adsorbent-silkworm pupa. *Journal of Hazardous Materials* B139: 167–174.

Noroozi, B., Sorial, G.A., Bahrami, H., and Arami, M. 2008. Adsorption of binary mixtures of cationic dyes. *Dyes and Pigments* 76: 784–791.

Novotný, Č., Dias, N., Kapanen, A., Malachová, K., Vándrovcová, M., Itävaara, M., and Lima, N. 2006. Comparative use of bacterial, algal and protozoan tests to study toxicity of azo-and anthraquinone dyes. *Chemosphere* 63(9): 1436–1442.

Novotný, Č., Svobodová, K., Benada, O., Kofroňová, O., Heissenberger, A., and Fuchs, W. 2011. Potential of combined fungal and bacterial treatment for color removal in textile wastewater. *Bioresource Technology* 102(2): 879–888.

Oladoja, N.A., Aliu, Y.D., and Ofomaja, A.E. 2011. Evaluation of snail shell as a coagulant aid in the alum precipitation of aniline blue from aqueous solution. *Environmental Technology* 32(6): 639–652.

Oliveira, D.L.R.O., Bazo, A.P., Salvadori, D.M.F.., Rech, C.M., de Palma Oliveira, D, and de Aragão Umbuzeiro, G. 2007. Mutagenic and carcinogenic potential of a textile azo dye processing plant effluent that impacts a drinking water source. *Mutation Research/ Genetic Toxicology and Environmental Mutagenesis* 626(1): 53–60.

Osman, G., Aysenur, K., Fatih, M., and Burhan, A. 2004. Sorption of basic dyes from aqueous solution by activated sludge. *Journal of Hazardous Materials* B108: 183–188.

Pajot, H.F., Figueroa, L.I., Spencer, J.F., and Farina, J.I. 2008. Phenotypical and genetic characterization of *Trichosporon* sp. HP-2023. A yeast isolate from Las Yungas rainforest (Tucumán, Argentina) with dye-decolorizing ability. *Antonie van Leeuwenhoek* 94(2): 233–244.

Parikh, A., and Madamwar, D. 2005. Textile dye decolorization using cyanobacteria. *Biotechnology Letters* 27(5): 323–326.

Parsons, S. 2004. *Advanced oxidation processes for water and wastewater treatment.* IWA Publishing, UK.

Parsons, S.A., and Jefferson, B. 2006. *Introduction to potable water treatment processes.* Blackwell Publishing.

Pavko, A., and Novotný, Č. 2008. Induction of ligninolytic enzyme production by *Dichomitus squalens* on various types of immobilization support. *Acta Chimica Slovenica* 55(3): 648–652.

Pavon-Silva, T., Pacheco-Salazar, V., Carlos Sánchez-Meza, J., Roa-Morales, G., and Colín-Cruz, A. 2009. Physicochemical and biological combined treatment applied to a food industry wastewater for reuse. *Journal of Environmental Science and Health, Part A* 44(1): 108–115.

Pazarlioglu, N.K., Akkaya, A., Akdogan, H.A., and Gungor, B. 2010. Biodegradation of Direct Blue 15 by free and immobilized *Trametes versicolor. Water Environment Research* 82(7): 579–585.

Pearce, C., Lloyd, J., and Guthrie, J. 2003. The removal of colour from textile wastewater using whole bacterial cells: A review. *Dyes and Pigments* 58(3): 179–196.

Pera-Titus, M., García-Molina, V., Baños, M.A., Giménez, J., and Esplugas, S. 2004. Degradation of chlorophenols by means of advanced oxidation processes: A general review. *Applied Catalysis B: Environmental* 47(4): 219–256.

Perkowski, J., and Kos, L. 2003. Decolouration of model dyehouse wastewater with advanced oxidation processes. *Fibres and Textiles in Eastern Europe* 11(3): 67–71.

Phugare, S.S., Waghmare, S.R., and Jadhav, J.P. 2011. Purification and characterization of dye degrading of veratryl alcohol oxidase from Pseudomonas aeruginosa strain BCH. *World Journal of Microbiology and Biotechnology* 27(10): 2415–2423.

Pignatello, J.J., Oliveros, E., and MacKay, A. 2006. Advanced oxidation processes for organic contaminant destruction based on the Fenton reaction and related chemistry. *Critical Reviews in Environmental Science and Technology* 36(1): 1–84.

Porter, J.F., McKay, G., and Choy, K.H. 1999. The prediction of sorption from a binary mixture of acidic dyes using single- and mixed-isotherm variants of the ideal adsorbed solute theory. *Chemical Engineering Science* 54: 5863–5885.

Poulopoulos, S.G., Arvanitakis, F., and Philippopoulos, C.J. 2006. Photochemical treatment of phenol aqueous solutions using ultraviolet radiation and hydrogen peroxide. *Journal of Hazardous Materials* 129(1): 64–68.

Qiao, S., Hu, Q., Haghseresht, F., Hu, X., and Lu, G.Q. 2009. An investigation on the adsorption of acid dyes on bentonite based composite adsorbent. *Separation and Purification Technology* 67: 218–225.

Rai, H.S., Bhattacharyya, M.S., Singh, J., Bansal, T., Vats, P., and Banerjee, U. 2005. Removal of dyes from the effluent of textile and dyestuff manufacturing industry: A review of emerging techniques with reference to biological treatment. *Critical Reviews in Environmental Science and Technology* 35(3): 219–238.

Rajeswari, K.R. 2000. Ozonation treatment of textile dyes wastewater using plasma ozonizer, Ph. D thesis, University Malaysia. Malaysia.

Ramakrishna, K.R., and Viraraghavan, T. 1996. Dye removal using peat. *American Dyestuff Reporter*, October, pp. 28–34.

Ramalho, P.A., Cardoso, M.H., Cavaco-Paulo, A., and Ramalho, M.T. 2004. Characterization of azo reduction activity in a novel ascomycete yeast strain. *Applied and Environmental Microbiology* 70(4): 2279–2288.

Ramos, B., Farah, J.P.S., and Teixeira, A. 2012. Estimating reaction constants by ab initio molecular modeling: A study on the oxidation of phenol to catechol and hydroquinone in advanced oxidation processes. *Brazilian Journal of Chemical Engineering* 29(1): 113–120.

Ranjan Jena, P., Kumar Basu, J., and De, S. 2004. A generalized shrinking core model for multicomponent batch adsorption processes. *Chemical Engineering Journal* 102: 267–275.

Rehn, L. 1895. Bladder tumours in fuchsin workers. *Arch Fuer Klin Chirurgie* 50: 588–600.

Reyes, P., Pickard, M.A., and Vazquez-Duhalt, R. 1999. Hydroxybenzotriazole increases the range of textile dyes decolorized by immobilized laccase. *Biotechnology Letters* 21(10): 875–880.

Robinson, T., McMullan, G., Marchant, R., and Nigam, P. 2001. Remediation of dyes in textile effluent: A critical review on current treatment technologies with a proposed alternative. *Bioresource Technology* 77(3): 247–255.

Rodríguez, M., Malato, S., Pulgarin, C., Contreras, S., Curcó, D., Giménez, J., and Esplugas, S. 2005. Optimizing the solar photo-Fenton process in the treatment of contaminated water. Determination of intrinsic kinetic constants for scale-up. *Solar Energy* 79(4): 360–368.

Sanghi, R., and Bhattacharya, B. 2002. Review on decolorisation of aqueous dye solutions by low cost adsorbents. *Coloration Technology* 118: 256–269.

Sanroman, M., Pazos, M., Ricart, M., and Cameselle, C. 2004. Electrochemical decolourisation of structurally different dyes. *Chemosphere* 57(3): 233–239.

Saratale, R., Saratale, G., Chang, J.-S., and Govindwar, S. 2009. Decolorization and biodegradation of textile dye Navy blue HER by *Trichosporon beigelii* NCIM-3326. *Journal of Hazardous Materials* 166(2): 1421–1428.

Saratale, R.G., Saratale, G., Chang, J., and Govindwar, S. 2011. Bacterial decolorization and degradation of azo dyes: A review. *Journal of the Taiwan Institute of Chemical Engineers* 42(1): 138–157.

Sauleda, R., and Brillas, E. 2001. Mineralization of aniline and 4-chlorophenol in acidic solution by ozonation catalyzed with Fe 2+ and UVA light. *Applied Catalysis B: Environmental* 29(2): 135–145.

Scholz, M. 2015. Wetlands for water pollution control. Elsevier, The Netherlands.

Scotter, M., and Castle, L. 2004. Chemical interactions between additives in foodstuffs: A review. *Food Additives and Contaminants* 21(2): 93–124.

Seesuriyachan, P., Takenaka, S., Kuntiya, A., Klayraung, S., Murakami, S., and Aoki, K. 2007. Metabolism of azo dyes by *Lactobacillus casei* TISTR 1500 and effects of various factors on decolorization. *Water Research* 41(5): 985–992.

Sevimli, M.F., and Sarikaya, H.Z. 2002. Ozone treatment of textile effluents and dyes: Effect of applied ozone dose, pH and dye concentration. *Journal of Chemical Technology and Biotechnology* 77(7): 842–850.

Sharma, D.K., Saini, H.S., Singh, M., Chimni, S.S., and Chadha, B.S. 2004. Isolation and characterization of microorganisms capable of decolorizing various triphenylmethane dyes. *Journal of Basic Microbiology* 44(1): 59–65.

Silva, S.S., Magalhães, F., and Sansiviero, M.T.C. 2010. Nanocompósitos semicondutores ZnO/TiO2. Testes fotocatalÃticos. *Quim. Nova* 33(1): 85–89.

Sincero, A.P and Sincero, G.A. 2002. *Physical-chemical treatment of water and wastewater*. CRC press, Boca Raton, FL.

Singh, K., and Arora, S. 2011. Removal of synthetic textile dyes from wastewaters: A critical review on present treatment technologies. *Critical Reviews in Environmental Science and Technology* 41(9): 807–878.

Sivaraj, R., Namasivayam, C., and Kadirvelu, K. 2001. Orange peel as an adsorbent in the removal of acid violet 17 (acid dye) from aqueous solutions. *Waste Management* 21: 105–110.

Solozhenko, E.G., Soboleva, N.M., and Goncharuk, V.V. 1995. Decolourization of azodye solutions by Fenton's oxidation. *Water Research* 29(9): 2206–2210.

Srivastava, S., Singh, N., Srivastava, A.K., and Sinha, R. 1995. Acute toxicity of malachite green and its effects on certain blood parameters of a catfish, *Heteropneustes fossilis*. *Aquatic Toxicology* 31(3): 241–247.

Staehelin, J., and Hoigne, J. 1982. Decomposition of ozone in water: Rate of initiation by hydroxide ions and hydrogen peroxide. *Environmental Science & Technology* 16(10): 676–681.

Stalidis, G.A., Matis, K.A., and Lazaridis, N.K. 1988. A statistical approach to precipitate flotation of CuSZnS. *International Journal of Mineral Processing* 24(3–4): 203–216.

Stalidis, G.A., Matis, K.A., and Zouboulis, A.I. 1986. Flotation techniques for the separation of trace pollutants. *Chimika Chronika (New Series)* 15: 133–146.

Stammati, A., Nebbia, C., De Angelis, I., Albo, A.G., Carletti, M., Rebecchi, C., Zampaglioni, F., and Dacasto, M. 2005. Effects of malachite green (MG) and its major metabolite, leucomalachite green (LMG), in two human cell lines. *Toxicology In-Vitro* 19(7): 853–858.

Stolz, A. 2001. Basic and applied aspects in the microbial degradation of azo dyes. *Applied Microbiology and Biotechnology* 56(1–2): 69–80.

Szyguła, A., Guibal, E., Ariño Palacín, M., Ruiz, M., and Sastre, A.M. 2009. Removal of an anionic dye (Acid Blue 92) by coagulation-flocculation using chitosan. *Journal of Environmental Management* 90(10): 2979–2986.

Tanaka, H., Koike, K., Itakura, S., and Enoki, A. 2009. Degradation of wood and enzyme production by Ceriporiopsis subvermispora. *Enzyme and Microbial Technology*) 45: 384–390.

Tang, W.Z. 2016. *Physicochemical treatment of hazardous wastes*. CRC Press.

Tchobanoglous, G., and Burton, F.L. 1991. Wastewater engineering. *Management* 7: 1–4.

Tchobanoglous, G., Burton, F.L., Stensel, H.D., and Metcalf & Eddy, Inc. 2003. *Wastewater engineering: Treatment and reuse*. McGraw-Hill, New York.

Teh, C.Y., Wu, T.Y., and Juan, J.C. 2014. Optimization of agro-industrial wastewater treatment using unmodified rice starch as a natural coagulant. *Industrial Crops and Products* 56: 17–26.

Teixeira, C., and Jardim, W.d.F. 2004. Processos oxidativos avançados: Conceitos teóricos. *Caderno Temático* 3: 83.

Thomas, A., and Yeh, R.Y.-L. 1995. Color removal from dye wastewaters by adsorption using powdered activated carbon: Mass transfer studies. *Journal of Chemical Technology and Biotechnology* 63: 48–54.

Tobaldi, D.M., Tucci, A., Camera-Roda, G., Baldi, G., and Esposito, L. 2008. Photocatalytic activity for exposed building materials. *Journal of the European Ceramic Society* 28(14): 2645–2652.

Tokumura, M., Znad, H.T., and Kawase, Y. 2008. Decolorization of dark brown colored coffee effluent by solar photo-Fenton reaction: Effect of solar light dose on decolorization kinetics. *Water Research* 42(18): 4665–4673.

Tsuboy, M., Angeli, J., Mantovani, M., Knasmüller, S., Umbuzeiro, G., and Ribeiro, L. 2007. Genotoxic, mutagenic and cytotoxic effects of the commercial dye CI Disperse Blue 291 in the human hepatic cell line HepG2. *Toxicology In-Vitro* 21(8): 1650–1655.

Tsui, L.S., Roy, W.R., and Cole, M.A. 2003. Removal of dissolved textile dyes from wastewater by compost sorbent. *Coloration Technology* 119: 14–18.

Tunay, O., Kabdasli, I., Eremektar, G., and Orhon, D. 1996. Color removal from textile wastewaters. *Water Science and Technology* 34(11): 9–16.

Turabik, M. 2008. Adsorption of basic dyes from single and binary component systems onto bentonite: Simultaneous analysis of Basic Red 46 and Basic Yellow 28 by first order derivative spectrophotometric analysis method. *Journal of Hazardous Materials* 158: 52–64.

Uday, U.S.P., Bandyopadhyay, T.K., and Bhunia, B. 2016a. Bioremediation and detoxification technology for treatment of dye(s) from textile effluent. In *Textile wastewater treatment,* Kumbasar, E.P.A., and Körlü, A.E. (eds.), Intechopen, Croatia.

Velic, D., Knapp, M., and Köhler, G. 2001. Supramolecular inclusion complexes between a coumarin dye and β-cyclodextrin, and attachment kinetics of thiolated β-cyclodextrin to gold surface. *Journal of Molecular Structure* 598(1): 49–56.

Verma, P., and Madamwar, D. 2005. Decolorization of azo dyes using Basidiomycete strain PV 002. *World Journal of Microbiology and Biotechnology* 21(4): 481–485.

Vijayaraghavan, K., Lee, M.W., and Yun, Y.-S. 2008. A new approach to study the decolorization of complex reactive dye bath effluent by biosorption technique. *Bioresource Technology* 99(13): 5778–5785.

Vilhunen, S., Vilve, M., Vepsäläinen, M., and Sillanpää, M. 2010. Removal of organic matter from a variety of water matrices by UV photolysis and UV/H2O2 method. *Journal of Hazardous Materials* 179(1): 776–782.

Vitor, V., and Corso, C.R. 2008. Decolorization of textile dye by *Candida albicans* isolated from industrial effluents. *Journal of Industrial Microbiology & Biotechnology* 35(11): 1353–1357.

Wang, H., Su, J.Q., Zheng, X.W., Tian, Y., Xiong, X.J., and Zheng, T.L. 2009. Bacterial decolorization and degradation of the reactive dye Reactive Red 180 by Citrobacter sp. CK3. *International Biodeterioration & Biodegradation* 63(4): 395–399.

Wang, S., and Ariyanto, E. 2007. Competitive adsorption of malachite green and Pb ions on natural zeolite. *Journal of Colloid and Interface Science* 314: 25–31.

Wang, S., Li, H., and Xu, L. 2006. Application of zeolite MCM-22 for basic dye removal from wastewater. *Journal of Colloid and Interface Science* 295: 71–78.

Wang, Y., and Hong, C. 1999. Effect of hydrogen peroxide, periodate and persulfate on photocatalysis of 2-chlorobiphenyl in aqueous TiO 2 suspensions. *Water Research* 33(9): 2031–2036.

Waranusantigul, P., Pokethitiyook, P., Kruatrachue, M., and Upatham, E.S. 2003. Kinetics of basic dye (methylene blue) biosorption by giant duckweed (*Spirodela polyrrhiza*). *Environmental Pollution* 125: 385–392.

Watharkar, A.D., and Jadhav, J.P. 2014. Detoxification and decolorization of a simulated textile dye mixture by phytoremediation using *Petunia grandiflora* and, *Gailardia grandiflora*: A plant–plant consortial strategy. *Ecotoxicology and Environmental Safety* 103: 1–8.

Wesley, E., and Eckenfelder, J.R. 1989. *Industrial water pollution control*. Editorial. Mc Graw Hill, New York.

Wilcoxon, J.P., Thurston, T.R., and Martin, J.E. 1999. Applications of metal and semiconductor nanoclusters as thermal and photo-catalysts. *Nanostructured Materials* 12(5): 993–997.

Wong, Y., and Yu, J. 1999. Laccase-catalyzed decolorization of synthetic dyes. *Water Research* 33(16): 3512–3520.

Wright, L.D. 1976. Effect of malachite green and formalin on the survival of largemouth bass eggs and fry. *The Progressive Fish-Culturist* 38(3): 155–157.

Wu, J.J., Chen, S.H., and Muruganandham, M. 2008. Catalytic ozonation of oxalic acid using carbon-free rice husk ash catalysts. *Industrial & Engineering Chemistry Research* 47(9): 2919–2925.

Wu, J.J., Muruganandham, M., Chang, L.T., Lee, G.J., Batalova, V.N., and Mokrousov, G.M. 2011. Catalytic ozonation of oxalic acid using SrTiO3 catalyst. *Ozone: Science & Engineering* 33(1): 74–79.

Yang, Q., Yang, M., Pritsch, K., Yediler, A., Hagn, A., Schloter, M., and Kettrup, A. 2003. Decolorization of synthetic dyes and production of manganese-dependent peroxidase by new fungal isolates. *Biotechnology Letters* 25(9): 709–713.

Yang, Y.-I., Hong, H.-Y., Lee, I.-S., Bai, D.-G., Yoo, G.-S., and Choi, J.-K. 2000. Detection of DNA using a visible dye, Nile blue, in electrophoresed gels. *Analytical Biochemistry* 280(2): 322–324.

Yahagi, T., Degawa, M., Seino, Y., Matsushima, T., Nagao, M., Sugimura, T., and Hashimoto, Y. 1975. Mutagenicity of carcinogenic azo dyes and their derivatives. *Cancer Letters* 1: 91–96.

Yeber, M., RodrÃguez, J., Freer, J., Baeza, J., Duran, N., and Mansilla, H.D. 1999. Advanced oxidation of a pulp mill bleaching wastewater. *Chemosphere* 39(10): 1679–1688.

Yu, Z., and Wen, X. 2005. Screening and identification of yeasts for decolorizing synthetic dyes in industrial wastewater. *International Biodeterioration & Biodegradation* 56(2): 109–114.

Yujie, J., Ruofei, J., Jing,W., and Jiti, Z. 2005. Researches on the Fermentation of Genetic Engineering Microorganism and the Decolorization of Azo-dyes [J]. Industrial Safety and Dust Control 9: 004.

Zazo, J.A., Casas, J.A., Mohedano, A.F., Gilarranz, M.A., and Rodriguez, J.J. 2005. Chemical pathway and kinetics of phenol oxidation by Fenton's reagent. *Environmental Science & Technology* 39(23): 9295–9302.

Zazo, J.A., Casas, J.A., Mohedano, A.F., and Rodriguez, J.J. 2006. Catalytic wet peroxide oxidation of phenol with a Fe/active carbon catalyst. *Applied Catalysis B: Environmental* 65(3): 261–268.

Zollinger, H. 2003. *Color chemistry: Syntheses, properties, and applications of organic dyes and pigments*. Wiley, Weinheim, Germany.

Zouboulis, A.I., and Matis, K.A. 1989. Electrolytic flotation of chromium from dilute solutions. *Environmental Technology* 10(6): 601–612.

Zouboulis, A.I., Matis, K.A., and Stalidis, G.A. 1990. Parameters influencing flotation in removal of metal ions. *International Journal of Environmental Studies* 35(3): 183–196.

Zouboulis, A.I., Matis, K.A., and Stalidis, G.A. 1992. Flotation techniques in waste water treatment. In *Innovations in flotation technology,* Springer, Germany, pp. 475–497.

Zouboulis, A.I., and Tzoupanos, N.D. 2009. Polyaluminium silicate chloride—A systematic study for the preparation and application of an efficient coagulant for water or wastewater treatment. *Journal of Hazardous Materials* 162(2): 1379–1389.

# 6 Lindane Contamination in the Environment

## *Toxicological Effects and Bioremediation Approaches*

*Siddharth Boudh and Jay Shankar Singh*

## CONTENTS


6.1 Introduction ... 178
6.2 Lindane ... 178
  6.2.1 Structure of Lindane ... 179
  6.2.2 Physical and Chemical Properties ... 179
  6.2.3 Commercial Production of Lindane and Its Applications ... 180
    6.2.3.1 Production of Lindane ... 180
    6.2.3.2 Application of Lindane ... 181
6.3 Sources of Lindane Contamination in Environment ... 183
6.4 Toxicological Effects of Lindane Contamination in Environments ... 185
  6.4.1 Physiological Effect of Lindane ... 185
  6.4.2 Health Hazards due to Lindane ... 186
    6.4.2.1 Acute Health Effects ... 186
    6.4.2.2 Chronic Health Effects ... 187
    6.4.2.3 Other Long-Term Effects ... 187
6.5 Bioremediation Approaches to Remove Lindane from Contaminated Environments ... 188
  6.5.1 Role of Bacteria in Lindane Remediation ... 188
  6.5.2 Lindane Mineralization ... 191
  6.5.3 Role of Fungal Extracellular Oxidative Enzymes in Lindane Decontamination ... 192
  6.5.4 Role of Plant–Microbe Association in Lindane Remediation ... 193
6.6 Government and Private Initiatives on Lindane Concern ... 194
6.7 Conclusions ... 196
Acknowledgments ... 197
References ... 197

## 6.1 INTRODUCTION

Lindane is the common name for gamma-hexachlorocyclohexane (γ-HCH) and has been listed as a persistent organic pollutant (POP) under Stockholm Convention on June 29, 2005. The prohibited dumping of waste hexachlorocyclohexane (HCH) isomers (called "muck") at sites all over the world is creating problems for the environment and soil systems. Lindane is a member of the organochlorine (OC) pesticide class. This is one of the oldest classes of pesticides, and few OCs are still in use. OC pesticides are so named because they include carbon, hydrogen, and chlorine. Three major subclasses of OC pesticides are diphenyl aliphatics, cyclodienes, and HCH. More than 52 countries have banned the use of lindane and production of any kind of its formulations (Vijgen 2011). The lindane-formulating countries are also one of the major problems as the formulations are accumulating at dumping sites and are toxic.

Among the five isomers of HCH (alpha [α], beta [β], gamma [γ], delta [δ], and epsilon [ε]), only the γ isomer has insecticidal properties (Hardie 1964). The γ isomer has been manufactured as lindane. Lindane has not been produced in the United States since 1977, but it is imported in multiple forms for pharmacological and industrial purposes. The use of lindane has been restricted by the U.S. Environmental Protection Agency (USEPA); presently it is used only by a few certified pesticide stakeholders. During the past, huge amounts of lindane were deposited in the soil, water, and other ecosystems; therefore, there is a need to remove this lethal compound from polluted environments. Microbial bioremediation of lindane deposited at various dumping sites may be one of the effective strategies to decontaminate affected sites (Boudh et al. 2016). Heritage and MacRae (1977) first reported anaerobic degradation of lindane by bacterial cell–free preparation, and Senoo and Wada (1989) first reported aerobic degradation of lindane by *Pseudomonas paucimobilis* bacteria. Bacteria have evolved at such contaminated sites that have the ability to degrade HCH. Degradation of various HCH isomers in bacterial strains is mediated primarily by two genes: linA and linB, which encode dehydrochlorinase and haloalkane dehalogenase, respectively (Nagata et al. 1993, 1994). Lindane is a persistent OC compound (produced by Faraday in 1825) which is widely distributed in the environment (IARC 1973). Long distance transport of lindane is evidenced by its presence in the Arctic Ocean, where it has never been used (Iwata et al. 1993; UNECE 2000). Most of the lindane present in the environment is in water, although a significant amount is also found in the soil/sediment and some in air (Willett et al. 1998). Lindane has also been shown to bioaccumulate in the fatty tissue of organisms (Bro-Rasmussen 1996). This chapter briefly acknowledges HCH properties, health hazards, its bioremediation prospects, and government and private organization's awareness regarding this chemical.

## 6.2 LINDANE

Lindane belongs to the OC class of pesticides that have been banned in most of the developed countries in the 1970s. Lindane, or HCH, is a halogenated organic insecticide that has been used worldwide. Most countries have prohibited the use of this xenobiotic compound, but some are still using γ-HCH for economic reasons, and new sites are consequently being contaminated. In addition, HCH is used against lice infestations

(Nagata et al. 1997). HCH has eight possible stereoisomers, of which four (α, β, γ, and δ-HCH) predominate in the technical product. γ-HCH is the best known and effective insecticide component of HCH.

For many years, we have enjoyed the benefits of using pesticides to control weeds, insects, pathogenic fungi, parasites, and rodent pests. Pesticides are the only effective means of controlling pathogens, weeds, or insect pests in many circumstances (USEPA 2012). Apart from the beneficial role, the use of pesticides also has some drawbacks or side effects, such as potential toxicity to humans, birds, and other aquatic animals. The HCH isomers are one of the most abundant OC contaminants in the Arctic Ocean (USEPA 2006). Production of 1 ton of lindane generates waste of around 6–10 tons of other isomers (Vijgen 2006). A massive 10 million tons of the technical HCH have been used worldwide from 1948 to 1997 (Li 1999; Walker et al. 1999). India started the production of technical HCH in the year 1952 and perhaps was the largest user of technical HCH and DDT in the world. Technical HCH and DDT amounted to 70% of total insecticide production in the 1980s.

Lindane has been recently listed as a POP under the United Nations' Economic Commission for Europe Convention on Long-range Transboundary Air Pollution (LRTAP POPs Protocol) and the Great Lakes Binational Toxics Strategy between the United States and Canada. It is also the subject of a joint reevaluation in the United States and Canada under the North American Free Trade Agreement (NAFTA) Technical Working Group on Pesticides. In the environment, lindane is potentially transformed into a variety of chemicals, most of which are volatile. These include γ-pentachlorocyclohexene, γ-3,4,5,6-tetrachlorocyclohexene, α-HCH (Cornacoff et al. 1988; Bintein and Devillers 1996). The ratio of α-HCH to γ–HCH concentration in air has been used as an indicator to estimate the possible origin of the air mass in the long-range transportation of contaminants (Iwata et al. 1993). Earlier, bioisomerization of γ-HCH to α-HCH was thought to be the principal route for long-range contamination; however, current field studies have found that only a small percentage of γ-HCH is converted to α-HCH as a result of biological activities (Singh et al. 1991; Waliszewski 1993).

### 6.2.1 Structure of Lindane

Lindane is the common name for γ-hexachlorocyclohexane. The chemical formula for lindane is $C_6H_6Cl_6$. It has a molecular weight of 290.83 g/mol (Figure 6.1).

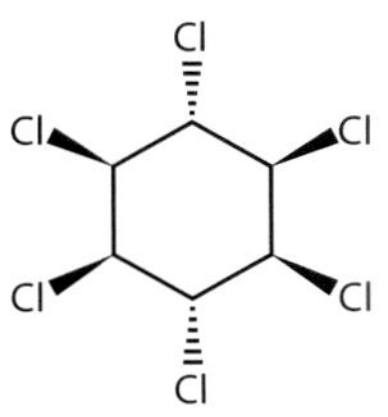


**FIGURE 6.1** Structure of lindane.

### 6.2.2 Physical and Chemical Properties

Lindane (γ-1,2,3,4,5,6,-hexachlorocyclohexane; γ-HCH) is an insecticide and fumigant that has been utilized for the control of an extensive variety of soil-dwelling and plant-eating (phytophagous) insects. It is commonly utilized on a wide variety of crops, in stockrooms, in public health to control insect-borne diseases, and with fungicides for the seed treatment. Lindane is also presently used in lotions, creams, and shampoos for the control of lice and mites (scabies) in humans. Trade or other names for lindane include Aphitiria, Agrocide, Aparasin, Ambrocide, Benesan, Benexane, benhexachlor, benzene hexachloride, BHC, gamma-BHC, Borer-Tox, BoreKil, Exagama, Gallogama, Gamaphex, Gamasan, Gammex, Gammexane, Gexane, Isotox, Jacutin, Kwell, Lorexane, Lindafor, Lintox, Lindagronox, Lindaterra, Lindatox, Noviagam, Quellada, Steward, Streunex, and Tri-6 (USEPA 1983). Lindane is a white powder that evaporates in air, with a musty (odorless when pure) odor at concentrations of 12 ppm and more. Lindane is known to be steady in air, light, heat, carbon dioxide, and strong acids.

Dehydrochlorination of lindane takes place in the presence of an alkali, forming trichlorobenzene and hydrochloric acid. Dechlorination of lindane occurs on exposure to ultraviolet radiation forming gamma pentachlorocyclohexenes and tetrachlorocyclohexenes. The half-life for the environmental degradation of lindane differs relying on the elements (e.g., climate, type of soil, and depth of application). Under humid and field conditions, the half-life of lindane is between a few days to 3 years (IPCS 1998). Harner et al. (1999) estimated the half-life of lindane in the Arctic Ocean as 19 years. Half-life for lindane in soil and water is accounted for as 708 days and 2292 days individually (Beyer and Matthies 2001).

Lindane is incompatible with strong bases and powdered metals such as iron, zinc, and aluminum. It can undergo oxidation when it comes in contact with ozone (Chalivendra 2011). Table 6.1 shows the physical and chemical properties of lindane.

### 6.2.3 Commercial Production of Lindane and Its Applications

#### 6.2.3.1 Production of Lindane

After World War II, numerous organizations started to produce HCH. They use various methods to produce technical HCH. These methods are by means of a catalyzer, butylacetate-method, benzine-dioxane method, modified benzine-dioxane method, benzine method, methanol-circulation-method, etc. Table 6.2 presents lindane-producing companies.

Technical-grade HCH is produced as a mixture of isomers (primarily the $\alpha$, $\beta$, $\gamma$, $\delta$, and $\varepsilon$ isomers) by photochlorination of benzene, a response that can be begun by free-radical initiators such as visible or ultraviolet light, x-rays, or $\gamma$-rays (ATSDR 2005). The active $\gamma$-HCH (lindane) can be concentrated by treatment with methanol or acetic acid, followed by fractional crystallization, which produces technical grade lindane containing 99.9% $\gamma$ isomer. Commercial production of lindane in the United States began in 1945 and peaked in the 1950s, when 17.6 million pounds were manufactured. Lindane is no longer produced commercially in the United States, but it is produced by 13 manufacturers worldwide, including 7 in India and 4 in China (SRI 2009), and is available from 42 suppliers, including 19 U.S. suppliers (ChemSources 2009). U.S. imports of HCH increased from 310,000 lb to 1.4 million pounds between 1989 and 1999 imports, declining to zero in 2005, and remaining zero through 2008 except in 2006, when 73,000 lb were imported.

**TABLE 6.1**
**Physical and Chemical Properties of Lindane**

| Common Name | Lindane |
|---|---|
| Chemical class | Organochlorine |
| Chemical name | 1,2,3,4,5,6-hexachlorocyclohexane, γ-isomer, γ-HCH |
| CAS Registry number | 58-89-9 |
| Chemical formula | $C_6H_6Cl_6$ |
| Molecular weight | 290.83 |
| Melting point | 112.5°C |
| Boiling point | 323.4°C |
| Solubility in water at 25°C | 7.52 mg/L |
| Partition coefficients | |
| Log KOW | 3.3, 3.61 |
| Log KOC | 3.0, 3.57 |
| Bioaccumulation factor in human fat | $19 \pm 9$ |
| Bioaccumulation factor in aquatic animals | $2.5 \pm 0.4$ |
| Vapor pressure at 20°C or 25°C | $5.3 \pm 1.4 \times 10^{-3}$ Pa |
| Henry's law constant at 25°C | $3.2 \times 10^{-4}$ KPa $m^3$/mol |
| Conversion factors | |
| Air 1 ppm | 1.18 $mg/m^3$ |
| Water 1 ppm | 1 mg/L |
| Carcinogenicity classification | |
| ACGIH | A3 |
| EPA | NA |
| IARC | 2B |

*Note:* NA, not available.

U.S. exports of HCH increased from zero in 1990 to 1.5 million pounds in 2005, declining to 154,000 lb in 2008 (USITC 2009).

#### 6.2.3.2 Application of Lindane

Before getting banned, lindane was used in various fields, like agriculture, forestry, household uses, seed treatment, etc.

1. Lindane is utilized for louse control and as a wood preservative in few nations.
2. Lindane is registered for the control of carpet beetles and clothes moths in the home.
3. It is reregistered by the EPA for the control of aphids, cucumber beetles, cutworms, melon worm, pickleworm, squash bug, squash vine borer, white grub, and other insects on cucurbits.
4. It is used for control of the following insects found in the home: ants, centipedes, clothes moths, house fly, odorous house ant, cockroaches, bed bugs,

**TABLE 6.2**
**Brief Introduction of Lindane-Producing Countries**

| Country/Producer | Period of Production | Reference |
|---|---|---|
| Albania | 1982–1990 | UNEP 2002 |
| Argentina | 1947–1949 | UNEP 2002 |
| Austria | 1965–1990 | Questionnaire, Department of the Environment and Heritage 2004; Li 1999 |
| Azerbaijan | 1960–1985 | |
| Belgium | — | Stoffbericht 1993 |
| Brazil | 1962–1985 | UNEP 2002; Lopez 1999 |
| Bulgaria | To 1966 | Hauzenberger 2004 |
| Canada | To 1974 | CEC 2000; NRTEE 2000 |
| China | 1995–2000 | Li 2001, 2004 |
| Czech Republic | 1954–April 1977 | Hauzenberger 2004 |
| Croatia | — | UNECE 2000 |
| Denmark | To 1977 | De Bruin 1997 |
| Egypt | — | Li 1999 |
| France | 1977–1979 | Heinisch 1994; De Bruin 1979 |
| Eastern Germany | 1967–1982: | Heinisch 1994 |
| West Germany | 1955–1971 Discontinued 1972 | De Bruin 1979 |
| Ghana | — | UNEP 2002 |
| Hungary | 1953–1964 | Bárczi 1994 |
| India | To 2003 | Rup Lal et al. 2005; Fact Sheet 2005 |
| Japan | 1948–1987 | UNEP 2002; Questionnaire Ministry of Agriculture October 2004 |
| Poland | 1956–1982 | Questionnaire September 2004; Heinisch 1994 |
| Romania | — | Nafta 2003; Hauzenberger April 2004 |
| (Former) Soviet Union | To 1990 | Li et al. 2005 |
| Switzerland | 1935–1965 | Bentz 2004 |
| Turkey | To 1982 | Türkman et al. 1993 |
| United States | 1950s–1970s | Stoffbericht 1993 |

chats, little black ants, sow bugs, mosquitoes, scorpions, spiders, water bugs, and silverfish.

5. It is also used for controlling wood boring beetles generally called powder post beetles and for dry wood termites.
6. It is used against all major arthropod pests that attack pets, including ticks, flies, lice, mites, and for pet premises. Ohio is the only reported state that recommended lindane for control of fleas on cats.
7. It has also been used as pharmaceutical treatment for lice and scabies, formulated as a shampoo or lotion.
8. It is used for control of phylloxera and shoot curculio on pecans.

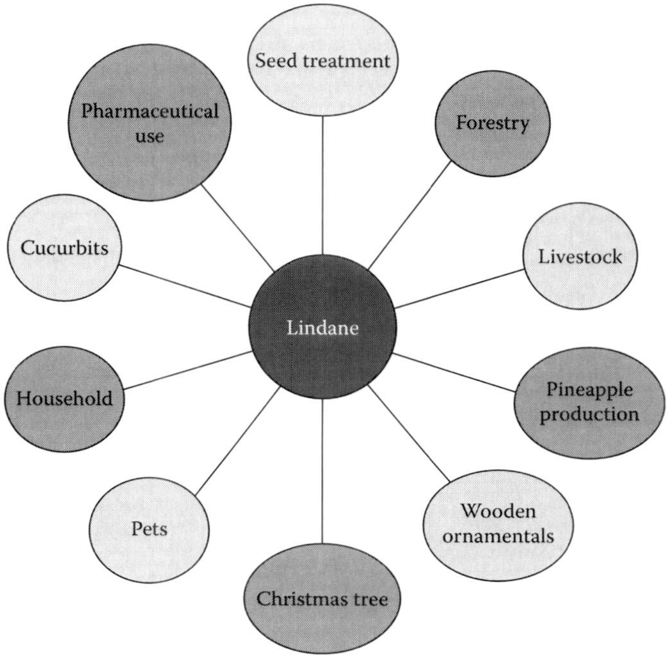


**FIGURE 6.2** Various applications of lindane.

9. Lindane is the only pesticide registered for the control of all the insect pests attacking Christmas trees.
10. It is used for all the major insects and leaf miners on woody ornamentals, floral, and foliage plants.
11. Lindane requires special EPA registration for the control of symphylans on Hawaiian pineapple production. It is used in conjugation of four soil fumigants.
12. It is used on a variety of livestock arthropod pests and on many classes of livestock—beef, cattle, hogs, pigs, sheep, goats, and horses—and livestock premises—barns, pens, sleeping quarters, and shelters.
13. In forest lands (commercial, seed orchards, and naval stores), lindane is the only registered chemical for use on living trees for insect control.
14. Lindane is an insecticide for use as a seed treatment on minor crops such as peas, beans, sunflower, lentils, and vegetables, and major crops including sorghum and sugar beets (Figure 6.2).

## 6.3 SOURCES OF LINDANE CONTAMINATION IN ENVIRONMENT

The routes of potential human exposure to lindane and other HCH isomers are ingestion, inhalation, and dermal contact (HSDB 2009). The general population potentially is exposed through consumption of foodstuffs contaminated with pesticide residues. According to the U.S. Food and Drug Administration's Total Diet Survey, lindane was detected in 279 of 2168 samples and in at least one sample of all 54 different food items analyzed (FDA 2006). Most of the food items in which lindane was detected had significant fat content; however, the highest lindane concentrations were in pickles and raw mushrooms, which have low fat content. Daily dietary intake of HCH isomers by the adult U.S. population was estimated at 0.010 μg/kg (10 ng/kg) of body weight for all isomers and 0.002 μg/kg (2 ng/kg) for lindane. For 1982 to 1984, the estimated dietary intake of lindane was 1.9 ng/kg of body weight for infants aged 6 to 11 months and 7.9 ng/kg for toddlers aged 2 years, who had the highest average daily intake. By 1986 to 1991, daily intake had fallen to 0.8 ng/kg for infants and 3.2 ng/kg for toddlers (ATSDR 2005).

Dermal introduction happens when shampoos and creams containing lindane are utilized for the treatment of lice and scabies (FDA 2009). The highest average blood concentration of lindane measured in children after scabies treatment with one of these products was 0.028 μg/mL (ATSDR 2005). According to the USEPA's Toxics Release Inventory, environmental releases of lindane ranged from 314 to 2118 lb between 1988 and 1997. In 1998, over 25,000 lb was sent to a hazardous-waste landfill. By 2006, releases had declined to 10 lb. In 2007, five facilities released a total of 1555 lb of lindane, mostly off site for unspecified management (TRI 2009). Lindane was found in at least 189 hazardous-waste sites currently or formerly on the National Priorities List; it occurred in air at 9 sites, surface water at 33 sites, sediment at 36 sites, and soil at 90 sites. The Non-Occupational Pesticide Exposure Study, published in 1990, collected personal air samples at one U.S. location with high pesticide usage and one with low to medium usage. The range of mean γ-HCH concentration was 7–22 ng/m$^3$ at the high-usage site and 0.7 to 5 ng/m$^3$ at the low- to medium-usage site (ATSDR 2005).

HCH isomers have been detected in human fatty tissue, blood, and breast milk. The National Human Adipose Tissue Survey (NHATS), conducted in 1982, found β-HCH in 87% of composite postmortem samples of fatty tissue. According to NHATS data, the mean concentration of β-HCH in fat decreased from 0.45 ppm in 1970 to 0.16 ppm in 1981. The levels were highest in the southern United States. In the 1970s, the National Health and Nutrition Examination Survey (NHANES) found β-HCH in blood at a median concentration of 1.7 ppb. When the NHANES was repeated in 1999–2000, the geometric mean concentration of β-HCH and γ-HCH in serum lipid was 9.68 ng/g (ppb) for individuals over 12 years of age (ATSDR 2005). β-HCH was measured in breast milk at a concentration of 0.6 ng/g in Canadian populations living near the Great Lakes. In the Netherlands, concentrations of γ-HCH in breast-milk fat in 1988 ranged from 0.01 to 0.24 mg/kg (HSDB 2009). Many other studies in populations throughout the world, especially Arctic populations, have found HCH isomers in blood, fat, and breast-milk samples. HCH isomers have been measured at higher concentrations in all types of samples in areas of the world where lindane is still extensively used for pest control, such as India and Africa (Figure 6.3).

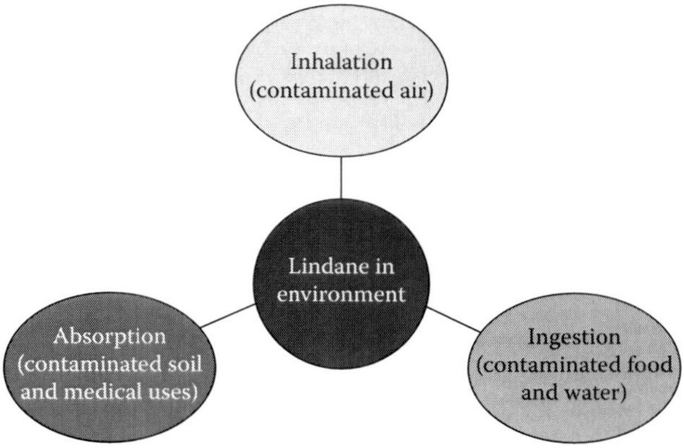


**FIGURE 6.3** Sources of lindane contamination in the environment.

## 6.4 TOXICOLOGICAL EFFECTS OF LINDANE CONTAMINATION IN ENVIRONMENTS

### 6.4.1 Physiological Effect of Lindane

Lindane produces central nervous system interruption through a variety of mechanisms, the most important being its ability to act as a noncompetitive GABA antagonist interacting with the picrotoxin site, both in membranes and in intact cultured neurons, thereby inhibiting the GABA-induced C1-flux following activation of either the GABA(A) or GABA(C) receptor (Pomes et al. 1994; Nagata et al. 1996; Aspinwall et al. 1997). Rosa et al. (1997) demonstrated that lindane induces significant changes in the intracellular $Ca^{2+}$ homeostasis of central neurons and has a significant impact in the cerebellum. Lindane principally influences the dantrolene-sensitive intracellular $Ca^{2+}$ stores, causing a release of calcium from these stores and altering the sensitivity of membranes.

Sunol et al. (1988) demonstrated that the neurotoxic effects of lindane are predominantly mediated through its noncompetitive antagonism of the GABA(A) receptor. It was noted that following poisoning and seizure, the concentrations of dopamine and its primary metabolite (DOPAC) were increased throughout the mesencephalon and the striatum. In cases of chronic intoxication, dopamine levels fluctuated. These fluctuations, following repetitive exposure to lindane, may account for some of the parkinsonian-like symptoms our patient experienced. Attia et al. (1991) have shown that chronic lindane poisoning causes an enhanced nighttime rise in pineal *N*-acetyltransferase (NAT) activity and increased central melatonin secretion, as well as increased serum melatonin levels. Chronic lindane poisoning also caused a significant reduction in central serotonin (5-HT) and 5-hydroxyindole acetic acid (5-HIAA). This change in central serotonin may well be the factor that mediates the anxiety and depressive changes seen in our patient, as well as the disruption of her sleep cycle. Nedkova-Bratanova et al. (1979) have shown that chronic lindane poisoning, even at low levels, diminishes dipeptidase activity and disrupts the activity of the intestinal disaccharidases for up to 90 days following low-level poisoning. Sucrose is the most effected disaccharide. This disruption of the disaccharidases may be responsible for the persistent diarrhea experienced by our patient.

## 6.4.2 Health Hazards due to Lindane

### 6.4.2.1 Acute Health Effects

The following acute (short-term) health effects may occur immediately or shortly (14 days or later) after exposure to lindane.

Lindane can cause headache, nausea, vomiting, dizziness and seizures, irritability, restlessness, muscle weakness and twitching, convulsions, and coma (Nantel et al. 1977). Vomiting and nausea are common signs of γ-HCH ingestion and furthermore have been accounted for after dermal presentation to γ-HCH. There are additionally reports of unfavorable hematological effects in individuals exposed to γ-HCH taken after inward breath or potentially dermal introduction to local items containing γ-HCH and following chronic occupational exposure. Immunotoxic impacts have been seen in other oral investigations of γ-HCH. Immunosuppression in the form of reduced antibody responses to *Salmonella* and typhoid vaccines happened in rats exposed to 6.25 mg/kg/day for up to 5 weeks (Dewan et al. 1980). Introduction to 10 mg/kg/day in mice for 10 days brought about lingering bone marrow harm and stifled granulocyte-macrophage ancestor cells, and decay of the thymus was seen in mice after 40 mg/kg/day for 3 days (Hong and Boorman 1993).

Neurological impacts after intense inward breath exposure to γ-HCH have included excitation, sedation, ataxia, and spasms. Neurological impacts of γ-HCH included improved susceptibility to kindling taking after a single 5 mg/kg dose (Gilbert and Mack 1995) or 3 mg/kg/day for 4 days (Joy et al. 1982), decreased brain serotonin level following 3 mg/kg/day for 6 days (Attia et al. 1991), and decreased brain barrier permeability in 10-day-old pups exposed to 2 mg/kg as a single dose or 8 daily doses (Gupta et al. 1999).

Studies regarding the inward breath lethality of HCH in people are constrained. The vast majority of the accessible data is from animal testing and experimental data. Rats breathing in up to 603 $mg/m^3$ γ-HCH vaporized for 4 hours in entire body exposure chambers showed no mortality all through the 14-day perception period (Oldiges et al. 1980). In an intermediate-duration study on CD-1 mice which is exposed to a lindane dust aerosol six hours per day, five days per week, reported a 22 percent mortality rate with exposure of 5 $mg/m^3$ for up to 20 weeks and a 2 percent mortality rate when the exposure was reduced to 1 $mg/m^3$ (Klonne and Kintigh, 1988). Other hematological anomalies, including isolated instances of leukopenia, leukocytosis, granulocytopenia, granulocytosis, eosinophilia, monocytosis, and thrombocytopenia, have been accounted for taking after chronic human occupational exposure to γ-HCH (Jedlicka et al. 1958; Brassow et al. 1981).

Kashyap (1986) reported significant increments in blood levels of the enzymes lactate dehydrogenase (33%), leucine aminopeptidase (45%), and γ-glutamyl transpeptidase (174%) in 19 individuals who worked in a HCH-producing plant for more than 10 years. The HCH isomer concentration in serum exhibited a 10-fold increase compared to the workers not exposed to lindane. A significant increase (approximately 18%) in the level of immunoglobulin M (IgM) was also observed in individuals affected by lindane compared to workers not affected. Kashyap (1986) also reported electrocardiogram (ECG) irregularities in 15% of 45 people involved in the production of technical-grade HCH. Weakness of the left and right limbs, dysarthria, and dysphagia was found in an agricultural worker revealed by internal breath and dermal contact to unspecified levels of a couple of OC pesticides, including γ-HCH (Fonseca et al. 1993).

#### 6.4.2.2 Chronic Health Effects

The following chronic (long-term) health effects can happen sooner or later after exposure to lindane and can keep going for a considerable length of time or years.

##### *6.4.2.2.1 Cancer Hazard*

Lindane may be a carcinogen in humans since it has been shown to cause liver, lung, endocrine gland, and other types of cancer in animals. Four cases of leukemia were reported in men exposed to lindane with or without other chemicals (IARC 1979; Sidi et al. 1983). Technical-grade α- and β-HCH and the γ isomer (lindane) produced liver tumors in mice when administered orally (Kashyap et al. 1979; Munir et al. 1983).

##### *6.4.2.2.2 Reproductive Hazard*

Lindane may damage the developing fetus and decrease fertility in females. Beard and Rawlings (1998) do three-generation study of mink and observed reductions in whelping rate, litter size, and testicular size when they exposed them to 1 mg/kg/day dietary γ-HCH. Dalsenter et al. (1997) studied reproductive toxicity in male offspring of rats that were exposed during lactation and found significant changes in testis and epididymis weights, sperm number, and hormone levels. Statistically significant increases in the levels of serum luteinizing hormone were reported by Tomczak et al. (1981) in a group of 54 persons who are occupationally exposed to unspecified concentrations of γ-HCH for approximately 8 years in a γ-HCH-producing factory.

#### 6.4.2.3 Other Long-Term Effects

High exposure to lindane may damage the blood cells causing anemia. Exposure to lindane has been linked with blood disorders known as blood dyscrasias, and in particular the disorder aplastic anemia where the formation of platelets and white cells is disrupted (Morgan et al. 1980). Lindane can cause abnormal heart rhythm (arrhythmia). Lindane is fat soluble, and this contributes to its tendency to bioaccumulate through the food chains. Lindane residues have been detected in the kidneys, livers, and adipose tissue of a wide variety of wild animals and birds. It is also highly toxic to aquatic invertebrates and fish.

## 6.5 BIOREMEDIATION APPROACHES TO REMOVE LINDANE FROM CONTAMINATED ENVIRONMENTS

There are several dumping sites present in the world from where this hazardous chemical easily comes into contact with biotic and abiotic components of environment. Lindane residues from "muck" due to rain, wind, and because of illegal discharge into water bodies, dry deposition, and snow can enter the environment and cause toxicity (Willett et al. 1998; ATSDR 1999). Lindane residues are readily found in milk, meat (Aulakh et al. 2006), water (Kaushik et al. 2007), soil (Abhilash and Singh 2008), packed water bottles, fish (Amaraneni and Pillala 2001), human blood (Dhananjayan et al. 2012), butter and ghee, honey samples (Choudhary et al. 2008), vegetables, breast milk (Kalra et al. 2003), and maternal and cord blood (Pathak et al. 2008).

For this reason, several laboratories have isolated and characterized the microorganisms that can cause the degradation of lindane and other HCH-isomers under anaerobic (MacRae et al. 1969; Jagnow et al. 1977; Van Eekert et al. 1998; Van Doesburg et al. 2005) and aerobic conditions (Senoo and Wada 1989; Sahu et al. 1990; Thomas et al. 1996; Gupta et al. 2000; Manonmani et al. 2000; Okeke et al. 2002; Boltner et al. 2005; Kumar et al. 2005) (Figure 6.4).

### 6.5.1 Role of Bacteria in Lindane Remediation

In the beginning it was believed that lindane biodegradation is largely an anaerobic process, and various levels of anaerobic degradation of α, β, γ, and δ-HCH have been observed. MacRae et al. (1969) reported degradation of lindane by anaerobic *Clostridium* sp. Initial studies suggested that the anaerobic microorganisms use lindane as a sole carbon and energy source. Later studies demonstrated that lindane is used as an electron acceptor under anaerobic conditions rather than as a carbon source. Other isolates capable of degrading one or more of the other four HCH isomers under anaerobic conditions include *C. sphenoides* (Heritage and MacRae 1977), *C. butyricum,* and *C. pasteurianum* (Jagnow et al. 1977). Ohisa and Yamaguchi (1978)

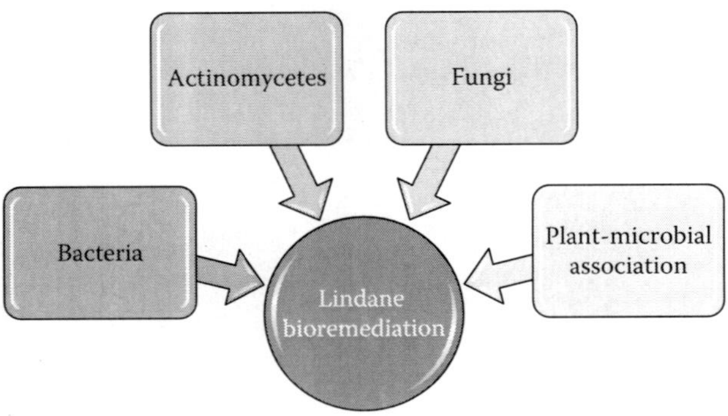


**FIGURE 6.4** Various approaches to lindane bioremediation.

reported that under anaerobic pure culture condition, strains of *C. rectum* degrade lindane optimally at pH 7–8.

Lindane degradation can also be done by *Escherichia coli,* which was isolated from rat feces. About 10% of the added lindane was metabolized by the bacterium in trypticase soy broth containing the pesticide lindane. A single metabolite, 2,3,4,5,6-pentachloro-1-cyclohexene (PCCH) was detected and identified by gas chromatography and mass spectrometry (Francis et al. 1975).

Members of the family Sphingomonadaceae appear to have an important role in aerobic lindane degradation. For example, three lindane-degrading species of *Sphingobium—S. japonicum* UT26 (Imai et al. 1989), *S. francense* Sp+ (Dogra et al. 2004), and *S. indicum* B90A (Sahu et al. 1990) (initially classified as three strains of *Sphingomonas paucimobilis*)—were isolated from lindane-contaminated soils from Japan, France, and India, respectively.

The pathway for the degradation of lindane has been comprehensively worked out in the bacterium *S. paucimobilis* UT26, and the genes for its different enzymes have been characterized (Nagata et al. 1999).The catabolic genes associated with the degradation of lindane were initially discovered in *S. japonicum* UT26, and termed *lin* genes (Nagata et al. 1999). Six structural *lin* genes (linA–linF) (Imai et al. 1991; Nagata et al. 1993; Nagata et al. 1994; Miyauchi et al. 1998; Miyauchi et al. 1999; Endo et al. 2005) and one regulatory gene (*linR*) (Miyauchi et al. 2002) are involved in the complete mineralization of lindane in *S. japonicum* UT26. In addition, a *linX* gene, encoding a protein that has activity similar to that of *linC*, was also characterized (Nagata et al. 1994). The *linA*-encoded HCH dehydrochlorinase (linA) mediates the first two steps of dehydrochlorination of lindane. In addition to mediating the second step in the degradation of lindane in *S. japonicum* UT26, *linB* has been reported, recently, to transform β-HCH to 2,3,4,5,6-pentachlorocyclohexanol (PCHL) (Nagata et al. 2005). PCHL has lower hydrophobicity and lower chemical stability than β-HCH, and the bacteria that degrade and use it might exist in the polluted environment, enabling the complete degradation of β-HCH by a combination of biological pathways. The location and stability of *lin* genes encoding the lindane catabolic pathway have been explored in the lindane-degrading *S. francense* strain Sp1+, and in two non-lindane-degrading mutants (Sp1- and Sp2-) (Ceremonie et al. 2006).

*Flavobacterium* sp., *Pseudomonas* sp., and *Achromobacter* sp. isolated from the gut of earthworms treated with lindane were capable of degrading α, β, and γ isomers of HCH (Ramteke and Hans 1992). Other lindane-degrading bacteria such as *Citrobacter freundii* (Jagnow et al. 1977), *Desulfovibrio gigas, D. africanus, Desulfococcus multivorans* (Boyle et al. 1999), and *Dehalobacter* sp. (van Doesburg et al. 2005) were also isolated. Nalin et al. (1999) isolated a new strain of *Rhodanobacter lindanclasticus* which degraded technical-grade HCH under aerobic conditions. Datta et al. (2000) reported the growth characteristics and degradation of the aerobic bacterial strain *A. citreus* BI-100 in mineral salts medium with lindane (100 mg/L) as the sole source of carbon. Gupta et al. (2001) reported the degradation of lindane by *Alcaligenes faecalis*, isolated from agricultural fields.

*Bacillus circulans* and *B. brevis* were isolated from soil contaminated with lindane and acclimatized to different concentrations of lindane for more than 2 years. These bacteria degraded 80% of γ-HCH at 5 μg $mL^{-1}$ lindane within 8 days (Gupta et al. 2000).

Abou-Arab (2002) studied the effect of starter on lindane in tryptone soya broth (TSB), mineral salt medium (MSM), and fermented sausage, and found the role of *Lactobacillus plantarum* on the degradation of lindane. The reductions of lindane were 27.9% and 40% in TSB and MSM without nitrite addition, or 38.4% and 48.4% in the same media with nitrite addition. He also found that *Micrococcus varians* had the ability to metabolize lindane. If the sausage mixture were cultured with meat starter (*L. plantarum* and *M. varians*) for 72 hours, then 18% of lindane was degraded. Gram-negative bacteria *Pandoraea* sp. substantially degraded lindane under aerobic conditions at concentrations of 10 to 200 mg $L^{-1}$ in liquid cultures. After 8 weeks of incubation in liquid culture, 89.9% of the lindane declined at an initial concentration of 150 mg $L^{-1}$. The optimum pH for growth and biodegradation of α and γ-HCH in soil slurries was 9 (Okeke et al. 2002). The aerobic biodegradation of lindane by *B. thiooxidans* bacteria from sediment at a polluted site on the Suquia river, Cordoba, Argentina, was reported by Pesce and Wunderlin (2004). *B. thiooxidans* were able to degrade lindane after 3 days of growth. About 12 novel lindane-degrading bacterial strains have been isolated from lindane-contaminated sites at Chemnitz in Germany (DS2, DS2–2, and DS3–1) (Boltner et al. 2005) and Bilbao in northern Spain (γ12–7, γ16–1, γ1–7, α1–2, α4–5, α16–10, α16–12, γ16–9, and α4–2), and all these isolates were also found to be represented by the members of the family *Sphingomonadaceae*.

A strain of *Pseudomonas paucimobilis* isolated from paddy field rhizosphere soil was demonstrated to degrade lindane (Sahu et al. 1990). About 98% of lindane was aerobically degraded by *S. paucimobilis* after 12 days of incubation (Johri et al. 2000). *S. ummariense* sp. nov. was isolated from an lindane dump site located in the northern part of India (Singh and Lal 2009). *Pseudomonas aeruginosa* ITRC-5 can degrade α-, β-, γ-, and δ-HCH, in both liquid-culture and contaminated soils, and the degradation of β- and δ-isomers is enhanced in the presence of α- or γ-HCH (Kumar et al. 2005). Incubation of "muck" with the isolated bacterium *P. aeruginosa* ITRC-5 under optimized conditions (i.e., 1.7 mM input concentration, pH 9, and temperature 20–30 ±1°C) causes substantial degradation of HCH-isomers, which is accompanied with reduction of their toxicity (Chaudhary et al. 2006). Bioconversion and biological growth kinetics of *P. aeruginosa* degrading technical HCH was investigated in batch process under aerobic conditions by Lodha et al. (2007). At lower technical HCH concentrations (1–10 mg/L), degradation (above 99%) was observed, whereas at higher concentration (20–50 mg/L), the degradation efficiency was reduced. A gram-positive *Microbacterium* sp. strain, ITRC1, has the capacity to degrade all four major isomers of HCH present in both liquid cultures, and aged contaminated soil was isolated and characterized by Manickam et al. (2006). For the first time a *Xanthomonas* sp. was isolated from a contaminated soil that utilized lindane as sole carbon and energy source by successive dechlorination (Manickam et al. 2008).

A yellow-pigmented, lindane-degrading bacterial strain, *S. quisquiliarum* P25 (T) was isolated from a lindane dumping site located in the northern part of India (Bala et al. 2009). Dadwal et al. (2009) reported a lindane degradation capability of *S. chinhatense* IP26, *S.* UM2, *S.* HDU05 and *S.* UM1, *S.* F2, *S.* HDIP04. *Azotobacter chroococcum* JL 102 was screened for lindane degradation by a chloride estimation method. Maximum degradation of lindane was recorded at 10 ppm concentration in

Jensen's medium. A pot culture experiment conducted to study the *in situ* degradation potential of this strain for a period of 8 weeks and showed increased degradation over the days with maximum degradation observed on the eighth week of incubation (Anupama and Paul 2010). In experiments with marine sediments collected from around burrows of *Saccoglossus kowalevskii* (a tribromopyrrole-producing marine hemichordate), Boyle et al. (2010) demonstrated lindane biodegradation under sulfate-reducing conditions with short-chain fatty acids serving as the carbon and energy source. Molybdate inhibited lindane biodegradation, suggesting that sulfate reduction plays a central role in this activity. Chlorobenzene (CB) and benzene were detected as products. Pure cultures of *Desulfovibrio gigas*, *D. africanus,* and *D. multivorans* were also able to dehalogenate lindane to benzene and CB. Nevertheless, no attempt was made to determine if lindane was being used as a terminal electron acceptor. Also, no attempt was made to reconcile potential competition between use of sulfate and lindane as terminal electron acceptors. Although sulfate-reducing bacteria clearly played an important role in lindane biodegradation in the study by Boyle et al. (2010), what that role is remains to be determined.

Zheng et al. (2011) reported degradation of lindane by *Sphingobium* strains (*S. indicum* B90A, *S. japonicum* UT26, *S. francense Sp+*) at low temperature (4°C). Further, *S. lactosutens* sp. nov. (Kumari et al. 2009), *Novosphingobium lindaniclasticum* sp. nov. (Saxena et al. 2012) were isolated from Lucknow. *S. baderi* LL03(T) sp. nov. (Kaur et al. 2012) and *S. czechense* sp. nov. *N. barchaimii* sp. nov. (Niharika et al. 2012a) were also isolated from lindane-contaminated soil at Spolana Neratovice, Czech Republic, where lindane was formerly produced. Also *Pseudoxanthomonas indica* sp. nov. (Kumari et al. 2011), *Flavobacterium ummariense* sp. nov. (Lata et al. 2011), *Pontibacter lucknowensis* sp. nov. (Dwivedi et al. 2012), *Pontibacter ramchanderi*, and *P. indicus* sp. nov. (Singh et al. 2013a, b), were isolated from a lindane dump site in Lucknow, India.

### 6.5.2 Lindane Mineralization

Actinomycetes for bioremediation are an attractive approach, since these microorganisms have already adapted to various habitats (Shelton et al. 1996). Pesticide-degrading actinomycetes belong to *Arthrobacter, Brevibacterium, Corynebacterium, Clavibacter Micromonospora, Mycobacterium, Nocardia, Nocardioides, Rhodococcus,* and *Streptomyces* genera (De Schrijver and De Mot 1999).

Actinomycetes have great potential for bioremediation of toxic compounds (Ravel et al. 1998). Benimeli et al. (2003, 2006, 2007) and Benimeli (2004) isolated and selected wild-type *Streptomyces* strains that were able to tolerate and remove lindane from river sediments and other local contaminated sites. Benimeli et al. (2007) reported removal of lindane without intracellular accumulation or biotransformation products of lindane using *Streptomyces* sp. M7 grown in sterile soil with different initial pesticide concentrations (100, 150, 200, and 300 μg $kg^{-1}$), when a decrease of the residual lindane concentration was detected in soils samples (29.1%, 78.03%, 38.81%, and 14.42%, respectively) (Benimeli et al. 2008).

Cuozzo et al. (2009) detected dechlorinase activity and lindane catabolism products as a result of microbial lindane degradation by *Streptomyces* sp. M7, isolated in Tucuman, Argentina. He also demonstrated that synthesis of dechlorinase in *Streptomyces* sp. M7 was induced when the microorganism was grown in the presence of lindane as only carbon source. He also reported the release of c-2,3,4,5,6-pentachlorocyclohexene and 1,3,4,6-tetrachloro-1,4-cyclohexadiene, the first and second products in lindane degradation by *Streptomyces* sp. M7 according to the catabolic pathway proposed by Nagata et al. (2007).

Fuentes et al. (2010, 2011) isolated actinomycetes, belonging to the *Streptomyces* and *Micromonospora* genera (*Streptomyces* sp. C39, *S.* MC1, *S.* M15, *S.* M30, *S.* M50, *S.* A1, *S.* A2, *S.* A3, *S.* A5, *S.* A6, *S.* A7, *S.* A8, *S.* A11, *S.* A12, *S.* A13, *S.* A14, and *Micromonospora* sp. A10) from sites contaminated with OC pesticides in Argentina and were able to grow in the presence of chlordane, lindane, and methoxychlor. They were also able to remove these compounds from the culture medium or degrade them and release chloride ions as a result of degradation. These results favor application of actinomycetes as potential agents for bioremediation of polluted environments with different OC pesticides.

De Paolis et al. (2013) isolate two *Arthrobacter* strains (*A. fluorescens* and *A. giacomelloi*), ability to grow in a mineral salt medium containing α-, β-, or γ-HCH (100 mg/L) as sole source of carbon. Although both bacteria are able to metabolize the HCHs but *A. giacomelloi* is the most effective one, as after 72 hours of incubation it produces 88% degradation of α-, 60% of β-, and 56% of γ-HCH (lindane). Pentachlorocyclohexenes and tetrachlorocyclohexenes have been detected as metabolites, which are almost completely eliminated after 72 hours of incubation.

### 6.5.3 Role of Fungal Extracellular Oxidative Enzymes in Lindane Decontamination

Most of the lindane-degrading fungi known to date are members of the family of white-rot fungi, and a very few non-white-rot fungi have been noted to degrade the lindane. The lindane biodegradation is accomplished with the action of extracellular oxidative enzymes, produced by the fungus to decompose woody substrates, such as laccase, manganese peroxidise and lignin peroxidase (Rigas et al. 2005; McErlean et al. 2006). Bumpus and Aust (1987) published the first report on the biodegradation of lindane by white-rot fungi *Phanerochaete chrysosporium*. Mougin et al. (1997) reported enhanced mineralization in soils supplemented with lindane by *Phanerochaete* sp., and the fungus seemed to modify the lindane degradation pathway by increasing the conversion of volatile intermediates to $CO_2$. Biodegradation of lindane up to 85%–95% by white-rot fungi such as *Pleurotus ostreatus, P. sajor-caju,* and *Trametes hirsutus* has been reported (Arisoy and Kolankaya 1997; Singh and Kuhad 1999; Papadopoulou et al. 2006). Singh and Kuhad (1999) investigated the lindane degradation ability of the white-rot fungus *T. hirsutus* in liquid culture, which was compared with *P. chrysosporium*. It was shown that *T. hirsutus* degraded lindane faster than *P. chrysosporium*. Singh and Kuhad (2000) studied the lindane degradation capability of two white-rot fungi, *Cyathus bulleri* and *P. sordida*, and reported that *C. bulleri* degraded lindane more efficiently than *P. sordida*.

The degradation of lindane at various concentrations by a subtropical white-rot fungus was studied in batch and packed bed bioreactor systems. About 82 ± 6% degradation of lindane was achieved in batch cultures and 81% degradation was noted in packed bed reactor (Tekere et al. 2002). The bioremediation process was evaluated in polypore fungus *Ganoderma australe* in mixtures of a sandy soil and wheat straw doped with lindane (Rigas et al. 2007). Maximal degradation of 94.5% was attained after 30 days for lindane by the white-rot fungus *Bjerkandera adusta* in a slurry batch bioreactor (Quintero et al. 2007). Biodegradation of lindane by *Phycomycetes* and *Conidiobolus*, a non-white-rot fungus was reported by Nagpal et al. (2008). The fungus completely degraded lindane on the fifth day in the culture medium. Quintero et al. (2008) observed lindane degradation by nine fungal species out of which only six fungal species (*B. adusta, Poliporus ciliatus, Lentinus tigrinus, Stereum hirsutum, Pleurotus eryngii,* and *Irpex lacteus*) show lindane degradation between 15.1% and 70.8%. The bracket-like polypore fungus *Ganoderma australe* was selected for its potential to degrade lindane in liquid-agitated sterile cultures. Maximum lindane biodegradation (3.11 mg $g^{-1}$ biomass) was obtained with the addition of nitrogen supplements during 5 days of cultivation time (Dritsa et al. 2009). Two *Fusarium* species (*F. poae* and *F. solani*) isolated from pesticide-contaminated soil showed better degradability of lindane used as a sole carbon source (Sagar and Singh 2011). The fungal strain *Fusarium verticillioides* AT-100 isolated from *Agave tequilana* leaves can degrade lindane (50 mg/L) after 7 days of incubation and utilize lindane as the sole source of carbon and energy (Guillen-Jimeneza et al. 2012).

Four yeast strains were isolated on the basis of their lindane degradation ability. Among them *Candida* sp. VITJzN04 showed the best potentiality for lindane degradation in solid as well as liquid media followed by *Rhodotorula* sp. VITJzN03, *Pseudozyma* sp. VITJzN01, and *Cintractia sorghi* VITJzN02. Initial concentration of 400 mg/L of lindane was found to be optimum for VITJzN01 and VITJzN02, whereas a higher concentration of 600 mg/L was optimum for the growth of VITJzN03 within a period of 10 days and VITJzN04 and complete removal of lindane. The calculated degradation rate and half-life of lindane were found to be 0.416 $day^{-1}$ and 1.66 days for VITJzN03 and 1.17 days, 0.588 $day^{-1}$ for VITJzN04. This is the first report of complete degradation of lindane at a high concentration such as 600 mg/L by yeasts *Candida* VITJzN04 within the shortest period of 6 days. The addition of $H_2O_2$ in the mineral medium showed 32% enhancement of lindane degradation within 3 days. In addition, involvement of the enzymes viz. dechlorinase, dehalogenase, dichlorohydroquinone reductive dechlorinase, lignin peroxidase, and manganese peroxidase was noted during lindane degradation (Salam and Das 2014).

### 6.5.4 Role of Plant–Microbe Association in Lindane Remediation

Rhizoremediation, the degradation of toxicants by microorganisms in the rhizosphere, holds incredible potential for the remediation of contaminated soil (Kuiper et al. 2004a,b). In the "rhizosphere effect," plants provide nutrients in the form of root exudates, oxygen, and favorable redox conditions to soil microorganisms, and this in turn results in increased bacterial diversity, population density, and activity compared with bulk soil (Molina et al. 2000; Vílchez et al. 2000; Espinosa-Urgel and Ramos 2001).

Presence of chili, corn, and coriander root exudates can increase lindane degradation efficiency of the three bacterial culture (*Klebsiella* sp., *Pseudomonas* sp., *Pseudoarthrobacter* sp.) up to ~10%–15% (Nagpal and Paknikar 2006). Boltner et al. (2007) used a two-step enrichment approach to isolate five root-colonizing HCH degrading *Sphingomonas* strains (DS-204B, OF-178A, GOF-203, Ans-PL0, Ans-PL2). Out of them two HCH-degrading sphingomonads (strains GOF-203 and Ans-PL0) show high colonization rate and were attached to corn seeds and used for rhizoremediation purposes. After 25 days about a 30% reduction of initial concentration 0.5 mg of γ-HCH per gram of soil was achieved.

A comparative study on lindane remediation potential of four rhizospheric bacterial species viz. *Kocuria rhizophila, Microbacterium resistens, Staphylococcus equorum,* and *S. cohnii* was reported by Abhilash et al. (2011). Abhilash et al. (2011) tested the combined rhizoremediation potential of *S. cohnii* sub sp. *urealyticus* in the presence of *Withania somnifera* grown in lindane-spiked soil. The results showed that the integrated use of stress-tolerant plant species and rhizospheric microbial inoculation enhanced the dissipation of lindane.

Alvarez et al. (2012) studied the dissipation of lindane by native *Streptomyces* strains in the presence of root exudates of *Zea mays* and reported enhanced dissipation of lindane by the microbes when grown on the root exudates. Abhilash and Singh (2013) demonstrated rhizoremediation potential of biodiesel plant, *Jatropha curcas* L. against lindane. *Jatropha* was proved to enhance the dissipation of lindane in simulated soil. Becerra-Castro et al. (2013a) improved the performance of leguminous shrub *Cytisus striatus* on substrates contaminated with HCH isomers using microbial inoculants. The plants were inoculated with endophyte *Rhodococcus erythropolis* ET54b and the HCH-degrader *Sphingomonas* sp. D4, singly or in combination. The outcome of the study revealed that inoculating *C. striatus* with the combination of bacterial strains is an effective method for the remediation of HCH contaminated sites. Kurashvili et al. (2014) used joint application of plants chickling vetch (*Lathyrus sativum*), soybean (*Glycine max*), maize (*Zea mays*), alfalfa (*Medicago sativa*), chickpea (*Cicer arietinum*), and lettuce (*Lactuca sativa*), and microorganisms (11 strains from genera *Pseudomonas*) to degrade lindane. They also studied the tolerance of plants to tested pesticides and induction degree of plant detoxification enzymes like phenoloxidase, peroxidase, glutathione S-transferase, and cytochrome P450, by these compounds. Obtained results showed that alfalfa, maize, and soybean among tested six plant species have highest tolerance to pesticides.

## 6.6 GOVERNMENT AND PRIVATE INITIATIVES ON LINDANE CONCERN

Before getting banned, many registered companies used lindane in various fields, which include direct treatment of livestock, pet products, ornamentals, home lawns, fallow areas, commercial food processing facilities and storage areas, greenhouses, wood treatment, forestry, Christmas tree plantations, military use on human skin and clothing (USEPA, fact sheet 2006). Lindane is classified by the US EPA as one of 22 "Bioaccumulative Chemicals of Concern" in the Great Lakes. New releases of these chemicals are restricted into "mixing zones" because of "proceeding with confirmation that the exceptionally bioaccumulative nature of these lethal chemicals displays a huge potential hazard to human well-being, aquatic life and wild life" (Great Lakes Initiative, Fact Sheet). The US EPA likewise arranged lindane as an "Extremely Hazardous Substance" in Section 302 of the Emergency Planning and Community Right-to-Know Act. 11.12. Under the Clean Water Act, the US EPA characterizes lindane as a "Priority Pollutant" "for the security of sea-going life and human well-being in surface water" (USEPA Water Quality Criteria). Lindane is incorporated into the US EPA's Toxic Release Inventory (TRI) program, which "requires facilities in specific enterprises, which produce, process, or utilize critical measures of harmful chemicals, to report every year on their release of these chemicals" (TRI actuality sheet).

Lindane and the other HCH isomers are mobile in the environment, and through long-extended environmental transport, are deposited in the Arctic, where they have been recognized in air, surface water, groundwater, residue, soil, ice, snowpack, fish, wildlife, and human beings. All utilizations with the exception of pharmaceutical uses are currently limited. "On August 2, 2006, EPA reported that registrants Chemtura USA Corporation, trailed by AGSCO Inc, Drexel Chemical Co. furthermore, JLM Industries, Inc. asked for to intentionally wipe out all residual pesticide enlistments of the OC pesticide lindane. EPA also made an assurance that the rest of the employments of lindane are not qualified for re-enlistment" (Lindane RED Addendum 2006).

The US FDA (2003a) published a general well-being consultative for lindane. It referred to an assortment of concerns, including the following: "In post-promoting reports, neurologic reactions happened in patients who misused lindane, and in addition in patients who utilized lindane according to labelled instructions. Among adverse events reported in the FDA database, 70% reported neurologic events including seizure, dizziness, headache and paresthesia." The US FDA (2003b) issued indistinguishable Black Box Warnings for creams and shampoos containing lindane: "Lindane moisturizer ought to be utilized with alert for babies, kids, the elderly, and people with other skin conditions (e.g. atopic dermatitis, psoriasis) and in the individuals who weigh <110 lbs (50 kg) as they might be at danger of genuine neurotoxicity." According to an appraisal by the Center for Drug Evaluation and Research (2003), "Lindane was marked a second line treatment in 1995 in light of the fact that, while it is comparative in real life to other endorsed treatments, it has a higher percutaneous absorption than other affirmed scabicides and pediculicides. This more prominent systemic exposure may mean a more noteworthy potential for genuine adverse events."

The U.S. Agency for Toxic Substances and Disease Registry (ATSDR) (2005) ranks lindane 32nd of the 275 substances on its list of CERCLA (Superfund) "Priority Pollutants":

> This list reflects a prioritization of substances in view of a mix of their recurrence, poisonous quality, and potential for human introduction at NPL (National Priorities List) sites. Beginning January 1, 2002, any item utilized for the treatment of lice or scabies in individuals that contains the pesticide Lindane should not be utilized or

> sold as a part in state of California (California Assembly Bill 2000). Lindane is banned for use in 52 nations, and limited or extremely confined in 33 nations. (EPA Risk Assessment Fact Sheet 2006)

The International Agency for Research on Cancer (IARC), the chief office on cancer-causing agent classification, at present considers HCHs, the class of chemicals to which lindane has a place, as "possibly carcinogenic to humans" (IARC 1998). The UN Environment Programme (2005) reported the selection of lindane to become one of five new contaminants to be added to the first 12 in the Stockholm Convention on Persistent Organic Pollutants (POPs): "Mexico is naming the pesticide lindane together with a related gathering of chemicals known as HCHs. It clarifies that creating the 99%-pure γ-HCHs required for each ton of lindane results in 6–10 tons of waste isomers. The subsequent waste isomer issue exacerbates the dangers postured by lindane itself. The Rotterdam Convention on Prior Informed Consent (PIC), a procedure for certain hazardous chemicals and pesticides in international trade, incorporates lindane among the 39 pesticides which are banned or extremely confined for human uses.

Several health professional organizations in Michigan have supported a lindane phase-out such as: Michigan Chapter of the American Academy of Pediatrics, Michigan Council of Nurse Practitioners, Michigan Nurses Association, Michigan Pharmacists Association, Michigan Association of School Nurses, Wayne County Medical Society of Southeast Michigan.

> On July 13, 2000, the European Union Standing Committee on Plant Health voted in favor of a restriction on lindane with the support from UK and other Member States. The European Commission confirmed the choice in December 2000, which ought to happen before the end of 2001. The boycott will cover all agricultural and amateur gardening uses of lindane. The 13th International HCH Pesticides Forum was organized in Zaragoza November 3–6, 2015 by International HCH and Pesticides Association (IHPA), the Aragon Government and SARGA to attract consideration regarding the huge ecological and money-related issues in the locale of Aragon (Spain) created by the former production of Lindane by the Inquinosa manufacturing plant and the legacies made by the wild dumping of gigantic amounts of HCH waste in the region encompassing Sabinanigo and the Gallego River.

## 6.7 CONCLUSIONS

Based on the above discussion it may be concluded that lindane is highly toxic to aquatic animals while moderately toxic to birds and mammals. It affects the central nervous system, liver, kidney, reproductive systems, and endocrine glands in humans. Several government and private organizations, environment agencies, and nongovernment organizations spread awareness about this chemical. The presence of lindane in the Arctic region is evidence of its long-range transport, as it has never been used or distributed in the Arctic regions.

Several soil microorganisms have been reported for degrading and using lindane as a carbon source. In selected bacterial strains, the genes encoding the catalysts required in the underlying degradation of lindane have been cloned, sequenced, and expressed, and the gene products have also been characterized. More research is expected to comprehend the essential component of associations of lindane-degrading microorganisms with the soil environment that directs lindane remediation.

In many cases, despite the fact that the additional microorganisms can degrade the target pollutants, bioremediation does not work effectively in field conditions. This could be due to the wide fluctuations in temperature, pH, and dampness content, under other ecological conditions. Constrained accessibility of supplements, the measure of the toxin, and its age nearness of inhibitory substances and rivalry with indigenous microflora are different determinants of the degradative action. In spite of these, reports related to bioremediation of lindane are not sufficient; therefore, much extensive work is required by isolation and identification of potential microbes that may solve the problem of lindane contamination.

## ACKNOWLEDGMENTS

Siddharth Boudh is thankful to the University Grants Commission (UGC) for awarding Rajiv Gandhi National Fellowship (Award Letter No: F.1-17.1/2013-14/RGNF-2013-14-SC-UTT-37387)/(SA-IIII/Website).

## REFERENCES

Abhilash, P.C., and Singh, N. 2008. Influence of the application of sugarcane bagasse on lindane (γ-HCH) mobility through soil column: Implication for biotreatment. *Bioresource Technology* 99: 8961–8966.

Abhilash, P.C., Srivastava, S., and Singh, N. 2011. Comparative bioremediation potential of four rhizospheric microbial species against lindane. *Chemosphere* 82: 56–63.

Abou-Arab, A.A.K. 2002. Degradation of organochlorine pesticides by meat starter in liquid media and fermented sausage. *Food Chemistry and Toxicology* 40(1): 33–41.

Alvarez, A., Yanez, M.L., Benimeli, C.S., and Amoroso, M.J. 2012. Maize plants (*Zea mays*) root exudates enhance lindane removal by native *Streptomyces* strains. *International Biodeterioration and Biodegradation* 66: 14–18.

Amaraneni, S.R., and Pillala, R.R. 2001. Concentrations of pesticide residues in tissues of fish from Kolleru Lake in India. *Environmental Toxicology* 16: 550–556.

Anupama, K.S., and Paul, S. 2010. Ex situ and in situ biodegradation of lindane by *Azotobacter chroococcum*. *Journal of Environmental Science and Health* 45: 58–66.

Arisoy, M., and Kolankaya, N. 1997. Biodegradation of lindane by *Pleurotus sajor-caju* and toxic effects of lindane and its metabolites on mice. *Bulletin of Environmental Contamination and Toxicology* 59: 352–359.

Aspinwall, L.S., Bermudez, I., and King, L.A. 1997. The interactions of hexachlorocyclohexane isomers with human gamma-aminobutyric acid (A) receptors expressed in *Xenopus* oocytes. *Journal of Pharmacology and Experimental Therapeutics* 282(3): 1557–1564.

ATSDR. 1999. *Toxicological profile for alpha-, beta-, gamma-, and delta hexachlorocyclohexane*. Agency for Toxic Substances and Disease Registry, US Department of Health and Human Services, Atlanta, GA.

ATSDR. 2005. *Toxicological profile for alpha-, beta-, gamma-, and delta hexachlorocyclohexane*. Agency for Toxic Substances and Disease Registry.

Attia, A.M., Richardson, B.A., and Rodriguez, C. 1991. Lindane may enhance nocturnal pineal N acetyltransferase activity via beta-adrenergic receptors. *Brain Resources* 554: 253–256.

Aulakh, R.S., Gill, J.P.S., Bedi, J.S., Sharma, J.K., Joia, B.S., and Ockerman, H.W. 2006. Organochlorine pesticide residues in poultry feed, chicken muscle and eggs at a poultry farm in Punjab, India. *Journal of Science and Food Agriculture* 86: 741–744.

Bala, K., Sharma, P., and Lal, R. 2009. *Sphingobium quisquiliarum* sp. nov., a hexachlorocyclohexane (HCH) degrading bacterium isolated from HCH contaminated soil. *International Journal of Systematic and Evolutionary Microbiology* 60(2): 429–433.

Bárczi, I. 1994. The greatest chlorinated hydrocarbon problem of Hungary, 3rd HCH and Halogenated Pesticides Forum, 19–20 September 1994, Bilbao. http://www.ihpa.info/library.htm 3rd Forum (accessed December 10, 2016).

Beard, A.P., and Rawlings, N.C., 1998. Reproductive effects in mink (*Mustela vison*) exposed to the pesticides lindane, carbofuran and pentachlorophenol in a multigenerational study. *Journal of Reproductive Fertility* 113: 93–104.

Benimeli, C.S., Amoroso, M.J., Chaile, A.P., and Castro, G.R. 2003. Isolation of four aquatic streptomycetes strains capable of growth on organochlorine pesticides. *Bioresource Technology* 89: 133–138.

Benimeli, C.S., Chaile, A.P., and Amoroso, M.J. 2004. Methods for determining lindane concentration in water and solid samples. In: *Environmental Microbiology: Methods and Protocols*, Spencer, J.F.T., Ragout de Spencer, Alicia, L. (eds.). Humana Press Inc., Totowa, NJ, pp. 279–282.

Benimeli, C.S., Castro, G.R., Chaile, A.P., and Amoroso, M.J. 2006. Lindane removal induction by *Streptomyces* sp. M7. *Journal of Basic Microbiology* 46(5): 348–357.

Benimeli, C.S., Castro, G.R., Chaile, A.P., and Amoroso, M.J. 2007. Lindane uptake and degradation by aquatic *Streptomyces* sp. strain M7. *International Biodeterioration and Biodegradation* 59: 148–155.

Benimeli, C.S., Fuentes, M.S., Abate, C.M., and Amarosoa, M.J. 2008. Bioremediation of Lindane contaminated soil by *Streptomyces* sp. M7 and its effects on *Zea mays* growth. *International Biodeterioration and Biodegradation* 61: 233–239.

Bentz, R. 2004. Remediation of the Maag-site in Dielsdorf (Switzerland), contaminated soil. *Proceedings of the Eighth International FZK/TNO Conference on contaminated Soil*, 12-16 May, Gent, Belgium. SpS-3899.

Beyer, A., and Matthie, M. 2001. Long-range transport potential of semi-volatile organic chemicals in coupled air-water systems. *Environmental Science and Pollution Research*, 8: 173–179.

Bintein, S., and Devillers, J. 1996. Evaluating the environmental fate of lindane in France. *Chemosphere* 32(12): 2427–2440.

Boltner, D., Godoy, P., Munoz-Rojas, J., Duque, E., Moreno-Morillas, S., Sanche Z. L., and Ramos, J.L. 2007. Rhizoremediation of lindane by root-colonizing *Sphingomonas*. *Microbial Biotechnology*, 1: 87–93.

Boltner, D., Moreno-Morillas, S., and Ramos, J. 2005. 16S rDNA phylogeny and distribution of lin genes in novel hexachlorocyclohexane-degrading *Sphingomonas* strains. *Environmental Microbiology* 7: 1329–1338.

Boudh, S., Tiwari, S., and Singh, J.S. 2016. Microbial mediated lindane bioremediation. In *Agro-environmental Sustainability-Managing Environmental Pollution*, Vol-II, Singh, J.S., and Seneviratne, G. (eds.). Springer Nature, Switzerland, pp. 213–233.

Boyle, A.W., Haggblom, M.M., and Young, L.Y. 1999. Dehalogenation of lindane (gamma-hexachlorocyclohexane) by anaerobic bacteria from marine sediments and by sulfate-reducing bacteria. *FEMS Microbiology Eco*logy 29: 379–387.

Brassow, H.L., Baumann, K., and Lehnert, G. 1981. Occupational exposure to hexachlorocyclohexane: II. Health conditions of chronically exposed workers. *International Archives of Occupational Environmental Health* 48: 81–87.

Bro-Rasmussen, F. 1996. Contamination by persistent chemicals in food chain and human health. *Science* of the *Total Environment* 188(Suppl 1): S45-S60.

Bumpus, J.A., and Aust, S.D. 1987. Biodegradation of Chlorinated Organic Compounds by *Phanerochaete chrysosporium*, a Wood-Rotting Fungus. *ACS Symposium Series* 338 (28): 340–349.

California Assembly Bill 2318. 2000. An act to add Section 111246 to the health and safety code, relating to environmental health. www.assembly.ca.gov/acs/acsframeset2text.htm (accessed December 9, 2016).

CEC. 2000. North American Commission for Environmental Cooperation (19th April, 2000): Draft Decision Document on Lindane, Under the Process for Identifying Candidate Substances for Regional Action under the Sound Management of Chemicals Initiative. Commission for Environmental Cooperation, Montreal, Canada.

Ceremonie, H., Boubakri, H., Mavingui, P., Simonet, P., and Vogel, T.M. (2006). Plasmid-encoded γ-hexachlorocyclohexane degradation genes and insertion sequences in *Sphingobium francense* (ex-*Sphingomonas paucimobilis* Sp+). *FEMS Microbiology Ecology* 257: 243–252.

Chaudhary, P., Kumar, M., Khangarot B.S., and Kumar, A. (2006). Degradation and detoxification of hexachlorocyclohexane isomers by *Pseudomonas aeruginosa* ITRC-5. *International Biodeterioration and Biodegradation* 57: 107–113.

ChemSources. 2009. *Chem Sources - Chemical Search*. Chemical Sources International. http://www.chemsources.com/chemonline.html and search on lindane (accessed December 13, 2016).

Choudhary, A., and Sharma, D.C. 2008. Pesticide residues in honey samples from Himachal Pradesh (India). *Bulletin of Environmental Contamination and Toxicology* 80: 417–422.

Cornacoff, J.B., Lloyd, D., Lauer, R.V., House, A.N., Tucker, L.M., Turmond, J.G., Vos, P.W., and Dean, J.H. 1988. Evaluation of the immunotoxicity of γ-hexachlorocyclohexane (γ-HCH). *Fundamental Applied Toxicology* 11: 293–99.

Cuozzo, S.A., Rollán, G.G., Abate, C.M., and Amoroso, M.J. 2009. Specific dechlorinase activity in lindane degradation by *Streptomyces* sp. M7. *The World Journal of Microbiology and Biotechnology* 25: 15391546.

Dadhwal, M., Jit, S., Kumari, H., and Lal, R. 2009. *Sphingobium chinhatense* sp. nov., a hexachlorocyclohexane (HCH)-degrading bacterium isolated from an HCH dumpsite. *International Journal of Systematic and Evolutionary Microbiology*, 59: 3140–3144.

Dalsenter, P.R., Faqi, A.S., and Webb, J. 1997. Reproductive toxicity and toxicokinetics of lindane in the male offspring of rats exposed during lactation. *Human & Experimental Toxiology* 16: 146–153.

Datta, J., Maiti, A.K., Modak, D.P., Chakrabartty, P.K., Bhattacharyya, P., and Ray, P.K. 2000. Metabolism of γ- hexachlorocyclohexane by *Arthrobacter citreus* strain BI-100: identification of metabolites. *Journal of General and Applied Microbiology*, 46: 59–67.

De Bruin, J. 1979. Final Report, Reduction of Chlordane, DDT, Heptachlor, Hexachlorobenzene and Hexachlorocyclohexane isomers, Prepared for the European Communities Environment and Consumer Protection Service, May 1979.

De Paolis, M.R., Lippi, D., Guerriero, E., Polcaro, C.M., and Donati, E. 2013. Biodegradation of α-, β-, and γ-hexachlorocyclohexane by *Arthrobacter fluorescens* and *Arthrobacter giacomelloi*. *Applied Biochemistry and Biotechnology*, 170(3): 514–524.

De Schrijver, A., and De Mot, R. 1999. Degradation of pesticides by actinomycetes. *Critical Reviews in Microbiology* 25: 85–119.

Dewan, A., Gupta, S.K., and Jani, J.P. 1980. Effect of lindane on antibody response to typhoid vaccine in weanling rats. *Journal of Environmental Science and Health*, B15: 395–402.

Dhananjayan, V., Ravichandran, B., and Rajmohan, H.R. 2012. Organochlorine pesticide residues in blood samples of agriculture and sheep wool workers in Bangalore (Rural), India. *Bulletin of Environmental Contamination and Toxicology* 88: 497.

Dogra, C., Raina, V., Pal, R., Suar, M., Lal, S., Gartemann, K.H., van der Meer, J.R., and Lal, R. (2004). Organization of lin genes and IS*6100* among different strains of hexachlorocyclohexane-degrading *Sphingomonas paucimobilis*: Evidence for horizontal gene transfer. *Journal of Bacteriology* 186: 2225–2235.

Dritsa, V., Rigas, F., Doulia, D., Avramides, E.J., and Hatzianestis, I. 2009. Optimization of culture conditions for the biodegradation of lindane by the polypore fungus *Ganoderma australe*. *Water, Air, & Soil Pollution* 204: 19–27.

Dwivedi, V., Niharika, N., and Lal, R. 2012. *Pontibacter lucknowensis* sp. nov., isolated from a hexachlorocyclohexane dump site. *International Journal of Systematic and Evolutionary Microbiology* 63(Pt 1): 309–13.

Endo, R., Kamakura, M., Miyauchi, K., Fukuda, M., Ohtsubo, Y., Tsuda, M., and Nagata, Y. (2005). Identification and characterization of genes involved in the downstream degradation pathway of γ-hexachlorocyclohexane in *Sphingomonas paucimobilis* UT26. *Journal of Bacteriology* 187: 847–853.

Espinosa-urgel, M., and Ramos, J.I. 2001. Expression of a *Pseudomonas putida* Aminotransferase Involved in Lysine Catabolism Is Induced in the Rhizosphere. *Applied and Environmental Microbiology*, 67(11): 5219–5224.

Fact Sheet. 2005. *Lindane's dirty secret: Indian facilities dump toxic waste*. CAPE (Community Action for Pesticide Elimination), Kerala, India.

FDA. 2006. *Total diet study market baskets 1991-3 through 2003-4*. U.S. Food and Drug Administration.

FDA. 2009. *The electronic orange book*. U.S. Food and Drug Administration. http://www.fda.gov/cder/ob/default.htm

Fonseca, R.G., Resende, L.A.L., and Silva, M.D. 1993. Chronic motor neuron disease possibly related to intoxication with organochlorine insecticides. *Acta Neurologica Scandinavica* 88: 56–58.

Francis, A.J., Spanggord, R.J., and Ouchi, G.I. 1975. Degradation of lindane by *Escherichia coli*. *Applied Microbiology* 29: 567–568.

Fuentes, M.S., Benimeli, C.S., Cuozzo, S.A., and Amoroso, M.J. 2010. Isolation of pesticide-degrading actinomycetes from a contaminated site: Bacterial growth, removal and dechlorination of organochlorine pesticides. *International Biodeterioration and Biodegradation* 64: 434–441.

Fuentes, M.S., Saez, J.M., Benimeli, C.S., and Amoroso, M.J. 2011. Lindane biodegradation by defined consortia of indigenous *Streptomyces strains*. *Water Air Soil Pollution*, 222: 217–231.

Gilbert, M.E., and Mack, C.M. 1995. Seizure thresholds in kindled animals are reduced by the pesticides lindane and endosulfan. *Neurotoxicology and Teratology* 17: 143–150.

Guillen-Jimeneza, F., Cristiani-Urbinab, E., Cancino-Díazc, J.C., Flores-Morenod, J.L., and Barragán-Huertaa, B.E. 2012. Lindane biodegradation by the Fusarium verticillioidesAT-100 strain, isolated from Agave tequilana leaves: Kinetic study and identification of metabolites. *International Biodeterioration and Biodegradation*, 74: 36–47.

Gupta, A., Agarwal, R., and Shukla, G.S. 1999. Functional impairment of blood–brain barrier following pesticide exposure during early development in rats. *Human and Experimental Toxicology* 18(3): 174–179.

Gupta, A., Kaushik, C.P., and Kaushik, A. 2000. Degradation of hexachlorocyclohexane (HCH; α-, β-, γ- and δ-) by *Bacillus circulans* and *Bacillus brevis* isolated from soil contaminated with HCH. *Soil Biology and Biochemistry* 32: 1803–1805.

Gupta, A., Kaushik, C.P., and Kaushik, A. 2001. Degradation of hexachlorocyclohexane isomers by two strains of *Alcaligenes faecalis* isolated from a contaminated site. *Bulletin of Environmental Contamination and Toxicology*, 66: 794–800.

Hardie, D.W.F. 1964. Benzene hexachloride. In *Encyclopedia of Chemical Technology,* vol. 5, Kirk, R.E. and Othmer, D.F. (eds.), 2nd edn., Wiley, New York, pp. 267–281.

Harner, T., Kylin, H., Bidleman, T.F., and Strachan, W.M. 1999. Removal of α-and γ-hexachlorocyclohexane and enantiomers of α-hexachlorocyclohexane in the eastern Arctic Ocean. *Environmental Science and Technology*, 33: 1157–1164.

Hauzenberger, I. 2004. Reports on substances scheduled for reassessments under the UNECE POPs protocol. *Technical review report on Lindane*, Federal Environment Agency, Wien, Austria.

Heritage, A.D., and MacRae, I.C. 1977. Identification of intermediates formed during the degradation of hexachlorocyclohexanes by *Clostridium sphenoides. Applied and Environmental Microbiology* 33: 1295–1297.

Hong, H.L., and Boorman, G.A. 1993. Residual myelotoxicity of lindane in mice. *Fundamental and Applied Toxicology* 21: 500–507.

HSDB. 2009. *Hazardous substances data bank.* National Library of Medicine. http://toxnet.nlm.nih.gov/cgi-bin/sis/htmlgen?HSDB and search on CAS number (accessed December 13, 2016).

IARC. 1973. Monographs on the evaluation of the carcinogenic risk of chemicals to man: some organochlorine pesticides. Volume 5. World Health Organization, Lyon, France, pp. 47–73.

IARC. 1979. Hexachlorocyclohexane (technical HCH and lindane). In *Some Halogenated Hydrocarbons. IARC Monographs on the Evaluation of Carcinogenic Risk of Chemicals to Humans*, vol. 20. International Agency for Research on Cancer, Lyon pp. 195–239.

IARC. 1998. IARC *Monographs on the evaluation of carcinogenic risks to humans. Overall evaluations of carcinogenicity: An updating of IARC monographs* Vol 1 to 42, Suppl 7. International Agency for Research on Cancer, Lyon, France.

Imai, R., Nagata, Y., Fukuda, M., Takagi, M., and Yano, K. 1991. Molecular cloning of a Pseudomonas paucimobilis gene encoding a 17-kilodalton polypeptide that eliminates HCl molecules from γ-hexachlorocyclohexane. *Journal of Bacteriology* 173: 6811–6819.

Imai, R., Nagata, Y., Senoo, K., Wada, H., Fukuda, M., Takagi, M., and Yano, K. 1989. Dehydrochlorination of γ-hexachlorocyclohexane (γ-BHC) by C-BHC assimilating Pseudomonas paucimobilis. *Agricultural and Biological Chemistry*, 53: 2015–2017.

IPCS. 1998. Hexachlorobenzene health and safety Guide. No.107. International Program on Chemical Safety, World Health Organization.

Iwata, H., Tanabe, S., Sakai, N., and Tatsukawa, R. 1993. Distribution of persistent organochlorines in the oceanic air and surface seawater and the role of ocean on their global transport and fate. *Environmental Science Technology* 27: 1080–1098.

Jagnow, G., Haider, K., and Ellwardt, P.C. 1977. Anaerobic dechlorination and degradation of hexachlorocyclohexane isomers anaerobic and facultative anaerobic bacteria. *Archives of Microbiology* 115: 285–292.

Jedlicka, V., Hermanska, Z., and Smida, I. 1958. Paramyeloblastic leukemia appearing simultaneously in two blood cousins after simultaneous contact with gammexane (hexachlorocyclohexane). *Acta Medica Scandinavica* 161: 447–451.

Johri, A.K., Dua, M., Saxena, D.M., and Sethunathan, N. 2000. Enhanced degradation of hexachlorocyclohexane isomers by *Sphingomonas paucimobilis. Current Microbiology* 41(5): 309–311.

Joy, R.M. 1982. Mode of action of lindane, dieldrin and related insecticides in the centralnervous system, Neurobeh. *Toxicology Teratology* 4: 813–823.

Kalra, R.L., Singh, B., and Battu, R.S. 2003. Organochlorine pesticide residues in human milk in Punjab, India. *Environmental Pollution* 85: 47–151.

Kashyap, S.K. 1986. Health surveillance and biological monitoring of pesticide formulators in India. *Toxicology Letters* 33: 107–114.

Kashyap, S.K., Nigam, S.K., Gupta, R.C., Karnik, A.B., and Chatterjee, S.K. 1979. Carcinogenicity of hexachlorocyclohexane (BHC) in pure inbred Swiss mice. *Journal of Environmental Science and Health* B14: 305–318.

Kaur, J., Moskalikova, H., Niharika, N., Sedlackova, M., Hampl, A., Damborsky, J., Prokop, Z., and Lal, R. 2013. *Sphingobium baderi* sp. nov., isolated from a hexachlorocyclohexane dump site. *International Journal of Systematic and Evolutionary Microbiology* 63: 673–678.

Kaushik, C.P., Sharma, H.R., Jain, S., Dawra, J., and Kaushik, A. 2007. Pesticide residues in river Yamuna and its canals in Haryana and Delhi, India. *Environmental Monitoring and Assessment* 144: 329–340.

Klonne, D.R., and Kintigh, W.J. 1988. *Lindane technical: Fourteen-week dust aerosol inhalation study on mice*. Bushy Run Research Center, Export, PA. Report no. 51-524.

Kuiper, I., Lagendijk, E.L., Bloemberg, G.V., and Lugtenberg, B.J.J. 2004. Rhizoremediation: a beneficial plant-microbe interaction. *Molecular Plant-Microbe Interactions*, 17: 6–15.

Kumar, M., Chaudhary, P., Dwivedi, M., Kumar, R., Paul, D., Jain, R.K., Garg, S.K., and Kumar, A. 2005. Enhanced biodegradation of β and δ -hexachlorocyclohexane in the presence of α- and γ-isomers in contaminated soils. *Environmental Science and Technology* 39: 4005–4011.

Kumari, H., Gupta, S.K., Jindal, S., Katoch, P., and Lal, R. 2009. *Sphingobium lactosutens* sp. nov., isolated from a hexachlorocyclohexane dump site and *Sphingobium abikonense* sp. nov., isolated from oil-contaminated soil. *International Journal of Systematic and Evolutionary Microbiology* 59(9): 2291–2296.

Kumari, K., Sharma, P., Tyagi, K., and Lal, R. 2011. Pseudoxanthomonas indica sp. nov., isolated from a hexachlorocyclohexane dumpsite. *International Journal of Systematic and Evolutionary Microbiology* 61(9): 2107–2111.

Kurashvili, M., Adamia, G., Ananiashvili, T., Amiranasvili, L., Varazi, T., Pruidze, M., Gordeziani, M., and Khatisashvili, G. 2014. Plants and microorganisms for phytoremediation of soils polluted with organochlorine pesticides. *International Journal of Biological, Biomolecular, Agricultural, Food and Biotechnological Engineering* 8(4): 382–384.

Lata, P., Lal, D., and Lal, R. 2011. Flavobacterium ummariense sp. nov., isolated from hexachlorocyclohexane-contaminated soil, and emended description of Flavobacterium ceti Vela et al. 2007. *International Journal of Systematic and Evolutionary Microbiology* 62(11): 2674–2679.

Lal, R., Suar, M., Raina, V., Dogra, C., Prakash, O., and Pal, R. 2004. Residues of hexachlorocyclohexane isomers in soil and water samples from Delhi and adjoining areas. *Current Science* 87: 73–77.

Li, Y.F. 1999. Global technical hexachlorocyclohexane usage and its contamination consequences in environment: From 1948 to 1997. *Science of the Total Environment* 232: 123–160.

Li, Y.F., Cai, D.J., Shan, Z.J., and Zhu, Z.L. 2001. Gridded usage inventories of technical hexachlorocyclohexane and lindane for China with 1/6° latitude by 1/4° longitude resolution. *Archives of Environmental Contamination and Toxicology* 41: 261–266.

Li, Y.F., Struger, J., and Waite, D., Ma, J. 2004. Gridded Canadian lindane usage inventories with 1/6° x 1/4° latitude and longitude resolution. *Atmospheric Environment* 38: 1117–1121.

Li, Y.F., Zhulidov, A.V., Robarts, R.D., and Korotova, L.G. 2005. Hexachlorocyclohexane use in the former soviet union. *The Archives of Environmental Contamination and Toxicology* 48: 10–15.

Lodha, B., Bhat, P., Kumar, M.S., Vaidya, A.N., Mudliar, S., Killedar, D.J., and Chakrabarti, T. 2007. Bioisomerization kinetics of γ-HCH and biokinetics of *Pseudomonas aeruginosa* degrading technical HCH. *Biochemical Engineering Journal* 35: 12–19.

Lopez, A.G., Deutcher, M.T., Aquino, W.F., and Marker, A. 1998. *HCH and pesticides contaminated sites in Brazil –Stat of the Art—5th International HCH and Pesticides Forum*. Forum Book, Bilbao.

MacRae, I.C., Raghu, K., and Bautista, E.M., 1969. Anaerobic degradation of the insecticide Lindane by *Clostridium* sp. *Nature* (*London*) 221: 859–860.

Manickam, N., Mau, M., and Schlomann, M. 2006. Characterization of the novel HCH-degrading strain, *Microbacterium* sp. ITRC1. *Applied Microbiology and Biotechnology* 69(5): 580–588.

Manickam, N., Reddy, M.K., Saini, H.S., and Shanker, R. 2008. Isolation of hexachlorocyclohexane- degrading *Sphingomonas* sp. by dehalogenase assay and characterization of genes involved in γ-HCH degradation. *Applied and Environmental Microbiology* 104: 952–960.

Manonmani, H.K., Chandrashekaraiah, D.H., Sreedhar, R.N., Elcey, C.D., and Kunhi, A.A.A. 2000. Isolation and acclimatization of a microbial consortium for improved aerobic degradation of α-hexachlorocyclohexane. *Journal of Agriculture and Food Chemistry* 48: 4341–4351.

McErlean, C., Marchant, R., and Banat, I.M. 2006. An evaluation of soil colonisation potential of selected fungi and their production of ligninolytic enzymes for use in soil bioremediation applications, *Antonie van Leeuwenhoek* 90: 147–158.

Miyauchi, K., Adachi, Y., Nagata, Y., and Takagi, M. 1999. Cloning and sequencing of a novel metacleavage dioxygenase gene whose product is involved in degradation of γ-hexachlorocyclohexane in *Sphingomonas paucimobilis*. *Journal of Bacteriology* 181: 6712–6719.

Miyauchi, K., Lee, H. S., Fukuda, M., Takagi, M., and Nagata, Y. 1998. Cloning and sequencing of a 2,5- dihydrochloroquinone reductive dehalogenase gene whose product is involved in the degradation of γ-hexachlorocyclohexane by Sphingomonas paucimobilis. *J. Bacteriol.* 180: 1354–1359.

Miyauchi, K., Lee, H.S., Fukuda, M., Takagi, M., and Nagata, Y. 2002. Cloning and characterization of linR, involved in regulation of downstream pathway for γ-hexachlorocyclohexane degradation in *Sphingomonas paucimobilis* UT26. *Applied and Environmental Microbiology* 68: 1803–1807.

Molina, L., Ramos, C., Duque, E., Ronchel, M.C., Garcia, J.M., Wyke, L., and Ramos, J.L. 2000. Survival of *Pseudomonas putida* KT2440 in soil and in the rhizosphere of plants under greenhouse and environmental conditions. *Soil Biology and Biochemistry* 32: 315–321.

Morgan, D.P., Roberts, R.J., Walter, A.W., and Stockdale, E.M. 1980. Anemia associated with exposure to lindane. *Archives of Environmental Health* 35: 307–310.

Mougin, C., Pericaud, C., Dubroca, J., and Asther, M. 1997. Enhanced mineralization of lindane in soils supplemented with the white rot basidiomycete *Phanerochaete chrysosporium*. *Soil Biology & Biochemistry* 29: 1321–1324.

Munir, K.M., Soman, C.S., and Bhide, S.V. 1983. Hexachlorocyclohexane-induced tumorigenicity in mice under different experimental conditions. *Tumorology* 69: 383–386.

NAFTA, North American Regional Action Plan on Lindane. 2000. *Background document on Lindane, North American Commission for Environmental Cooperation: Decision Document on Lindane*. http://www.cec.org (accessed December 15, 2016).

Nagata, K., Huang, C.S., and Hamilton, B.J. 1996. Differential effects of hexachlorocyclo hexane isomers on the GABA receptor subunits expressed in human embryonic kidney cell line. *Brain Research* 738(1): 131–137.

Nagata, Y., Endo, R., Ito, M., Ohtsubo, Y., and Tsuda, M. 2007. Aerobic degradation of lindane (γ-hexachlorocyclohexane) in bacteria and its biochemical and molecular basis. *Applied Microbiology and Biotechnology* 76: 741–752.

Nagata, Y., Miyauchi, K., and Takagi, M. 1999. Complete analysis of genes and enzymes for γ-hexachlorocyclohexane degradation in *Sphingomonas paucimobilis* UT26. *Journal of Industrial Microbiology and Biotechnology* 23: 380–390.

Nagata, Y., Miyauchi, K., Damborsky, J., Manova, K., Ansorgova, A., and Takagi, M. 1997. Purification and characterization of a haloalkane dehalogenase of a new substrate class from α-hexachlorocyclohexane-degrading bacterium, Sphingomonas paucimobilis UT26. *Applied and Environmental Microbiology* 63: 3707–3710.

Nagata, Y., Nariya, T., Ohtomo, R., Fukuda, M., Yano, K., and Takagi, M. 1993. Cloning and sequencing of a dehalogenase gene encoding an enzyme with hydrolase activity involved in the degradation of γ-hexachlorocyclohexane in *Pseudomonas paucimobilis*. *Journal of Bacteriology* 175: 6403–6410.

Nagata, Y., Nariya, T., Ohtomo, R., Fukuda, M., Yano, K., and Takagi, M. 1994. Cloning and sequencing of a 2, 5-dichloro-2,5-cyclohexadiene-1,4-diol dehydrogenase gene involved in the degradation of γ-hexachlorocyclohexane in *Pseudomonas paucimobilis*. *Journal of Bacteriology* 176: 3117–3125.

Nagata, Y., Prokop, Z., Sato, Y., Jerabek, P., Kumar, A., Ohtsubo, Y., Tsuda, M., and Damborsky, J. 2005. Degradation of β- hexachlorocyclohexane by haloalkane dehalogenase LinB from *Sphingomonas paucimobilis* UT26. *Applied and Environmental Microbiology* 71: 2183–2185.

Nagpal, V., and Paknikar, K.M. 2006. Integrated biological approach for the enhanced degradation of lindane. *Indian Journal of Biotechnology* 5: 400–406.

Nagpal, V., Srinivasan, M. C., and Paknikar, K. M. 2008. Biodegradation of γ-hexachlorocyclohexane (Lindane) by a non-white rot fungus Conidiobolus 03-1-56 isolated from litter. *Indian J. Microbiol.* 48: 134–141.

Nalin, R., Simonet, P., Vogel, T.M., and Normand, P. 1999. *Rhodanobacter lindaniclasticus* gen. nov., sp. nov., a lindane-degrading bacterium. *International Journal of Systematic Bacteriology* 49: 19–23.

Nantel, A.J., Ayotte, L., Beneditti, J.L., Savoie, J.Y., Tessier, L., Schwarz, T., and Weber, J.P. 1977. A group of adults acutely poisoned by food contaminated with lindane. *Acta Pharmacology Toxicology* 41(Suppl. 2): 250.

Nedkova-Bratanova, N., Ivanov, E., and Savov, G. 1979. Effect of the pesticides phosalone and l lindane on the activity of some dipeptidases and disaccharidases in rat intestinal mucosa. *Enzyme* 24(5): 281–284.

Niharika, N., Moskalikova, H., Kaur, J., Khan, F., Sedlackova, M., Hampl, A., Damborsky, J., Prokop, Z., and Lal, R., 2012a. *Sphingobium czechense* sp. nov., isolated from a hexachlorocyclohexane dump site. *International Journal of Systematic and Evolutionary Microbiology* 63(2): 723–728.

Niharika, N., Moskalikova, H., Kaur, J., Sedlackova, M., Hampl, A., Damborsky, J., Prokop, Z., and Lal, R. 2012b. *Novosphingobium barchaimii* sp. nov., isolated from hexachlorocyclohexane-contaminated soil. *International Journal of Systematic and Evolutionary Microbiology* 63(2): 667–672.

NRTEE (National Round Table on the Environment and the Economy). 2000. *Managing potentially toxic substances in Canada*. NRTEE Publication, Canada.

Ohisa, N., and Yamaguchi, M. 1978. Gamma-BHC degradation accompanied by the growth of *Clostridium rectum* isolated from paddy field soil. *Agricultural and Biological Chemistry* 42: 1819–1823.

Okeke, B.C., Siddique, T., Arbestain, M.C., and Frankenberger, W.T., 2002. Biodegradation of γ-hexachlorocyclohexane (lindane) and α-hexachlorocyclohexane in water and a soil slurry by a Pandoraea species. *Journal of Agricultural and Food Chemistry* 50: 2548–2555.

Oldiges, H., Takenaka, S., and Hochrainer, D. 1980. *Inhalation study with lindane (γ-hexachlorocyclohexane) to determine the LC50.* Fraunhofer-Institut, Institute for Toxicology and Aerosol Research, Schmallenberg, Germany. Celamerck document no. 111AA-423-002.

Papadopoulou, K., Rigas, F., and Doulia, D. 2006. Lindane degradation in soil by Pleurotus ostreatus. *WSEAS Transactions Topics Environment and sustainable development* 5: 489–496.

Pathak, R., Suke, S.G., Ahmed, R.S., Tripathi, A.K., Guleria, K., Sharma, C.S., Makhijani, S.D., Mishra, M., and Banerjee, B.D. 2008. Endosulfan and other organochlorine pesticide residues in maternal and cord blood in North Indian population. *Bulletin of Environmental Contamination and Toxicology* 81: 216–219.

Pesce, S.F., and Wunderlin, D.A. 2004. Biodegradation of lindane by a native bacterial consortium isolated from contaminated river sediment. *International Biodeterioration & Biodegradation,* 54, 255–260.

Pomes, A., Rodriguez-Farre, E., and Sunol, C. 1994. Effects of organochlorine pesticides on 36Cl-flux in primary neuronal cultures. *Neurotoxicology* 15(3): 745–749.

Questionnaire on HCH and Lindane. 2004a. Submitted by Japanese Ministry of Agriculture.

Questionnaire on HCH and Lindane. 2004b. Submitted by Lindita Tafaj, Albania.

Questionnaire on HCH and Lindane. 2004c. Submitted by Polish producers.

Quintero, J.C., Lu-Chau, T.A., Moreira, M.T., Feijoo, G., and Lema, J.M., 2007. Bioremediation of HCH present in soil by white rot fungus *Bjerkandera adusta* in a slurry batch bioreactor. *International Biodeterioration and Biodegradation* 60: 319–326.

Quintero, J.C., Moreira, M.T., Feijoo, G., and Lema, J.M. 2008. Screening of white rot fungal species for their capacity to degrade lindane and other isomers of hexachlorocyclohexane (HCH). *Cienciae Investigación Agraria* 35: 159–167.

Ramteke, P.W., and Hans, R.K. 1992. Isolation of hexachlorocyclohexane (HCH) degrading microorganisms from earthworm got. *Journal of Environmental Science and Health* 27: 2113–2122.

Ravel, J., Amoroso, M.J., Colwell, R.R., and Hill, R.T. 1998. Mercury-resistant actinomycetes from the Chesapeake Bay. *FEMS Microbiology Letters*, 162(1): 177–184.

Rigas, F., Dritsa, V., Marchant, R., Papadopouloua, K., Avramides, E.J., and Hatzianestis I. 2005. Biodegradation of lindane by *Pleurotus ostreatus* via central composite design. *Environment International* 31: 191–196.

Rigas, F., Papadopoulou, K., Dritsa, V., and Doulia, D. 2007. Bioremediation of a soil contaminated by Lindane utilizing the fungus *Ganoderma australe* via response surface methodology. *Journal of Hazardous Materials* 140: 325–332.

Rosa, R., Sanfeliu, C., and Sunol, C. 1997. The mechanism for hexachlorocyclohexane-induced cytotoxicity and changes in intracellular Ca2+ homeostasis in cultured cerebellar granule neurons is different for the gamma- and delta-isomers. *Toxicology and Applied Pharmacology* 142(1): 31–39.

Sagar, V., and Singh, D.P. 2011. Biodegradation of lindane pesticide by non white-rots soil fungus *Fusarium* sp. *World Journal of Microbiology and Biotechnology* 27: 1747–1754.

Sahu, S.K., Patnaik, K.K., Sharmila, M., and Sethunathan, N. 1990. Degradation of alpha-, beta-, and gamma-hexachlorocyclohexane by a soil bacterium under aerobic conditions. *Applied Environmental Microbiology* 56: 3620–3622.

Salam, J.A., and Das, N. 2014. Degradation of lindane by a novel embedded bio-nano hybrid system in aqueous environment. *Applied Microbiology and Biotechnology* 99: 2351–2360.

Saxena, A., Anand, S., Dua, A., Sangwan, N., Khan, F., and Lal, R. 2012. Novosphingobium lindaniclasticum sp. nov., a hexachlorocyclohexane (HCH)-degrading bacterium isolated from an HCH dumpsite. *International Journal of Systematic and Evolutionary Microbiology* 63: 2160–2167.

Senoo, K., and Wada, H., 1989. Isolation and identification of an aerobic gamma-HCH decomposing bacterium from soil. *Soil Science and Plant Nutrition* 35: 79–87.

Shelton, D.R., Khader, S., Karns, J.S., and Pogell, B.M. 1996. Metabolism of twelve herbicides by Streptomyces. *Biodegradation* 7: 129–136.

Sidi, Y., Kiltchevsky, E., Shaklai, M., and Pinkhas, J. 1983. Acute myeloblastic leukemia and insecticide. *New York State Journal of Medicine* 83: 161.

Singh, A., and Lal, R., 2009. Sphingobium ummariense sp. nov., a hexachlorocyclohexane (HCH)-degrading bacterium, isolated from HCH-contaminated soil. *International Journal of Systematic and Evolutionary Microbiology* 59: 162–166.

Singh, A.K., Garg, N., Lata, P., Kumar, R., Negi, V., Vikram, S., and Lal, R. 2013a. Pontibacter indicus sp. nov., isolated from hexachlorocyclohexane-contaminated soil. *International Journal of Systematic and Evolutionary Microbiology* 64(1): 254–259.

Singh, A.K., Garg, N., Sangwan, N., Negi, V., Kumar, R., Vikram, S., and Lal, R. 2013. *Pontibacter ramchanderi* sp. nov., isolated from hexachlorocyclohexane-contaminated pond sediment. *International Journal of Systematic and Evolutionary Microbiology* 63(8): 2829–2834.

Singh, B.K., and Kuhad, R.C. 1999. Biodegradation of lindane (γ-hexachlorocyclohexane) by the white-rot fungus *Trametes hirsutus*. *Letters in Applied Microbiology* 28: 238–241.

Singh, B.K., and Kuhad, R.C. 2000. Degradation of insecticide lindane (γ-HCH) by white-rot fungi *Cyathus bulleri* and *Phanerochaete sordida*. *Pest Management Science* 56: 142–146.

Singh, G., Kathpal, T.S., Spencer, W.F., and Dhankar, J.S. 1991. Dissipation of some organochlorine insecticides in cropped and uncropped soil. *Environmental Pollution* 70: 219–239.

SRI. 2009. *Directory of chemical producers*. SRI Consulting, Menlo Park, CA.

Sunol, C., Vale, C., and Rodriguez-Farre, E. 1998. Polychlorocycloalkane insecticide action on GABA- and glycine-dependent chloride flux. *Neurotoxicology* 19(4–5): 573–580.

Tekere, M., Ncube, I., Read, J.S., and Zvauya, R. 2002. Biodegradation of the organochlorine pesticide, lindane by a subtropical white rot fungus in batch and packed bed bioreactor system. *Environmental Technology* 23: 199–206.

Thomas, J.C., Berger, F., Jacquier, M., Bernillon, D., Baud-Grasset, F., Truffaut, N., Normand, P., Vogel, T.M., and Simonet, P. 1996. Isolation and characterization of a novel γ-hexachlorocyclohexane-degrading bacterium. *Journal of Bacteriology* 178: 6049–6055.

Tomczak, S, Baumann K, and Lehnert G. 1981. Occupational exposure to hexachlorocyclohexane: IV. Sex hormone alterations in HCH-exposed workers. *International Archives of Occupational and Environmental Health* 48: 283–287.

TRI. 2009. *TRI explorer chemical report*. U.S. Environmental Protection Agency. http://www.epa.gov/triexplorer and select Lindane.

Turkman, A., Nuri, A., and Nermin, A. 1993. An overview on HCH application in Turkey, A situation report, HCH and halogenated pesticides—State of the art for risk assessment and technology development. http://www.ihpa.info/library.htm 1st & 2nd Forum (accessed April 4, 2016).

UNECE. 2000. The 2000 review on strategies and policies for air pollution abatement, the 1998 protocol on persistent organic pollutants, replies to questions 37-49 of the 2000 questionnaire. *Executive body for the convention on long-range transboundary air pollution*, http://www.unece.org/env/documents/2000/eb/eb.air.2000.1.q.37-49.pdf (accessed October 24, 2016).

UNEP. 2002a. *Central and Northeast Asia regional report, regionally based assessment of persistent toxic substances.*

UNEP. 2002b. *Eastern and western South America regional report, regionally based assessment of persistent toxic substances.*

UNEP. 2002c. *Indian Ocean report, regional report, regionally based assessment of persistent toxic substances.*

UNEP. 2002d. *Mediterranean regional reports, regionally based assessment of persistent toxic substances.*

UNEP. 2006e. Stockholm convention on persistent organic pollutants. *Report of the persistent organic pollutants review committee on the work of its second meeting, risk profile on Lindane.* UNEP/POPS/POPRC.2/17/Add.4.

UNEP. 2002f. Sub-Saharan Africa, regional report. *Regional based assessment of persistent toxic substances.*

U.S. Environmental Protection Agency. *Toxic Release Inventory (TRI) Program, TRI Chemicals* www.epa.gov/triinter/chemical/index.htm (accessed December 18, 2016).

U.S. Environmental Protection Agency (USEPA). 1983. *Guidance for the re registration of pesticide products containing chlorobenzilate as the active ingredient.* Environmental Protection Agency. Washington, D.C., pp. 6–7.

U.S. Environmental Protection Agency (USEPA). 2006. Lindane and Other HCH Isomers—EPA Risk Assessment Fact Sheet.www.epa.gov/oppsrrd1/REDs/factsheets/lindane_isomers_fs.htm (accessed December 18, 2016).

U.S. Environmental Protection Agency (USEPA). 2012. Water Quality Criteria. www.epa.gov/waterscience/ criteria/wqcriteria.html (accessed December 18, 2016).

U.S. Food and Drug Administration (FDA). 2003a. Center for Drug Evaluation and Research. Lindane Lotion USP, 1% RX Only.

USITC. 2009. USITC Interactive Tariff and Trade DataWeb. United States International Trade Commission. http://dataweb.usitc.gov/scripts/user_set.asp and search on HTS no. 290351 (accessed March 25, 2016).

Van Doesburg, W., van Eekert, M.H.A., Middeldorp, P.J.M., Balk, M., Schraa, G., and Stams, A.J.M. 2005. Reductive dechlorination of β-hexachlorocyclohexane (β-HCH) by a Dehalobacter species in coculture with a Sedimentibacter sp. *FEMS Microbiology Ecology* 54: 87–95.

Van Eekert, M.H.A., van Ras, N.J.P., Mentink, G.H., Rijnaarts, H.H.M., Stams, A.J.M., Field, J.A., and Schraa, G., 1998. Anaerobic transformation of β-hexachlorocyclohexane by methanogenic granular sludge and soil microflora. *Environmental Science and Technology* 32: 3299–3304.

Vijgen, J. 2006. The legacy of lindane HCH isomer production: A global overview of residue management, formulation and disposal. In: Main report, International HCH and Pesticides Association (IHPA) Secretariat, Denmark, ISBN: 87-991210-1-8, pp. 1–22.

Vijgen, J., Abhilash, P.C., Li, Y.F., Lal, R., Forter, M., Torres, J., Singh, N., et al. 2011. HCH as new Stockholm Convention POPs, a global perspective on the management of Lindane and its waste isomers. *Environmental Science and Pollution Research* 18: 152–162.

Vilchez, S., Molina, L., Ramos, C., and Ramos, J.L. 2000. Proline catabolism by Pseudomonas putida: cloning, characterization, and expression of the put genes in the presence of root exudates. *Journal of Bacteriology* 182: 91–99.

Waliszewski, S.M. 1993. Residues of lindane, HCH isomers and HCB in the soil after lindane application. *Environmental Pollution* 82: 289–293.

Walker, K., Vallero, D.A., and Lewis, R.G. 1999. Factors influencing the distribution of lindane and other hexachlorocyclohexanes in the environment. *Environmental Science and Technology* 33: 4373–4378.

Weston. 1993. Phase I environmental conditions assessment chemical report.

Willett, K.L., Ulrich, E.M., and Hites, R.A. 1998. Differential toxicity and environmental fate of hexachlorocyclohexane isomers. *Environmental Science and Technology* 32: 2197–2207.

Zheng, G., Selvam, A., and Wong, J.W.C. 2011. Rapid degradation of lindane (γ-hexachlorocyclohexane) at low temperature by *Sphingobium* strains. *International Biodeterioration & Biodegradation* 65: 612–618.

# 7 Toxic Metals Contamination in the Environment

## *Toxicological Effects and Bioremediation Approaches for Environmental Cleanup*

*Janardhan Reddy Koduru, Shiv Shankar, Nandkishor S. More, Shikha, Lakshmi Prasanna Lingamdinne, and Jiwan Singh*

**CONTENTS**

7.1 Introduction .......... 210
7.2 Definition of Toxic Metals .......... 211
7.3 Origin and Nomenclature of Heavy Metals .......... 212
7.4 Sources of Toxic Metals Contamination in Environments .......... 213
7.5 Toxicological Effects and Health Hazards from Toxic Metals .......... 213
7.6 Approaches for the Remediation of Toxic Metals–Contaminated Environments .......... 214
7.6.1 Physicochemical Remediation Techniques .......... 214
7.6.1.1 Absorption .......... 215
7.6.1.2 Filtration and Membrane Technology .......... 215
7.6.1.3 Adsorption .......... 216
7.6.1.4 Electrokinetic Remediation .......... 217
7.6.2 Other Physicochemical Soil Treatment Processes .......... 218
7.6.2.1 Immobilization .......... 218
7.6.2.2 Vitrification .......... 218
7.6.2.3 Excavation .......... 219
7.6.2.4 Solidification/Stabilization .......... 219
7.6.2.5 Soil Leaching: Removal and Soil Isolation .......... 219
7.6.2.6 Capping .......... 220

7.6.2.7 Physical Separation ... 220
7.6.2.8 Limitation of Physicochemical Techniques ... 220
7.6.3 Biological Approaches for the Remediation of Toxic Metals–Contaminated Environments ... 220
7.6.3.1 Bioremediation and Its Types ... 220
7.6.3.2 Application of Bioremediation in Removal of Toxic Metals ... 222
7.6.3.3 Mechanism of Bioremediation of Heavy Metals ... 225
7.6.3.4 Advantages and Limitations of Biological Methods ... 229
7.7 Conclusion ... 230
Acknowledgments ... 231
References ... 231

## 7.1 INTRODUCTION

Heavy metals contamination in the environment is a major global concern, because of toxicity and threat to the human life and ecosystem. Industrial wastewater has been considered as the major sources of heavy metal pollution in the environment, which create serious toxic effects in living beings (Bharagava et al. 2008, 2014; Chandra et al. 2009, 2011, 2015; Saxena and Bharagava 2015, 2016; Saxena et al. 2016). The levels of metals in environments are increasing in some cases to toxic levels, with contributions from a broad range of industrial and domestic sources. Metal-contaminated environments pose a significant threat to health and ecosystems. Heavy metals including arsenic, cadmium, lead, mercury, silver, etc., cause hypophosphatemia, heart disease and liver damage, cancer and neurological and cardiovascular diseases, central nervous system damage, and sensory disturbances. One of the leading reasons for the abatement of the Roman Empire is believed to be lead poisoning that resulted because of the extensive use of wine stored in pottery lined with lead and lead water pipes. Another example of heavy metal poisoning is the Minamata Bay incident of Japan, which was caused by consumption of fishes and shellfishes contaminated with methylmercury in their body. Metals concentration is linked to congenital disabilities, cancers, skin lesions, retardation leading to disabilities, liver and kidney damage, and a host of other maladies (ATSDR 2001). Hence, almost all heavy metals are called toxic metals.

Rapid industrialization, development, and urbanization have resulted in severe environmental problems. The resultant degradation and contamination of the ecosystems have today become a significant threat to all life forms on earth in general and humanity (EPA 1990). Globally, land, air, and aquatic ecosystems have been contaminated with heavy metals through various human activities and because they move up through the food chain, they have become a significant human health hazard. Due to increased awareness of the harmful effects of pollutants, particularly heavy metals and hydrocarbons, there has been a marked growth in research on the effects of metal ions in the environment as well in animals, plants, and microbes, along with research on the exploration of various remediation strategies for the cleaning of contaminated environments (Pandey and Jain 2002). Remediation or cleanup deals with the removal of pollution or contaminants from environmental media such as soil, groundwater, sediment, or surface water. Remedial action is subject to an array of regulatory requirements and also can be based on assessments of human health and ecological risks where no legislated standards exist or where standards are advisory. Many treatment methods have been applied to remove toxic metals from the soil and the water system. In the chemical treatment method of toxic metals in wastewaters, a mixture of chemical compound was required to reduce the toxicity of metals; however, the removal could not be achieved as predicted in the guideline (Ismail et al. 2014). Some physicochemical treatment methods such as coagulation, flocculation, reverse osmosis, membrane separation, and ion exchange were carried out to remove the toxic metals from wastewaters (Prieto-Rodrıguez et al. 2013; Lim and Aris 2014). These methods may change the hazardous state of the toxic metals into a stable and safe form.

For the bioremediation of toxic metals from soil and aqueous systems, living cells, nonliving biomass, or biopolymers as biosorbents are used. A variety of microorganisms (fungi/algae/bacteria) have been used for the bioremediation of toxic metals from the contaminated sites (Saxena and Bharagava 2016). Bioremediation is a cost-effective process and has been widely accepted. Phytoremediation is also an emerging technology that uses higher plants for the cleanup of contaminated environments. It is an efficient, economical, and eco-friendly *in situ* remediation technology. This chapter presents a view on the toxicity of metals, their remediation techniques including physicochemical and biological methods, as well as the mechanism accountable for the removal of toxic metals.

## 7.2 DEFINITION OF TOXIC METALS

Metals are important not only for humankind but also for other living beings. However, its natural form and its actions classify them as useful and harmful. Heavy metals are defined as metals with relatively high densities, atomic weights, or atomic numbers. The criteria used and whether metalloids are included vary depending on the author and context. In metallurgy, for example, a heavy metal may be defined based on the density; in physics, the distinguishing criterion might be atomic number; and a chemist would likely be more concerned with chemical behavior. More specified definitions have been published, but none of these has been widely accepted. The definitions surveyed in this chapter encompass up to 96 out of the 118 chemical elements—only mercury, lead, and bismuth meet all of them. Despite this lack of agreement, the term (plural or singular) is widely used in science. A density of more than 5 g/cm$^3$ is a commonly used criterion.

The earliest metals are known as common metals, such as iron, copper, and tin; precious metals such as silver, gold, and platinum are heavy metals. From 1809 onward, light metals, such as magnesium, aluminum, and titanium, were discovered as well as less well-known heavy metals like gallium, thallium, and hafnium (Hawkes 1997). Physical and chemical characterizations of heavy metals need to be treated with caution, as the metals involved are not always consistently defined. As well as being relatively dense, heavy metals tend to be less reactive than lighter metals and have much less soluble sulfides and hydroxides. While it is relatively easy to distinguish a heavy metal such as tungsten from a lighter metal such as sodium, a few heavy metals such as zinc, mercury, and lead have some of the characteristics of lighter metals, and lighter metals such as beryllium, scandium, and titanium have some of the characteristics of heavier metals.

## 7.3 ORIGIN AND NOMENCLATURE OF HEAVY METALS

An early use of the term *heavy metal* dates from 1817, when the German chemist Leopold Gmelin divided the elements into nonmetals, light metals, and heavy metals. Light metals had densities of 0.860–5 g/cm$^3$; heavy metals 5.308–22 g/cm$^3$. The term later became associated with elements of high atomic weight or high atomic number. It is sometimes used interchangeably with the term *heavy element* as it was designated then. The heaviness of naturally occurring metals such as gold, copper, and iron may have been noticed in prehistory and, in light of their malleability, led to the first attempts to craft metal ornaments, tools, and weapons. All metals discovered from then until 1809 had relatively high densities; their heaviness was regarded as a singularly distinguishing criterion. From 1809 onward, light metals such as sodium, potassium, and strontium were isolated. Their low densities challenged conventional wisdom, and it was proposed to refer to them as metalloids (meaning "resembling metals in form or appearance"). This suggestion was ignored and the new elements came to be recognized as metals, and the term *metalloid* was then used to refer to nonmetallic elements and, later, elements that were hard to describe as either metals or nonmetals. Table 7.1 represents the metals of biological importance.

Trace amounts of some heavy metals, mostly in period 4, are required for certain biological processes. These are iron and copper (oxygen and electron transport); cobalt (complex syntheses and cell metabolism); zinc (hydroxylation); vanadium and manganese (enzyme regulation or functioning); chromium (glucose utilization); nickel (cell growth); arsenic (metabolic growth in some animals and possibly in humans); and selenium (antioxidant functioning and hormone production). Periods 5 and 6 contain fewer essential heavy metals, consistent with the general pattern

**TABLE 7.1**
**Metals of Biological Importance**

| S.No. | Element | Milligram |
|---|---|---|
| 1 | Iron | 4000 |
| 2 | Zinc | 2500 |
| 3 | Lead | 120 |
| 4 | Copper | 70 |
| 5 | Tin | 30 |
| 6 | Vanadium | 20 |
| 7 | Cadmium | 20 |
| 8 | Nickel | 15 |
| 9 | Selenium | 14 |
| 10 | Manganese | 12 |
| 11 | Other | 200 |
| 12 | Total | 7000 |

*Source:* Emsley, J. 2011. *Nature's Building Blocks*, new edition, Oxford University Press, Oxford, UK.

that heavier elements tend to be less abundant and that scarcer elements are less likely to be nutritionally essential. In period 5, molybdenum is required for the catalysis of redox reactions; cadmium is used by some marine diatoms for the same purpose; and tin may be required for growth in a few species. In period 6, tungsten is required by some bacteria for metabolic processes. An average human body (~70 kg) is about 0.01% heavy metals (~7 g, equivalent to the weight of two dried peas, with iron at 4 g, zinc at 2.5 g, and lead at 0.12 g comprising the three main constituents), 2% light metals (~1.4 kg, the weight of a bottle of wine), and nearly 98% nonmetals (mostly water) (Moore and Ramamoorthy 1984). A deficiency of any of these period 4–6 essential heavy metals may increase susceptibility to heavy metal poisoning. A few nonessential heavy metals have also been observed to have biological effects. Gallium, germanium (a metalloid), indium, and most lanthanides can stimulate metabolism, and titanium promotes growth in plants (though it is not always considered a heavy metal). Thus, there are many complexities with the deficiency and excess of them.

## 7.4 SOURCES OF TOXIC METALS CONTAMINATION IN ENVIRONMENTS

It is not a gross understatement that metals are predominantly found in nature, but their use after extraction and processing makes their way back to the environment in a form that is a serious concern. Heavy metals are relatively scarce in the Earth's crust but are present in many aspects of modern life. They are used in, for example, golf clubs, cars, antiseptics, self-cleaning ovens, plastics, solar panels, mobile phones, and particle accelerators. Some heavy metals are either essential nutrients (typically iron, cobalt, and zinc) or are relatively harmless (such as ruthenium, silver, and indium) but can be toxic in large amounts or certain forms. Other heavy metals, like chromium, cadmium, mercury, and lead, are notably toxic. Potential sources of heavy metal poisoning include mining and industrial wastes, agricultural runoff, occupational exposure, and contact with lead-based paints (Hubner et al. 2010).

## 7.5 TOXICOLOGICAL EFFECTS AND HEALTH HAZARDS FROM TOXIC METALS

Metal toxicity or metal poisoning is the toxic effect of certain metals in certain forms and doses on life. Some metals are toxic when they form poisonous soluble compounds. Certain metals have no biological role (i.e., are not essential minerals) or are toxic when in a certain form. In the case of lead, any measurable amount may have negative health effects. Often heavy metals are thought of as synonymous, but lighter metals may also be toxic in certain circumstances, such as beryllium and lithium. Not all heavy metals are particularly toxic, and some are essential, such as iron. The definition may also include trace elements when considered in abnormally high, toxic doses. An option for treatment of metal poisoning may be chelation therapy, which is a technique that involves the administration of chelation agents to remove metals from the body (Leeper 1978).

Toxic metals sometimes imitate the action of an essential element in the body, interfering with the metabolic process to cause illness. Many metals, particularly heavy metals, are toxic, but some heavy metals are essential, and some, such as bismuth, have a low toxicity. Most often, the definition of toxic metals includes at least cadmium, manganese, lead, mercury, and the radioactive metals. Metalloids (arsenic, polonium) may be included in the definition. Radioactive metals have both radiological toxicity and chemical toxicity. Metals in an oxidation state abnormal to the body may also become toxic: chromium (III) is an essential trace element, but chromium (VI) is a carcinogen.

Toxicity is a function of solubility. Insoluble compounds as well as the metallic forms often exhibit negligible toxicity. The toxicity of any metal depends on its ligands. In some cases, organometallic forms, such as methyl mercury and tetraethyl lead, have proven to be extremely toxic. In other cases, organometallic derivatives are less toxic such as the cobaltocenium cation. Decontamination for toxic metals is different from organic toxins, because toxic metals are elements and cannot be destroyed. Toxic metals can be made insoluble or collected, possibly by the aid of chelating agents. Alternatively, they can be diluted into a sufficiently large reservoir, such as the sea, because immediate toxicity is a function of concentration rather than amount. Toxic metals can bioaccumulate in the body and in the food chain. Therefore, a common characteristic of toxic metals is the chronic nature of their toxicity. This is particularly notable with radioactive heavy metals such as radium, which imitates calcium to the point of being incorporated into human bone, although similar health implications are found in lead or mercury poisoning. The exceptions to this are barium and aluminum, which can be removed efficiently by the kidneys (Leeper 1978).

## 7.6 APPROACHES FOR THE REMEDIATION OF TOXIC METALS–CONTAMINATED ENVIRONMENTS

### 7.6.1 Physicochemical Remediation Techniques

Globally, toxic metals–contaminated soils and waters caused significant environmental, agricultural, and human health problems. Cleanup to date has been in response to such historical chemical management practices as dumping, poor storage, and free release or spillage. Greater effort in recent years has been directed toward pollution prevention, which is more costeffective than remediation. Programs such as Superfund in the United States, as well as parallel state programs, represent a commitment of billions of dollars to the cleanup of contaminated sites. These contaminated sites can be cleaned by physical, chemical, and/or biological methods.

The methods are dependent on physical processes or activities such as the civil construction of barriers, physical adsorption or absorption, mass transfer, as well as harnessed chemical or biochemical processes. Most of the times, two or more processes are coupled to deal with the contamination problem. Many of the adsorbents and the filtration process described here have frequently been used in treating contaminated water or effluent but had been scarcely used for *in situ* groundwater treatment.

#### 7.6.1.1 Absorption

Inorganic surfactants initiate some exemplary desirable absorption processes, which help in contaminants removal (e.g., solubility enhancement, surface tension reduction, micellar solubilizing, wettability, and foaming capacity) (Mulligan et al. 2001). Doong et al. (1998) observed the combined effect of complexing agents and surfactants for Cd, Pb, and Zn removal from soil. Anionic (sodium dodecyl sulfate, SDS), cationic (cetyltrimethylammonium bromide, CTAB), and nonionic (Triton X-100) surfactants were evaluated in combination with complexing agents EDTA and diphenylthiocarbazone (DPC). Anionic and nonionic surfactants were enhanced desorption rates of Cd, Pb, and Zn, whereas cationic surfactant was most efficient at lower pH. In the surfactant/EDTA mixture system, the heavy metal desorption efficiency was in the order of Cd > Pb > Zn. However, surfactant/DPC system resulted in two to four times reduction in extraction efficiency. The results demonstrated that surfactants in combination with complexing agents could be used to flush heavy metal–contaminated soil (Doong et al. 1998). Zhang et al. (2007, 2008) also successfully removed Pb from contaminated soil by washing it with EDTA followed by SDS (Zhang et al. 2007, 2008).

#### 7.6.1.2 Filtration and Membrane Technology

Membrane technology can be used in conjunction with soil washing or flushing techniques for concentrating the contaminants from the wash liquid so that the raffinate can be treated accordingly. Membranes can be of several types such as electrodialytic membrane liquid membrane, polymer membrane, ultrafiltration membrane, nanofiber membrane, etc. An active membrane method should reduce the volume of contaminated water to be treated while producing clean water that meets the applicable effluent guidelines (Sikdar et al. 1998). Partitioning and transport of water and dissolved salts at the membrane and the bulk solution depend on the chemical properties of the membrane as well as the physical structures on nano- to micro-scales. From a filtration perspective, this intermediate range contains, for example, colloidal solids, primarily organic and biological molecules, polymers, and viruses. It also corresponds to the dimensions at which we recognize distinct modes of material transport across a membrane. For a larger size of porous membranes, transport is described regarding the convective flow through pores. On the other hand, carriage in a thick reverse osmosis membrane is typically defined regarding diffusive flow through a homogeneous material. Hansen et al. (1997) simultaneously applied cation and anion exchange membrane barriers in an electrokinetic method to remove Cu, Cr, Hg, Pb, and Zn (concentration several hundred ppm each) from soils contaminated by wood-preserving facilities, chloralkali plants, and Cu rolling mills.

Heavy metals such as Cu(II), Cr(VI), and As(V) might be removed from the aqueous phase by ultrafiltration, pretreating the phase with a functionalized polymer such as polyethylene imine or polyacrylic acid, which can form complex with metal ions. It was also shown that for low contaminant concentrations below 50 ppm, excellent binding and subsequent removal by polysulfone ultrafiltration membranes (mol. wt. cut-off 10 or 20 kDa) could be achieved in the presence of sodium salts, if the pH of the solution was controlled within a narrow range (Geckeler and Volchek 1996). A microfiltration process was developed to treat heavy metal–contaminated groundwater using DuPont's polyolefin Tyvek filter material best suited for suspended solid concentrations below 5000 ppm. An ultrafiltration technique, which employed chelating polymers for capturing dissolved metal ions, could be applied to the filtrate for further cleanup (James and Stacy 1993). Sirkar (1997) published a review of the developments in membrane separation technologies such as reverse osmosis, ultrafiltration, microfiltration, electrodialysis, dialysis, pervaporation, gas permeation, and emulsion liquid membrane.

#### 7.6.1.3 Adsorption

Adsorption, a surface-based process, is the adhesion of atoms, ions, or molecules from a gas, liquid, or dissolved solid to a surface. The nature of the bonding depends on the details of the species involved, but the adsorption process is classified as physiosorption (weak van der Waals forces or electrostatic attraction) or chemisorption (covalent bonding or chemical binding/ion exchange etc.). Adsorption is involved in many natural, physical, biological, and chemical systems. It is widely used in industrial applications such as activated charcoal, capturing and using waste heat to provide cold water for air conditioning and other process requirements (adsorption chillers), synthetic resins, increasing the storage capacity of carbide-derived carbons, and water purification. Adsorption, ion exchange, and chromatography are sorption processes in which individual adsorbates are selectively transferred from the fluid phase to the surface of insoluble, rigid particles suspended in a vessel or packed in a column.

Recently, the adsorption by commercial and synthetic activated carbon (Rajakovic 1992; Lorenzen et al. 1995; Gu et al. 2005; Gu and Deng 2006), adsorption in industrial by-products and wastes (lignite, peat charcoals) (Allen and Brown 1995; Allen et al. 1997; Mohan and Chander 2006), biochar (Fan et al. 2004; Mohan et al. 2007) and bone char (Sneddon et al. 2005), and use of ferrous materials as adsorbents and mixed composites were used in wastewater treatment. Singh et al. (2008) prepared activated carbon from tamarind wood material by chemical activation with sulfuric acid. The BET surface area of this material was found to be 612 $m^2\ g^{-1}$ and total pore volume of 0.508 $cm^3\ g^{-1}$. It was tested for Pb(II) adsorption from dilute aqueous solution, and the maximum removal rate of 97.95% (experimental) and 134.22 mg $g^{-1}$ (from Langmuir isotherm model) was obtained at initial concentration Pb(II) of 40 mg $L^{-1}$, adsorbent dose 3 g $L^{-1}$ and pH 6.5. Tamarind wood activated at pH 5.41 demonstrated high removal rate of Cr(VI) to >89% (Sahu et al. 2009a). High Pb(II) and Cr(VI) rates were also observed by zinc chloride activated carbon prepared from tamarind wood ash (Acharya et al. 2009a, 2009b; Sahu et al. 2010).

Biochar from fast wood/bark pyrolysis was adequately investigated as adsorbent for the heavy metals such as $As^{3+}$, $Cd^{2+}$, and $Pb^{2+}$ from water (Mohan et al. 2007). Maple wood ash without any chemical treatment could also be utilized to immobilize As(III) and As(V) from contaminated aqueous streams in low concentrations (Rahman et al. 2004). Some researchers proposed the use of Fe oxides, oxyhydroxides, and sulfides to sorb or immobilize a range of heavy metals from groundwater and wastes (Heal et al. 2003; Kumpiene et al. 2006; Contin et al. 2007; Mohan and Pittman 2007; Naveau et al. 2007). Mg–Al hydrotalcite was used, sometimes in conjugation with chelating ligand such as EDTA to remove As(III), As(V), Se, Cr, Cu, Cd, and Pb (Hu et al. 2005; Kameda et al. 2005; Norihiro et al. 2005; Pérez et al. 2006). Notably, Cr was the most researched heavy metal followed by As and Se.

#### 7.6.1.4 Electrokinetic Remediation

Electrokinetics remediation consists of electrodes that act as the anode and cathode, where oxidation and reduction take place. Many physicochemical processes such as oxidation, reduction, coagulation, and adsorption govern the electrokinetics. Electrokinetic remediation of soil, when a direct current electrical field is applied across a wet mass of contaminated soil, the migration of nonionic pore fluids by electro-osmosis, and the ionic migration of dissolved ions toward the electrodes take place. Combining these two removal mechanisms results in the electrokinetic extraction of metal contaminants from soils (Lestan et al. 2008). The process is schematically represented in Figure 7.1. This technique has been used for the treatment of dyes, heavy metals, nitrates, fluorides, and phenolic compounds from wastewater.

Virkutyte et al. (2002) considered electroremediation to be most effective in treating near saturated, clay soils polluted with metals, whereby removal is greater than 90%. Electromigration rates in the subsurface are dependent on the soil pore, density of water current, grain size, ionic mobility, the concentration of the contaminant, and total ionic concentration (Sims 1990; Cauwenberghe 1997). In turn, it is governed by advection, which is generated by electro-osmotic flow and externally applied hydraulic gradients, diffusion of acid from anode to cathode, and migration of cations and anions toward the respective electrodes (Zelina and Rusling 1999). Electrolysis of water is the dominant electron transfer reaction occurring at electrodes during the electrokinetic process:

$$H_2O \rightarrow 2H^+ + \frac{1}{2}\,O_2(g) + 2e^- \tag{7.1}$$

$$2H_2O + 2e^- \rightarrow 2OH^- + H_2(g) \tag{7.2}$$

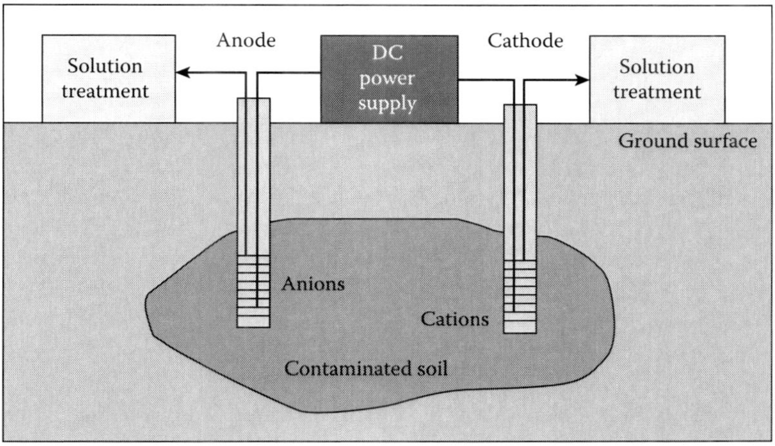


**FIGURE 7.1** Electrokinetic treatment of soil contaminants. (Adapted from Hashim et al. 2011.)

The hydrogen ions produced in the process decrease the pH near the anode causing desorption of metallic contaminants from the solid soil phases. The dissolved metallic ions are then removed from the soil solution by ionic migration and precipitation at the cathode (Acar and Alshawabkeh 1993). An increase in the hydroxide ion concentration causes an increase of the pH near the cathode. Three phenomena occurring during electrokinesis are electro-osmosis, electromigration, and electrophoresis (Virkutyte et al. 2002).

Electrokinetic remediation techniques demonstrated 85%–95% efficiency in removing As, Cd, Cr, Co, Hg, Ni, Mn, Mb, Zn, Sb, and Pb from low-permeability soils such as clay, peat, kaolinite, high-purity fine quartz, Na, and sand montmorillonite mixtures, as well as from argillaceous sand (Yeung et al. 1997). In addition, kaolinite showed more than 90% removal efficiencies of heavy metals (Pamukcu and Wittle 1992). However, the removal efficiency of porous, high permeability soils, such as peat and river sediment, was approximately 65% (Chilingar et al. 1997). Some researchers performed numerical simulations by a mathematical model for electrokinetic treatment of heavy metal–contaminated aquifer. They reported that electromigration was found to be useful for the local removal of heavy metals between the electrodes and heavy metal accumulated at the cathode region downstream after being removed from the upstream anode region. Also, they found that the hydraulic flow by purge water was essential to carry away the heavy metal from the aquifer (Shiba and Hirata 2002; Shiba et al. 2005).

### 7.6.2 Other Physicochemical Soil Treatment Processes

Some more *ex situ* and *in situ* physicochemical treatment processes that are primarily used for soil remediation can be applied in combination with some other methods to treat groundwater as well. These methods are solidification/stabilization (complete and partial vitrification), pyrometallurgical separation (*in situ* and *ex situ*), and physical separation processes (screening, gravity, magnetic separations).

#### 7.6.2.1 Immobilization

The immobilization method is mainly designed to reduce the mobility of contaminants by changing the physical or leaching characteristics of the contaminated matrix. Usually, the mobility is decreased by physically restricting contact between the contaminant and the surrounding groundwater, or by chemically altering the contaminant to make it more stable with dissolution in groundwater. The aqueous and solid phase chemistry of metals is conducive to immobilization by these techniques. A variety of methods is available for immobilization of metal contaminants, including those that use chemical reagents and thermal treatment to bind the contaminated soil or sludge physically. Most immobilization technologies can be performed *ex situ* or *in situ. In situ* processes are preferred due to lower labor and energy requirements, but the implementation *in situ* will depend on specific site conditions.

#### 7.6.2.2 Vitrification

The mobility of metal contaminants can be decreased by high-temperature treatment of the contaminated area that results in the formation of transparent material, usually an oxide solid. During this process, the increased temperature may also volatilize and destroy organic contaminants or volatile metal species (such as Hg) that must be collected for treatment or disposal. Most soils can be treated by vitrification, and a wide variety of inorganic and organic contaminants can be targeted. Vitrification may be performed *ex situ* or *in situ,* although *in situ* processes are preferred due to the lower energy requirements and cost (USEPA 1992).

#### 7.6.2.3 Excavation

Excavation and physical removal of the soil are perhaps the oldest remediation method for contaminated soil. It is still in use at many locations, including residential areas contaminated with lead in southwestern Missouri. Advantages of excavation include the complete removal of the contaminants and the relatively rapid cleanup of a contaminated site (Wood 1997). Disadvantages include the fact that the contaminants are only moved to a different place, where they must be monitored; the risk of spreading contaminated soil and dust particles during removal and transport of contaminated soil; and the relatively high cost. Excavation can be the most expensive option when large amounts of soil must be removed or disposal as hazardous or toxic waste is required.

#### 7.6.2.4 Solidification/Stabilization

Solidification and stabilization (S/S) immobilization technologies are the most common. Solidification involves the formation of a solidified matrix that physically binds the contaminated material. Stabilization, also referred to as fixation, usually utilizes a chemical reaction to convert the waste to a less mobile form. The general approach for solidification/stabilization treatment processes involves mixing or injecting treatment agents to the contaminated soils. Inorganic binders, such as cement, fly ash, or blast furnace slag, and organic binders such as bitumen are used to form a crystalline, glassy, or polymeric framework around the waste. The dominant mechanism by which metals are immobilized is by precipitation of hydroxides within the solid matrix (Bishop et al. 1982; Shively et al. 1986).

#### 7.6.2.5 Soil Leaching: Removal and Soil Isolation

The principle of soil leaching is to wash the heavy metal–contaminated soil with specific reagents and thus remove the heavy metal compound and soluble irons adsorbed on the solid-phase particles. By using this method, heavy metals are separated from the soil, and heavy metals are then recycled from the extracting solution. Removal and isolation of contaminated soil mean adding a significant amount of clean soil to cover the surface of the contaminated soil or to blend with the latter. Soil removal refers to removal of the contaminated soil and renewal of it with the clean soil, which is necessary for severely contaminated soil with a little area. Soil isolation means isolating the contaminated soil from the uncontaminated soil, but to completely remediate the soil it still needs other auxiliary engineering measures (Zheng et al. 2002). These methods will cost an enormous amount of human resources and material resources, so they can only be applied to a small area of soils.

#### 7.6.2.6 Capping

Capping systems are used to provide an impermeable barrier to surface water infiltration to the contaminated soil for prevention of further release of contaminants to the surrounding surface water or groundwater. Secondary objectives include controlling gas and odor emissions, improving esthetics, and providing a stable surface to a contaminated site. Capping also eliminates risks associated with dermal contact and incidental ingestion of surface soils, but if this is the primary goal for the location, and surface water infiltration is not a concern, a less expensive permeable cover may be preferred.

#### 7.6.2.7 Physical Separation

Physical separation is an *ex situ* process that attempts to separate the contaminated material from the rest of the soil matrix by exploiting certain characteristics of the metal and soil. Physical separation techniques are available that operate based on particle size, particle density, and surface and magnetic properties of the contaminated soil. These techniques are most useful when the metal is either in the form of discrete particles in the soil or if the metal is sorbed to soil particles that occur in a size fraction of the soil. Physical separation is often used as a form of pretreatment to reduce the amount of material requiring subsequent treatment (Rosetti 1993). Several techniques are available for physical separation of contaminated soils including screening, classification, gravity concentration, magnetic separation, and froth flotation.

#### 7.6.2.8 Limitation of Physicochemical Techniques

The commonly used physicochemical procedures for removing metal ions include chemical precipitation, ion exchange, reverse osmosis, electrodialysis, and ultrafiltration. These conventional physicochemical methods for heavy metal removal are economically expensive (Archana and Jailty 2014) and have disadvantages like incomplete metal removal, higher reagent, energy requirements, and generation of toxic sludge. Furthermore, physicochemical remediation approaches are unfit for very large areas of contaminated sites such as mining sites and industrially and agriculturally polluted soils. These procedures are improper for plant growth, and they are also damaging to almost all soil biological activities (Gaur and Adholeya 2004; Khan 2005). Hence, with these problems with physical and chemical approaches for remediation, a need for cost-effective alternative technologies is essential. In this endeavor, bioremediation has emerged as an option for developing economic and eco-friendly treatment processes for contaminated soil and water.

### 7.6.3 Biological Approaches for the Remediation of Toxic Metals–Contaminated Environments

#### 7.6.3.1 Bioremediation and Its Types

The technique involving the use of living organisms to decontaminate or neutralize environmental pollutants from a contaminated site is called bioremediation. The U.S. Environmental Protection Agency defines bioremediation as "treatment that uses naturally occurring organisms to break down hazardous substances into less toxic or nontoxic substances." Bioremediation techniques can be categorized into two basic types as *in situ* or *ex situ*. In *in situ* techniques, remediation of a contaminated site is carried out at the site, while in *ex situ* techniques contaminated material is treated elsewhere. Some examples of bioremediation techniques include bioventing, land forming, biobleaching, phytoremediation, composting, bioaugmentation, phytoremediation, and bioreactor composting.

#### *7.6.3.1.1* In Situ *Bioremediation Techniques*

In *in situ* bioremediation techniques, nutrients and oxygen are supplied by passing aqueous medium via contaminated soil sediment to enhance naturally surviving bacteria so that they may biodegrade environmental pollutants. *In situ* techniques are cost-effective and do not employ harmful microbes for the degradation of the environmental pollutants (Chauhan and Jain 2010; Rayu et al. 2012).

*In situ* bioremediation techniques can be further divided into "intrinsic bioremediation" and "engineered bioremediation" (Hazen 2010). Intrinsic bioremediation techniques involve stimulation of the naturally occurring microbial population by supplying them oxygen and gases to facilitate their metabolic activities. It is an unstimulated, unmanipulated, unenhanced bioremediation technique to treat a contaminated environment. In other words, it is a biological or natural process of decontamination of environmental pollutants. Engineered bioremediation techniques involve any type of deliberately altered or stimulated or enhanced biological remediation of an environment (Hazen 2010). The employment of a particular microbe to the contaminated environment facilitates the process of degradation, creating favorable physicochemical conditions (Kumar et al. 2011). The chief benefits of *in situ* bioremediation are its cost-effectiveness, no excavation, nominal site disruption, minimal production of dust, and the chances of simultaneous treatments of soil and groundwater in the future. However, the major problems associated with this technique are consumption of more time, seasonal variations affecting activities of the microbes, and problems encountered while applying treatment additives to the natural environment.

#### *7.6.3.1.2* Ex Situ *Bioremediation*

In *ex situ* bioremediation, the contaminated part of the environment is excavated and treated at a different site. *Ex situ* bioremediation techniques have been categorized into a solid-phase system (involves treatment of the land and piles of soil) and slurry-phase system, which involves liquid solid suspensions in bioreactors. Organic wastes like manures, agricultural wastes, leaves, domestic and industrial wastes, municipal solid wastes, and sewage sludge are treated under solid-phase treatment. The common practices of wastes treatment under solid phase include composting, land forming, and hydroponics (Kumar et al. 2011; Ramos et al. 2011; Rayu et al. 2012).

Slurry-phase bioremediation of a contaminated environment is a more rapid process wherein polluted soil sediments with water and other additives are transferred in a bioreactor and mixed properly to ensure proper exposure of pollutants with microbes. However, in the meantime optimum physical and chemical conditions required for the growth of microbes are maintained. Effective treatment of pollutants can be obtained by maintaining an adequate supply of nutrients and gases (Paliwal et al. 2012). In land forming, polluted soil is excavated and spread on a previously prepared platform. Consistent tilting of contaminated soil sediment is performed until pollutants are not degraded. Under this technique, a 10.35 cm soil layer can be treated effectively. In composting, contaminated soil is subjected to organic amendments such as manures or agricultural wastes. The incorporation of organic materials facilitates the growth and development of a microbial population (Paliwal et al. 2012).

#### 7.6.3.2 Application of Bioremediation in Removal of Toxic Metals

Bioremediation has potential to restore environments contaminated with heavy metals (Dowarah et al. 2009), but a lack of information about the factors controlling the growth and metabolism (Li et al. 2013) of microorganisms to be used for cleaning the polluted environment often limits its implementation.

##### *7.6.3.2.1 Removal of Toxic Metals Using Fungi*

*Mycoremediation* was a term coined by Stamet, which is a form of bioremediation, employing fungi to degrade or sequester contaminants from the polluted environments, thereby restoring the weakened immune system of the environment. In mycofiltration, fungal mycelia are employed to filter obnoxious waste and microbes from contaminated soil and water. In this process, mycorrhizal, endophytic, and saprophytic fungi are employed, which are capable of recovering the soil-water ecosystems and maintaining the microbial population. Extracellular enzymes and organic acids are secreted by fungi, which biologically degrade plant cell components like lignin and cellulose (main constituents of plant cell). In mycoremediation, selection of promising fungal strains for the detoxification of peculiar pollutant is essential (Stamets 2005). Dudhane et al. (2012) used mycorrhizal fungi for the stabilization of aluminum in soil. He reported fungi secrete glomaline (one of the glycoprotein) enzyme, which stabilized the aluminum in soil as well as in roots of Gmelina plants (Dudhane et al. 2012). For the removal of heavy metals from a polluted environment, several fungal species like *Trametes versicolor*, *Rhizopus arrhizus, Penicillium* spp., *Ganoderma lucidum*, *Funalia trogii*, *Cladosporium resinae*, *Aureobasidium pullulans*, and *Aspergillus niger*, etc., have been reported (Loukidou et al. 2003; Say et al. 2003). *Aspergillus versicolor* has drawn significant attention as a promising bioaccumulator of chromium from wastewaters. The bioaccumulation potential of *Aspergillus versicolor* for heavy metal was investigated by Tastan et al. (2010) at optimal pH 6 for 50 mg/L Cr(VI), Ni(II) and 5 for Cu (II). He obtained 99.89%, 30.05%, and 29.06% removal of chromium, respectively. Ramasamy et al. (2011) reported that the fungal strain (*Aspergillus fumigates*) is appropriate for removing Pb(II) ions from wastewater originating from the electroplating industry (containing Pb 100 mg/L) and reported its maximum adsorption of 85.41% during the batch sorption experiment (Ramasamy et al. 2011). Zeftawy and Mulligan (2011) reported that in a micellar-enhanced ultrafiltration (MEUF) of phosphorous-rich wastewater, it could remediate heavy metals with rejection ratios of 99% at an optimal MEUF conditions at biosurfactant to metal molar ratios of 2:1, a transmembrane pressure of 69 kPa, at 25°C and pH 6.9 (Zeftawy and Mulligan 2011). A statistically significant impact of pressure and cross-flow velocity on the flux was investigated by Hayrynen et al. (2012), who observed and achieved significant heavy metals (Cd and Cu) rejection coefficients (Landaburu-Aguirre et al. 2012), while P was not retained. Different fungi employed in removal of heavy metals and developments in context to mycoremediation are highlighted in Table 7.2.

**TABLE 7.2**
**Various Microbes Used in Removal of Toxic Metals from Different Contaminated Sites**

| Microbes | Species | Metals | References |
|---|---|---|---|
| Algae | *Ascophyllum nodosum, Bacillus firmus, Chlorella fusca, Oscillatoria anguistissima* | Pb, Zn, Cd, Cr, Cu, Ni | Ahluwalia and Goyal (2007) |
| | *Chlorella pyrenoidosa* | U | Singhal et al. (2004) |
| | *Saccharomyces cerevisiae* | Cr, Ni, Cu, Zn | Machado et al. (2010) |
| | *Ascophyllum nodosum, Fucus spiralis, Laminaria hyperborea*, and Pelvetia canaliculata | Zn, Cu, Ni | Cechinel et al. (2016) |
| | *Cladophora fascicularis* | Pb | Deng et al. (2007) |
| | *Spirogyra* sp. and *Spirulina* sp. | Cr, Cu, Fe, Mn, Zn | Mane and Bhosle (2012) |
| | *Spirogyra* sp. and *Cladophora* sp. | Pb, Cu | Lee and Chang (2011) |
| | *Hydrodictyon, Oedogonium,* and *Rhizoclonium* sp. | V, As | Saunders et al. (2012) |
| Bacteria | *Bradyrhizobium* sp. and *Rhizobacteria* sp. | Cd, Pb, Cu | Dary et al. (2010) |
| | *Bacillus* sp. | Cd, Pb, Cu | Guo et al. (2010) |
| | *Kocuria flava* | Cu | Achal et al. (2011) |
| | *Serratia marcescens* | U | Kumar et al. (2011) |
| | *Pseudomonas aeruginosa* | U | Choudhary and Sar (2011) |
| | *Bacillus cereus* | Cd, Zn | Hrynkiewicz et al. (2012) |
| | *Bacillus cereus* | Cr | Kanmani et al. (2012) |
| | *Halomonas* sp. | Sr | Achal et al. (2012a) |
| | *Bacillus, Streptococci, Salmonella, Pseudomonas, Micrococcus,* and *E. coli* | Cd, Cu, Fe | Fulekar et al. (2012) |
| | *Escherichia coli* | Pb, Cd, Zn | Akhigbe et al. (2016) |
| | *Cloacibacterium normanense* | Ni | Nouha et al. (2016) |
| | *Bacillus cereus* | Cr-6 Cr | Dong et al. (2013) |
| Fungi | *Penicillium canescens* | Cr | Say et al. (2003) |
| | *Ganoderma lucidum, Penicillium* sp. | Ar | Loukidou et al. (2003) |
| | *Aspergillus versicolor* | Cr, Ni, Cu | Tastan et al. (2010) |
| | *Aspergillus fumigatus* | Pb | Ramasamy et al. (2011) |
| | *Cladonia rangiformis* (Lichens) | Pb | Ekmekyapar et al. (2012) |
| | *Aspergillus, Mucor, Penicillium,* and *Rhizopus* | Cd, Cu, Fe | Fulekar et al. (2012) |
| | *Beauveria bassiana* | Zn, Cu, Cd, Cr, Ni | Gola et al. (2016) |
| | *Claroideoglomus claroideum* | Pb, Cd | Hristozkova et al. (2016) |

#### *7.6.3.2.2 Removal of Toxic Metals Using Algae*

The rate of release of heavy metals like Ni, Cd, Pb, and V in the atmosphere has increased many folds leading to their accumulation in marine and soil ecosystems (Shirdam et al. 2006). For the treatment of the environmental sphere with heavy metals, a wide spectrum of algae such as microalgae, green algae, and blue green algae has been found suitable (Norstrom et al. 2004; Singhal et al. 2004; Deng et al. 2007; Tripathi et al. 2008; Yin et al. 2012). The technique of removing environmental pollutants using algae, as wild type, mutant, or genetically engineered, is often referred to as cyanoremediation (Yin et al. 2012). A unicellular algae *Synechocystis* sp. PCC6803 has been reported to accumulate arsenic as much as 1 and 0.9 g/kg DW when exposed to 0.5 mM arsenate and arsenite for 14 days, respectively (Yin et al. 2012).

Algae *Synechocystis* quickly oxidized arsenite to arsenate and accumulated As rapidly through cellular oxidation when it was treated with 2.671 M arsenite. Results obtained from this study firmly establish that it can be employed for the efficient removal of arsenic from wastewater and contaminated environment.

Widespread application of cyanobacteria in the treatment of wastewaters and soil sediments contaminated with heavy metals has still not been well reported (Fiset et al. 2008). However, the roles of different cyanobacteria like *Anabaena* spp., *Nostoc* spp., *Oscillatoria* spp., *Calothrix* spp., and *Synechococcus* spp. in the treatment of heavy metals from wastewaters, wetland ecosystems, and agricultural fields has been very well reported (Fiset et al. 2008; Tripathi et al. 2008). Green marine algae *Cladophora fascicularis* was observed to be an efficient biosorbent material for removal of Pb(II) from wastewater (as a function of time, initial pH, initial Pb[II] concentrations, temperature, and co-existing ions) by Deng et al. (2007). In addition, the heavy metal removal potential of cyanobacterial species like *Oscillatoria* sp., *Synechococcus* sp., *Nodularia* sp., *Nostoc* sp., and *Cyanothece* sp., was evaluated by Dubey and Shiwani (2011). They found these algae suitable for metal removal, especially in biodegradation and biosorption of contaminants (ammonia, nitrate), either as single cell or in consortium, and the metal removal efficiency (RE) percentage of cyanobacterial species varied from 69.5% to 99.6%, while the mixed culture RE percentages ranged from 91.6% to 100%, at 5 ppm concentration of the metal taken for the experiment (Deng et al. 2007).

The biosorption potential of the green algae species *Spirogyra* and *Cladophora* was investigated by Lee and Chang (2011) for lead ($Pb_2^+$) and copper ($Cu_2^+$) from aqueous solutions. *Spirogyra* was found to result in the highest removal of Cu (89.6%), Cr (98.23%), Mn (99.6%), Fe (99.73%), Zn (81.53%), and Se (98.16%). The *Spirulina* was found to remove Cu (81.2%), Cr (98.3%), Mn (99.73%), Fe (98.93%), Zn (79%), and Se (98.83%) at 5 mg/L initial metal concentration (Mane and Bhosle 2012). Saunders and co-workers (2012) cultured three species of algae (*Rhizoclonium* sp., *Hydrodictyon,* and *Oedogonium*) to investigate their metal uptake and bioremediation potential in wastewater contaminated with heavy metals, which was released from coal-fired power generation. All algal species tested were found to be accumulating heavy metals up to 8% dry mass in general and accumulation of vanadium and arsenic concentration in particular, which reached remarkably up to 1543 and 137 mg/kg DW, respectively. Aforesaid results obtained by the researchers consolidate the facts that algae are promising agents of bioremediation of heavy metals from contaminated water and soil sediments. Algae are biological vectors, which may play a pivotal role in practical applications of metal removal from contaminated sites. The advances regarding cyanoremediation were highlighted under Table 7.2.

#### *7.6.3.2.3 Removal of Toxic Metals Using Bacteria*

Bacteria are the most versatile and abundant microorganisms, which constitute a significant portion of an entire living terrestrial biomass (Mann 1990). Earlier, bacteria are deemed as biosorbents on the basis of their small size, ubiquity and capability of proliferating under controlled conditions, and their resistance toward different environmental conditions (Urrutia 1997; Eman 2012; Kinoshita et al. 2013). Several bacterial species like *Micrococcus, Escherichia, Streptomyces, Pseudomonas,* and *Bacillus* have been successfully investigated for their heavy metal removal capacity (Duda-Chodak et al. 2008; Ma et al. 2013; Poirier et al. 2013). In addition, bacteria have slime layers composed of polysaccharide, which renders carboxyl, amino, phosphate, and sulfate groups for metals binding. Nevertheless, numerous anomalies in heavy metal uptake capacity have been reported in different bacterial genera (Mullen et al. 1989; Pandey et al. 2001). Generally, heavy metals bind to the surface of bacteria in two stages.

The first stage involves the interaction between metal ions and reactive groups on cell surfaces, and the second stage includes accumulation of target metal species in larger amounts (Vasudevan et al. 2001). Generally, the carboxyl groups of glutamic acid of the peptidoglycan layer are the principal sites of metal deposition (Chen and Wang 2007). Further, there are some bacteria that perform biosorption, a metabolism-independent major mechanism of heavy metal uptake. The process of biosorption is metabolism independent; metabolism-dependent processes may facilitate the deposition of metal on the cell wall of bacteria (Ahluwalia and Goyal 2007). However, different microbes employed for the removal of heavy metals from contaminated sites are given in Table 7.2.

### 7.6.3.3 Mechanism of Bioremediation of Heavy Metals

Microbes are present everywhere. They have the capability to change the toxic forms of heavy metals into nontoxic forms. During bioremediation, microbes perform mineralization of organic compounds into end products like water, carbon dioxide, and some metabolic intermediates, which may be utilized as primary substrates for microbial growth. Microorganisms produce degradative enzymes, which attack the pollutants as well as they have resistance toward heavy metals. Microorganisms decontaminate sites polluted with heavy metal by their enzyme systems and chemicals produced during their metabolism. They are able to dissolve, reduce, and oxidize transition metals through different methods and can restore the natural conditions. Bioremediation strategies can be successfully employed using a designer microbe approach and by knowing different environmental stresses and optimal physical and chemical conditions required for its growth. Some pollutants like organic solvents alter the cell membrane of the microbes, but microbes may develop defense mechanisms, which may involve formation of outer cell membrane protective material, often hydrophobic or solvent efflux pumps. For instance, plasmid-encoded and energy-dependent metal efflux systems involving ATPases and chemiosmotic ion/proton pumps have been reported for As, Cr, and Cd resistance in many bacteria (Roane and Pepper 2000).

#### *7.6.3.3.1 Physico-Bio-Chemical Mechanism*

In biosorption, biosorbents are employed that have strong affinity toward metal ions lasting until equilibrium is established between the two components (Das et al. 2008). *Saccharomyces cerevisiae* as a biosorbent has been reported to remove Zn(II) and Cd (II) through the ion exchange mechanism (Chen and Wang 2007; Talos et al. 2009). For the removal of heavy metals from textile mill effluent, *Cunninghamella elegans* has emerged as a promising sorbent (Tigini et al. 2010). The degradation of heavy metal involves energy for the cell metabolic cycle. The process of combined active and passive modes of toxic metal remediation can be called bioaccumulation (Brierley 1990). Significant heavy metal binding potential has been reported in some fungi like *Botryosphaeria rhodina, Pleurotus pulmonarius, Phlebia sp., Stachybotrys* sp., *Allescheriella* sp., and *Klebsiella oxytoca* (D'Annibale et al. 2007). Fungal species like *Aspergillus parasiticus* and *Cephalosporium aphidicola* have been reported to remove Pb(II) from soil by biodegradation and biosorption (Tunali et al. 2006; Akar et al. 2007).

Biotransformation of Hg(II) state to a nontoxic state by Hg-resistant fungi like *Verticillium terrestre, Neocosmospora vasinfecta,* and *Hymenoscyphus ericae* has been successfully reported. Many heavy metals and contaminants are hydrophobic and are taken up by microbes with the help of the secretion of some biosurfactant and direct cell-contaminant associations. Secreted biosurfactants form strong ionic linkage with heavy metals forming a metal complex before being desorbed from a soil matrix to an aqueous phase because of minimum interfacial tension (Thavasi 2011). Bioremediation of heavy metal can be both aerobic and anaerobic. In aerobic bioremediation, the atoms of the oxygen or chemically reactive oxygen are incorporated in reactions that are catalyzed by ligninases, peroxidases, oxidative dehalogenases, hydroxylases, dioxygenases, and monooxygenases. However, in anaerobic degradation of contaminants, initial activation reactions are involved followed by oxidative catabolism mediated by anoxic electron acceptors.

Immobilization is a process wherein reduction in the mobilization of the physical or chemical state of toxic heavy metals from contaminated sites is performed. The process of solidification treatment involves incorporation of chemical substances to the polluted sites or precipitation of hydroxides (Evanko and Dzombak 1997). Microbes mobilize heavy metals from contaminated sites by leaching, chelation, methylation, and redox transformation of toxic metals. Toxic heavy metals cannot be destroyed completely; however, with the help of bioremediation their oxidation state or organic complex can be changed so that they become water soluble, less toxic, and precipitated (Garbisu and Alkorta 2001). In microbial metabolism, heavy metals either can be utilized as terminal electron acceptors or may be reduced via detoxification reactions catalyzed by microbial enzyme systems. Microorganisms remove heavy metals through the mechanisms that they use to obtain energy from metals redox reactions, to deal with toxic metal through enzymatic and nonenzymatic processes.

Two chief mechanisms responsible for the development of resistance in bacteria are detoxification (conversion of the toxic metal state and making it unavailable) and active efflux pumping of the toxic heavy metals from bacterial cells (Silver 1996). The oxidation and reduction reactions occur in soil between toxic heavy metals and microorganisms; microorganisms serve as oxidizing agents for heavy metals and result in loss of electrons, which are gained by alternative electron acceptors (nitrate, sulfate, and ferric oxides).

Under aerobic conditions, the role of electron acceptor is performed by oxygen; under anaerobic conditions, organic contaminants are oxidized by reducing electron acceptors by microbes. The microorganism obtains energy for growth and metabolism by the oxidation of an organic compound with Fe(III) or Mn(IV) as an electron acceptor (Lovley and Phillips 1988). Anaerobic degradation of organic contaminants is facilitated under higher availability of Fe(III) for microbial reduction (Lovley et al. 1996; Spormann and Widdel 2000). The process of heavy metals being utilized as final electron acceptor is known as *dissimilatory metal reduction* (Lovely 2002). Microorganisms have been reported to perform the reduction of metals and change in their solubility pattern. Bacteria *Geobacter* reduce uranium from a soluble state ($U^{6+}$) to an insoluble state ($U^{4+}$) (Lovley et al. 1991). Microbes develop different defense strategies like complex formation and synthesis of binding protein and peptides, compartmentalization, and exclusion to overcome the stress exerted by toxic metals (Gómez Jiménez et al. 2011). By the expression of metal-binding protein and peptides, accumulation of heavy metals by microorganisms can be investigated (phytochelatins [PC] and metallothionein) (Cobbett and Goldsbrough 2002). The metal-binding protein transcription factors have been reported to mediate the process of hormone and redox signaling with respect to toxic metal (Cd, Zn, Hg, Cu, Au, Ag, Co, Ni, and Bi) exposure (Kagi and Schaffer 1988). *Synechococcus* sp. (cyanobacterial strains) has been reported with successful expression of the *smtA* gene and secretion of metal-binding protein (Huckle et al. 1993). In order to express mouse metallothionein on the cell surface and decrease the toxic effect of Cd(II) in the contaminated sites, *Ralstonia eutropha* has been genetically modified (Valls et al. 2000). Expression of several peptides and proteins in Escherichia coli regulate the accumulation of the cadmium (Mejare and Bulow 2001). Co-expression of PC along with precursor glutathione (GSH) has been reported to result in 10-fold increase in PC concentration, which finally increased accumulation of cadmium, by twofold (Kang et al. 2007).

#### *7.6.3.3.2 Molecular Mechanisms Involved in Bioremediation Process*

A wide spectrum of mechanisms has been suggested for the removal of heavy metals by microorganisms. Hg reduction has been demonstrated at high temperatures due to the expression of *mer* operon from *E. coli* coded for $Hg^{2+}$ reduction in a genetically engineered bacterium *Deinococcus geothemalis* (Brim et al. 2003). Bacterium *Cupriavidus metallidurans* strain MSR33 (mercury-resistant bacteria) was genetically modified by incorporating a pTP6 plasmid that provided genes (*merB* and *merG*) regulating biodegradation of Hg along with the synthesis of organomercurial lyase protein (MerB) and mercuric reductase (MerA).

Modification of bacterial strain *Pseudomonas* with pMR68 plasmid with novel genes (*mer*) made it resistant to mercury (Rojas et al. 2011). Mercury volatilization by reduction of Hg(II) to Hg(0) and mercury precipitation as insoluble Hg due to volatile thiol ($H_2S$), are two different mechanisms for the degradation of Hg bacteria (*Klebsiella pneumoniae* M426) (Essa et al. 2002). Genetic modification in *Deinococcus radiodurans* (radiation-resistant bacterium), which naturally reduces Cr(IV) to Cr(III) has been undertaken for complete toluene (fuel hydrocarbon) degradation by cloned genes of *Pseudomonas putida* (Brim et al. 2006). Microbial metabolites like metal-bound coenzymes and siderophores play an important role in the degradation pathway (Penny et al. 2010).

#### *7.6.3.3.3 Phytoremediation of Toxic Metals*

Phytoremediation is an emerging technology that employs the use of higher plants for the cleanup of contaminated environments. It is an efficient, economical, and eco-friendly *in situ* remediation technology. Fundamental and applied research have unequivocally demonstrated that selected plant species possess the genetic potential to remove, degrade, metabolize, or immobilize a broad range of pollutants. Despite this tremendous potential, phytoremediation is yet to become a commercial technology. Progress in the field is precluded by limited knowledge of central plant remedial mechanisms. Also, the effect of agronomic practices on these mechanisms is poorly understood. Another limitation lies within the biological nature of this novel approach. For example, the potential for phytoremediation depends on the interactions among soil, contaminants, microbes, and plants. This complex interplay, affected by a variety of factors, such as climatic conditions, soil properties, and site hydrogeology, argues against generalization, and for site-specific phytoremediation practices. Thus, an understanding of the underlying plant mechanisms and the effect of agronomic practices on plant/soil/contaminant interaction would allow practitioners to optimize phytoremediation by customizing the process to site-specific conditions. The technique of phytoremediation is classified as phytoextraction, phytofiltration (rhizofiltration), phytostabilization, phytovolatilization, and phytodegradation.

*7.6.3.3.3.1 Phytoextraction* Phytoextraction is the process of extraction of metals by plant roots and the translocation thereof to shoots. The roots and shoots are subsequently harvested to remove the contaminants from the soil. When compared with conventional soil remediation techniques, phytoextraction would be more than 10 times less cost (Salt et al. 1995). It is considered as a low-impact technology, which is its main environmental benefit. Moreover, during the phytoextraction procedure, plants cover the soil and erosion, and leaching will thus be reduced. Recently, many countries have been continuously cropping and harvesting by reducing the level of contaminants in soil (Schnoor 1997; Ma et al. 2001; Vandenhove et al. 2001).

*7.6.3.3.3.2 Phytostabilization* Phytostabilization is primarily used for the remediation of soil, sediment, and sludge (USEPA 2000). It involves the use of plant roots to limit contaminant mobility and bioavailability in the soil. The plants' primary purposes are to (1) decrease the amount of water percolating through the soil matrix, which may result in the formation of a hazardous leachate; (2) act as a barrier to prevent direct contact with the contaminated soil; and (3) prevent soil erosion and the distribution of the toxic metal to other areas (Raskin and Ensley 2000; Chhotu and Fulekar 2009). It can occur through sorption, precipitation, complexation, or metal valence reduction. It is useful for the treatment of toxic metals such as lead (Pb), arsenic (As), cadmium (Cd), chromium (Cr), and heavy metals, copper (Cu) and zinc (Zn). Some of the advantages associated with this technology are that the disposal of hazardous material/biomass is not required (USEPA 2000), and it is effective when rapid immobilization is needed to preserve groundwater and surface waters. The presence of plants also reduces soil erosion and decreases the amount of water available in the system (USEPA 2000). Mainly, it has been used to treat contaminated land areas affected by mining activities and Superfund sites.

*7.6.3.3.3.3 Rhizofiltration* Rhizofiltration is used to remediate extracted groundwater, surface water, and wastewater with low contaminant concentrations (Ensley 2000). It is defined as the use of plants, both terrestrial and aquatic, to absorb, concentrate, and precipitate contaminants from polluted aqueous sources in their roots. Rhizofiltration can be employed for the treatment of Pb, Cd, Cu, Ni, Zn, and Cr, which are primarily retained within the roots (USEPA 2000). Sunflower, Indian mustard, tobacco, rye, spinach, and corn have been studied for their ability to remove lead from water, with sunflower having the greatest ability. Indian mustard has a bioaccumulation coefficient of 563 for lead and has also proven to be effective in removing a wide concentration range of lead (4–500 mg/L) (Raskin and Ensley 2000; USEPA 2000). The advantages associated with rhizofiltration are the ability to use both terrestrial and aquatic plants for either *in situ* or *ex situ* applications. Another advantage is that contaminants do not have to be translocated to the shoots. Thus, species other than hyperaccumulators may be used. Terrestrial plants are preferred because they have a fibrous and much longer root system, increasing the amount of root area (Raskin and Ensley 2000).

*7.6.3.3.3.4 Phytovolatilization* Phytovolatilization is the use of plants to take up contaminants from the soil, transforming them into volatile forms and transpiring them into the atmosphere (USEPA 2000). So far, this process has been used for mercury as the primary metal contaminant. The benefits of this process are that the contaminant, mercuric ion, may be transformed into a less toxic substance (that is, elemental Hg). The main disadvantage of this process is that the mercury released into the atmosphere is likely to be recycled by precipitation and then redeposited back into lakes and oceans, repeating the production of methyl-mercury by anaerobic bacteria.

#### 7.6.3.4 Advantages and Limitations of Biological Methods

Bioremediation, the use of microorganisms or microbial processes to degrade environmental contaminants, is among these new technologies. Bioremediation has numerous applications, including cleanup of groundwater, soils, lagoons, sludges, and process-waste streams. Bioremediation has been used on enormous-scale applications, as demonstrated by the shoreline cleanup efforts in Prince William Sound, Alaska, after the Exxon oil spill. Although the Alaskan oil spill cleanup represents the most extensive use of bioremediation on any one site, there has been much other successful application on a smaller scale. Bioremediation frequently must address multiphasic, heterogeneous environments, such as soils in which the contaminant is present in association with soil particles, dissolved in soil liquids and in soil atmosphere. Because of these complexities, successful bioremediation is dependent on an interdisciplinary approach involving such disciplines as microbiology, engineering, ecology, geology, and chemistry.

For bioremediation to be successful, the bioremediation methods depend on having the right microbes in the right place with the right environmental factors for degradation to occur. The right microbes are bacteria or fungi, which have the physiological and metabolic capabilities to degrade the pollutants. Bioremediation offers several advantages over conventional techniques such as landfilling or incineration. Bioremediation can be done on-site, it is often less expensive, site disruption is minimal, it eliminates waste permanently, it eliminates long-term liability, it has greater public acceptance, with regulatory encouragement, and it can be coupled with other physical or chemical treatment methods. Bioremediation also has its limitations. Some chemicals are not amenable to biodegradation, for instance, heavy metals, radionuclides, and some chlorinated compounds. In some cases, microbial metabolism of contaminants may produce toxic metabolites. Bioremediation is a scientifically intensive procedure, which must be tailored to the site-specific conditions, which means one must do treatability studies on a small scale before the actual cleanup of the sites. Some of the questions one must answer before using bioremediation techniques are as follows: Is the contaminant biodegradable? Is biodegradation occurring in the site natural? Are environmental conditions appropriate for biodegradation? If the waste does not completely biodegrade, where will it go? These questions can be answered by doing site characterization and treatability studies.

Bioremediation is influenced by many factors such as microbial properties and growth (biomass concentration, population diversity, enzyme activities), environmental factors (pH, temperature, moisture content, Eh, availability of electron acceptors and carbon and energy sources), nature of the biological process (aerobic versus anaerobic process), and physicochemical characteristics of pollutants (molecular structure, and concentration). These parameters affect the acclimation period of the microbes to the substrate. The molecular structure and contaminant concentration have been shown to strongly affect the feasibility of bioremediation and the type of microbial transformation occurring, and whether the compound will serve as a primary, secondary, or co-metabolic substrate (Boopathy 2000).

## 7.7 CONCLUSION

The contamination of soils and waters by toxic metals is a serious environmental problem globally; hence, active remediation techniques are essential. Physicochemical methods for cleanup of toxic metal–contaminated soils have some serious limitations such as high cost and generation of secondary pollution problems. However, low-cost adsorbents can be applied with the further modification to change the materials into valuable and marketable products. Heavy metals can also be successfully removed by employing low-cost and efficient microbial biomass from wastewater. The process of biosorption offers several advantages including elimination of metals ions under varying pH and temperature. Biosorption efficiency of different microbial biomass has been undertaken by several research groups, which advocate microbial biomass–based removal of heavy metals in a cost-effective manner. Consequently, with unrelenting research and developmental efforts, biosorption using microbes can be performed at pilot and industrial levels, thereby offering promising alternatives to physicochemical technologies of heavy metal removal. Research is being carried out on genetically modified plants, which are better suited for phytoremediation of toxic heavy metals. Many studies are also being completed to identify the proteins that participate the in cross-membrane transport and vacuolar sequestration of toxic metals. Phytoextraction of toxic metals is predicted to be a commercially feasible scheme for phytoremediation of toxic metals in the future.

## ACKNOWLEDGMENTS

This work was supported by Young Scientist Fellowship (Award No: CST/242/, dated 12/05/2015, Council of Science and Technology Uttar Pradesh, Lucknow, India) granted to co-author Dr. Shiv Shankar.

## REFERENCES

Acar, Y.B., and Alshawabkeh, A. 1993. The principle of electrokinetic remediation. *Journal of Environmental Science and Technology* 27: 2638–2647.

Achal, V., Pan, X., Fu, Q., and Zhang, D. 2012. Biomineralization based remediation of As (III) contaminated soil by *Sporosarcina ginsengisoli*. *Journal of Hazardous Materials* 201: 178–184.

Acharya, J., Sahu, J.N., Mohanty, C.R., and Meikap, B.C. 2009a. Removal of lead(II) from wastewater by activated carbon developed from Tamarind wood by zinc chloride activation. *Chemical Engineering Journal* 149: 249–262.

Acharya, J., Sahu, J.N., Sahoo, B.K., Mohanty, C.R., and Meikap, B.C. 2009b. Removal of chromium(VI) from wastewater by activated carbon developed from Tamarind wood activated with zinc chloride. *Chemical Engineering Journal* 150: 25–39.

Ahluwalia, S.S., and Goyal, D. 2007. Microbial and plant derived biomass for removal of heavy metals from wastewater. *Bioresource Technology* 98: 2243–2257.

Akar, T., Tunali, S., and Cabuk, A. 2007. Study on the characterization of lead (II) biosorption by fungus *Aspergillus parasiticus*. *Applied Biochemistry and Biotechnology* 136: 389–406.

Akhigbe, L., Ouki, S., and Saroj, D. 2016. Disinfection and removal performance for *Escherichia coli* and heavy metals by silver-modified zeolite in a fixed bed column. *Chemical Engineering Journal* 295: 92–98.

Allen, S.J., and Brown, P. 1995. Isotherm analysis for single component and multicomponent metal sorption onto lignite. *Journal of Chemical Technology & Biotechnology* 62: 17–24.

Allen, S.J., Whitten, L.J., Murray, M., and Duggan, O. 1997. The adsorption of pollutants by peat, lignite and activated chars. *Journal of Chemical Technology & Biotechnology* 68: 442–452.

Archana, H.S., and Jaitly, A.K. 2014. Bioremediation: Environmental biotechnology for heavy metal decontamination of soil and water. *Biochemical and Cellular Archives* 14(2): 259–281.

ATSDR (Agency of Toxic Substances and Disease Registry). 2001. *CERCLA priority list of hazardous substances*. http://www.atsdr.cdc..gov/clist.html.

Bharagava, R.N., Chandra, R., and Rai, V. 2008. Phytoextraction of trace elements and physiological changes in Indian mustard plants (*Brassica nigra* L.) grown in post methanated distillery effluent (PMDE) irrigated soil. *Bioresource Technology* 99: 8316–8324.

Bharagava, R.N., Yadav, S., and Chandra, R. 2014. Antibiotic and heavy metal resistance properties of bacteria isolated from the aeration lagoons of common effluent treatment plant (CETP) of tannery industries (Unnao, India). *Indian Journal of Biotechnology* 13(4): 514–519.

Bishop, P., Gress, D., and Olafsson, J. (1982), Cement stabilization of heavy metals: Leaching rate assessment, industrial wastes. *Proceedings of the 14th Mid-Atlantic Industrial Waste Conference.* Technomics, Lancaster, PA.

Boopathy, R. 2000. Factors limiting bioremediation technologies. *Bioresource Technology* 74: 63–67.

Brierley, C.L. 1990. Bioremediation of metal-contaminated surface and groundwater. *Journal of Geomicrobiology* 8: 201–223.

Brim, H., Osborne, J.P., Kostandarithes, H.M., Fredrickson, J.K., Wackett, L.P., and Daly, M.J. 2006. *Deinococcus radiodurans* engineered for complete toluene degradation facilities Cr (IV) reduction. *Microbiology* 152: 2469–2477.

Brim, H., Venkateshwaran, A., Kostandarithes, H.M., Fredrickson, J.K., and Daly, M.J. 2003. Engineering *Deinococcus geothermalis* for bioremediation of high temperature radioactive waste environments. *Applied and Environmental Microbiology* 69: 4575–4582.

Cauwenberghe, L.V. 1997. *Electro kinetics: Technology overview report.* Groundwater Remediation Technologies Analysis Centre, Pittsburgh, PA, pp. 1–17.

Cechinel, M.A.P., Diego, A., Mayer, D.A., Pozdniakova, TA., Mazur, L.P., Boaventura, R.A.R., de Souza, A.A.U., Guelli, S.M.A., de Souza, U., and Vilar, V.J.P. 2016. Removal of metal ions from a petrochemical wastewater using brown macro-algae as natural cation-exchangers. *Chemical Engineering Journal* 286: 1–15.

Chandra, R., Bharagava, R.N., Kapley, A., and Purohit, H.J. 2011. Bacterial diversity, organic pollutants and their metabolites in two aeration lagoons of common effluent treatment plant (CETP) during the degradation and detoxification of tannery wastewater. *Bioresource Technology* 102(3): 2333–2341.

Chandra, R., Bharagava, R.N., Yadav, S., and Mohan, D. 2009. Accumulation and distribution of toxic metals in wheat (*Triticum aestivum* L.) and Indian mustard (*Brassica campestris* L.) irrigated with distillery and tannery effluents. *Journal of Hazardous Material* 162: 1514–1521.

Chandra, R., Saxena, G., and Kumar, V. 2015. Phytoremediation of environmental pollutants: An eco-sustainable green technology to environmental management. In *Advances in biodegradation and bioremediation of industrial waste,* Chandra, R. (ed.), CRC Press, Boca Raton, FL, pp. 1–30.

Chauhan, A., and Jain, R.K. 2010. Biodegradation: Gaining insight through proteomics. *Biodegradation* 21: 861–879.

Chen, C., and Wang, J.L. 2007. Characteristics of $Zn^{2+}$ biosorption by *Saccharomyces cerevisiae. Biomedical and Environmental Sciences* 20: 478–482.

Chhotu, D.J., and Fulekar, M.H., 2009. Phytoremediation of heavy metals: Recent techniques. *African Journal of Biotechnology* 8(6): 921–928.

Chilingar, G.V., Loo, W.W., Khilyuk, L.F., and Katz, S.A. 1997. Electrobioremediation of soils contaminated with hydrocarbons and metals: Progress report. *Energy Sources, Part A: Recovery, Utilization, and Environmental Effects* 19: 129–146.

Cobbett, C., and Goldsbrough, P. 2002. Phytochelatins and metallothioneins: Role in heavy metals detoxification and homeostatis. *Annual Review in plant Biology* 53: 159–182.

Conner, J.R., 1990. *Chemical fixation and solidification of hazardous wastes.* Van No- strand Reinhold, New York.

Contin, M., Mondini, C., Leita, L., and Nobili, M.D. 2007. Enhanced soil toxic metal fixation in iron (hydr)oxides by redox cycles. *Geoderma* 140: 164–175.

D'Annibale, A., Leonardi, V., Federici, E., Baldi, F., Zecchini, F., and Petruccioli, M. 2007. Leaching and microbial treatment of a soil contaminated by sulphide ore ashes and aromatic hydrocarbons. *Applied Microbiology and Biotechnology* 174: 1135–1144.

Dary, M., Chamber-Pe´rez, M.A., Palomares, A.J., and Pajuelo, E. 2010. "In situ" phytostabilisation of heavy metal polluted soils using Lupinus luteus inoculated with metal resistant plant-growth promoting rhizobacteria. *Journal of Hazardous Materials* 177: 323–330.

Das, N., Vimala, R., and Karthika, P. 2008. Biosorption of heavy metals—An overview. *Indian Journal of Biotechnology* 7: 159–169.

Deng, L., Su, Y., Su, H., Wang, X., and Zhu, X. 2007. Sorption and desorption of lead (II) from wastewater by green algae *Cladophora fascicularis*. *Journal of Hazardous Materials* 143 (1–2): 220–225.

Dong, G., Wang, Y., Gong, L., Wang, M., Wang, H., He, N., Zheng, Y., and Li, Q. 2013. Formation of soluble Cr (III) end-products and nanoparticles during Cr (VI) reduction by *Bacillus cereus* strain XMCr-6. *Journal of Biochemical Engineering* 70: 166–172.

Doong, R.A., Wu, Y.W., and Lei, W.G. 1998. Surfactant enhanced remediation of cadmium contaminated soils. *Water Science and Technology* 37: 65–71.

Dowarah, J., Boruah, H.P.D., Gogoi, J., Pathak, N., Saikia, N., and Handique, A.K. 2009. Eco-restoration of a high-sulphur coal mine overburden dumping site in northeast India: A case study. *Journal of Earth System and Science* 118(5): 597–608.

Dubey, A., and Shiwani, S. 2012. Adsorption of lead using a new green material obtained from *Portulaca plant. International Journal of Environmental Science and Technology* 9: 15–20.

Duda-Chodak, A., and Baszczyk, U. 2008. The impact of nickel on human health. *Journal of Elementology*. (13): 685–696.

Dudhane, M., Borde, M., and Jite, P.K. 2012. Effect of aluminium toxicity on growth responses and antioxidant activities in *Gmelina arborea* inoculated with AM fungi. *International Journal of Phytoremediation* 14(7): 643–655.

Ekmekyapar, F., Aslan, A., Bayhan, Y.K., and Cakici, A. 2012. Biosorption of Pb(II) by nonliving lichen biomass of *Cladonia rangiformis* (Hoffm). *International Journal of Environmental Science Research* 6(2): 417–424.

Eman. 2012. Production and characteristics of a heavy metals removing bioflocculant produced by Pseudomonas aeruginosa. *Polish Journal of Microbiology* 61(4): 281–289.

Ensley, B.D. 2000. Rationale for the Use of Phytoremediation. In *Phytoremediation of toxic metals: Using plants to clean-up the environment*. Raskin, I., and Ensely, B.D. (eds.). John Wiley Publishers, New York, pp. 205–210.

Environmental Protection Agency (EPA). 1990. Draft Report, Govt. of the Punjab, Lahore, Pakistan, pp. 51.

Essa, A.M.M., Macaskie, L.E., and Brown, N.L. 2002. Mechanisms of mercury bioremediation. *Biochemical Society Transactions* 30: 672–674.

Evanko, C.R., and Dzombak, D.A. 1997. Remediation of metals-contaminated soil and groundwater. *Environmental Science* 412: 1–45.

Fan, M., Marshall, W., Daugaard, D., and Brown, R.C. 2004. Steam activation of chars produced from oat hulls and corn stover. *Bioresource Technology* 93: 103–107.

Fiset, J.F., Blais, J.F., and Riverso, P.A. 2008. Review on the removal of metal ions from effluents using seaweeds, alginate derivatives and other sorbents. *Revue des sciences de l'eau* 21(3): 283–308.

Fulekar, M.H., Sharma, J., and Tendulkar, A. 2012. Bioremediation of heavy metals using biostimulation in laboratory bioreactor. *Environmental Monitoring and Assessment* 184(12): 7299–7307.

Garbisu, C., and Alkorta, I. 2001. Phytoextraction: A cost-effective plant-based technology for the removal of metals from the environment. *Bioresource Technology* 77: 229–236.

Gaur, A., and Adholeya, A. 2004. Prospects of arbuscular mycorrhizal fungi in phytoremediation of heavy metal contaminated soils. *Current Science* 86: 528–534.

Geckeler, K.E., and Volchek, K. 1996. Removal of hazardous substances from water using ultrafiltration in conjunction with soluble polymers. *Environmental Science and Technology* 30: 725–734.

Gola, D., Dey, P., Bhattacharya, A., Mishra, A., Malik, A., Namburath, M., and Ahammad, S.Z. 2016. Multiple heavy metal removal using an entomopathogenic fungi *Beauveria bassiana*. *Bioresource Technology* 218: 388–396.

Gu, Z., and Deng, B. 2006. Use of iron-containing mesoporous carbon (IMC) for arsenic removal from drinking water. *Environmental Engineering Science* 24: 113–121.

Gu, Z., Fang, J., and Deng, B. 2005. Preparation and evaluation of GAC-Based iron-containing adsorbents for arsenic removal. *Environmental Science and Technology* 39: 3833–3843.

Guo, H., Luo, S., Chen, L., Xiao, X., Xi, Q., Wei, W., Zeng, G., et al. 2010. Bioremediation of heavy metals by growing hyperaccumulator endophytic bacterium *Bacillus* sp. L14. *Bioresource Technology* 101: 8599–8605.

Hansen, H.K., Ottosen, L.M., Laursen, S., and Villumsen, A. 1997. Electrochemical analysis of ion-exchange membranes with respect to a possible use in electrodialytic decontamination of soil polluted with heavy metals. *Separation Science and Technology* 32: 2425–2444.

Hashim, M.A., Mukhopadhyay, S., Sahu, J.N., and Sengupta, B. 2011. Remediation technologies for heavy metal contaminated groundwater. *Journal of Environmental Management* 92: 2355–2388.

Hawkes, S.J. 1997. What is a "heavy metal"? A chemist's perspective. *Journal of Chemical Education* 74: 1369–1374.

Häyrynen, P., Landaburu-Aguirre, A., Pongrácz, E., and Keiski, R.L. 2012. Study of permeate flux in micellar-enhanced ultrafiltration on a semi-pilot scale: Simultaneous removal of heavy metals from phosphorous rich real wastewaters. *Separation and Purification Technology* 93(1): 59–66.

Hazen, T.C. 2010. In situ: Groundwater bioremediation. In *Handbook of hydrocarbon and lipid microbiology,* Timmis, K.N. (ed.), Springer, Berlin, pp. 2583–2594.

Heal, K., Younger, P.L., Smith, K., Glendinning, S., Quinn, P., and Dobbie, K. 2003. Novel use of ochre from mine water treatment plants to reduce point and diffuse phosphorus pollution. *Land Contamination and Reclamation* 11: 145–152.

Hristozkova, M., Geneva, M., Stancheva, I., Boychinova, M., and Djonova, E. 2016. Contribution of arbuscular mycorrhizal fungi in attenuation of heavy metal impact on *Calendula officinalis* development. *Applied Soil Ecology* 101: 57–63.

Hu, J., Chen, G., and Lo, I.M.C. 2005. Removal and recovery of Cr(VI) from wastewater by maghemite nanoparticles. *Water Research* 39: 4528–4536.

Hübner, R., Astin, K.B., and Herbert, R.J.H. 2010. "Heavy metal"—Time to move on from semantics to pragmatics? *Journal of Environmental Monitoring* 12: 1511–1514.

Huckle, J.W., Morby, A.P., Turner, J.S., and Robinson, N.J. 1993. Isolation of a prokaryotic metallothionein locus and analysis of transcriptional control by trace metal ions. *Molecular Microbiology* 7: 177–187.

Ismail, F.A., Aris, A.Z., and Latif, P.A. 2014. Dynamic behaviour of $Cd^{2+}$ adsorption in equilibrium batch studies by CaCO3–rich *Corbicula fluminea* shell. *Environmental Science and Pollution Research* 21(1): 344–354.

James, S.V., and Stacy, G.L. 1993. *Overview of separation technologies for EPA's superfund innovative technology evaluation program.* Gulf Publishing, Houston, TX.

Jiménez-T, R.G., Moliterni, E., Rodríguez, L., Fernández, F.J., and Villaseñor, J. 2011. Feasibility of mixed enzymatic complexes to enhanced soil bioremediation processes. *Procedia Environmental Science* 9: 54–59.

Kagi, J.H.R., and Schaffer, A. 1988. Biochemistry of metallothionein. *Biochemistry* 27: 8509–8515.

Kameda, T., Saito, S., and Umetsu, Y. 2005. Mg-Al layered double hydroxide intercalated with ethylene-diaminetetra acetate anion: Synthesis and application to the uptake of heavy metal ions from an aqueous solution. *Separation and Purification Technology* 47: 20–26.

Kang, S.H., Singh, S., Kim, J.Y., Lee, W., Mulchandani, A., and Chen, W. 2007. Bacteria metabolically engineered for enhanced photochelatin production and cadmium accumulation. *Applied and Environmental Microbiology* 73: 6317–6320.

Khan, A.G. 2005. Role of soil microbes in the rhizospheres of plants growing on trace metal contaminated soils in phytoremediation. *Journal of Trace Elements in Medicine and Biology* 18: 355–364.

Kinoshita, H., Sohma, Y., Ohtake, F., and Kimura, K. 2013. Bio sorption of heavy metals by lactic acid bacteria and identification of mercury binding protein. *Research in Microbiology* 164(7): 701–709.

Kumar, R., Acharya, C., and Joshi, S.R. 2011. Isolation and analyses of uranium tolerant *Serratia marcescens* strains and their utilization for aerobic uranium U (VI) bioadsorption. *Journal of Microbiology* 49(4): 568–574.

Kumpiene, J., Ore, S., Renella, G., Mench, M., Lagerkvist, A., and Maurice, C. 2006. Assessment of zerovalent iron for stabilization of chromium, copper, and arsenic in soil. *Environmental Pollution* 144: 62–69.

Landaburu-Aguirre, J., Pongrácz, E., Sarpola, A., and Keiski, R.L. 2012. Simultaneous removal of heavy metals from phosphorous rich real wastewaters by micellar-enhanced ultrafiltration. *Separation and Purification Technology* 88: 130–137.

Lee, Y.C., and Chang, S.P. 2011. The biosorption of heavy metals from aqueous solution by *Spirogyra* and *Cladophora* filamentous macroalgae. *Bioresource Technology* 102(9): 5297–5304.

Leeper, G.W. 1978. *Managing the heavy metals on the land*. Marcel Dekker, New York, NY.

Lestan, D., Luo, C.L., and Li, X.D. 2008. The use of chelating agents in the remediation of metal-contaminated soils: A review. *Environmental Pollution* 153: 3–13.

Li, M., Cheng, X., and Guo, H. 2013. Heavy metal removal by biomineralization of urease producing bacteria isolated from soil. *International Journal of Biodeterioration and Biodegradation* 76: 81–85.

Lim, A.P., and Aris, A.Z. 2014. A novel approach for the adsorption of cadmium ions in aqueous solution by dead calcareous skeletons. *Desalination and Water Treatment* 52: 16–18.

Lorenzen, L., van Deventer, J.S.J., and Landi, W.M. 1995. Factors affecting the mechanism of the adsorption of arsenic species on activated carbon. *Minerals Engineering* 8: 557–569.

Loukidou, M.X., Matis, K.A., Zouboulis, A.I., and Liakopoulou-Kyriakidou, M. 2003. Removal of As (V) from wastewaters by chemically modified fungal biomass. *Water Research* 37(18): 4544–4552.

Lovely, D.R. 2002. Dissimilatory metal reduction: From early life to bioremediation. *ASM News* 68: 231–237.

Lovley, D.R., Coates, J.D., Blunt-Harris, E.L., Philips, E.J.P., and Woodward, J.C. 1996. Humic substances as electron acceptors for microbial respiration. *Nature* 382: 445–448.

Lovley, D.R., Philips, E.J., Gorby, Y.A., and Landa, E.R. 1991. Microbial reduction of uranium. *Nature* 350: 413–416.

Lovley, D.R., and Phillips, E.J.P. 1988. Novel mode of microbial energy metabolism: Organic carbon oxidation to dissimilatory reduction of iron or manganese. *Applied and Environmental Microbiology* 54: 1472–1480.

Ma, L.Q., Komar, K.M., Tu, C., Zhang, W., Cai, Y., and Kennelley, E.D. 2001. A fern that hyperaccumulate arsenic. *Nature* 409: 579.

Ma, Y., Rajkumar, M., Zhang, C., and Freitas, H. 2016. Beneficial role of bacterial endophytes in heavy metal phytoremediation. *Journal of Environmental Management* 174 (1): 14–25.

Machado, M.D., Santos, M.S.F., Gouveia, C., Soares, H.M.V.M., and Soares, E.V. 2008. Removal of heavy metal using a brewer's yeast strain of *Saccharomyces cerevisiae*: The flocculation as a separation process. *Bioresource Technology* 99: 2107–2115.

Mane, P.C., and Bhosle, A.B. 2012. Bioremoval of some metals by living algae Spirogyra sp. and *Spirullina* sp. from aqueous solution. *International Journal of Environmental Research* 6(2): 571–576.

Mann, M. 1990. Removal and recovery of heavy metals by biosorption. In *Biosorption of heavy metals,* Volesky, B. (ed.), CRC press, Boca Raton, FL, pp. 93–137.

Mejare, M., and Bulow, L. 2001. Metal binding proteins and peptides in bioremediation and phytoremediation of heavy metals. *Trends in Biotechnology* 19: 67–73.

Mohan, D., and Chander, S. 2006. Removal and recovery of metal ions from acid mine drainage using lignite: A low cost sorbent. *Journal of Hazardous Materials* 137: 1545–1553.

Mohan, D., and Pittman, J.C.U. 2007. Arsenic removal from water/wastewater using adsorbent. *Journal of Hazardous Materials* 142: 1–53.

Mohan, D., Pittman, J.C.U., Bricka, M., Smith, F., Yancey, B., Mohammad, J., Steele, P.H., Alexandre-Franco, M.F., Serrano, V.G., and Gong, H. 2007. Sorption of arsenic, cadmium, and lead by chars produced from fast pyrolysis of wood and bark during bio-oil production. *Journal of Colloid and Interface Science* 310(1): 57–73.

Moore, J.W., and Ramamoorthy, S. 1984. *Heavy metals in natural waters: Applied monitoring and impact assessment*. Springer Verlag, New York, NY.

Mullen, M.D., Wolf, D.C., Ferris, F.G., Beveridge, T.J., Flemming, C.A., and Bailey G.W. 1989. Bacterial sorption of heavy metals. *Applied and Environmental Microbiology* 55(12): 3143–3159.

Mulligan, C.N., Yong, R.N., and Gibbs, B.F. 2001. Remediation technologies for metal contaminated soils and groundwater: An evaluation. *Engineering Geology* 60: 193–207.

Naveau, A., Monteil-Rivera, F., Guillon, E., and Dumonceau, J. 2007. Interactions of aqueous selenium (II) and (IV) with metallic sulfide surfaces. *Environmental Science & Technology* 41: 5376–5382.

Norihiro, M., Mitsuaki, T., Ryo, S., Hideki, Y., and Junji, S. 2005. Removal of toxic heavy metal ions in aqueous solution with Mg/Al type hydrotalcite derived from wastes. *Kagaku Kogaku Ronbunshu* 31: 285–290.

Norstrom, A., Larsdotter, K., Gumaelius, L., Jansen, J.L.C., and Dalhammar, G. 2004. A small scale hydroponics wastewater treatment system under Swedish conditions. *Water Science and Technology* 48(11–12): 161–167.

Nouha, K., Kumar, R.S., and Tyagi, R.D. 2016. Heavy metals removal from wastewater using extracellular polymeric substances produced by *Cloacibacterium normanense* in wastewater sludge supplemented with crude glycerol and study of extracellular polymeric substances extraction by different methods. *Bioresource Technology* 212: 120–129.

Paliwal, V., Puranik, S., and Purohit, H.J. 2012. Integrated perspective of effective bioremediation. *Applied Biochemistry and Biotechnology* 166: 903–924.

Pamukcu, S., and Wittle, J.K. 1992. Electrokinetic removal of selected heavy metals from soil. *Environmental Progress* 11: 241–250.

Pandey, A., Nigam, P., and Singh, D. 2001. Biotechnological treatment of pollutants. *Chemical Industry Digest* (14): 93–95.

Pandey, G., and Jain, R.K. 2002. Bacterial chemotaxis towards environmental pollutants: Role in bioremediation. *Applied Environmental Microbiology* 68(12): 5789–5795.

Penny, C., Vuilleumier, S., and Bringel, F. 2010. Microbial degradation of tetrachloromethane: Mechanisms and perspectives for bioremediation. *FEMS Microbiology and Ecology* 74(2): 257–275.

Pérez, M.R., Pavlovic, I., Barriga, C., Cornejo, J., Hermosín, M.C., and Ulibarri, M.A. 2006. Uptake of $Cu^{2+}$, $Cd^{2+}$ and $Pb^{2+}$ on Zn-al layered double hydroxide intercalated with EDTA. *Applied Clay Science* 32: 245–251.

Poirier, I., Hammann, P., Kuhn, L., and Bertrand, M. 2013. Strategies developed by the marine bacterium *Pseudomonas fluorescens* BA3SM1 to resist metals: a proteome analysis. Aquatic Toxicology 128(129): 215–232.

Prieto-Rodríguez, L., Oller, I., Klamerth, N., Aguera, A., Rodríguez, E.M., and Malato, S. 2013. Application of solar AOPs and ozonation for elimination of micro pollutants in municipal wastewater treatment plant effluents. *Water Research* 47(4): 1521–1528.

Rahman, M.H., Wasiuddin, N.M., and Islam, M.R. 2004. Experimental and numerical modeling studies of arsenic removal with wood ash from aqueous streams. *The Canadian Journal of Chemical Engineering* 82: 968–977.

Rajakovic, L.V. 1992. The sorption of arsenic onto activated carbon impregnated with metallic silver and copper. *Separation Science and Technology* 27: 1423–1433.

Ramasamy, R.K., Congeevaram, S., and Thamaraiselvi, K. 2011. Evaluation of isolated fungal strain from e-waste recycling facility for effective sorption of toxic heavy metals Pb (II) ions and fungal protein molecular characterization: A Mycoremediation approach. *Asian Journal of Experimental Biology* 2(2): 342–347.

Ramos, J.L., Marques, S., Dillewijn, P.V., Espinosa-Urgel, M., Segura, A., and Duque, E. 2011. Laboratory research aimed at closing the gaps in microbial bioremediation. *Trends in Biotechnology* 29(12): 641–647.

Raskin, I., and Ensley, B.D. 2000. *Phytoremediation of toxic metals: Using plants to clean up the environment*. Wiley, Inc., New York, NY.

Rayu, S., Karpouzas, D.G., and Singh, B.K. 2012. Emerging technologies in bioremediation: Constraints and opportunities. *Biodegradation* 23: 917–926.

Roane, T.M., and Pepper, I.L. 2000. Microorganisms and metal pollution. In *Environmental microbiology,* Maier, R.M., Pepper, I.L., and Gerba, C.B. (eds.), Academic Press, London, UK, p. 55.

Rojas, L.A., Yanez, C., Gonzalez, M., Lobos, S., Smalla, K., and Seeger, M. 2011. Characterization of the metabolically modified heavy metal-resistant *Cupriavidus metallidurans* strain MSR33 generated for mercury bioremediation. *PLoS One* 6: 17555.

Rosetti, P.K. 1993. Possible methods of washing fine soil particles contaminated with heavy metals and radionuclides. M.S. Thesis. Carnegie Mellon University, Pittsburgh, PA.

Sahu, J.N., Acharya, J., and Meikap, B.C. 2009. Response surface modeling and optimization of chromium (VI) removal from aqueous solution using tamarind wood activated carbon in batch process. *Journal of Hazardous Materials* 172: 818–825.

Sahu, J.N., Acharya, J., and Meikap, B.C. 2010. Optimization of production conditions for activated carbons from tamarind wood by zinc chloride using response surface methodology. *Bioresource Technology* 101: 1974–1982.

Salt, D.E., Blaylock, M., Kumar, N.P.B.A., Dushenkov, V., Ensley, B.D., Chet, I., and Raskin, I. 1995. Phytoremediation: A novel strategy for the removal of toxic metals from the environment using plants. *Biotechnology* 13: 468–475.

Saunders, R.J., Paul, N.A., Hu, Y., and de Nys, R., 2012. Sustainable sources of biomass for bioremediation of heavy metals in waste water derived from coal-fired power generation. *PLoS One* 7: e36470.

Saxena, G., and Bharagava, R.N. 2015. Persistent organic pollutants and bacterial communities present during the treatment of tannery wastewater. In *Environmental waste management,* Chandra, R. (ed.), CRC Press, Boca Raton, FL, pp. 217–247.

Saxena, G., and Bharagava, R.N. 2016. Ram Chandra: Advances in biodegradation and bioremediation of industrial waste. *Clean Technology and Environmental Policy* 18: 979–980. doi: http://dx.doi.org/10.1007/s10098-015-1084-9.

Saxena, G., Chandra, R., and Bharagava, R.N. 2016. Environmental pollution, toxicity profile and treatment approaches for tannery wastewater and its chemical pollutants. *Reviews of Environmental Contamination and Toxicology* 240: 31–69. doi: http://dx.doi.org/10.1007/398_2015_5009.

Say, R., Yimaz, N., and Denizli, A. 2003. Removal of heavy metal ions using the fungus *Penicillium canescens. Adsorption Science and Technology* 21: 643–650.

Schnoor, J.L. 1997. *Phytoremediation.* Department of Civil and Engineering, The University of Iowa, Iowa, IA, p. 62.

Shiba, S., and Hirata, Y. 2002. In-situ electrokinetic remediation of groundwater contaminated by heavy metal. In *Developments in water science,* Hassanizadeh, S.M., Schotting, R.J., Gray, W.G., and George, F.P. (eds.), Elsevier, Amsterdam, The Netherlands, pp. 883–890.

Shiba, S., Hirata, Y., and Seno, T. 2005. Mathematical model for hydraulically aided electrokinetic remediation of aquifer and removal of non-anionic copper. *Engineering Geology* 77: 305–315.

Shirdam, R., Khanafari, A., and Tabatabaee, A. 2006. Cadmium, nickel and vanadium accumulation by three strains of marine bacteria. *Iranian Journal of Biotechnology* 4(3): 180–187.

Shively, W., Bishop, P., Gress, D., and Brown, T. 1986. Leaching tests of heavy metals stabilized with portland cement. *Journal of Water Pollution Control Federation* 38: 234–241.

Sikdar, S.K., Grosse, D., and Rogut, I. 1998. Membrane technologies for remediating contaminated soils: A critical review. *Journal of Membrane Science* 151: 75–85.

Silver, S. 1996. Bacterial heavy metal resistance: New surprises. *Annual Reviews in Microbiology* 50: 753–789.

Sims, R.C. 1990. *Soil remediation techniques at uncontrolled hazardous waste sites.* Utah State University, Logan, UT, pp. 704–732.

Singh, C.K., Sahu, J.N., Mahalik, K.K., Mohanty, C.R., Mohan, B.R., and Meikap, B.C. 2008. Studies on the removal of Pb (II) from wastewater by activated carbon developed from tamarind wood activated with sulphuric acid. *Journal of Hazardous Materials* 153: 221–228.

Singhal, R.K., Joshi, S., Tirumalesh, K., and Gurg, R.P. 2004. Reduction of uranium concentration in well water by Chlorella (*Chlorella pyrenoidosa*) fresh water algae immobilized in calcium alginate. *Journal of Radioanalytical and Nuclear Chemistry* 261: 73–78.

Sirkar, K.K. 1997. Membrane separation technologies: Current developments. *Chemical Engineering Communications* 157: 145–184.

Sneddon, I.R., Garelick, H., and Valsami-Jones, E. 2005. An investigation into arsenic (V) removal from aqueous solutions by hydroxyl apatite and bone-char. *Mineralogical Magazine* 69: 769–780.

Spormann, A.M., and Widdel, F. 2000. Metabolism of alkyl benzenes, alkanes, and other hydrocarbons in anaerobic bacteria. *Biodegradation* 11: 85–105.

Stamets, P. 2005. *Mycelium running: How mushroom can help save the world.* Ten Speed Press, Crown Publishing Group, New York, NY.

Talos, K., Pager, C., Tonk, S., Majdik, C., Kocsis, B., Kilar, F., and Pernyeszi, T. 2009. Cadmium biosorption on native *Saccharomyces cerevisiae* cells in aqueous suspension. *Acta Universitatis Sapientiae* 1: 20–30.

Tastan, B.E., Ertuğrul, S., and Dönmez, G. 2010. Effective bioremoval of reactive dye and heavy metals by *Aspergillus versicolor. Bioresource Technology* 101(3): 870–876.

Thavasi, R. 2011. Microbial biosurfactants: From an environment application point of view. *Journal of Bioremediation and Biodegradation* 2: 104.

Tigini, V., Prigione, V., Giansanti, P., Mangiavillano, A., Pannocchia, A., and Varese, G.C. 2010. Fungal biosorption, an innovative treatment for the decolourisation and detoxification of textile effluents. *Water* 2: 550–565.

Tripathi, R.D., Dwivedi, S., Shukla, M.K., Mishra, S., Srivastava, S., Singh, R., Rai, U.N., and Gupta, D.K. 2008. Role of blue green algae biofertilizer in ameliorating the nitrogen demand and fly-ash stress to the growth and yield of rice (*Oryza sativa L*) plants. *Chemosphere* 70: 1919–1929.

Tunali, S., Akar, T., Oezcan, A.S., Kiran, I., and Oezcan, A. 2006. Equilibrium and kinetics of biosorption of lead (II) from aqueous solutions by *Cephalosporium aphidicola*. *Separation and Purification Technology* 47: 105–112.

Urrutia, M.M. 1997. General Bacterial Sorption Processes. In: *Biosorbents for metal ions*, Wase, J., Forster, C., (eds). CRC Press, London, UK, pp. 39–66.

USEPA (United States Environmental Protection Agency). 1992. *Vitrification technologies for treatment of hazardous and radioactive waste*. Handbook EPA/625/R-92/002. Office of Research and Development, U.S. Environmental Protection Agency, Washington, DC.

USEPA (United States Environmental Protection Agency). 2000. *Introduction to phytoremediation*. EPA 600/R-99/107. Office of Research and Development, U.S. Environmental Protection Agency, Cincinnati, OH.

Valls, M., Atrian, S., de Lorenzo, V., and La, F. 2000. Engineering a mouse metallothionein on the cell surface of *Ralstonia eutropha* CH34 for immobilization of heavy metals in soil. *Nature Biotechnology* 18: 661–665.

Vandenhove, H., Van Hees, M., and Van Winkel, S. 2001. Feasibility of phytoextraction to clean up low-level uranium-contaminated soil. *International Journal of Phytoremediation* 3: 301–320.

Vasudevan, P.V., Padmavathy, Tewari, N., and Dhingra. S.C. 2001. Biosorption of heavy metal ions. *Journal of Scientific and Industrial Research* 60(2): 112–120.

Virkutyte, J., Sillanpää, M., and Latostenmaa, P. 2002. Electrokinetic soil remediation critical overview. *The Science of the Total Environment* 289: 97–121.

Wood, P. 1997. Remediation methods for contaminated sites. In *Contaminated land and its reclamation,* Hester, R., and Harrison, R. (eds.), The Royal Society of Chemistry, Milton, Cambridge, pp. 47–71.

Yeung, A.T., Hsu, C.N. and Menon, R.M. 1997. Physicochemical soil-contaminant interactions during electrokinetic extraction. *Journal of Hazardous Materials* 55: 221–237.

Yin, X., Yao, C., Song, J., Li, Z., and Zhang, C. 2009. Mercury contamination in vicinity of secondary copper smelters in Fuyang, Zhejiang Province, China: Levels and contamination in top soils. *Environmental Pollution* 157(6): 1787–1793.

Zeftawy, M.A.M.E., and Mulligan, C.N. 2011. Use of rhamnolipid to remove heavy metals from wastewater by micellar-enhanced ultrafiltration (MEUF). *Separation and Purification Technology* 77(1): 120–127.

Zelina, J.P., and Rusling, J.F. 1999. Electrochemical remediation of soils. *Encyclopedia of Environmental Pollution and Cleanup* 11: 532–539.

Zhang, W., Tsang, D.C.W., and Lo, I.M.C. 2007. Removal of Pb and MDF from contaminated soils by EDTA and SDS-enhanced washing. *Chemosphere* 66: 2025–2034.

Zhang, W., Tsang, D.C.W., and Lo, I.M.C. 2008. Removal of Pb by EDTA washing in the presence of hydrophobic organic contaminants or anionic surfactant. *Journal of Hazardous Materials* 155: 433–439.

Zheng, X., Lu, A., Gao, X., Zhao, J., and Zheng, D.S. 2002. Contamination of heavy metals in soil present situation and method. *Soils and Environmental Sciences* 11(l): 79–84.

# 8 Bioremediation of Nonaqueous Phase Liquids (NAPLs)-Polluted Soil-Water Resources

*Pankaj Kumar Gupta and Brijesh Kumar Yadav*

**CONTENTS**

8.1 Introduction ........ 241
8.2 NAPLs Movement in Subsurfaces ........ 242
8.3 Engineered Bioremediation Techniques: An Overview ........ 243
8.4 Enhanced Biostimulation Strategies ........ 244
8.5 Bioaugmentation ........ 246
8.6 Plant-Assisted Bioremediation Strategies ........ 247
8.7 Constructed Wetland Treatment ........ 248
8.8 Integrated Bioremediation Strategies ........ 249
8.9 Multielectron Acceptors/Microbial Seeding Strategies ........ 250
8.10 Bioremediation Kinetics Models ........ 250
8.11 Conclusion and Recommendations ........ 252
References ........ 252

## 8.1 INTRODUCTION

Pollution of natural resources due to release of several pollutants including hydrocarbons is a key human and ecological health concern. The hydrocarbons, like nonaqueous phase liquids (NAPLs), are of particular concern under varying environmental conditions because of their high sensitivity to subsurface variability, which enables them to spread widely. The main sources of NAPLs contaminants are above-ground storage tanks (AST) or underground storage tanks (UST) holding oil and/or petrochemical products, generally found near petroleum and natural gas production sites, petrol stations (USEPA 2005). Such types of pollution from (non)point sources transformed the quality of (sub)surface water to a toxic stream; thus, the remediation of soil-water resources is needed to reduce the hazards to humans and/or environment.

Various physical, chemical, and biological practices were used to remediate NAPLs pollutants in soil-water resources (Nadim et al. 2000; Farhadian et al. 2008). Bioremediation is an emerging cost-effective technique that causes no destruction to the ecosystem as linked to the other physicochemical methods (Yang et al. 2009). Also, these practices significantly work to remediate the NAPLs pollutants from soil-water systems under a wide range of subsurface conditions (Yadav and Hassanizadeh 2011). Thus, bioremediation of natural resources polluted by NAPLs pollutants is getting increasing attention and, where relevant, can assist as a gainful decontamination alternative. Therefore, the main focus of this chapter is to present the different approaches of bioremediation of NAPLs-polluted soil-water resources. The outcomes of this chapter are of direct use in applying bioremediation techniques in the field and for the decision-making related to planning of NAPLs polluting industrial locations under varying environmental conditions.

## 8.2 NAPLs MOVEMENT IN SUBSURFACES

The NAPLs are mainly classified as light and dense nonaqueous phase liquids (LNAPLs and DNAPLs, respectively) based on their density to water and present a possible risk to soil-water systems having dynamic environmental conditions. When released in sufficient amount at the (sub)surface, the LNAPLs moves downward through the unsaturated zone and are generally retained by the water table due to their lighter density to soil water, while DNAPLs penetrate the water table and move downward until they are retained by an impermeable layer (USEPA 1995). The movement of NAPLs in a partially saturated zone becomes even more complex when a third fluid phase as an immiscible organic fluid is involved and creates a multiphase system. The NAPLs in partially saturated zones fractionize and cause multiphase partitioning (i.e., air phase, aqueous phase, solid phase, and pure phase NAPLs). The LNAPLs pool retained at the water table starts dissolving with groundwater and subsequently moves to surrounding locations due to advection, diffusion, and dispersion mechanisms of mass transport (Powers et al. 1991; Dobson et al. 2007). Similarly, the DNAPLs also form the dissolved plume, but most of the pure phase mass penetrates the water table and moves downward due to gravity.

Environmental variability and changes affect groundwater resources considerably and are responsible for frequent fluctuations in the groundwater table and its flow velocities, particularly in shallow unconfined aquifers. Rapid groundwater table fluctuations along with high pore-water velocities are expected in shallow aquifers to cause the enhanced mobilization of NAPLs. Water table fluctuations cause the (up)downward movement of the plume, causing the entrapment of NAPLs in pore space, which increases the wide coverage of the NAPLs masses. Thus, NAPLs entrapped in the form of isolated blobs or ganglia increased the water interfacial area, which enhanced NAPLs dissolution (Soga et al. 2004; Dobson et al. 2007). NAPLs trapped in the porous media become the residual and long-lasting pollution sources, which very slowly attenuated (Yadav and Hassanizadeh 2011). Bioremediation of NAPLs-polluted sites has recognized the need to improve a wide variety of ground-breaking advance techniques to eliminate NAPLs without affecting further environmental damage.

## 8.3 ENGINEERED BIOREMEDIATION TECHNIQUES: AN OVERVIEW

Bioremediation is a grouping of biological processes using (native)microorganisms under favorable conditions that act on pollutants to reduce the mass, toxicity, mobility, volume, or concentration of contaminants in soil or groundwater resources. Various studies evaluated the bioremediation techniques and a combination of these techniques and showed the significant removal rate of NAPLs from the soil-water system. The primary metabolism/secondary co-metabolism of the potential microbes causes the degradation or transformation of pollutants in nontoxic end products, which is environmentally friendly and self-sustainable.

Bioremediation practices can be categorized as *in situ* and *ex situ*. The *in situ* methods remediate soils and groundwater in place, whereas *ex situ* practices include the elimination of the contaminated soil-water resources away from the challenging site. *In situ* bioremediation of aquifer contaminated with NAPLs has been in use for more than four decades and is mostly reliant on native microbes to reduce pollutants. It does not require any mining; therefore, it is supplemented by slight or no disruption to soil morphology. Preferably, these methods should be less expensive compared to *ex situ* bioremediation, due to no additional cost prerequisite for excavation practices; nevertheless, cost of plan and in-site setting up of some sophisticated tools to increase microbial growth during bioremediation is of major concern. Some *in situ* bioremediation might be enhanced as biostimulation and phytoremediation, while others influence keeps without any form of enrichment (i.e., intrinsic bioremediation or natural attenuation). The native microbial population is familiarized to degrade the pollutant being exposed for a long time, but it takes a significantly long time for cleanup.

The engineered bioremediation emerges to accelerate the bioremediation using modification of the environmental condition and native microflora. This advanced bioremediation influences microbial activities and their neighboring environmental conditions for accelerating the practice of biodegradation and is categorized as biostimulation and bioaugmentation. The biostimulation is enhanced by the adding of nutrients, electron acceptors, oxygen, and other relevant compounds to the polluted sites, which enhanced the (co) metabolic actions of the microflora. Bioaugmentation is a microorganism seeding practice for cultivating the volume of a NAPLs degrader by adding potential microbial cultures, which are grown independently in well-defined conditions. Furthermore, the plants also accelerate the NAPLs removal by promoting the microbial re-establishment in polluted soils and water due to constant delivering of oxygen by root zone aeration and nutrients for microbial development by fixation and exudation. Similarly, the constructed wetlands techniques are concurrent treatments to polluted soil-water resources. The nature, depth, and degree of pollution, type of situation, and location are among the selection measures that are reflected when selecting any bioremediation strategies (Azubuike et al. 2016). In Figure 8.1, the conceptualized different bioremediation techniques are presented according to their application to different levels of NAPLs-polluted sites.

Furthermore, performance measures like amount of nutrient, etc., that control the success of bioremediation are also given major attention prior to beginning a bioremediation

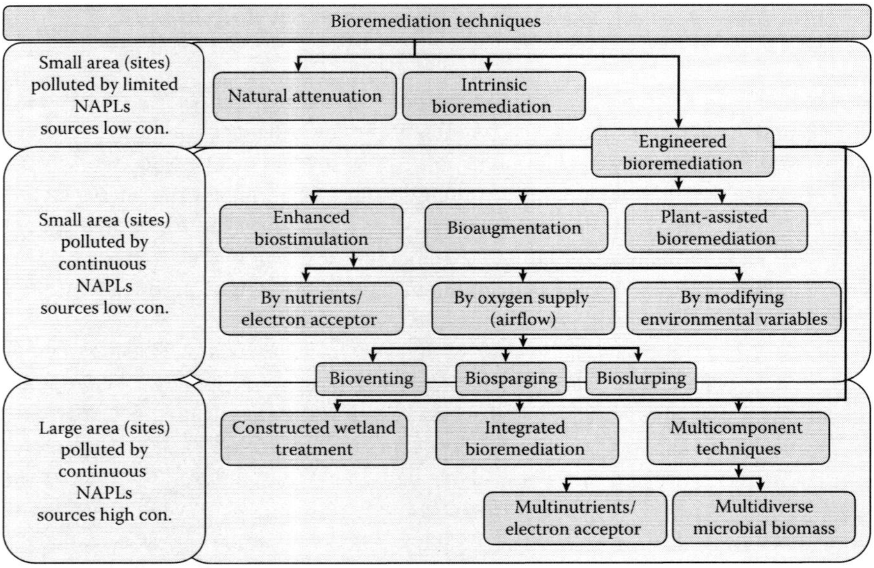


**FIGURE 8.1** Various bioremediation techniques and area/sources of NAPLs-polluted sites.

project. In this chapter, the art pertaining to bioremediation of NAPLs-polluted soil-water resources is presented with special emphasis on the engineered bioremediation strategies.

## 8.4 ENHANCED BIOSTIMULATION STRATEGIES

Enhanced biostimulation is a special case of engineered bioremediation, where the existing conditions of polluted sites may be modified by providing a favorable environment and/or nutrients for growth of potential microbial populations. The introduction of pollutants in the subsurface causes a reduction in oxygen level, nutrients, etc., which directly affects the microbial activities. Thus, native microbes need more (micro)-nutrients, electron acceptors, and favorable environmental conditions to achieve complete degradation of pollutants. Therefore, modification in polluted sites by providing such essential components stimulates microbial growth and ultimately removal of pollutants. Generally, the NAPLs-polluted sites were modified by providing (1) oxygen sources, (2) nutrients, (3) electron acceptors, (4) commercial products, and (5) favorable environmental conditions.

In order to enhance pollutant removal by increasing activities of indigenous microbes, delivering oxygen to an unsaturated (vadose) zone is the most effective technique for NAPLs-polluted sites. The controlled injection of airflow in an unsaturated zone to simulate the oxygen level in soil (water)air and to maintain aerobic conditions causes increases in microbial activities. These techniques are generally referred to as bioventing and gained popularity among other *in situ* biostimulation techniques, especially in restoring sites polluted with low concentration LNAPLs.

Sui and Li (2011) investigated the effects of air injection rate on the fate of a LNAPL-polluted site by bioventing and observed that higher airflow rate resulted in enhanced toluene removal compared to a lower rate. Frutos et al. (2010) investigated the effectiveness of bioventing treatment in stimulation of NAPLs-polluted soil and showed greater than 93% NAPLs mass removal. Similarly, *in situ* air sparging is injected below the water table so that the NAPLs pollutants move in an unsaturated zone, where a dominating microbial population promotes the removal rate. The air sparging technique is used generally for deep-penetrating pollutants like DNAPLs in contaminated aquifer systems.

Biostimulation by nutrient amendment is a practical approach to improve the rate of biodegradation of NAPLs from (un)saturated zones. Roling et al. (2002) investigated the impact of nutrient amendment on dynamics of bacterial communities along with the NAPLs biodegradation and found that the nutrient amendment significantly increased the bacterial population and improved biodegradation up to 92% of NAPLs mass removal. Yadav et al. (2013) conducted a series of microcosm experiments, in which initially the natural biodegradation was investigated, and they subsequently added domestic wastewater as nutrients. The results showed the addition of wastewater significantly enhanced the biodegradation rate at room temperature. Similarly, Macnaughton et al. (1999) investigated the impact of nutrients on removal rate in batches having NAPLs-polluted soil and showed the higher degradation rate in batches having additional nutrients. The (micro) nutrients are also available as commercial products, which are used mostly in critical cases like heavy pollution load cases. Franzetti et al. (2008) showed the increased removal of NAPLs mass enhanced by Brij 56 and Tween 80. Similarly, Asquith et al. (2012) showed 60%–69% and 69%–80% additional removal of NAPLs mass in RemActive and Daramend amended batch systems, respectively.

The addition of electron acceptors is biochemically complementary to nutrient amendment for the enhancement of the NAPLs degradation. Generally, the oxygen-releasing compounds like $H_2O_2$. $MgO_2$, $O_2$, $NO_3$, $SO_4$, Mn(IV), and Fe(III) are used to stimulate the NAPLs-polluted sites. These electron acceptors significantly increased the oxygen level, which helped in maintaining the aerobic condition. Alvarez and Vogel (1995) used nitrate as an electron acceptor as well as nutrient to incubate NAPLs degrader in a batch system. The addition of nitrate to soil is an enhancement, as the denitrification causes increased oxygen levels and significantly degrades the NAPLs mass. It is shown in the literature that the application of electron acceptors and nutrient amendments accelerates the aerobic condition at polluted sites, causing high NAPLs removal rates.

Other than the nutrient amendments and addition of electron acceptors, maintaining the environmental variables like temperature, moisture conditions, pH, salinity, etc., is the most important strategy to enhance the NAPLs' removal from polluted sites. Yadav et al. (2012) examined the role of varying temperature on biodegradation of LNAPLs under aerobic environments. To see the seasonal impact of temperature, a series of microcosm experiments was conducted at three different constant temperatures: 10°C, 21°C, and 30°C. The results of constant and fluctuating temperatures in microcosm show that toluene degradation is strongly dependent on temperature level. An almost two-time growth in LNAPLs degradation time was shown for every 10°C decrease in temperature. Similarly, Coulon et al. (2003) investigated the impact of temperature on the biodegradation of NAPLs under increasing temperature from 4°C to 20°C. The results showed the increasing temperature accelerate the degradation rate in batch system. Similarly, Dibble and Bartha (1979) determined that optimal soil moisture contents between 30% and 90% of soil field capacity are suitable for complete NAPLs removal from polluted soil-water systems. In the context of pH, microorganisms survived in a certain range of pH, and for biodegradation of NAPLs, the optimal range of pH is 6–8.

The substrate concentration is also an important variable for the stimulation of biodegradation rate, as the high concentration leads to toxicity risk. Gupta et al. (2013) investigated biodegradation having different substrate concentrations of NAPLs and showed an increased removal rate with increasing initial substrate concentration up to 100 ppm that starts decreasing with higher concentration. Research has proven that biostimulation is a cost-effective technique, but the sites' orientated nutrients, electron acceptors, under optimal environmental variables are needed to investigate for effective implementation of bioremediation in field conditions.

## 8.5 BIOAUGMENTATION

The continuous release of NAPLs causes adverse effects on microbial activities and a decline in microbial population in (sub)surface. To maintain the optimal population of such potential microbes, it is necessary to seed the native potential microbes. This microbial seeding technique to the polluted sites to achieve the maximum removal of NAPLs mass is known as a bioaugmentation technique (Atlas 1991; Sarkar et al. 2005). This technique is preferably used in the polluted area having continuous release with high substrate concentrations, less microbial counts, and dynamic environmental variables. A list of potential microbial species used in bioaugmentation of NAPLs-polluted sites is presented in Table 8.1.

**TABLE 8.1**
**Potential Microbial Species Used in Bioaugmentation of NAPLs-Polluted Sites**

| Microbial Species | References |
|---|---|
| *Pseudomonas* sp.<br>*Acinetobacter* sp.<br>*Alcaligenes* sp.<br>*Mycobacterium* sp.<br>*Xanthomonas* sp.<br>*Rhodococcus* sp.<br>*Achromobacter* sp.<br>*Pseudoxanthomonas* sp.<br>*Methanogenic* sp.<br>*Bacillus* sp. | Mishra et al. 2001; Samenta et al. 2002; Margesin et al. 2003; Bento et al. 2005; Vinas et al. 2005; Witzig et al. 2006; Popp et al. 2006; Farhadiam et al. 2008; Kim et al. 2008; Roy et al. 2014 |

Da Silva and Alvarez (2004) investigated enhanced biodegradation of NAPLs in microbial seeded aquifer columns and showed removal of NAPLs mass increased up to 88%. Joo et al. (2008) used the microbial stain of *Candida catenulate* (CM1) to contaminated soil and showed that 84% NAPLs mass removal took place in 13 days. Similarly, toluene degradation by seeding of *Corynebacterium varialilis* (SVB74) and *Acinetobacter radioresistens* (SVB65) shows the highest rate in the vadose zone.

## 8.6 PLANT-ASSISTED BIOREMEDIATION STRATEGIES

Plant-assisted bioremediation refers to the use of selective plant species for the targeted pollutant to mitigate the toxic effects and removal of pollutant mass from the (sub)surface. This technique used the plant~geochemical interaction to modify the polluted site and also supply (micro)nutrients, oxygen, etc., into the subsurface for better performance of NAPLs degrader on targeted pollutants (Susarla et al. 2002). NAPLs are mostly removed by degradation, rhizoremediation, stabilization, and volatilization, with mineralization being possible when some plants such as *Canna generalis* are used (Yadav et al. 2013). The plant–geochemical interaction enhances the (1) physical and chemical properties of sites, (2) nutrient supply by releasing root exudates (Shimp et al. 1993), (3) aeration by transfer of oxygen (Burken and Schnoor 1996), (4) interception and retardation of the movements of chemicals, (5) the plant enzymatic transformation, and (6) resistance to the vertical and lateral migration of pollutants (Narayanan et al. 1998a, 1998b). Similarly, the plant–microbe's interaction increased mineralization in the rhizosphere and the numbers of degraders and shortened the lag phase until disappearance of the compound. Some key factors to consider when choosing a plant include root system, which may be fibrous or tap subject to the depth of contaminant, toxicity of pollutant to plant, plant survival and its adaptability to prevailing environmental conditions, plant growth rate, and resistance to diseases and pests.

The deep root systems of the plant improve aeration in the subsurface, which maintains the oxygen level in the deep vadose zone. The root exudates, dead root hair, and fine root serve as important sources of the carbon for microbial growth (Shimp et al. 1993). The root exudates also accelerate the enzyme synthesis of microbial metabolisms (Dzantor et al. 2007). The role of plants on LNAPL removal using a pot scale setup planted with *Canna generalis* under controlled conditions was investigated by Basu et al. (2015). The total removal time of the LNAPLs was found to be uppermost in the unplanted followed by planted mesocosm, and degradation rates were found to be higher in planted pot setups. A three-dimensional model has been developed for the plant-assisted remediation of NAPLs-polluted sites (Narayanan et al. 1998a, 1998b). A list of potential plants species mostly used in plant-assisted bioremediation of NAPLs-polluted sites is presented in Table 8.2.

Overall, plants play a crucial role in removal of NAPLs mass, but many issues related to plant application to NAPLs-polluted sites require investigation before implementation of such techniques. Further, the impacts of static and dynamic environmental variables on pollutant removal, the combination of other

**TABLE 8.2**
**Potential Plant Species Used in Plant-Assisted Bioremediation of NAPLs-Polluted Sites**

| Plant Species | References |
|---|---|
| *Phragmites australis* | Ranieri et al. 2013; Vymazal et al. 2009 |
| *Typha latifolia* | Imfeld et al. 2008 |
| *Typha domingensis* | Shehzadi et al. 2014 |
| *Juncus subsecundus* | Zhang et al. 2010 |
| *Lolium perenne* | Tsao et al. 1998 |
| *Bassia indica* | |
| *Canna generalis* | Boonsaner et al. 2011; Yadav et al. 2013 |
| *Baumea juncea, Baumea articulata, Schoenoplectus validus, Juncus subsecundus* | Zhang et al. 2010 |
| *Juncus roemerianus* | Lin et al. 2009 |
| *Cyperus alternifolius* | Langwaldt et al. 2000 |
| *Scirpus grossus* | Bedessem et al. 2007 |
| *Polygonum aviculare* | Mohsenzadeh et al. 2010 |

bioremediation techniques to plant-assisted bioremediation, and multiscale investigation are needed.

## 8.7 CONSTRUCTED WETLAND TREATMENT

Constructed wetlands (CWs) are treatment systems that use natural processes involving wetland vegetation, soils, and their associated microbial assemblages to remediate the polluted sites. CWs are a low-cost treatment option that has been formerly assessed mainly at the mesocosm or pilot level for their capability to eliminate NAPLs. In plant-based systems, complex physical, chemical, and biological processes may occur simultaneously, including volatilization, sorption, phytodegradation, plant uptake, and accumulation, as well as microbial degradation.

A considerable amount of work has been done in the last decade with respect to application of constructed wetlands for remediation of NAPLs in pilot and field-scale systems. Treatment wetlands can support a mixed redox environment that may offer diverse biochemical pathways for pollutant removal. This shows that the rhizospheric soil has 10–100 times more microbes in wetland setups (Gerhardt et al. 2009; Imfeld et al. 2009). Various aerobic microbial groups (heterotrophs, methanotrophs, and ammonia oxidizers) are active in the wetland rhizosphere and can potentially degrade various NAPLs by (co)metabolism (Powell et al. 2011).

## 8.8 INTEGRATED BIOREMEDIATION STRATEGIES

Individual application of the above-mentioned bioremediation techniques shows effective strategies to NAPLs-polluted sites, but is limited for sites having continuous release of pollutants with high concentration levels under faster subsurface flow and dynamic environmental conditions. Therefore, the integrated or combined application of these bioremediation techniques is needed to achieve a higher degree of removal of NAPLs mass from realistic conditions. Zu and Lu (2010) integrated the bioaugmentation, biostimulation, and phytoremediation techniques and found 100% removal of NAPLs from a polluted soil-water system. The different integrated bioremediation strategies and associated dominant bioagents for bioremediation of NAPLs-polluted sites are listed in Table 8.3.

**TABLE 8.3**
**Different Integrated Bioremediation Strategies and Associated Dominant Bioagents for Remediation of NAPLs-Polluted Sites**

| Integrated Strategy | Dominant Bioagents | References |
|---|---|---|
| Bioaugmentation | *Dehalococcoides* sp., *Azoarcus* sp, *Corynebacterium variabilis, Pseudomonas aeruginosa, Pseudomonas putida, Rhodococcus* sp., *Mycobacterium* sp., *Pseudomonas* sp., *Pseudoxanthomonas spadix, Cladophialophora* sp. | Fan and Scow 1993; Romine et al. 1999; Margesin et al. 2003; Kim et al. 2008; Karamalidis et al. 2010 |
| Phytoremediation | *Juncus subsecundus, Canna generalis, Scirpus grossus Polygonum aviculare, Mirabilis jalapa* | Mohsenzadeh et al. 2010; Boonsaner et al. 2011; Zhang et al. 2010; Yadav et al. 2013 |
| Biostimulation along with microbial seeding | *Bacillus* sp., *Pseudomonas* sp., *Trichoderma* sp., *Candida catenulate, Dehalococcoides* sp., *Achromabacter* sp., *Xanthomonas* sp., *Alcaligenes* sp., *Acinetobacter* sp., *Baumannii* sp. | Mishra et al. 2001; Lendvay et al. 2003; Van Gestel et al. 2003; Huang et al. 2004; Vinas et al. 2005; Bento et al. 2005; Sarkar et al. 2005; Joo et al. 2008 |
| Plant-enhanced biostimulation | *Galega orientalis* sp., *Rhizobium* sp., *Scorzonera mongolica, Atriplex centralasiatica, Limonium bicolor, Lolium perenne, Typha domingensis, Vetiver grass, Achromobacter xylosoxidans* | Yang et al. 2010; Shehzadi et al. 2014 |

## 8.9 MULTIELECTRON ACCEPTORS/MICROBIAL SEEDING STRATEGIES

The fate and transport of the NAPLs plumes in the subsurface originate from different biochemical zones, dominated by aerobic respiration (AR), denitrification (D), iron III reduction (IR), sulfate reduction (SR), and methanogenic (M). Therefore, polluted sites having large coverage area and high pollution load, the zones-specific multielectron acceptors and or microbial seeding are applied to remediate them completely. Alvarez and Illman (2006) investigated the application of multielectron acceptor to a NAPLs-polluted site and showed first-order degradation of NAPLs mass, which means that the rate is proportional to NAPLs concentration. Cunningham et al. (2001) describes the zones of electron acceptors during bioremediation of NAPLs-polluted sites by collective addition of nitrate and sulfate.

Further, the specified zones–based application of electron acceptors along with microbes for the bioremediation is an advanced strategy for NAPLs-polluted subsurface. Generally, the microbes are associated favorable electron acceptors, thermodynamically. The zone or electron acceptor–based application of microbe diversity results in more effective and efficient bioremediation (e.g., application of anaerobic microbes in the methanogenic zone, etc.). This technique requires a better understanding of the polluted sites including the geo-hydrological assessment, fate and transport, pollution source monitoring, and many site-specific investigations for the effective implementation of the remediation project.

## 8.10 BIOREMEDIATION KINETICS MODELS

Bioremediation kinetics is one of the major decision-making factors for the remediation and better management of NAPLs-polluted site. The kinetic model helps in selection of different best techniques for polluted sites under hostile environmental conditions. Furthermore, the understanding of microbial degradation kinetics will promote the multimicrobial/electron acceptors strategy (Powell et al. 2014). Therefore, the focus of this section is to present an overview of bioremediation kinetics models associated with NAPLs-polluted sites.

The relationship between the specific growth rate ($\mu$) of the microbial population and the subtract concentration (S) is called *microbial growth kinetics* (Kovarova-kovar and Egli 1998). Microbial growth kinetics meet mass transfer kinetics and enzymatic kinetics which results in the ultimate biodegradability of substrates. Generally, the NAPLs degradation is expressed as follows:

$$-\frac{\partial C}{\partial t} = \mu_{max} C\left(\frac{C_0 + X_0 - C}{K_s + C}\right) \tag{8.1}$$

where $\mu_{max}$ is the maximum specific growth rate, $C$ is the pollutant concentration at time $t$, $C_0$ is the initial concentration and $X_0$ corresponding to contaminant required to produce initial microbial density, and $K_s$ is the substrate affinities constant (Yadav and Hassanizadeh 2011). The growth and degradation phenomena can be described satisfactorily with four parameters; the two kinetics parameters are $\mu_{max}$ and $K_s$ and

**TABLE 8.4**
**Summary of Fundamental Microbial Growth Kinetic Models Generally Referred for Site-Specific Bioremediation of NAPLs-Polluted Sites**

| Kinetics Model | Condition | Equation | Rate Constant |
|---|---|---|---|
| Constant or zero order | $X_0 >> C_0$; $C_0 >> K_s$ | $-\frac{\partial C}{\partial t} = k_0$ | $k_0 = \mu_{max} X_0$ |
| Linear of first order | $X_0 >> C_0$; $K_s >> C_0$ | $-\frac{\partial C}{\partial t} = k_1 C$ | $k_1 = \mu_{max} X_0 / K_s$ |
| Monod | $X_0 >> C_0$ | $-\frac{\partial C}{\partial t} = k_m C/(K_s + C)$ | $k_m = \mu_{max} X_0$ |
| Logistic | $K_s >> C_0$ | $-\frac{\partial C}{\partial t} = k_1 C/(C_0 + X_0 - C)$ | $k_1 = X_0/K_s$ |
| Logarithmic | $K_s >> C_0$ | $-\frac{\partial C}{\partial t} = k/(C_0 + X_0 - C)$ | $k = \mu_{max}$ |

*Source*: Adapted from Yadav, B. K., and Hassanizadeh, S. M., *Water Air Soil Pollut.*, 220(1–4), 225–239, 2011.

the other two stoichiometric parameters $Y_{s/x}$ and $S_{min}$. The fundamental biodegradation kinetics models are listed in Table 8.4, which have been generally used to describe microbial growth in a soil-water system.

Monod's kinetics, in which degradation rates resulting from zero-order to first-order kinetics to target substrate concentration, is broadly used to define biodegradation rates of NAPLs during remediation of soil-water systems (Alvarez et al. 1994; Kovarova-kovar and Egli 1998; Littlejohns and Daugulis 2008; Trigureros et al. 2010; Yadav and Hassanizadeh 2011). The equation is written as $k_m C/(K_s + C)$, where $k_m = \mu_{max} X_0$.

Alvarez et al. (1994) calculated Monod's coefficient and $K_s$ for benzene and toluene during aerobic biodegradation in a sandy aquifer system. The Monod coefficients were calculated as $k = 8.3$ g benzene/g-cells/day and $k_s = 12.2$ mg/L for benzene and $k = 9.9$ g benzene g-cells/day and $K_s = 17.4$ mg/L for toluene.

The zero-order kinetics represents the oversimplification in NAPLs degradation and mostly used low concentration degradation (Yadav and Hassanizadeh 2011; Datta et al. 2014). In zero-order kinetics, the rate of depletion of pollutant is taken as constant under conditions $X_0 >> C_0$ and $C_0 >> K_s$. In the first-order model the necessary conditions are $X_0 >> C_0$ and $K_s >> C_0$, and the model represented as $k_1 C$, where $k_1 = \mu_{max} X_0/K_s$. Gupta et al. (2013) investigated the first-order model for the NAPLs degradation up to 100 ppm substrate concentration. Similarly, Alvarez et al. (1994) investigated the first-order kinetics model during the biodegradation of NAPLs and calculated the value kinetic parameter $K_s$ is equal to 0.01 for BTEX compounds in sandy aquifer. In the logistics model, the half saturation constant is greater than the initial contaminate concentration. The differential form of equation is written as $k_1 C/(C_0 + X_0 - C)$, where $k_1 = X_0/K_s$. Similarly, the logarithmic model has lacked a horizontal asymptote as time becomes larger. In the logarithmic model, the rate constant $K$ is equal to the maximum growth rate $\mu_{max}$. These models are relevant to soils having nonlimiting nutrients, absence of mass transfer limitations, optimal soil moisture content, and constant physicochemical factors (i.e., temperature, salinity, and pH). Out of these factors, some may affect the rate of substrate uptake by microbial assemblage, and others may alter the rate of contaminant transport/supply to the microorganisms.

## 8.11 CONCLUSION AND RECOMMENDATIONS

Soil-water resources contamination resulting from releases of NAPLs in (sub) surface has long been known to pose threats to human health and the environment through multiple transport pathways. Based on their density to water, NAPLs are classified as light and dense nonaqueous phase liquids (LNAPLs and DNAPLs, respectively, and move downward through the unsaturated zone. LNAPLs are generally retained by the water table due to their lighter density to soil water, while DNAPLs penetrate the water table and move downward until they are retained by an impermeable layer. The groundwater table fluctuations along with high pore-water velocities are expected in shallow aquifers, causing enhanced mobilization of NAPLs, and (up) downward movement of the plume causes the entrapment of NAPLs in pore space, which increased the wide coverage of the NAPLs masses. Thus, the remediation of soil-water resources is needed to eliminate hazards to humans and/or the environment. Therefore, the main focus of this chapter is to present different aspects of bioremediation for NAPLs-polluted sites.

Engineered bioremediation emerges as fast-tracking methods to the bioremediation using modifications to the environmental condition and potential (native) microflora. This advance bioremediation influences microbial activities and their neighboring environmental conditions for accelerating the practice of NAPLs removal and is categorized as biostimulation and bioaugmentation. The biostimulation is enhanced by adding nutrients, electron acceptors, oxygen, and other relevant compounds to the polluted sites, which enhanced the (co) metabolic actions of the microflora. Bioaugmentation is a microorganism seeding practice for cultivating the volume of a NAPLs degrader by adding potential microbial cultures that are grown independently in well-defined conditions. Furthermore, the plants also accelerate the NAPLs removal by promoting microbial re-establishment in contaminated soils and water due to constant delivery of oxygen by root zone aeration and nutrients for microbial development by fixation and exudation. Subsequently, the integrated strategies, advanced multimicrobial and multielectron acceptor's approaches of bioremediation are discussed.

Geological characteristics of polluted site(s) including soil type, pollutant depth and type, site location relative to human habitation, and performance characteristics of each bioremediation technique should be incorporated in deciding the most suitable and efficient method to effectively treat polluted sites. Further, investigation of the multiscale modeling of NAPLs bioremediation is needed, considering site heterogeneity and environmental/climatic changes.

## REFERENCES

Alvarez, P.J.J., Anid, P.J., and Vogel, T.M. 1994. Kinetics of toluene degradation by denitrifying aquifer microorganisms. *Journal of Environmental Engineering* 120: 1327–1336.

Alvarez, P.J.J., and Illman, W.A. 2006. *Bioremediation and natural attenuation, process fundamentals and mathematical models*. Wiley-Interscience, NJ, ISBN-10 0-471-65043-9.

Alvarez, P.J.J., and Vogel, T.M. 1995. Biodegradation of BTEX and their aerobic metabolize by indigenous microorganisms under nitrate reducing conditions. *Water Sciences Technology* 3(1): 15–28.

Asquith, E.A., Geary, P.M., Nolan, A.L., and Evans, C.A. 2012. Comparative bioremediation of petroleum hydrocarbon-contaminated soil by biostimulation, bioaugmentation and surfactant addition. *Journal of Environmental Science and Engineering* 1: 637–650.

Atlas, R.M. 1991. Microbial hydrocarbon degradation and bioremediation of oil spills. *Journal of Chemical Technology and Biotechnology* 52: 149–156.

Azubuike, C.C., Chikere, C.B., and Okpokwasili, G.C. 2016. Bioremediation techniques-classification based on site of application: Principles, advantages, limitations and prospects. *World Journal of Microbiology and Biotechnology* 32: 180. DOI: http://dx.doi.org/10.1007/s11274-016-2137-x.

Basu, S., Yadav, B.K., and Mathur, S. 2015. Enhanced bioremediation of BTEX contaminated groundwater in pot-scale wetlands. *Environmental Science and Pollution Research* 22(24): 20041–20049.

Bedessem, M.E., Ferro, A.M., and Hiegel, T. 2007. Pilot-scale constructed wetlands for petroleum-contaminated groundwater. *Water Environment Research* 79(6): 581–586.

Bento, F.M., Camargo, F.A.O., Okeke, B.C., and Frankenberg, W.T. 2005. Comparative bioremediation of soils contaminated with diesel oil by natural attenuation, biostimulation and bioaugmentation. *Bioresource Technology* 96(9): 1049–1055.

Boonsaner, M., Borrirukwisitsak, S., and Boonsaner, A. 2011. Phytoremediation of BTEX contaminated soil by *Canna generalis*. *Ecotoxicology and Environmental Safety* 74(6): 1700–1707.

Burken, J.G., and Schnoor, J.L. 1998. Predictive relationships for uptake of organic contaminants by hybrid poplar trees. *Environmental Science & Technology* 32(21): 3379–3385.

Coulon, F., McKew, B.A., Osborn, A.M., McGenity, T.J., and Timmis, K.N. 2007. Effects of temperature and biostimulation on oil-degrading microbial communities in temperate estuarine waters. *Environmental Microbiology* 9(1): 177–186.

Cunningham, J.A., Rahme, H., Hopkins, G.D., Lebron, C., and Reinhard, M. 2001. Enhanced in-situ bioremediation of BTEX-contaminated groundwater by combined injection of nitrate and sulfate. *Environmental Science & Technology* 35(8): 1663–1670.

Da Silva, M.L.B., and Alvarez, P.J.J. 2004. Enhanced anaerobic biodegradation of benzene-toluene-ethylbenzene-xy-lene-ethanol mixtures in bioaugmented aquifer columns. *Applied and Environmental Microbiology* 70: 4720–4726.

Datta, A., Philip, L., and Murty, B.S. 2014. Modeling the biodegradation kinetics of aromatic and aliphatic volatile pollutant mixture in liquid phase. *Chemical Engineering Journal* 241: 288–300.

Dibble, J.T., and Bartha, R. 1979. Effect of environmental parameters on parameters on the biodegradation of oil sludge. *Applied and Environmental Microbiology* 37: 729–739.

Dobson, R., Schroth, M.H., and Zeyer, J. 2007. Effect of water-table fluctuation on dissolution and biodegradation of a multi-component, light non-aqueous-phase liquid. *Journal of Contaminant Hydrology* 94: 235–248.

Dzantor, E.K. 2007. Phytoremediation: The state of rhizosphere "engineering" for accelerated rhizodegradation of xenobiotic contaminants. *Journal of Chemical Technology and Biotechnology* 232: 228–232.

Fan, S., and Scow, K.M. 1993. Toluene by indigenous microbial populations in biodegradation of trichloroethylene and toluene by indigenous microbial populations in soil. *Applied and Environmental Microbiology* 59(6): 1911–1918.

Farhadian, M., Vachelard, C., Duchez, D., and Larroche, C. 2008. In situ bioremediation of monoaromatic pollutants in groundwater: A review. *Bioresource Technology* 99(13): 5296–5308.

Franzetti, A., Di Gennaro, P., Bestetti, G., Lasagni, M., Pitea, D., and Collina, E. 2008. Selection of surfactants for enhancing diesel hydrocarbons-contaminated media bioremediation. *Journal of Hazardous Materials* 152(3): 1309–1316.

Frutos, F.J.G., Escolano, O., García, S., Babín, M., and Fernández, M.D. 2010. Bioventing remediation and ecotoxicity evaluation of phenanthrene-contaminated soil. *Journal of Hazardous Materials* 183: 806–813.

Gerhardt, K.E., Huang, X.-D., Glick, B.R., and Greenberg, B.M. 2009. Phytoremediation and rhizoremediation of organic soil contaminants: Potential and challenges. *Plant Science* 176(1): 20–30.

Gupta, P.K., Ranjan, S., and Yadav, B.K. 2013. BTEX biodegradation in soil-water system having different substrate concentrations. *International Journal of Engineering Research & Technology* (IJERT) 2(12): 1765–1772. ISSN: 2278-0181.

Huang, X.D., El-Alawi, Y., Penrose, D.M., Glick, B.R., and Greenberg, B.M. 2004. A multi-process phytoremediation system for removal of polycyclic aromatic hydrocarbons from contaminated soils. *Environmental Pollution* 130(3): 465–476.

Imfeld, G., Braeckevelt, M., Kuschk, P., and Richnow, H.H. 2009. Monitoring and assessing processes of organic chemicals removal in constructed wetlands. *Chemosphere* 74(3): 349–362.

Joo, H.S., Ndegwa, P.M., Shoda, M., and Phae, C.G. 2008. Bioremediation of oil-contaminated soil using *Candida catenulata* and food waste. *Environmental Pollution (Barking, Essex: 1987)* 156(3): 891–896.

Karamalidis, A.K., Evangelou, A.C., Karabika, E., Koukkou, A.I., Drainas, C., and Voudrias, E.A. 2010. Laboratory scale bioremediation of petroleum-contaminated soil by indigenous microorganisms and added *Pseudomonas aeruginosa* strain Spet. *Bioresource Technology* 101(16): 6545–6552.

Kim, J.M., Le, N.T., Chung, B.S., Park, J.H., Bae, J.W., Madsen, E.L., and Jeon, C.O. 2008. Influence of soil components on the biodegradation of benzene, toluene, ethylbenzene, and o-, m-, and p-xylenes by the newly isolated bacterium *Pseudoxanthomonas spadix* BD-a59. *Applied and Environmental Microbiology* 74(23): 7313–7320.

Kovarova-kovar, K., and Egli, T. 1998. Growth kinetics of suspended microbial cells: From single-substrate-controlled growth to mixed-substrate kinetics. *Microbiology and Molecular Biology Reviews* 62(3): 646.

Langwaldt, J.H., and Puhakka, J.A. 2000. On-site biological remediation of contaminated groundwater: A review. *Environmental Pollution* 107(2): 187–197. DOI: http://dx.doi.org/10.1016/S0269-7491(99)00137-2.

Lendvay, J.M., and Lo, F.E. 2003. Bioreactive barriers: A comparison of bioaugmentation and biostimulation for chlorinated solvent remediation. *Environmental Science and Technology* 37(7): 1422–1431.

Lin, Q., and Mendelssohn, I.A. 2009. Potential of restoration and phytoremediation with *Juncus roemerianus* for diesel-contaminated coastal wetlands. *Ecological Engineering* 35(1): 85–91.

Littlejohns, J.V., and Daugulis, A.J. 2008. Kinetics and interactions of BTEX compounds during degradation by a bacterial consortium. *Process Biochemistry* 43(10): 1068–1076.

Macnaughton, S.J., Stephen, J.R., Venosa, A.D., Davis, G.A., Chang, Y., and White, D.C. 1999. Microbial population changes during bioremediation of an experimental oil spill. *Applied and Environmental Microbiology* 65(8): 3566.

Margesin, R., and Schinner, F. 2001. Biodegradation and bioremediation of hydrocarbons in extreme environments. *Applied Microbiology and Biotechnology* 56(5–6): 650–663.

Mishra, S., Jyot, J., Kuhad, R.C., and Lal, B. 2001. Evaluation of inoculum addition to stimulate in-situ bioremediation of oily-sludge-contaminated soil evaluation of inoculum addition to stimulate in-situ bioremediation of oily-sludge-contaminated soil. *Applied and*

*Environmental Microbiology* 67(4): 1675–1681. DOI: http://dx.doi.org/10.1128/AEM.67.4.1675.

Mohsenzadeh, F., Nasseri, S., Mesdaghinia, A., Nabizadeh, R., Zafari, D., Khodakaramian, G., and Chehregani, A. 2010. Phytoremediation of petroleum-polluted soils: Application of Polygonum aviculare and its root-associated (penetrated) fungal strains for bioremediation of petroleum-polluted soils. *Ecotoxicology and Environmental Safety* 73(4): 613–619.

Nadim, F., Hoag, G.E., Liu, S., Carley, R.J., and Zack, P. 2000. Detection and remediation of soil and aquifer systems contaminated with petroleum products: An overview. *Journal of Petroleum Science and Engineering* 26(1–4): 169–178.

Narayanan, M., Tracy, J.C., Davis, L.C., and Erickson, L.E. 1998a. Modeling the fate of toluene in a chamber with alfalfa plants 1. Theory and modeling concepts. *Journal of Hazardous Substance Research* 1: 5a-1–5a-30.

Narayanan, M., Tracy, J.C., Davis, L.C., and Erickson, L.E. 1998b. Modeling the fate of toluene in a chamber with alfalfa plants 2. Numerical results and comparison study. *Journal of Hazardous Substance Research* 1: 5b-1–5b-28.

Popp, N., Schlömann, M., and Mau, M. 2006. Bacterial diversity in the active stage of a bioremediation system for mineral oil hydrocarbon-contaminated soils. *Microbiology (Reading, England)* 152(Pt 11): 3291–3304.

Powell, C.L., Goltz, M.N., and Agrawal, A. 2014. Degradation kinetics of chlorinated aliphatic hydrocarbons by methane oxidizers naturally-associated with wetland plant roots. *Journal of Contaminant Hydrology* 170: 68–75.

Powell, C.L., Nogaro, G., and Agrawal, A. 2011. Aerobic co-metabolic degradation of trichloroethene by methane and ammonia oxidizing microorganisms naturally associated with *Carex comosa* roots. *Biodegradation* 22: 527–538.

Powers, S.E., Loureiro, C.O., Abriola, L.M., and Weber, W.J. 1991. Theoretical study of the significance of non-equilibrium dissolution of non-aqueous phase liquids in subsurface systems. *Water Resources Research* 27(4): 463–477.

Ranieri, E., Gikas, P., and Tchobanoglous, G. 2013. BTEX removal in pilot-scale horizontal subsurface flow constructed wetlands. *Desalination and Water Treatment* 51(13–15): 3032–3039.

Röling, W.F., Milner, M.G., Jones, D.M., Lee, K., Daniel, F., Swannell, R.J., and Head, I.M. 2002. Robust hydrocarbon degradation and dynamics of bacterial communities during nutrient-enhanced oil spill bioremediation. *Applied and Environmental Microbiology* 68(11): 5537–5548.

Romantschuk, M., Sarand, I., Petanen, T., Peltola, R., Jonsson-Vihanne, M., Koivula, T., Yrjala, K., and Haahtela, K. 2000. Means to improve the effect of in situ bioremediation of contaminated soil: An overview of novel approaches. *Environmental Pollution* 107: 179–185.

Roy, A.S., Baruah, R., Borah, M., Singh, A.K., Deka Boruah, H.P., Saikia, N., and Chandra Bora, T. 2014. Bioremediation potential of native hydrocarbon degrading bacterial strains in crude oil contaminated soil under microcosm study. *International Biodeterioration & Biodegradation* 94, 79–89.

Samanta, S.K., Singh, O.V., and Jain, R.K. 2002. Polycyclic aromatic hydrocarbons: Environmental pollution and bioremediation. *Trends in Biotechnology* 20(6): 243–248.

Sarkar, D., Ferguson, M., Datta, R., and Birnbaum, S. 2005. Bioremediation of petroleum hydrocarbons in contaminated soils: Comparison of biosolids addition, carbon supplementation, and monitored natural attenuation. *Environmental Pollution* 136(1): 187–195.

Shehzadi, M., Afzal, M., Khan, M.U., Islam, E., Mobin, A., Anwar, S., and Khan, Q.M. 2014. Enhanced degradation of textile effluent in constructed wetland system using *Typha domingensis* and textile effluent-degrading endophytic bacteria. *Water Research* 58: 152–159.

Shimp, J.F., Tracy, J.C., Davis, L.C., Lee, E., Huang, W., Erickson, L.E., and Schnoor, J.L. 1993. Beneficial effects of plants in the remediation of soil and groundwater contaminated with organic materials. *Critical Reviews in Environmental Science and Technology* 23(1): 41–77.

Soga, K., Page, J.W.E., and Illangasekare, T.H. 2004. A review of NAPL source zone remediation efficiency and the mass flux approach. *Journal of Hazardous Materials* 110(1–3): 13–27.

Sui, H., and Li, X. 2011. Modeling for volatilization and bioremediation of toluene-contaminated soil by bioventing. *Chinese Journal of Chemical Engineering* 19: 340–348.

Susarla, S., Medina, V.F., and McCutcheon, S.C. 2002. Phytoremediation: An ecological solution to organic chemical contamination. *Ecological Engineering* 18(5): 647–658.

Trigueros, D.E.G., Modenes, A.N., Kroumov, A.D., and Espinoza-Quinones, F.R. 2010. Modeling of biodegradation process of BTEX compounds: Kinetic parameters estimation by using particle swarm global optimizer. *Process Biochemistry* 45(8): 1355–1361.

Tsao, C., Song, H., and Bartha, R. 1998. Metabolism of benzene, toluene, and xylene hydrocarbons in soil. *Applied and Environmental Microbiology* 64(12): 4924.

USEPA (U.S. Environmental Protection Agency). 1995. *Light non-aqueous phase liquids.* EPA/540/S-95/500. Office of solid waste and emergency response, Washington, DC.

USEPA. 2005. *A decision-making framework for cleanup of sites impacted with light non aqueous phase liquids (LNAPL)*. EPA 542-R-04-011. U.S. Environmental Protection Agency, Washington, D.C.

Van Gestel, K., Mergaert, J., Swings, J., Coosemans, J., and Ryckeboer, J. 2003. Bioremediation of diesel oil-contaminated soil by composting with biowaste. *Environmental Pollution* 125(3): 361–368.

Vinas, M., Sabate, J., Espuny, M.J., Anna, M., and Vin, M. 2005. Bacterial community dynamics and polycyclic aromatic hydrocarbon degradation during bioremediation of heavily creosote-contaminated soil. *Applied and Environmental Microbiology* 71(11): 7008.

Vymazal, J. 2009. The use constructed wetlands with horizontal sub-surface flow for various types of wastewater. *Ecological Engineering* 35(1): 1–17.

Witzig, R., Junca, H., Hecht, H., and Pieper, D.H. 2006. Assessment of toluene/biphenyl dioxygenase gene diversity in benzene-polluted soils: Links between benzene biodegradation and genes similar to those encoding isopropyl benzene dioxygenases assessment of toluene/biphenyl dioxygenase. *Applied and Environmental Microbiology* 72(5): 3504–3514.

Yadav, B.K., Ansari, F.A, Basu, S., and Mathur, A. 2013. Remediation of LNAPL contaminated groundwater using plant-assisted biostimulation and bioaugmentation methods. *Water, Air, & Soil Pollution* 225(1): 1793–1799.

Yadav, B.K., and Hassanizadeh, S.M. 2011. An overview of biodegradation of LNAPLs in coastal (semi)-arid environment. *Water, Air, & Soil Pollution* 220(1–4): 225–239. DOI: http://dx.doi.org/10.1007/s11270-011-0749-1.

Yadav, B.K., Hassanizadeh, S.M., and Rajbhandari, S. 2012. Biodegradation of toluene under seasonal and diurnal fluctuations of soil-water temperature conditions. *Water, Air, & Soil Pollution* 223(7): 3579–3588.

Yang, S.Z., Jin, H.J., Wei, Z., He, R.X., Ji, Y.J., Li, X.M., and Yu, X.P. 2009. Bioremediation of oil spills in cold environments: A review. *Pedosphere* 19: 371–381.

Zhang, B.Y., Zheng, J.S., and Sharp, R.G. 2010. Phytoremediation in engineered wetlands: Mechanisms and applications. *Procedia Environmental Sciences* 2: 1315–1325.

Zu, Y., and Lu, M. 2010. Bioremediation of crude oil-contaminated soil: Comparison of different biostimulation and bioaugmentation treatments. *Journal of Hazardous Materials* 183(1–3): 395–401.

# 9 Microbial Strategies for Enhanced Phytoremediation of Heavy Metal-Contaminated Soils

*Jyoti Singh and Ajay Veer Singh*

## CONTENTS

9.1 Introduction ........ 257
9.2 Phytoremediation Technologies ........ 258
  9.2.1 Phytoextraction/Phytoaccumulation ........ 259
  9.2.2 Phytostabilization/Phytoimmobilization ........ 259
  9.2.3 Phytovolatilization/Rhizovolatilization ........ 259
  9.2.4 Rhizofiltration ........ 259
9.3 Remediation of Organic Contaminants by Plant Growth-Promoting Rhizobacteria (PGPR) ........ 260
9.4 Impact of Plant, Soil, and Microbe Interaction ........ 261
  9.4.1 Positive Interactions ........ 262
  9.4.2 Negative Interactions ........ 263
9.5 Heavy Metal and Microbe Interaction ........ 264
9.6 Internal Physiological Mechanisms ........ 264
9.7 Genetically Engineered Strategies ........ 266
9.8 Conclusion ........ 267
References ........ 268

## 9.1 INTRODUCTION

Heavy metals are recognized relatively as high-density inorganic pollutants, which comprise 75% known elements and are toxic at low concentrations. Industrial wastewater has been considered as the major source of heavy metal pollution in the environment (Chandra et al. 2011; Saxena and Bharagava 2015; Saxena et al. 2016a, 2016b). These metals have been used since the beginning of ancient human civilization and are constantly being added to soil, water, and eventually the biosphere through rapid modernization as well as industrialization around the world via smelting of metalliferous surface finishing industry, combustion of fossils, electroplating, photography, electrical devices, manufacturing, aerospace, waste incineration, transportation, mining, agricultural applications, etc. The accrual of heavy metals, as nonbiodegradable material, not only decreases soil microbial activity along with crop production but also threatens human health due to their persistence in the course of the food chain. Various noxious heavy metals present in different oxidation states such as zinc (Zn), arsenic (As), chromium (Cr), cadmium (Cd), mercury (Hg), copper (Cu), nickel (Ni), and lead (Pb); radioactive elements such as uranium and strontium; as well as organic compounds like trinitrotoluene, 1,3,5-trinitro-1,3,5-hexahydrotriazine, petroleum hydrocarbons (benzene, toluene, xylene, etc.) due to their nonbiodegradable nature, all are difficult to eliminate from the environment, and they become acutely toxic if their concentration exceeds certain threshold (Glick 2010; Ma et al. 2011; Shinwari et al. 2015).

A number of heavy metals are required by plants in traces as micronutrients, but uneven accretion of heavy metals is detrimental to the majority of plants. Heavy metals, when present at a prominent level in the environment, are absorbed by plant roots and translocated to shoots, leading to impaired metabolism, reduced growth, and even plant death (Tak et al. 2013). In addition, cadmium, for example, is a toxic heavy metal to plants, which can inhibit root and shoot growth, affect nutrient uptake, and affect homeostasis. On conditions, the contamination of soil with Cd can negatively influence biodiversity as well as the activities of soil microbial communities (Jing et al. 2007).

## 9.2 PHYTOREMEDIATION TECHNOLOGIES

To avoid the toxicity associated with these hazardous compounds, a number of technologies and methods have been developed to eliminate them from polluted soil. These heavy metals cannot be degraded easily but are barely transformed from one oxidation state to another or from one organic complex to a different complex. Until now, some physical methods have been developed and designed for their remediation, such as exclusion of soil to landfill sites or mining through chemical or physical means. These techniques are prompt, but regrettably, they are expensive from both economic and environmental points of view, and they potentially have adverse impacts on soil physical, chemical, and biological properties (Glick et al. 2007). Traditional methods of environmental remediation do not resolve the issue; somewhat they only transfer it to next generation. Evidently, there is a vital need for alternative, inexpensive, and resourceful methods to clean up sites infected with toxic metals.

Phytoremediation is a plant-based alternative green technology that is a cost-effective, environmentally friendly, esthetically satisfying approach in support of the remediation of toxic metals (Bharagava et al. 2008; Chandra et al. 2009; Chandra et al. 2015; Laghlimi et al. 2015). Due to the level of contaminated areas, approaches of phytoremediation have already received noteworthy scientific and commercial consideration. Two approaches have been projected for the phytoremediation of toxic metals from soils as well as wetlands: natural phytoremediation and induced phytoremediation.

Natural phytoremediation makes use of hyperaccumulating vegetation along with associated soil microbes, whereas induced phytoremediation employs chemicals, especially artificial chelating ligands, for the augmentation of metal bioavailability as well as uptake in plants (Rahman et al. 2016). Depending on the methods used and the nature of the concerned contaminant, phytoremediation of heavy metals may exploit one of several techniques: phytoextraction, phytovolatilization, phytostabilization, as well as rhizofiltration.

### 9.2.1 Phytoextraction/Phytoaccumulation

This technique eliminates metals by taking advantage of the unusual ability of some plants to absorb and accumulate metals, by concentrating them within the biomass. The principle of this kind of remediation is to diminish the metal concentration in contaminated soils so that they can be used advantageously for agriculture, forestry, horticulture, grazing, etc. Hyperaccumulation of these metal ions can be induced by adding a chemical amendment, such a ethylenediaminetetraacetic acid (EDTA), in the direction of a plant substrate to formulate a soluble an otherwise insoluble target metal, such as lead (Christopher et al. 1998).

### 9.2.2 Phytostabilization/Phytoimmobilization

This technique utilizes plants, frequently blended with soil additives, to assist in mechanically stabilizing sites intended for reducing pollutant transfer to other compartments of the ecosystem and to the network of links in a food web; the stabilized organic or inorganic composite is generally integrated into plant lignin or soil humus. The foundation for phytostabilization is that metals do not degrade, and thus capturing them *in situ* is often the best alternative. This approach is pertinent when small concentrations, diffused areas, and huge areas of contamination are to be treated. Plants confine pollutants through creating a zone around the roots where the pollutant is precipitated and stabilized. Phytostabilization is undertaken when the plants used do not absorb the targeted pollutants into plant tissue.

### 9.2.3 Phytovolatilization/Rhizovolatilization

These techniques employ the metabolic capabilities of plants as well as the associated rhizospheric microorganisms to modify pollutants into volatile compounds to release into the atmosphere. Some ions are absorbed by roots and are transformed into less toxic forms and subsequently released.

### 9.2.4 Rhizofiltration

Plant roots absorb, concentrate, as well as precipitate contaminants in the aqueous system. This approach is also used to moderately treat industrial and agricultural runoff (Tak et al. 2013). Rhizofiltration is chosen due to its ability to be used for *in situ* or *ex situ* application and because species other than hyperaccumulators may be used in the process. Plants like *Helianthus*, *Brassica*, *Nicotiana*, *Secale*, *Spinacia,* and *Zea mays* have been studied for their ability to eliminate lead from industrial effluents (Bharti et al. 2014).

As compared in the midst of chemical and physical remediation, phytoremediation has the following properties:

1. It preserves the natural properties of the soil.
2. It obtains energy primarily from sunlight.
3. Eminent levels of microbial biomass within the rhizosphere can be achieved.
4. It is economical.
5. It has the potential to be expeditious (Huang et al. 2004a).

Unfortunately, many of the plants effective in removing metals from soil (e.g., hyperaccumulators such as *Thlaspi caerulescens* as well as *Alyssum bertolonii*) are small and slow growing, thus minimizing their potential for phytoextraction of metal from soil on a large scale (Khan et al. 2000). To be effectual in the remediation of soil polluted with metals, plants must be tolerant to one or more metals, highly competitive, fast growing, and have a widespread root system; trees are considered for phytoremediation, but the accumulation of metals by trees is generally low (Glick 2012).

## 9.3 REMEDIATION OF ORGANIC CONTAMINANTS BY PLANT GROWTH-PROMOTING RHIZOBACTERIA (PGPR)

Heavy metals are associated with a number of soil components, which determines the behavior in the soil and influences their bioavailability (Boruvka and Drabek 2004). There is a way to maximize the chances of success of phytoremediation by exploiting plant growth-promoting rhizobacteria (PGPR), which are soil bacteria that reside in the rhizosphere. When PGPR are introduced to a contaminated site, they augment the potential for plants that sequester heavy metals, recycle various nutrients, maintain the structure of soil, detoxify chemicals, and control diseases along with pests. PGPR also decrease the toxicity of metals by altering their bioavailability within plants. The plants provide root exudates to the microorganisms in turn (Tak et al. 2013). Under metal stress, PGPR have developed various strategies to escape the toxicity generated by various heavy metals, which include exclusion of metal species outside the microbial cell surface, bioaccumulation of metal ions in the cell actively or passively, biotransformation of toxic metals into less toxic metals, and adsorption of metal on the cell wall (Ahemad 2014).

As stated by Huang et al. (2004a, 2004b), the addition of PGPR improve seed germination, growth, and survival, and also stimulate faster plant growth in soils contaminated with organic compounds such as polycyclic aromatic hydrocarbons and creosote (Zhuang et al. 2007). PGPR are reported to secrete some extracellular metabolites such as siderophores that play an important role in sequestering iron, and have more affinity toward plants, as they form a ferric-siderophore complex that can move via diffusion and be returned back to the cell surface by solubilizing iron from their surroundings (Singh et al. 2010a). Most of the catecholates are bacterial siderophores, and few are hydroxamates and carboxylates (Rachna and Snehlata 2013). A load of bacterial species belonging to genera *Azospirillum, Arthrobacter, Flavobacterium, Bacillus, Erwinia, Alcaligenes, Pseudomonas, Acinetobacter, Burkholderia, Enterobacter,*

*Chryseobacterium,* and *Rhizobium* are associated with plant rhizosphere. These bacteria are able to put forth an advantageous effect on plant growth (Tilak et al. 2005; Egamberdieva 2009; Singh et al. 2013; Singh and Goel 2015). It has been suggested that Indole-3-acetic acid (IAA) synthesized by PGPR may prevent the detrimental effects of environmental stresses. As IAA, the lengthening of root and shoot of wheat seedlings exposed to high levels of salt is stimulated (Lindberg et al. 1985).

PGPR also facilitate plant growth through $N_2$ fixation, insoluble phosphorus solubilization (Singh et al. 2010b, 2010c; Ahemad and Khan 2012a), production of phytohormones, lowering of ethylene concentration, synthesis of antibiotics and antifungal metabolites, and induced systemic resistance (Glick 2012). In this way, Plant growth promotory bacteria (PGPB) are also known to boost soil fertility, and in return, the plant acquiesces by supplying essential nutrients and growth regulators (Ahemad and Khan 2012b) and enhancing the ethylene-mediated stress by 1-aminocyclopropane-1-carboxylate (ACC) deaminase production along with improving plant stress tolerance to drought, salinity, metals, and toxicity of pesticide (Glick 2012; Ahemad and Khan 2012a).

By observing a variety of contaminants in the environment such as total petroleum hydrocarbons, remediation expertise even among both PGPR and plants may still be low in efficiency. The assortment of PGPR and specific bacteria, which are capable of degrading contaminants, was found to be effectual (Huang et al. 1990; Ajithkumar et al. 1998; Huang et al. 2005). With the discovery of a number of a soil microorganisms that are able to degrade xenobiotic chemicals together with herbicides, pesticides, refrigerants, solvents, and other organic compounds, the concept that microbial degradation might provide a rational and effectual means of disposing toxic chemical wastes gained credibility (Glick 2010). Specific bacteria that are capable of degrading contaminants can be screened out according to the contaminants' properties. They can rapidly metabolize several available compounds, whereas the part of PGPR is still prompting plant growth and enhancing plant tolerance to pollutants (Zhuang et al. 2007).

## 9.4 IMPACT OF PLANT, SOIL, AND MICROBE INTERACTION

The phenomenon of plant–microbe interaction is usually accepted as being of undisputed noteworthy significance both for the macro-partner (higher plants) and for the plant-associated bacteria (micro-partners). The synergy of soil microorganisms with plants has been increasingly attracting consideration throughout the past decade in basic research as well as in applied fields (Alexander 2008). Phytoremediation potential relies on the interaction among soil, microbes, plants, as well as heavy metals. As depicted in Figure 9.1, these complex interactions are pretentious by various biotic and abiotic factors, such as characteristics along with activities of plant and rhizobacteria, soil properties, as well as climatic condition, etc.

Plant roots interrelate with a large number of different microorganisms, with these interactions being the main determinants of the extent of phytoremediation. The symbiosis interaction between plant and bacteria in heavy metal–contaminated soil can be affected by the sides of both the mineral micro-partner and host plant. Bacteria can supplement the remediation aptitude of plants or trim down the phytotoxicity of the contaminated soil (Jing et al. 2007). Generally, there are two types of interactions, as shown in Figure 9.2.

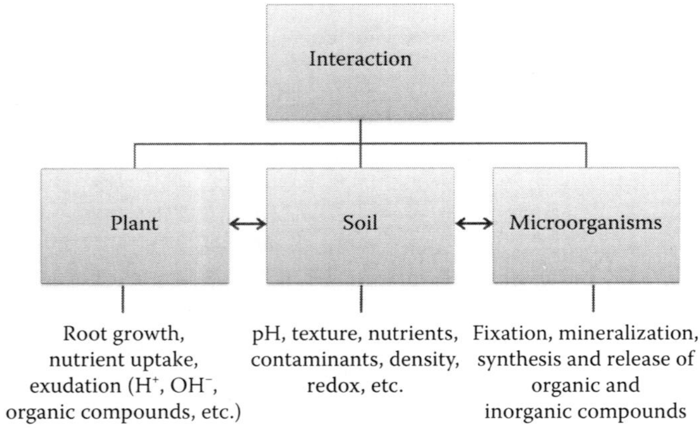


**FIGURE 9.1** Plant–soil–microbial interactions within the rhizosphere.

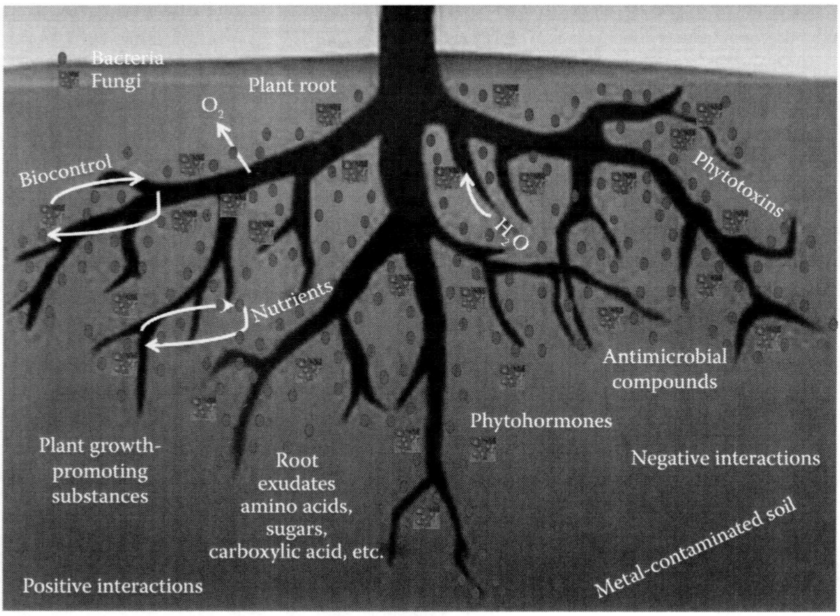


**FIGURE 9.2** Plant–microbe interactions.

### 9.4.1 Positive Interactions

Throughout plant growth, roots either actively or passively liberate an array of organic compounds. Among them are exudates, mostly sugars, amino acids, and carboxylic acids, which are passively released along concentration gradients with the aim of being provided as nutrients for microorganisms in the rhizosphere. Microbes colonize the rhizosphere of various plants and usually have beneficial effects, such as promoting plant growth along with reducing susceptibility to diseases engendered by plant pathogens such as fungi, bacteria, nematodes, and viruses. Bacterial plant growth-promoting mechanisms incorporate the atmospheric nitrogen fixation, 1-aminocyclopropane-1-carboxylic acid utilization as a sole source of N, siderophores production, or plant growth regulators production (Rajkumar et al. 2009; Weyens et al. 2009). The strategies with the intent for controlling disease vulnerability include competition of rhizosphere microbes intended for nutrients, niche exclusion, antifungal metabolites production, in addition to induced systemic resistance (Compant et al. 2005; Miethke and Marahiel 2007).

### 9.4.2 Negative Interactions

Rhizosphere microorganisms are also capable of having detrimental effects on plant health and survival by means of increasing the risk intended for infection with plant pathogen. Root exudates containing toxic substances, such as antimicrobials, phytotoxins, and nematicidal as well as insecticidal compounds might also hinder the growth of beneficial as well as pathogenic organisms within the rhizosphere (Bais et al. 2006).

Moreover, it is well known that heavy metals can even be toxic for metal-accumulating as well as metal-tolerant plants, if the metal concentration in the environment is adequately high. Elevated levels of metals in soil not only decrease soil microbial activity but also lessen the rate of crop production by accrual in plant organs. The metal ions are excessively absorbed by the roots and translocated to different parts of the plants (Figure 9.3).

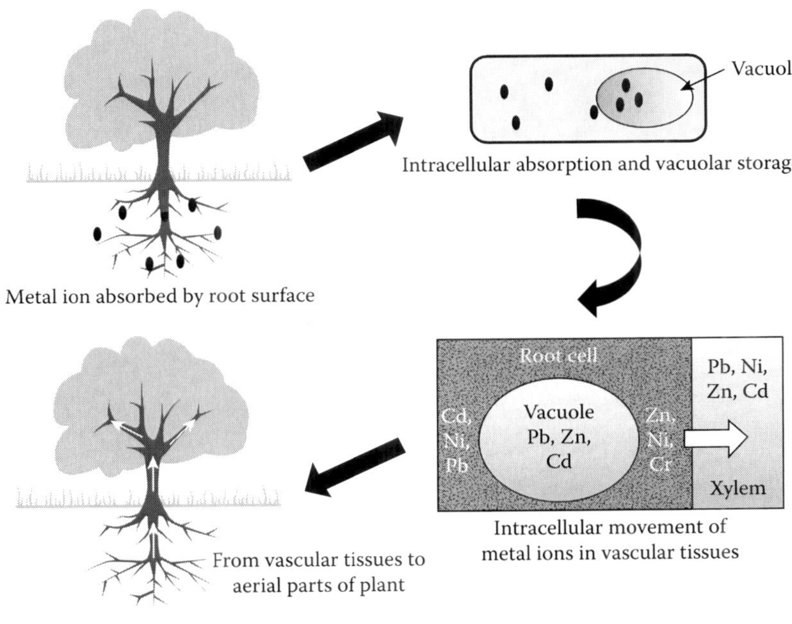


**FIGURE 9.3** Movement of heavy metals within the plants.

## 9.5 HEAVY METAL AND MICROBE INTERACTION

Plant growth-promoting rhizobacteria include a vast group of free-living soil bacteria. They have the ability to improve growth of the host plant in addition to development in heavy metal–contaminated soils by extenuating toxic effects of heavy metals on plants (Belimov et al. 2004). Rhizobacteria have been shown to acquire various traits that can change heavy metal bioavailability through releasing chelating substances, causing acidification of the microenvironment, and influencing changes in redox potential. The adaptation capabilities of both partners of the associative symbiosis as well as the micro-symbiont's bioremediation potential are significant in minimizing the detrimental effect of heavy metal pollution (Jing et al. 2007). Moreover, rhizobacteria also play a significant role in recycling plant nutrients, soil structure maintenance, detoxification of noxious chemicals, and plant growth (Giller et al. 1998; Elsgaard et al. 2001; Filip 2002). Thus, rhizobacteria can augment the remediation capacity of plants or lessen the phytotoxicity of contaminated soil. In addition, bacteria and plants can form explicit associations in which the plants bestow bacteria with a specific carbon source with the addition of root exudates, increase ion solubility that induces the bacteria to reduce the phytotoxicity of the contaminated soil, and increase the remediation activity of bacteria allied with plant roots. These multifaceted interactions are affected by an array of factors, such as characteristics and activity of rhizobacteria and plant, climatic conditions, properties of soil, etc. (Jing et al. 2007).

In general, metal's low bioavailability in soils reduces their uptake by organisms (Whiting et al. 2001; Braud et al. 2006). The bioavailability is influenced by loads of edaphic and ecological factors, such as:

1. Soil properties together with soil pH, cation exchange capacity, organic matter content, content of clay minerals, moisture content, hydrous metal oxides, buffering capacity, redox potential, and temperature
2. Chemical properties of metal
3. Biological properties of soil, which includes exudation by plant roots along with microbial activities in soil
4. Climate (Roane et al. 2000; Fischerova et al. 2006)

These biogeochemical processes are catalyzed mainly by sulfate-reducing bacteria such as *Desulfovibrio* or *Desulfotomaculum* (Rachna and Snehlata 2013).

## 9.6 INTERNAL PHYSIOLOGICAL MECHANISMS

The majority of metal ions have to enter the bacterial cell in order to have physiological or toxic effects. The soil bacteria encompass a broad kind of mechanism for surviving in environments under elevated heavy metals concentrations through various biological processes, such as physical sequestration, detoxification, exclusion (Figure 9.4), transportation over the cell membrane, entrapment in extracellular capsules, biosorption to the cell walls and production of chelators known to detoxify

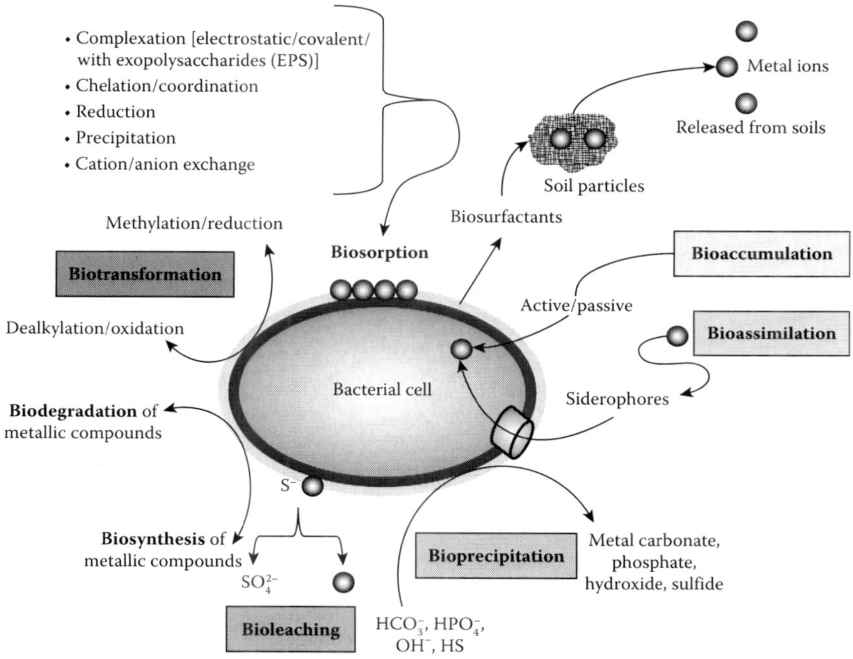


**FIGURE 9.4** Depiction of different kinds of bacterial interaction with heavy metals in metal-contaminated soils. (From Ahemad, M., *Arab. J. Chem.*, 2014, http://dx.doi.org/10.1016/j.arabjc.2014.11.020. With permission.)

heavy metals in bacteria, precipitation, complexation, and oxidation–reduction (Ahemad 2014; Singh and Lal 2015).

Recently, investigation designed for novel and innovative technologies has centered on the biological treatment methods. Bioremediation is the use of microorganisms to break down or remove toxic and hazardous compounds in the environment. It commonly utilizes microbes (bacteria, fungi, algae, and yeast), although higher plants are also used in some applications (Acquaah 2004; Morales-Barrera et al. 2008).

In spite of the diverse types of toxic metal ions and the various methods by which they interact with microorganisms, many bacteria have developed diverse methods of resistance to the noxious metal ions, and this factor makes microbial cells extraordinary. Bacterial cells usually code for proteins and enzymes that perform explicit functions either to protect the bacterial cell, block or change the incoming toxic metal, or both (Cavicchioli and Thomas 2002). As stated by Nies and Silver (1995), heavy metals cannot be degraded like toxic organic compounds, so there are only few possible mechanisms for heavy metal ion resistance. However, bacteria generally have two types of uptake systems for heavy metal ions: one is rapid, unspecific, and driven by the gradient of chemiosmotic pressure across the cytoplasmic membrane. In which first, the accrual of the respective ion can be diminished by efflux, an active exclusion of the heavy metal ion from the cell. Cations can also be segregated into intricate compounds by thiol-containing molecules, while several heavy metal ions may be reduced to less toxic oxidation states, and another type exhibits high substrate specificity by which they expel an elevated amount of toxic metals to the outer surface of the cell and are coupled with ATP hydrolysis (Nies 1999).

Moreover, several studies reveal that some bacteria have developed a cytosolic sequestration mechanism for protection against heavy metal toxicity. In this course of action, metal ions might also turn out to be compartmentalized or transformed into more innocuous forms and then enter the bacterial cell. In addition, some different metal-reducing bacteria can eliminate several heavy metals from the metal-polluted soils by reducing them to a lower oxidation state, making use of metals as terminal electron acceptors in anaerobic respiration, even though most of them utilize $Fe^{3+}$ and $S^0$ as terminal electron acceptors. Metal-chelating agents such as siderophores secreted by a number of bacteria have an important role in the acquisition of some heavy metals (Lovley 1995; Jing et al. 2007; Rajkumar et al. 2010). These resistance mechanisms explain why several microorganisms are able to survive in metal-polluted environments.

Bacteria including PGPR are able to immobilize, mobilize, or transform metals through the binding of heavy metals to extracellular materials and consequently immobilize the metal, avoiding its intake into the bacterial cell, and as a result reducing their toxicity to tolerate heavy metal ion uptake (Ahemad 2014). Likewise, bacterial extracellular macromolecules, such as polysaccharides, proteins, along with humic substances, also adeptly bind with heavy metals (Ahemad and Kilbret 2013). These substances therefore detoxify metals merely via complex formation or via forming an effectual barrier surrounding the cell (Rajkumar et al. 2010). *Thlaspi caerulescens,* a member of the *Brassicaceae* family, is one of the best known Cd and Zn hyperaccumulator plants (Rachna and Ṣnehlata 2013).

Microbial biomass can also passively bind to large amounts of metals during biosorption, hence providing a commercial resolution for management. Biosorption does not consume cellular metabolic energy; metal ions carrying positive charge are sequestered, primarily in the course of the adsorption of metal ions to the negative ionic groups on cell surfaces, and on the polysaccharide coating found on most bacterial cell walls, or other extracellular structures such as capsules. However, on long-term contact with the metal carriage solution, the living biomass may also be able to sequester the metal intracellularly by an active process known as bioaccumulation (Malik 2004).

## 9.7 GENETICALLY ENGINEERED STRATEGIES

PGPR effective in metal bioremediation of a particular environment or coupled by means of a specific host plant cannot be effectively utilized in other contaminated environments. However, it is achievable to modify PGPR according to the polluted environment or it can be feasible that the less effective existing microbes interacting in the midst of plants in an environment of metal pollution can be modified via genetic engineering development. With the progression of molecular biology, genetically engineered rhizobacteria with the pollutant-degrading gene are constructed to perform bioremediation in the rhizosphere. Barac et al. (2004) formulated the engineered endophytic bacteria to ameliorate the phytoremediation of water-soluble, volatile, organic pollutants.

The genetically constructed endophytic strain showed improved degradation and lessens evapotranspiration of toluene and a sort of hydrophobic volatile compound. They hypothesized that the aim of endophytic bacteria possessing the genetic information requisite to resourcefully degrade the organic contaminant, promoted its breakdown at the same time it moved through the plant's vascular system. Due to the extended transportation period of the contaminant in the system, there was ample time for proficient degradation by endophytic bacteria in xylem.

Moreover, gene transition between bacteria and transgenic plants has been constructed for advanced remediation effectiveness (Grckho et al. 2000; Nie et al. 2002). The ACC deaminase expression in the plant exhibits some advantages against bacterial ACC deaminase:

1. In primary stages of seed germination, the activity of bacterial ACC deaminase is probably much lower than the transgenic plant's activity (Nie et al. 2002).
2. It can persistently arouse plant growth, which leads to a higher metal accumulation in several cases and an increase in the shoot/root ratio (Grckho et al. 2000).
3. Encourage metal uptake of certain fast-growing plants for the replacement of sluggish growing hyperaccumulators (Stearns et al. 2005).

For PGPR improvements, genetic manipulation needs incessant research and development of molecular techniques for determining an effectual remediation role of PGPR in natural environment. However, it has beneficial consequences for protecting along with promoting development of plants under stresses. The expression of target genes or the production of target enzymes or proteins of interest by PGPR involved in metal bioremediation can be improved, and among these proteins, phytochelatins (PCs) as well as metalloproteins (MTs) are the main focus. PCs and MTs bind a broad range of heavy metals with large affinity (Zhuang et al. 2007). In bacterial and plant root cells, MTs on the outer membrane unite with metal ions and make sure they carry in the cytosol, where specific protein chelators (i.e., metallochaperons) transfer these combined metals to the related receptor proteins. Both approaches can be efficient against heavy metal tolerance.

Sriprang et al. (2003) revealed that by inserting novel PC-synthesizing genes into *Mesorhizobium huakuii* subsp. *rengei* strain B3, the bacteria were able to produce more phytochelatins, which were functional for accumulation of Cd ions subsequent to establishing symbiosis with the plant *Astragalus sinicus*.

## 9.8 CONCLUSION

Phytoremediation through rhizobacteria is a promising move to re-establish metal-contaminated sites. A sustainable and economical strategy is quickly becoming a viable substitute for conventional remediation methods and will be more appropriate in developing countries. Microbial activities in the rhizosphere contribute appreciably to the sustainability of agriculture and forestry and to the remediation of dispersed soils. Studies on PGPR necessitate building up innovative strategies by inoculating plants with or without genetically modified rhizospheric microbes for the purpose of removing metals from polluted soils in addition to enhancing plant growth effectively. Results from the literature imply that the heavy metals may be removed from polluted soil either by escalating the metal-accumulating capability of plants or by increasing the amount of plant biomass. Finally, it is important to keep in mind that a range of remediation approaches may necessitate accomplishing all recovery goals at a contaminated site.

## REFERENCES

Acquaah, G. 2004. *Understanding biotechnology: An integrated and cyber-based approach.* (1st edn.). Pearson Education, Inc, NJ, pp. 294–295.

Ahemad, M. 2014. Remediation of metalliferous soils through the heavy metal resistant plant growth promoting bacteria: Paradigms and prospects. *Arabian Journal of Chemistry* doi: http://dx.doi.org/10.1016/j.arabjc.2014.11.020.

Ahemad, M., and Khan, M.S. 2012a. Evaluation of plant growth promoting activities of rhizobacterium *Pseudomonas putida* under herbicide-stress. *Annals of Microbiology* 62(4): 1531–1540.

Ahemad, M., and Khan, M.S. 2012b. Alleviation of fungicide-induced phytotoxicity in greengram [*Vigna radiata* (L.) Wilczek] using fungicide-tolerant and plant growth promoting *Pseudomonas* strain. *Saudi Journal of Biological Sciences* 19(4): 451–459.

Ahemad, M., and Kilbret, M. 2013. Recent trends in microbial biosorption of heavy metals: A review. *Biochemistry and Molecular Biology* 1(1): 19–26.

Ajithkumar, P.V., Gangadhara, K.P, Manilal, P., and Kunhi, A.A.M. 1998. Soil inoculation with *Pseudomonas aeruginosa* 3MT eliminates the inhibitory effect of 3-chloroand 4-chlorobenzoate on tomato seed germination. *Soil Biology and Biochemistry* 30(8): 1053–1059.

Alexander, A. K. 2008. Metals in soil versus plant-microbe interactions: Biotic and chemical interferences. *Plant Microbe Interactions* 291–318.

Bais, H.P., Tiffany, L.W., Laura, G.P., Simon, G., and Jorge, M.V. 2006. The role of root exudates in rhizosphere interactions with plants and other organisms. *Annual Review of Plant Biology* 57: 233–266.

Barac, T., Taghavi, S., Borremans, B., Provoost, A., Oeyen, L., and Colpaert, J.V. 2004. Engineered endophytic bacteria improve phytoremediation of water-soluble, volatile, organic pollutants. *Nature Biotechnology* 22(5): 583–588.

Belimov, A.A., Kunakova, A.M., Safronova, V.I., Stepanok, V.V., Yudkin, L.Y., Alekseev, Y.V., and Kozhemyakov, A.P. 2004. Employment of rhizobacteria for the inoculation of barley plants cultivated in soil contaminant with lead and cadmium. *Microbiology* 73(1): 99–106.

Bharagava, R.N., Chandra, R., and Rai, V. 2008. Phytoextraction of trace elements and physiological changes in Indian mustard plants (*Brassica nigra* L.) grown in post methanated distillery effluent (PMDE) irrigated soil. *Bioresource Technology* 99: 8316–8324.

Bharti, R.P., Vastava, A.S., Soni, N., Tiwari, A., More, S., and Choudhary, J.R. 2014. Phytoremediation of heavy metal toxicity and role of soil in rhizobacteria. *International Journal of Scientific and Research Publications* 4(1): 2250–3153.

Boruvka, L., and Drabek, O. 2004. Heavy metal distribution between fractions of humic substances in heavily polluted soils. *Plant, Soil and Environment 50*(8); 339–345.

Braud, A., Jezequel, K., Vieille, E., Tritter, A., and Lebeau, T. 2006. Changes in extractability of Cr and Pb in a polycontaminated soil after bioaugmentation with microbial producers of biosurfactants, organic acids and siderophores. *Water, Air, Soil Pollution: Focus* 6(3–4): 261–279.

Cavicchioli, R., and Thomas, T. 2002. Extremophiles. In *Encyclopedia of Microbiology,* (2nd edn.), Joshua Lederberg, Martin Alexander, Barry bloom, David hopwood, Roger hull, Barbara Iglewski. (eds.). Academic Press, San Diego, CA, 2: pp. 317–337.

Chandra, R., Bharagava, R.N., Kapley, A., and Purohit, H.J. 2011. Bacterial diversity, organic pollutants and their metabolites in two aeration lagoons of common effluent treatment plant (CETP) during the degradation and detoxification of tannery wastewater. *Bioresource Technology* 102(3): 2333–2341.

Chandra, R., Bharagava, R.N., Yadav, S., and Mohan, D. 2009. Accumulation and distribution of toxic metals in wheat (*Triticum aestivum* L.) and Indian mustard (*Brassica campestris* L.) irrigated with distillery and tannery effluents. *Journal of Hazardous Material* 162: 1514–1521.

Chandra, R., Saxena, G., and Kumar, V. 2015. Phytoremediation of environmental pollutants: An eco-sustainable green technology to environmental management. In *Advances in biodegradation and bioremediation of industrial waste,* Chandra, R. (ed.), CRC Press, Boca Raton, FL, pp. 1–30.

Christopher, W.N.A., Brooks, R.R., Stewart, R.B., and Simcock, R. 1998. Harvesting a crop of gold in plants. *Nature 395*(6702): 553–554.

Compant, S., Duffy, B., Nowak, J., Clement, C., and Ait, B.E. 2005. Use of plant growth-promoting bacteria for biocontrol of plant diseases: principles, mechanisms of action, and future prospects. *Applied and Environmental Microbiology* 71(9): 4951–4959.

Egamberdieva, D. 2009. Allevation of salt stress by plant growth regulators and IAA producing bacteria in wheat. *Acta Physiologiae Plantarum* 31(4): 861–864.

Elsgaard, L., Petersen, S.O., and Debosz, K. 2001. Effects and risk assessment of linear alkylbenzene sulfonates in agricultural soil. 1. Short-term effects on soil microbiology. *Environmental Toxicology and Chemistry* 20(8): 1656–1663.

Filip, Z. 2002. International approach to assessing soil quality by ecologically related biological parameters. *Agriculture Ecosystem and Environment* 88(2): 689–712.

Fischerova, Z., Tlustos, P., Szakova, J., and Sichorova, K. 2006. A comparison of phytoremediation capability of selected plant species for given trace elements. *Environmental Pollution* 144(1): 93–100.

Giller, K.E, Witter, E., and McGrath, S.P. 1998. Toxicity of heavy metals to microorganisms and microbial processes in agricultural soils. *Soil Biology and Biochemistry* 30(10–11): 1389–1414.

Glick, B.R. 2010. Using soil bacteria to facilitate phytoremediation. *Biotechnology Advances* 28: 367–374.

Glick, B.R. 2012. Plant growth-promoting bacteria: Mechanisms and applications. *Scientifica* 2012: 963401.

Glick, B.R., Todorovic, B., Czarny, J., Cheng, Z., Duan, J., and McConkey, B. 2007. Promotion of plant growth by ACC deaminase containing containing soil bacteria. *European Journal of Plant Pathology* 119: 329–339.

Grckho, V.P., Filby, B., and Glick, B.R. 2000. Increased ability of transgenic plants expressing the bacterial enzyme ACC deaminase to accumulate Cd, Co, Cu, Ni, Pb, and Zn. *Journal of Biotechnology* 81(1): 45–53.

Huang, C., Huang, C., and Morehart, A.L. 1990. The removal of copper from dilute aqueous solutions by *Saccharomycetes cerevisiae. Water Resource* 24(4): 433–439.

Huang, X.D., El-Alawi, Y., Gurska, J, Glick, B.R., and Greenberg, B.M. 2005. A multi-process phytoremediation system for decontamination of persistent total petroleum hydrocarbons (TPHs) from soils. *Microchemical Journal* 81(1): 139–147.

Huang, X.D., El-Alawi, Y., Penrose, D.M., Glick, B.R., and Greenberg, B.M. 2004a. Responses of three grass species to creosote during phytoremediation. *Environmental Pollution* 130(3): 453–463.

Huang, X.D., El-Alawi Y., Penrose, D.M., Glick, B.R., and Greenberg, B.M. 2004b. A multi-process phytoremediation system for removal of polycyclic aromatic hydrocarbons from contaminated soils. *Environmental Pollution* 130(3): 465–476.

Jing, Y., He, Z., and Yang, X. 2007. Role of soil rhizobacteria in phytoremediation of heavy metal contaminated soils. *Journal of Zhejiang University Science* B 8(3): 192–207.

Khan, A.G., Kuek, C., Chaudhry, T.M., Khoo, C.S., and Hayes, W.J. 2000. Role of plants, mycorrhiza and phytochelators in heavy metal contaminated land remediation. *Chemosphere* 41: 197–207.

Laghlimi, M., Baghdad, B., Hadi, H.E., and Bouabdli, A. 2015. Phytoremediation mechanism of heavy metal contaminated soils: a review. *Open Journal of Ecology* 5: 375–388.

Lindberg, T., Granhall, U., and Tomenius, H. 1985. Infectivity and acetylene reduction of diazotrophic rhizosphere bacteria in wheat (*Triticum aestivum*) seedlings under xenobiotic conditions. *Biology and Fertility of Soils* 1(3): 123–129.

Lovley, D.R. 1995. Bioremediation of organic and metal contaminants with dissimilatory metal reduction. *Journal of Industrial Microbiology and Biotechnology* 14(2): 85–93.

Ma, Y., Prasad, M.N.V., Rajkumar, M., and Freitas, H. 2011. Plant growth promoting rhizobacteria and endophytes accelerate phytoremediation of metalliferous soils. *Biotechnology Advances* 29(2): 248–258.

Malik, A. 2004. Metal bioremediation through growing cells. *Environment International* 30(2): 261–278.

Miethke, M., and Marahiel, M.A. 2007. Siderophore-based iron acquisition and pathogen control. *Microbiology and Molecular Biology Reviews* 71(3): 413–451.

Morales-Barrera, L., de Maria Guillen Jimenez, F., Ortiz Moreno, A., Villegas Garrido. T.L., Sandoval-Cabrera, A., Hernande Rodriguez, C.H., and Cristiani Urbina, E. 2008. Isolation, identification and characterization of a *Hypocrea tawa* strain with high Cr(VI) reduction potential. *Biochemical Engineering* Journal 40(2): 284–292.

Nies, D.H. 1999. Microbial heavy metal resistance. *Applied Microbiology and Biotechnology* 51(6): 730–750.

Nies, D.H., and Silver, S. 1995. Ion efflux systems involved in bacterial metal resistances. *Journal of Industrial Microbiology* 14(2): 186–199.

Nie, L., Shah S., Burd, G.I., Dixon, D.G., and Glick, B.R. 2002. Phytoremediation of arsenate contaminated soil by transgenic canola and the plant growth-promoting bacterium *Enterobacter cloacae* CAL2. *Plant Physiology and Biochemistry* 40(4): 355–361.

Rachna, B., and Snehlata. 2013. Role of rhizospheric bacteria in phytoremediation of heavy metal contaminated soil. *International Journal of Current Research* 2(12): 3897–3907.

Rahman, M.A., Reichman, S.M., De Filippis, L., Sany, S.B.T., and Hasegawa, H. (2016). Phytoremediation of toxic metals in soils and wetlands: Concepts and applications. In *Environmental Remediation Technologies for Metal-Contaminated Soils,* Hiroshi Hasegawa, Ismail Md. Mofizur Rahman, Mohammad Azizur Rahman. (eds.), Springer, Japan. pp. 161–195.

Rajkumar, M., Ae, N., Prasad M.N.V., and Freitas, H. 2010. Potential of siderophore producing bacteria for improving heavy metal phytoextraction. *Trends in Biotechnology* 28(3): 142–149.

Rajkumar, M., Prasad M.N.V., Freitas, H., and Ae, N. 2009. Biotechnological applications of serpentine soil bacteria for phytoremediation of trace metals. *Critical Reviews in Biotechnology* 29(2): 120–130.

Roane, T.M., and Pepper, I.L. 2000. Microorganisms and metal pollution. In *Environmental Microbiology,* Maier, I.L., Pepper, C.B., (eds.), Gerba, Academic Press: London, UK, pp. 403–423.

Saxena, G., and Bharagava, R.N. 2015. Persistent organic pollutants and bacterial communities present during the treatment of tannery wastewater. In *Environmental waste management,* Chandra, R. (ed.), CRC Press, Boca Raton, FL, pp. 217–247.

Saxena, G., Chandra, R., and Bharagava, R.N. 2016a. Environmental pollution, toxicity profile and treatment approaches for tannery wastewater and its chemical pollutants. *Reviews of Environmental Contamination and Toxicology*, 240: 1–39.

Saxena, G., and Bharagava, R.N. 2016b. Ram Chandra: Advances in biodegradation and bioremediation of industrial waste. *Clean Technology and Environmental Policy* 18: 979–980. doi: http://dx.doi.org/10.1007/s10098-015-1084-9.

Shinwari, K.I., Shah, A., Afridi, M.I., Zeeshan, M., Hussain, H., Hussain, J., Ahmad, O., and Jamil, M. 2015. Application of plant growth promoting rhizobacteria in bioremediation of heavy metal polluted soil. *Asian Journal of Multidisciplinary Studies* 3(4): 179–185.

Singh, A.V., Agarwal, A. and Goel, R. 2010b. Comparative phosphate solubilization efficiency of two bacterial isolates and their effect on *Cicer arietinum* seeds in indigenous and alternative soil system. *Environment and Ecology* 28: 1979–1983.

Singh, A.V., Chandra, R., and Goel, R. 2013. Phosphate solubilization by *Chryseobacterium* sp. and their combined effect with N and P fertilizers on plant growth promotion. *Archives of Agronomy and Soil Science* 59(5): 641–651.

Singh, A.V., and Goel, R. 2015. Plant growth promoting efficiency of *Chryseobacterium* sp. PSR 10 on finger millet (*Eleusine coracana*). *Journal of Global Biosciences* 4(6): 2569–2575.

Singh, A.V., Prasad B., and Shah, S. 2010a. Screening plant growth promotory rhizobacteria for improving seed germination and seedling vigor of lentil (*Lens culinaris* Medik). *Environment and Ecology* 28 (3B): 2055–2058.

Singh, Y., and Lal, N. 2015. Investigation of the heavy metal resistant bacterial isolates in vitro from industrial effluent. *World Journal of Pharmacy and Pharmaceutical Sciences* 4(2): 343–350.

Singh, A.V., Shah, S., and Prasad, B. 2010c. Effect of phosphate solubilizing bacteria on plant growth promotion and nodulation in soybean (Glycine max (L.) Merr.). *Journal of Hill Agriculture* 1(1): 35–39.

Sriprang, R., Hayashi, M., Ono, H., Takagi, M., Hirata, K., and Murooka, Y. 2003. Enhanced accumulation of $Cd^{2+}$ by a *Mesorhizobium* sp. transformed with a gene from *Arabidopsis thaliana* coding for phytochelatin synthase. *Applied and Environmental Microbiology* 69(3): 1791–1796.

Stearns, J.C., Shah, S., Dixon, D.G., Greenberg, B.M., and Glick, B.R. 2005. Tolerance of transgenic canola expressing 1-aminocyclopropane-carboxylic acid deaminase to growth inhibition by nickel. *Plant Physiology and Biochemistry* 43(7): 701–708.

Tak, H.I., Ahmad, F., and Babalola, O.O. 2013. Advances in the application of plant growth promoting rhizobacteria in phytoremediation of heavy metals. *Reviews of Environmental Contamination and Toxicology* 223: 33–52.

Tilak, K.V.B.R., Ranganayaki, N., Pal, K.K., De, R., Saxena, A.K., Nautiyal, C.S., Mittal, S., Tripathi, A.K., and Johri, B.N. 2005. Diversity of plant growth and soil health supporting bacteria. *Current Science* 89(1): 136–150.

Weyens, N., Van der lelie, D., Taghavi, S., and Vangronsveld, J. 2009. Phytoremediation: plant endophyte partnerships take the challenge. *Current Opinion in Biotechnology* 20(2): 248–254.

Whiting, S.N., de Souza, M.P., and Terry, N. 2001. Rhizosphere bacteria mobilize Zn for hyperaccumulation by *Thlaspi caerulescens*. *Environmental Science and Techno*logy 35(15): 3144–3150.

Zhuang, X., Chen, J., Shim, H., and Bai, Z. 2007. New advances in plant growth promoting rhizobacteria for bioremediation. *Environment International* 33(3): 406–413.

# 10 Microbial Cellulases and Their Applications in the Pulp and Paper Industry

## *An Emerging Paradigm*

*Shiv Shankar, M. Gupta, and Shikha*

**CONTENTS**

10.1 Introduction....274
10.2 Cellulase: A Promising Enzyme for Cellulose Degradation....275
  10.2.1 Cellulases....275
  10.2.2 Occurrence and Physiological Role of Cellulase Enzyme....275
10.3 Structure of Cellulase Enzyme....276
10.4 Classification of Cellulase Enzymes....277
10.5 Mode of Action of Cellulase Enzyme....278
10.6 Mechanism of Action of Cellulase Enzyme....280
  10.6.1 Retaining Mechanism....280
  10.6.2 Inverting Mechanism....282
10.7 Production of Cellulase Enzyme....282
  10.7.1 Production of Cellulase Enzyme under Solid-State Fermentation (SSF)....283
  10.7.2 Production of Cellulase Enzyme under Submerged Fermentation (SmF)....283
10.8 Factors Affecting Cellulase Production....287
  10.8.1 Temperature....287
  10.8.2 pH....288
  10.8.3 Chemical Composition of Cellulose....288
  10.8.4 Concentration of Cellulase....288
  10.8.5 Carbon and Nitrogen Sources....288
10.9 Cellulolytic Microorganisms....288
  10.9.1 Fungal Cellulases....288
  10.9.2 Bacterial Cellulases....291
  10.9.3 Application of Microbial Cellulases in the Pulp and Paper Industry....292
    10.9.3.1 Cellulases in Biobeating and Pulp Refining....292
    10.9.3.2 Cellulases in the Improvement of Pulp Drainage Rate....293
    10.9.3.3 Cellulases in Deinking....293
    10.9.3.4 Cellulases in the Improvement of Pulp Brightness....294

10.10 Role of Microbial Cellulases in the Biodegradation of Lignocellulosic Waste into Ethanol ..... 294
10.10.1 Lignocellulosic Biomass Composition ..... 295
10.10.2 Role of Cellulase in Hydrolysis of Cellulose ..... 295
10.10.3 Overview of the Conversion of Lignocellulose into Bio-Ethanol ..... 295
10.11 Problems Associated with the Use of Cellulases ..... 296
10.12 Conclusion and Future Prospects ..... 297
Acknowledgments ..... 297
References ..... 297

## 10.1 INTRODUCTION

Cellulose is earth's major biopolymer and is of immense economic importance across the globe. It is the primary structural component of the plant cell wall accounting for over half of the carbon in the biosphere. Synthesis and degradation of approximately $1.5 \times 10^{12}$ tons of cellulose were estimated annually, and it is considered an almost inexhaustible source of raw material for the synthesis of different products. Hence, the conversion of cellulosic biomass to fermentable sugars and alcohol through cellulase produced by various cellulolytic organisms as "biological catalysts" has attracted worldwide attention.

Cellulases are inducible enzymes, composed of structurally and functionally discrete units called domains, which can fold independently, making cellulases modules (Henrissat et al. 1998). A large diversity of microorganisms including both fungi and bacteria synthesize cellulase during their growth on cellulosic materials (Wilson 2011). These microorganisms can be aerobic, anaerobic, mesophilic, or thermophilic. Complete enzymatic hydrolysis of cellulose is achieved by synergistic action of three cellulase enzymes: endoglucanase, exoglucanase, and β-glucosidases (Asztalos et al. 2012). Extensive basic and applied research carried out on cellulases has revealed their widespread commercial and industrial application. However, the commercial application of this enzyme is cost intensive, which has hindered the industrial application of cellulose bioconversion. Therefore, there is an ever-increasing demand for more stable, highly active, specific enzymes reduced at a nominal cost. In order to increase the production and applicability of microbial cellulases, various biotechnological measures were adopted. Biotechnology of cellulases started in the early 1980s, initially in animal feed followed by food applications, and later in textile laundry, as well as in the pulp and paper industry.

During the last two decades, the use of cellulases has considerably increased in the pulp and paper industry (Bajpai 2012). The mechanical pulping processes such as grinding and refining of the wood-based raw materials lead to pulp having a high content of bulk and stiffness. On the contrary, biomechanical pulping uses cellulases, which not only saves energy substantially (20%–40%) during the refining process but also improves the hand-sheet strength properties (Lecourt et al. 2010). Cellulases have also been employed for biomodification of fiber properties with an aim of improving drainage and beatability in the paper mills prior to or after the beating of pulp (Dienes et al. 2004). Cellulases can also be used to enhance the bleaching ability of softwood kraft pulp producing a final brightness score comparable to that of xylanase treatment.

It has been experimentally documented that improvements in dewatering and deinking of various pulps result in the peeling of the individual fibrils and bundles, which have high affinity for the surrounding water and ink particles (Das et al. 2013). These enzymes are also employed in the preparation of easily biodegradable cardboard; the manufacturing of soft paper, paper towels, and sanitary paper; and the removal of adhered paper (Acharya and Chaudhary 2012). Although a number of review works are available on the production and application of microbial cellulases, the increasing demand for cellulases in various sectors of the pulp and allied industries requires improved production of cellulases. In the previously mentioned context, the present book chapter intends to highlight advancements in the biotechnology of microbial cellulases of commercial significance and their applications in pulp and paper industries.

## 10.2 CELLULASE: A PROMISING ENZYME FOR CELLULOSE DEGRADATION

### 10.2.1 Cellulases

Cellulase (EC 3.2.1.4) is currently the third largest industrial enzyme worldwide and finds extensive use in paper recycling, as a detergent enzyme, in juice extraction, and as an animal feed additive (Wilson 2011). Therefore, if ethanol (or another fermentation product of sugars), produced from biomass by enzymes, becomes a major transportation fuel, cellulases will certainly occupy the largest volume of industrial enzyme (Wilson 2011). Subsequent to enzymatic hydrolysis, cellulose releases soluble sugars like glucose, xylose, and other hexoses and pentoses. In order to make the sugar monomers available for fermentation, the cellulose and hemicellulose chains need to be hydrolyzed. The hydrolysis of hemicelluloses is catalyzed by xylanases, together with other accessory enzymes (α-L-arabinofuranosidases, feruloyl and acetylxylan esterases, β-xylosidases, etc.), while the hydrolysis of cellulose can be undertaken by microorganisms that produce enzymes known as the cellulase systems (Gusakov et al. 2007).

### 10.2.2 Occurrence and Physiological Role of Cellulase Enzyme

Cellulases are produced in nature by various organisms, such as plants, animals, and microorganisms, and are distributed throughout the world. Chiefly microorganisms, like bacteria, actinomycetes, filamentous fungi, and protozoans, produce cellulase enzymes. The most important sources for industrial cellulases production are filamentous fungi, distributed in the soil, plants, and in marine environments (Eriksson et al. 1990). Although microorganisms having potential for complete degradation of native cellulose are widespread in the soil environment, this ability appears to be confined to a small number of species, which are predominantly fungi.

Cellulose is the most abundant organic compound in the biosphere and comprises almost 50% of the biomass synthesized by photosynthetic fixation of carbon dioxide

(Eriksson et al. 1990). Growth and survival of microorganisms identified as important in most of the agricultural soils, depend on the carbon source locked in the cellulose occurring in the soils. However, for carbon to be released as a source of energy for use by the microorganisms, cellulose in plant debris has to be degraded into glucose, cellobiose, and high molecular weight oligosaccharides by cellulase enzymes (White 1982). During recent years, one of the most important biotechnological applications of cellulases is the conversion of agricultural wastes and all lignocellulosics into products of commercial interest such as ethanol, glucose, and single-cell products. The bioconversions of cellulosic materials by the hydrolytic enzymes are now a subject of intensive research as a contribution to the development of a large-scale conversion process beneficial to mankind (Foust et al. 2009). Moreover, such a process would help alleviate shortages of food and animal feeds, solve modern waste disposal problems, and reduce man's dependence on fossil fuels by providing a convenient and renewable source of energy in the form of glucose.

## 10.3 STRUCTURE OF CELLULASE ENZYME

The modular structures of all cellulases have two domains, which are structurally and functionally independent, including one catalytic domain (CD) linked to one or more noncatalytic modules called cellulose-binding domains (CBDs) via a flexible linker (Figure 10.1). The linker is usually very rich in amino acids like threonines, serines, and prolines, and it is heavily glycosylated (Tomme et al. 1995). The probable role of the linker is to keep the two domains apart and to restrict their movements with respect to each other. Glycosylation of the linker probably makes it less flexible, and probably decreases its sensitivity to proteolytic enzymes. The catalytic domain brings hydrolysis of β-1,4-glycosidic bonds in cellulose, while CBD mediates the attachment of cellulases to cellulose (Watanabe and Tokuda 2010).

The noncatalytic module is also involved in protein–carbohydrate (ligands binding) and protein–protein (cellulosomes) interactions. CBD domains are ubiquitous and show a wide spectrum of substrate-binding properties that include cellulose, xylan, chitin,

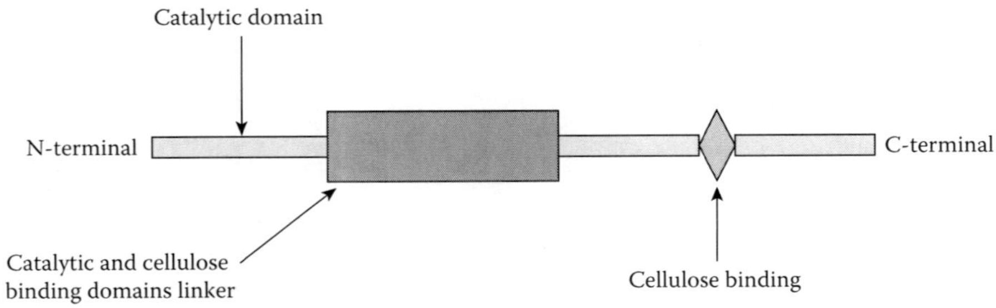


**FIGURE 10.1** Architectural primary sequence organization of cellulase domains of glycosyl hydrolase family-5A. (Adapted from Tomme, P., et al., *Adv. Microb. Physiol.*, 37, 1–81, 1995.)

and cellulose derivatives. Structurally the cellulase enzymes are either all α-helix domains or a combination of α/β barrels or all β-sheet domains (Henrissat 1991).

## 10.4 CLASSIFICATION OF CELLULASE ENZYMES

Cellulases produced in nature can be classified in three main categories: (1) endoglucanases or 1,4-β-D-glucan-4-glucanohydrolases (EC 3.2.1.4; 52,000 m.wt.); (2) exoglucanases, including 1,4-β-D-glucan glucanohydrolases (also known as cellodextrinases, CBHI) (EC 3.2.1.74) and 1,4-β-D glucan cellobiohydrolases (cellobiohydrolases, CBHII) (EC 3.2.1.91; 60,500–62,000 m.wt.); and (3) β-glucosidases or β-glucoside glucohydrolases (EC 3.2.1.21; 76,000 m.wt.) (Lynd et al. 2002). Endoglucanases act randomly at internal amorphous sites within the cellulose polysaccharide chain, generating oligosaccharides of different lengths, and consequently producing new chain ends. Although most endoglucanases are nonprocessive, a new class of processive endoglucanases has been located in bacteria (Wilson 2011).

Exoglucanases act in a processive manner either on the reducing or nonreducing ends of cellulose polysaccharide chains, liberating either glucose (glucanohydrolases) or cellobiose (cellobiohydrolase) as major products (Figure 10.2). Exoglucanases enzymes remain bound to the glucose chain even after cleaving a cellobiose molecule and will continue to cleave cellobiose units until a minimum chain length is reached (Barr et al. 1996). Exoglucanases can also result in splits/peeling of cellulose chains from microcrystalline structure by acting on microcrystalline cellulose. β-Glucosidases perform hydrolysis of soluble cellodextrins and cellobiose to glucose. Cellulases enzymes

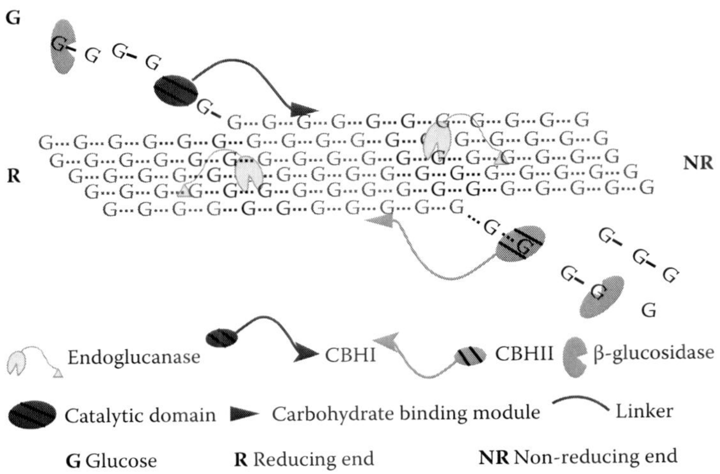


**FIGURE 10.2** Mode of action of various cellulases enzymes on the surface layer of cellulose. Glucose molecules in gray color represent the crystalline region, and glucose molecules in black color are in the amorphous region. (Adapted from Kumar, D., and Murthy, G.S., *Biotechnol. Biofuels,* 6(1), 63, 2013.)

differ from other glycoside hydrolases in respect to their ability to hydrolyze β-1,4-glucosidic bonds between glucosyl residues. Cellulase systems are not simply an agglomeration of enzymes representing the three enzyme groups (endoglucanases, exoglucanases, and β-glucosidases, with or without carbohydrate-binding module [CBM]) but rather act jointly to efficiently hydrolyze cellulose. Different approaches have been adopted by microorganisms to effectively hydrolyze cellulose, naturally occurring in insoluble particles or embedded within hemicellulose and lignin biopolymers (Tomme et al. 1995). Cellulolytic filamentous fungi and actinomycete bacteria are efficient in penetrating into cellulosic substrates via hyphal extensions, thereby presenting their cellulase systems in confined cavities within cellulosic particles (Eriksson et al. 1990).

The production of "free" cellulases, with or without CBM, may be sufficient for the efficient hydrolysis of cellulose under these conditions. The enzymes in these cellulase systems do not synthesize stable high molecular weight complexes and therefore are called "noncomplexed" systems. By contrast, anaerobic bacteria perhaps use alternative mechanisms for cellulose degradation due to a lack of ability to effectively penetrate cellulolytic material. Anaerobic bacteria get access to the products of cellulosic hydrolysis in competition with other microorganisms and with limited ATP available for cellulase synthesis (Bayer et al. 1994). This could have led to the development of "complexed" cellulase systems (called "cellulosomes"), which keeps cellulase-producing cells at the site of hydrolysis, as observed for clostridia and ruminal bacteria (Figure 10.3). The cellulosome is thought to synchronize concerted enzyme activity in close proximity to the bacterial cell, enabling optimum synergism between the cellulases presented on the cellulosome. Concomitantly, the cellulosome also minimizes the distance over which cellulose hydrolysis products must diffuse, allowing efficient uptake of these oligosaccharides by the host cell (Schwarz 2001).

## 10.5 MODE OF ACTION OF CELLULASE ENZYME

In order to gain an understanding about the application of cellulase enzyme, one needs to know the mode of action of the enzyme and the actual role of each component of the enzyme in performing any specific reaction. Cellobiohydrolases (exoglucanases) have been classified as exo-acting based on the assumption that they all cleave β-1,4-glycosidic bonds from chain ends. The truly exo-acting enzymes often have a tunnel-shaped, closed active site that retains a single glucan chain and prevents it from re-adhering to the cellulose crystal (Rouvinen et al. 1990). Endoglucanases are classified as endo-acting, because they are thought to cleave glycosidic bonds internally and have cleft-shaped, open active sites. Since endoglucanases are active on the amorphous region of cellulose, their activity can be assayed using soluble cellulose substrates. Subsequent to the evidence that some cellulases display both modes of action, endo- and exo-function (Davies and Henrissat 1995), the classification has changed; whereas exoglucanases are described as active on the crystalline regions of cellulose, endoglucanases are typically active on the more soluble amorphous region of the cellulose crystal. Zhang and Lynd (2004) observed cellulase enzymes possess different modes of actions and, hence, cannot be strictly

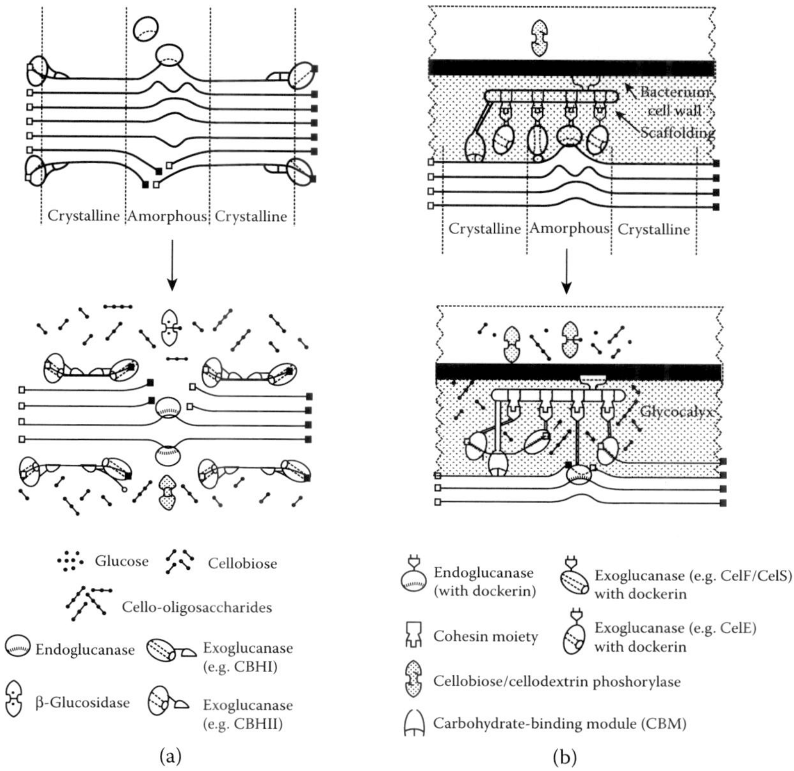


**FIGURE 10.3** Hydrolysis of amorphous and microcrystalline cellulose by noncomplexed (a) and complexed (b) cellulase systems. The solid squares represent reducing ends, and the open squares represent nonreducing ends. Amorphous and crystalline regions are indicated. (Based on Lynd, L.R., et al., *Microbiol. Mol. Biol. Rev.*, 66(3), 506–577, 2002.)

titled under broad groups such as endoglucanases and exoglucanases (Zhang and Lynd 2004).

According to stochastic molecular modeling (SMM) as proposed by Kumar and Murthy (2013), enzymes have been categorized into eight classes on the basis of their molecular structure and mode of action (Table 10.1) (Kumar and Murthy 2013). Each class of enzyme may have multiple enzymes identified from different organisms; however, they differ only in terms of their hydrolytic efficiency (activity). Therefore, modeling the action of these eight classes of enzymes will sufficiently uncover the dynamics of individual cellulases. The important aspect of this classification is that a novel enzyme can be used in the model after characterization of its action patterns and placing it into one or multiple groups with its specific activity/activities, without making any change to the present model.

**TABLE 10.1**
**Mode of Action of Different Cellulases: Hydrolysis of Cellulose**

| Different Cellulose | Hydrolytic Mechanism |
|---|---|
| Endocellulase (EG, nonprocessive with CBM) | • Carbohydrate-binding module (CBM) binds to surface chains randomly and breaks bonds at catalytic domain (CD). |
| Endocellulase (EG, nonprocessive without CBM) | • They break bonds on surface chains in a random pattern. |
| Endocellulase (EG, processive with CBM) | • CBM binds to surface chains randomly and breaks bonds at CD.<br>• The enzyme moves along the chains (toward nonreducing end or reducing end) and cuts every alternate bond releasing cellobiose unit and a minimum chain length is achieved. |
| Exocellulase (CBHI, processive) | • CBM attaches from the reducing end only on surface chains and pulls the chain toward CD.<br>• The chain passes through CD (tunnel-like shape), and every alternate bond is broken to produce cellobiose.<br>• The enzyme moves along the chain (toward nonreducing end) and cuts every alternate bond until a minimum chain length is reached. |
| Exocellulase (CBHI, nonprocessive) | • This attacks from the reducing end on surface chains and cuts every alternate bond to produce cellobiose. |
| Exocellulase (CBHII, processive) | • CBM attaches from the nonreducing end on surface chains and pulls the chain toward CD.<br>• The chain passes through the tunnel-shaped CD, and every alternate bond is broken to produce cellobiose.<br>• Enzyme moves along the chain (toward reducing end) and cuts every alternate bond until a minimum chain length is reached. |
| Exocellulase (CBHII, nonprocessive) | • Attack is from the nonreducing end on surface chains and breaks alternate bonds to produce cellobiose. |
| β-Glucosidase | • This acts on cellobiose and soluble oligomers (DP $\leq$ 6) and produces glucose by breaking bonds. |

*Source*: Kumar, D., and Murthy, G.S., *Biotechnol. Biofuels*, 6(1), 63, 2013.
CBH, cellobiohydrolases; CBHI, cellobiohydrolase-I; CBHII, cellobiohydrolase-II; CBM, carbohydrate-binding module; CD, catalytic domain.

## 10.6 MECHANISM OF ACTION OF CELLULASE ENZYME

The enzymatic cleavage of 1,4-glucosidic bonds in cellulose proceeds via acid hydrolysis using a proton donor and nucleophile or base. The products of hydrolysis can result in either the inversion or retention (double replacement mechanism) of the anomeric configuration of carbon-1 at the reducing end (Davies and Henrissat 1995).

### 10.6.1 Retaining Mechanism

The retaining glycoside hydrolase mechanism results in a net retention of the configuration at the anomeric carbon (C1) of the substrate after cleavage. This is performed following a double displacement mechanism (i.e., the hydrolysis of a glycosidic bond

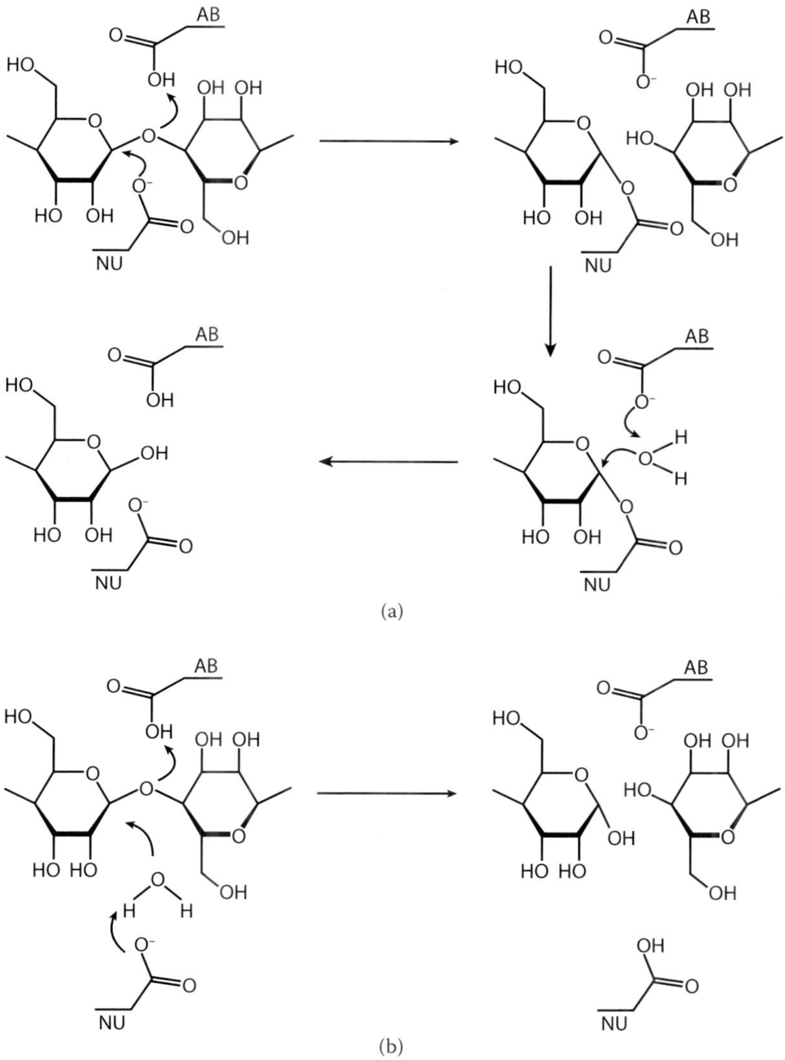


**FIGURE 10.4** Hydrolytic mechanism of cellulase. The two enzyme mechanisms observed for cellulases: (a) the retaining mechanism and (b) the inverting mechanism. In the retaining mechanism the configuration at the anomeric carbon will be in (b) configuration after hydrolysis (i.e., the configuration is retained). The distance between the two catalytic carboxylates in the retaining enzyme is ~5.5Å. In the inverting mechanism, the configuration at the anomeric carbon gets changed from (b) to (a) configuration upon hydrolysis. The distance between the two catalytic carboxylates in the inverting enzymes varies between 6.5 and 9.5Å. (Based on Davies, G., and Henrissat, B., *Structure*, 3(9), 853–859, 1995.)

end in a product with the same configuration at the anomeric carbon as the substrate had prior to hydrolysis) (Figure 10.4a). The catalytic machinery of these enzymes consists of two catalytic carboxylate residues that usually sit at opposite sides of the sugar plane originating subsequent to glycosylation and deglycosylation. Glycosylation being a double displacement reaction is generally an acid-catalyzed leaving group, from a glycosyl-enzyme intermediate. In deglycosylation, the first carboxylate residue now functions as a general base that activates an incoming nucleophile by stealing a proton from it, which subsequently hydrolyzes the glycosyl-enzyme intermediate.

### 10.6.2 Inverting Mechanism

The inverting glycoside hydrolase mechanism (Figure 10.4b) leads to a net inversion of the configuration at the anomeric carbon (C1) of the substrate subsequent to cleavage. This inversion is performed via a single nucleophilic displacement mechanism, which involves the hydrolysis of a beta-glycosidic bond creating a product with the alpha-configuration, and vice versa. The catalytic machinery of these enzymes requires two catalytic carboxylates. These two carboxylate residues lead to a general acid-catalyzed leaving group departure and a general base-assistance to nucleophilic attack by a water molecule from the opposite side of the sugar ring. The mechanism of cellulose degradation by aerobic bacteria is similar to that of aerobic fungi. However, anaerobic celluloytic bacteria must attach themselves to cellulose to effect degradation. Bayer et al. (1994) coined the term *cellulosomes* to describe the multicomponent cellulolytic complex produced by *Clostridium thermocellum* (Bayer et al. 1994). Aggregates of these cellulosomes (i.e., polycellulosomes) located on the cell surface are responsible for attachment of the cell to the substrate, with this adherence being specific for cellulose (Jindou et al. 2006).

## 10.7 PRODUCTION OF CELLULASE ENZYME

Research on cellulase production has rejuvenated globally due to its emerging applications in ligno-cellulose conversion. The production of low titers of cellulase has always been a major concern; thus, several workers are attempting to improve production titers by adopting multifaceted approaches, which include the use of better bioprocess technologies, using cheaper or crude raw materials as substrates for enzyme production, bioengineering the microorganisms, etc. (Lynd et al. 2002). A significant portion of research attempts to address the bioprocess improvement strategies for enhancing the yield and specific activities of cellulases. The majority of reports on microbial production of cellulases utilize the submerged fermentation technology (SmF). However, in nature, the growth and cellulose utilization of aerobic microorganisms elaborating cellulases probably resembles solid-state fermentation (SSF) more than a liquid culture (Zhu et al. 2009).

During the last two decades, solid-state fermentation has regained interest due to the high titers of enzyme production employing fungal cultures (Singhania et al. 2009). Different types of lignocellulosic substrates have the greatest impact on cellulase secretion. Some of the substrates significantly stimulated lignocellulolytic enzyme synthesis without supplementation of the culture medium with specific inducers (Elisashvili et al. 2009). Nevertheless, the advantages of better monitoring and handling are associated with the submerged cultures. Currently used bioprocesses and the status of cellulase production based on literature are given in Tables 10.2 and 10.3.

### 10.7.1 Production of Cellulase Enzyme under Solid-State Fermentation (SSF)

Solid-state fermentation (SSF) is defined as the fermentation in absence or near absence of free water. SSF for production of cellulases is rapidly gaining interest as a cost-effective technology as the microorganisms, especially fungal cultures, produce comparatively high titers of cellulase due to the conditions of fermentation, which show similarity to the natural environment (Singhania et al. 2007, 2009). Filamentous fungi such as *Trichoderma reesei, Aspergillus niger, Penicillium* sp., etc., have been employed for cellulase production using SSF where a basal mineral salts medium was used for moistening the substrate. Solid-state cultures are strongly recommended as systems for producing cellulases at lower price than submerged cultures, as the product concentration remains higher, thereby reducing the steps in downstream processing, and in turn reducing the cost of operation.

Nigam and Singh (1996) have reviewed the use of agricultural wastes as substrates for cellulolytic enzyme production under SSF and strongly believe that with the appropriate technology, improved bioreactor design, and operation controls, SSF may become a competitive method for the production of cellulases. SSF can thus be considered as a future technology for commercial production of cellulases considering the low cost input and ability to utilize naturally available sources of cellulose as substrate. Cellulases produced in solid-state culture show remarkable stability toward temperature, pH, metal ions, etc. It has also been reported that the enzyme produced using the same biomass as to be used for bioconversion proves more efficient than the one produced on other cellulosic substrate. Optimization of SSF conditions for cellulase production can always be improved, and overall production economics eventually makes it an attractive technology for cellulase production. It offers many advantages over SmF, including high volumetric productivity, higher concentration of products, less effluent generation, and low catabolic repression, which makes it a promising technology (Singhania et al. 2007, 2009).

### 10.7.2 Production of Cellulase Enzyme under Submerged Fermentation (SmF)

Submerged fermentation (SmF) has been defined as fermentation in the presence of water. Almost all the large-scale enzyme-producing facilities have adopted the proven technology of SmF due to ease in both monitoring and handling. Though bacteria and actinomycetes are also reported for cellulase production under SmF, the titers are very low to make the technology economically feasible. Most of the commercial cellulases are produced by the filamentous fungi *Trichoderma reesei*

**TABLE 10.2**
**Solid-State Fermentation (SSF) and Substrate Involved in Cellulase Production**

| Microorganism | Substrate | Magnitude | Enzyme Activity | References |
|---|---|---|---|---|
| *Aspergillus fumigatus* SK1 | Oil palm trunk | Flask | CMCase 54.27 U/g | Ang et al. (2013) |
| *Aspergillus fumigatus* ABK 9 | Wheat bran and rice straw (1:1) | Flask | Endogluconase 826 U/g, BGL 102 U/g FPase 1130 U/g | Das et al. (2013) |
| *Aspergillus terreus* | Rice straw | Flask | CMCase 676 mg/g | Narra et al. (2012) |
| *Aspergillus niger* NS2 | Wheat bran | Flask | CMCase 310 U/g | Bansal et al. (2012) |
| Mixed culture: *Trichoderma reesei, Aspergillus niger* | Rice chaff/wheat bran (9:1) | Flask | FPase 5.64 IU/g | Yang et al. (2009) |
| *Aspergillus niger* FGSCA733 | *Jatropha curcas* seed cake | Flask | CMCase 3974 U/g | Ncube et al. (2012) |
| *Fomitopsis* sp. RCK2010 | Wheat bran | Flask | CMCase 71.526 IU/g FPase 3.268 IU/g BGA 50.69 IU/g | Deswal et al. (2011) |
| *Penicillum decumbens* | Rice bran | Flask | CMCase 5.76 IU/g | Liu et al. (2011) |
| *Bacillus subtilis* | Banana waste | Shake flask | FPase 2.8 IU/gds, CMCase 9.6 IU/gds, Cellobiase 4.5 IU/gds | Krishna (1999) |
| *Thermoascus aurantiacus* | Wheat straw | Perforated-drum bioreactor | FPase 4.4 U/gds, CBH 2.8 U/gds, Endoglucanase 987 U/gds, BGL 48.8 U/gds | Kalogeris et al. (1999) |
| *Trichoderma reesei* ZU 02 | Corn cob residue | Tray fermentor | FPase 158 U/gds | Xia and Cen (1999) |
| *Aspergillus* sp. SU14 | Wheat bran | Shake flask | 19.2 (U/g) | Vu et al. (2011) |

*(Continued)*

**TABLE 10.2** ***(CONTINUED)***
**Solid-State Fermentation (SSF) and Substrate Involved in Cellulase Production**

| Microorganism | Substrate | Magnitude | Enzyme Activity | References |
|---|---|---|---|---|
| *Trichoderma viride* FCBP-142 | Wheat bran | Shake flask | 122.66 IU/mL | Shazia Shafique (2011) |
| *Aspergillus awamori* IOC-3914 | Babassu cake | Fixed-bed SSF | 55.7 U/g | Castro et al. (2015) |
| *Kluyveromyces marxianus* | Sugarcane bagasse | Rotary drum bioreactor (RDB) to 100-L capacity | 24.6 g/L | Lin et al. (2013) |

BGL, β-glucosidases; CBH, cellobiohydrolases; CMCase, carboxymethyl cellulase; FPase, filter paper activity; RDB, rotary drum bioreactor; SSF, solid-state fermentation.

**TABLE 10.3**
**Submerged Fermentation (SmF) and Substrate Involved in Cellulase Production**

| Microorganism | Substrate | Magnitude | Enzyme Activity | References |
|---|---|---|---|---|
| *Acremonium cellulolyticus* (CF-2612) | Solka-folc | 2-L jar fermenter | Cellulase 240.3 FPU/L | Fujii et al. (2009) |
| *Aspergillus niger* A 20 | Cellulose | Shake flask | Cellobiase 27.5 U/mL | Abdel-Fattah et al. (1997) |
| *Chaetomium thermophilum* CT2 | Cellulose (sigma cell) | SF | CMCase 2.7 IU/mL | Li et al. (2003) |
| *Mucor circinelloides* | Lactose | Shake flask | EGL 0.25 U/mL | Saha (2004) |
| *Neurospora crassa* | Wheat straw | Shake flask | FPase 1.33 U/mL, CMCase19.7 U/mL, BGL 0.58 U/mL | Romero et al. (1999) |
| *Penicillium janthinellum* | Sugar cane bagasse | Shake flask | FPase 0.55 U/mL, CMCase 21.5 U/mL, BGL 2.3 IU/mL | Adsul et al. (2004) |
| *Phanerochaete chrysosporium* | Cellulose (Avicell) | 100-L fermenter | 29 mg/g cellulose | Szabó et al. (1996) |
| *Thermotoga maritima* | Xylose | Shake flask | Cellobiase 11 mU/mL, Avicellase 0.3 mU/mL β-glucosidase 30 mU/mL | Bronnenmeier et al. (1995) |
| *Trichoderma reesei* | Steam-treated willow | 22-l fermenter | FPase 108 U/g | Reczey et al. (1996) |
| *Trichoderma viride* | Sugar cane bagasse | Shake flask | FPase 0.88 U/mL, CMCase 33.8 U/mL, BGL 0.33 U/mL | Adsul et al. (2004) |
| *Agaricus bisporus* | Lignocellulosic wastes | Shake flask | 0.26 U/mL | Yang et al. (2009) |
| *Trichoderma reesei* | Bagasse | Shake flask | 0.163 IU/mL | Aiello et al. (1996) |

BGL, β-glucosidases; CMCase, carboxymethyl cellulase; FPase, filter paper activity.

or *Aspergillus niger* under SmF (Cherry and Fidantsef 2003). Cellulase production in cultures under SmF is highly influenced by various physicochemical parameters including the nature of the cellulosic substrate, pH of the medium, nutrient availability, inducer supplementation, fermentation temperature, etc., and a large-scale production of cellulases requires thorough understanding and proper control of the growth and enzyme production capabilities of the producer.

Synthesis of microbial cellulases is subject to induction and repression mechanisms; therefore, the process design and media formulation for cellulase production have to take care of these aspects. Out of the known inducers of cellulase genes, lactose has been considered to be the only economically feasible additive in industrial fermentation media (Aro et al. 2005). Although the majority of the processes described for cellulase production under SmF are batch processes, there have been attempts to produce cellulase in fed-batch with improved enzyme production (Belghith et al. 2001; Wen et al. 2005) or continuous mode (Schafner and Toledo 1992; Ju and Afolabi 1999), which supposedly helps to overrule the repression caused by the accumulation of reducing sugar. Prolonged fermentation time with low productivity has long been the major technical limitation in fermentative production of cellulases. To overcome this limitation, a two-stage continuous process for cellulase production could be employed in which the growth phase and production phase could be separated by different pH and temperature optima.

Mostly, pure cellulose preparations like Solka-Floc and Avicell have been used in the liquid cultures of cellulolytic microbes for production of the enzymes, while natural cellulosic materials when used as the carbon source gave poor enzyme yields. When using soluble substrates, there is a chance that the breakdown products may hamper cellulase synthesis by promoting catabolite repression due to accumulation of free sugars. The carbon sources in the majority of commercial cellulase fermentations are cellulosic biomass ranging from pure cellulose to straw, spent hulls of cereals and pulses, rice or wheat bran, bagasse, paper industry waste, dairy manure, and various other lignocellulosic residues (Romero et al. 1999; Heck et al. 2002). Improved production of cellulase in fermenters may be achieved by a gradient feed of a suitable cellulose and maintenance of optimal process conditions.

## 10.8 FACTORS AFFECTING CELLULASE PRODUCTION

Cellulose degradation is affected by several physicochemical conditions, such as temperature, pH, chemical composition and concentration of cellulose, the nature of the microorganism, and nutritional parameters such as the presence of carbon and nitrogen sources, etc.

### 10.8.1 Temperature

Cellulolytic microorganisms grow well in a wide range of temperatures (for bacteria 45°C–82°C and for fungi 50°C–70°C). Most cellulase-producing microorganisms have their own optimum temperature for cellulase production.

### 10.8.2 pH

Medium pH is the key factor that affects the production of cellulases. Fungi mostly prefers acidic medium, whereas bacteria require neutral and alkaline pH for enzyme production (Gautam et al. 2011).

### 10.8.3 Chemical Composition of Cellulose

Cellulose occurs in nature in association with other plant substances. Such substances may affect its natural degradation. Cellulose fibrils are embedded in a matrix of other polymers, including hemicelluloses, pectin, and proteins (Yamada et al. 2005). Mechanisms of cellulose degradation by cellulases have not been fully described because of the high crystallinity and water insolubility of cellulose fibers.

### 10.8.4 Concentration of Cellulase

Cellulose concentration in the medium affects the growth and production of cellulase enzymes by microorganisms. Mostly, the optimum cellulase production is obtained when the concentration of cellulose is approximately 1%. Beyond this level, the production of cellulase declines. Most cellulolytic microorganisms lack efficient ligninase systems and have problems degrading lignin. However, some basidiomycetes (white-rot fungi) have efficient ligninase systems making lignin degradation feasible (Leschine 1995).

### 10.8.5 Carbon and Nitrogen Sources

Carbon and nitrogen are essential sources for the growth and production of enzymes. Cellulases are inducible enzymes that often require the presence of an inducer. There are other carbon sources that induce cellulases production (e.g., sophomers, lactose, CMC, and sucrose), whereas glucose and end products repress the production of cellulase enzymes. Successful scale-up strategy necessitates the optimization of physicochemical parameters affecting microbial growth and product formation (Pothiraj and Eyini 2007).

## 10.9 CELLULOLYTIC MICROORGANISMS

Cellulase, an industrially important enzyme, can be produced by fungi, bacteria, or actinomycetes.

### 10.9.1 Fungal Cellulases

Fungi are the main producers of cellulases because of their high enzyme yield and capacity to produce complete cellulase complex. A number of the studied fungi that produce cellulase include *Streptomyces, Actinomadura, Saccharomonospora, Trichoderma, Neurospora, Penicillium, Candida, Piromyces, Aspergillus,* and *Rhizopus* strains (Table 10.4).

**TABLE 10.4**
**Cellulase-Producing Microorganisms**

| Microorganisms | Enzyme: Activity | Cultivation Condition | Media | References |
|---|---|---|---|---|
| *Aspergillus sydowii* | FPase 1.33 U/mL<br>Endogluconase 1.32 U/mL<br>Exoglucanase 3.99 U/mL<br>β-glucosidases 9.24 U/mL | Temp: 40°C<br>pH: 5.5<br>Moisture: 70% | Potato dextrose agar (PDA) | Matkar et al. (2013) |
| *Chrysoporthe cubensis* | Endo 33.84 U/g<br>FPase 2.52 U/g<br>β-glucosidases 21.55 U/g | Temp: 50°C–60°C<br>pH: 4<br>Moisture: 70% | PDA | Falkoski et al. (2013) |
| *Aspergillus terreus* | FPase 16.96 U/g | Temp: 50°C–60°C<br>pH: 4<br>Moisture: 70% | PDA | Bansal et al. (2012) |
| *Trichoderma harzianum* | Cellulases 121 FPU/g<br>β-glucosidases 1730 IU/g | Temp: 29°C<br>pH: 5 | PDA | Delabona et al. (2012) |
| *Fusarium oxysporum* | Cellulases 1.43 U/mL | Temp: 60°C–65°C<br>pH: 4.5–5.5 | PDA | Yuan et al. (2012) |
| *Aspergillus niger* | FPase 17 U/g<br>CMCase 310 U/g<br>β-glucosidases 33 U/g | Temp: 20°C–25°C<br>pH: 3–8<br>Substrate and moisture ratio: 1:1.5–1:1.75 | PDA | Yang et al. (2009) |
| *Aspergillus heteromorphus* | Cellulase 13.05 U/mL | Temp: 60°C<br>pH: 5.5<br>Moisture content: 86% | Nutrient media | Singh et al. (2009) |
| *Fomitopsis* sp. RCK | CMCase 71.526 IU/g<br>FPase 3.268 IU/g<br>B 50.696 IU/g | Temp: 60°C<br>pH: 5.5<br>Substrate and moisture ratio: 1:3.5 | PDA | Deswal et al. (2011) |

*(Continued)*

**TABLE 10.4** *(CONTINUED)*
**Cellulase-Producing Microorganisms**

| Microorganisms | Enzyme: Activity | Cultivation Condition | Media | References |
|---|---|---|---|---|
| Mixed culture of *Aspergillus niger* and *Trichoderma reesei* | FPase 10.7 U/gds<br>β-glucosidases 10.71 IU/gds | Temp: 70°C<br>pH: 5<br>Moisture ratio: 70% | PDA | Brijwani et al. (2010) |
| *Rhodococcus* sp. | FPase 5.6 mg/mL<br>CMCase 4.5 mg/mL | Temp: 50°C<br>pH: 6.5 | Starch casein agar and nutrient agar | Tanskul et al. (2013) |
| *Streptomyces griseorubens* | Cellulase 143.05 U/mL | Temp: 55°C<br>pH: 6 | Enrichment media | Prasad et al. (2013) |
| *Sporotrichum pulverulentum* | Endoglucanase 1.4 glucose/mL | pH: 6 | H324 media | Tirado et al. (2010) |
| *Bacillus* sp. | CMCase 5.2 U/mL | pH: 6 | Carboxy methyl cellulose agar | Paudel and Qin (2015) |
| *Aspergillus terreus* UniMAP AA-6 | Cellulase 0.0625 U/mL | — | Enrichment medium | Gunny et al. (2015) |
| *Bacillus subtilis* | Endoglucanase, 0.658 U/mg protein | pH: 7.2 | Modified DSMZ medium | Taechapoempol et al. (2011) |

CMCase, carboxymethyl cellulase; DSMZ, Deutsche Sammlung von Mikroorganismen und Zellkulturen (German collection of microorganisms and cell cultures).

Fungal cellulase production is controlled by induction and catabolic repression by cellulase degradation products.

The cellulase system of *Trichoderma reesei* (teleomorph: *Hypocrea jecorina*, initially called *Trichoderma viride*) has been a focus of research for the last 50 years (Takashima et al. 1999). *T. reesei* produces at least two exoglucanases (CBHI and CBHII), five endoglucanases (EGI, EGII, EGIII, EGIV, and EGV), and two glucosidases (BGLI and BGLII) (Medve et al. 1998). The necessity for the two exoglucanases (cellobiohydrolases) has been attributed to their particular preferences for the reducing (CBHI) and nonreducing (CBHII) ends of cellulose chains of microcrystalline cellulose. This notion has also been supported by the exo–exo synergy observed between these two enzymes (Rouvinen et al. 1990; Nidetzky et al. 1994). Crystallography has elucidated the three-dimensional (3-D) structures of the two cellobiohydrolases (Kleywegt et al. 1997). CBHI contains four surface loops that give rise to a tunnel with a length of 50Å, while CBHII contains two surface loops that give rise to a tunnel of 20Å. These tunnels prove to be essential to the cellobiohydrolases for the processive cleavage of cellulose chains from the reducing or nonreducing ends. The 3-D structure of CBHI confirmed that cellobiose is the major hydrolytic product as the cellulose chain passes through the tunnel.

CBHI and CBHII are the principal components of the *T. reesei* cellulase system, representing 60% and 20%, respectively, of the total cellulase protein produced by the fungus on a mass basis (Wood 1992). Endoglucanases are thought to be primarily responsible for a decreasing degree of polymerization by internally cleaving cellulose chains at relatively amorphous regions, thereby generating new cellulose chain ends susceptible to the action of cellobiohydrolases (Teeri 1997; Teeri et al. 1998). The need for five endoglucanase species in the *T. reesei* cellulase system has not been clearly explained, particularly considering that endoglucanases (with EGI and EGII as major species) represent less than 20% of the total cellulase protein of *T. reesei*. Cellobiose, the major product of CBHI and CHBII activity, inhibits the activity of the cellobiohydrolases and endoglucanases (Medve et al. 1994, 1998).

The production of at least two β-glucosidases by *T. reesei* facilitates the hydrolysis of cellobiose and small oligosaccharides to glucose. Both BGLI and BGLII have been isolated from culture supernatants, but a large fraction of these enzymes remains cell wall bound (Usami et al. 1990). The presence of β-glucosidases in close proximity to the fungal cell wall may limit loss of glucose to the environment following cellulose hydrolysis. Likewise, the cellulase system of the thermophilic fungus *Humicola insolens* is homologous to the *T. reesei* system and contains at least seven cellulases (two cellobiohydrolases [CBHI and CBHII] and five endoglucanases [EGI, EGII, EGIII, EGV, and EGVI]). However, differences exist, such as the absence of a CBM in EGI of *H. insolens*.

### 10.9.2 Bacterial Cellulases

Cellulase-producing bacteria have been isolated from environments such as rumen, soil, compost, manure, municipal solid waste, brewery sludge, sewerage sludge, wood material, hot springs, marine sediments, alkaline environments, and geysers. Some common cellulase-producing bacteria are *Bacillus, Clostridium, Cellulomonas,*

*Acetovibrio, Bacteroides, Microbiospora, Ruminococcus, Streptomyces, Thermomonospora, Cytophaga, Micrococcus, Sporocytophaga,* and *Pseudomonas* species (Kuhad et al. 2011). The mechanism for cellulase action in bacteria may differ from that of fungi as observed by the differences of their optima of pH and temperature. A special feature of a number of bacterial endoglucanases is that they ineffectively hydrolyze crystalline cellulose.

Recently, few researchers have purified and characterized *Bacillus* endoglucanase capable of hydrolyzing crystalline cellulose. The best-studied species of cellulolytic aerobic bacteria belong to the genera *Cellulomonas* and *Thermobifida* (formerly *Thermomonospora*). *Cellulomonas* species are coryneform bacteria that produce at least six endoglucanases and at least one exoglucanase (Cex) (Chaudhary et al. 1997). The individual cellulases of *Cellulomonas* resemble the cellulase systems of aerobic fungi and contain CBMs; however, cellulosome-like protuberant structures have been noted on *Cellulomonas* cells grown with cellulose and cellobiose as carbon sources (Vladut-Talor et al. 1986). The thermophilic filamentous bacterium *Thermobifida fusca* (formerly *Thermomonospora fusca*) is a major cellulose degrader in soil. Six cellulases, three endoglucanases (E1, E2, and E5), two exoglucanases (E3 and E6), and an unusual cellulase with both endoglucanase and exoglucanase activity (E4) have been isolated.

### 10.9.3 Application of Microbial Cellulases in the Pulp and Paper Industry

The pulp and paper industry processes utilize huge quantities of lignocellulosic biomass annually. The technology for pulp manufacture is very diverse, and numerous opportunities exist for the application of microbial enzymes. Historically, enzymes have found some applications as in the paper industry, but such applications have been mainly confined to areas such as modifications of raw starch. However, a wide range of applications of microbial enzymes has been identified in the pulp and paper industry. The application of cellulases in the pulp and paper industry has grown rapidly since 1980.

#### 10.9.3.1 Cellulases in Biobeating and Pulp Refining

Paper production is a complex process that aims at producing paper sheets out of fibers derived from various origins having maximum strength at the lowest cost. Bleached softwood kraft pulp is the most common produce in the pulp industry. After pulping, these fibers cannot retain enough bonding strength. Therefore, a mechanical treatment (i.e., refining) is required to increase fiber bonding and improve paper strength. Refining is a high energy-consuming process where the cellulosic fibers are mechanically treated in water, which results in morphological (internal and external fibrillation, fines formation, fiber shortening or cutting, and fiber curling or straightening) and structural changes. Because of the cost reduction and environmental regulations, the biggest challenge before the paper producers is to reduce energy consumption in refining. To meet the challenge, different solutions have been developed, such as adapting plate pattern to modify fiber treatment or varying consistency.

Nevertheless, these methods require important investments. Enzyme treatments on cellulose fibers have been documented to have several positive effects in papermaking, drainage improvement, deinking, and reduced refining energy. Enhancement of pulp drainage by applying cellulases on recycled fibers has been well documented, which also aims at improved beatability and runnability of the paper mills. Enzymatic treatment can be performed either before or after beating of the pulps. Cellulase treatment prior to the refining process is aimed either at improving the beatability response or modifying the fiber properties. Cellulase is usually added after beating to improve the drainage properties of pulps. A commercial cellulase preparation, named Pergalase-A40, from *Trichoderma* has been used by many paper mills around the world for the production of wood-containing printing papers (Gil et al. 2009) revealed that enzymatic treatment of pulp using cellulase enzyme mixtures gave rise to better properties, even without mechanical refining, which is in agreement with the bio-refining concept.

#### 10.9.3.2 Cellulases in the Improvement of Pulp Drainage Rate

Cellulases have been successfully applied to improve drainage rates of thermomechanical pulp and secondary fibers (Oksanen et al. 2011). In addition, they counter the adverse effects on mechanical properties of fiber horrification during drying. Improved drainage favors smooth working of machines leading to energy savings and, thus, making the entire process cost intensive. The possibility of improvement in the drainage rates of recycled fibers by cellulase, especially by endoglucanase, was identified in the late 1980s, which was subsequently confirmed by purified endoglucanases from *Trichoderma*.

Both endoglucanases I and II from *Trichoderma* were found equally effective in decreasing the Schopper-Riegler (SR) value (provide a measure of the rate at which a dilute suspension of pulp may be dewatered) of recycled soft wood craft pulp and showed improved drainage, while cellobiohydrolase did not affect SR. Xylanase and mannanase treatment resulted in only a marginal improvement of the SR value. The cellulase treatment can make drainability and brightness better and can improve strength to a certain extent. A combination treatment of hemicellulases with endoglucanase was found to increase the positive effects of the endoglucanases on pulp drainage (Oksanen et al. 2000; Oksanen et al. 2011). The use of cellulase enzyme modifies the interfacial properties of fibers, increasing the affinity for water, which in turn promotes changes in the technical properties of the pulp and paper, such as pulp drainability and paper strength (Dienes et al. 2004).

#### 10.9.3.3 Cellulases in Deinking

The chemical deinking process in the pulp and paper industry employs large quantities of chemicals, requiring costly water treatment systems (Lee et al. 2007). To overcome these disadvantages, the process of enzymatic deinking has attracted a great deal of attention owing to its high efficiency and low environmental impact. Bio-deinking as an alternative to the use of chemical products uses a neutral medium, which allows reduction of the contaminant charge and is comparatively more suitable for maintenance of the environment. Cellulases and hemicellulases can dislodge inks by peeling off fibers or fines on paper surfaces. Cellulases find significant commercial importance due to their potential application in the deinking of mixed office waste, magazines and newspapers, and denim stone washing. Cellulases along with hemicellulases in the enzyme cocktails have proven potential in the saccharification of lignocellulosics to fermentable sugars. Enzymatic deinking can also contribute to improving the strength of paper as well as freeness of the recycled fibers.

Generally, three approaches are available for the use of enzymes in the deinking: (1) the fiber surface can be attacked by the action of cellulases, hemicellulases, and pectinases to detach the ink particles from their surfaces; (2) starch-based coating can be solubilized by amylase; and (3) vegetable oil–based or soya-based ink carriers can be hydrolyzed by lipases. Endoglucanase, a major component of cellulases, plays a crucial and significant role in the deinking process in the pulp and paper industry. Several mechanisms have been proposed for the removal of ink by the action of cellulases. Cellulases bind and alter the fiber surfaces in the vicinity of the ink particles loading for removal of small fibers form the surface of ink particles. Endoglucanase initiates their action by splitting the cellulose fibers into several amorphous sites and results in generation of numerous reducing ends, which favors the loosening and releasing of the ink particles from the recycled papers (Vyas and Lachke 2003). The main advantage of enzymatic deinking is the avoidance of the use of alkali. Enzymatic deinking at acidic pH also prevents alkaline yellowing, makes the deinking process simple, changes the ink particle size distribution, and reduces environmental pollution. Enzymatic deinking is a promising technology and has great potential from both commercial and environmental standpoints.

#### 10.9.3.4 Cellulases in the Improvement of Pulp Brightness

Cellulases improve fiber brightness, strength properties, pulp freeness, and cleanliness as well as reduce fine particles in the pulp. A detailed study of the modes of action of different cellulases on different types of pulps is imperative for the development of enzymatic modification of fibers. Rahkamo et al. (1996) investigated the effect of major cellulase components from *T. reesei* on the fiber properties of unbleached soft wood kraft and dissolving pulps (Rahkamo et al. 1996). They found that the cellobiohydrolases had moderate effect on fiber viscosity, while endoglucanases, especially endoglucanase II, dramatically reduced the pulp viscosity even at a low concentration. Cellobiohydrolase I treatment did not affect hand-sheet properties even after pulp refiner (PFI) refining, indicating that this enzyme did not cause any structural damage to the fibers. Endoglucanase II treatment resulted in damage of the strength properties and indicated that this enzyme attacked cellulose fibers at sites where even low contamination of hydrolysis resulted in a large decrease in viscosity and deterioration in the tensile index.

## 10.10 ROLE OF MICROBIAL CELLULASES IN THE BIODEGRADATION OF LIGNOCELLULOSIC WASTE INTO ETHANOL

Agricultural activities across the world generate huge amounts of agricultural waste like bagasse, rice straw, wheat straw, and corn straw throughout the year, which can be utilized for the production of useful products like bio-ethanol. A major proportion of the agricultural waste generated is disposed of as waste. For example, 600–900 million tons of rice straw is generated globally as waste, but a very little amount is used as animal feed. The major proportion of this waste is removed from the agricultural field by burning. Burning of lignocellulosic waste is practiced worldwide, but it causes air pollution and reduction in fertility of the soil (Kumar et al. 2008). Considering the enormity and abundance of agricultural waste, microbial cellulases can be employed to treat lignocellulosic waste and transform it into useful products like bioethanol.

### 10.10.1 Lignocellulosic Biomass Composition

The lignocellulosic biomass is composed of cellulose (30%–50%), hemicelluloses (15%–35%), and lignin (10%–20%) (Chandel et al. 2015). Cellulose and hemicellulose constitute up to 70% of total biomass. Lignin is the third most abundant component tightly linked to cellulose and hemicellulose by covalent and hydrogen bonds, rendering the structure robust and resistant to microbial attack (Buaban et al. 2010).

### 10.10.2 Role of Cellulase in Hydrolysis of Cellulose

Cellulose is a crystalline, fibrous, and insoluble biopolymer. It is the principal component of the cell wall, which is composed of D-glucose units connected by β-1,4-glucosidic linkage (Shahzadi et al. 2014). Cellulases can be utilized for the conversion of lignocellulosic biomass into bio-ethanol as a cost-effective agent and can help in minimizing the volume of agricultural, industrial, and municipal cellulosic waste (Wu et al. 2010). It is of economic and ecological interest to explore cost-effective ways to utilize and treat lignocellulosic waste.

A wide spectrum of organisms including bacteria, fungi, and protists as well as invertebrate animals like crustaceans, insects, mollusks, annelids, and nematodes use cellulose as their source of energy (Kuhad et al. 2011).

A typical cellulase enzyme system comprises endo-β-1,4-glucanases (EC 3.2.1.4); exo-β-1,4-glucanases (EC 3.2.1.91); and β-1,4-glucosidase (EC 3.2.1.21) (Lynd et al. 2005). The combined actions of all three components act sequentially and result in the breakdown of cellulose into fermentable sugar. Endoglucanases randomly break the crystalline region of cellulose. Each cleavage results in formation of two new ends in the cellulose chain. The exoglucanases bind to the free end of the chain and release cellobiose and glucose units. Cellobiose is composed of two molecules of glucose linked with each other that are subsequently cleaved by β-glucosidase into glucose monomers (Pandey et al. 2013).

### 10.10.3 Overview of the Conversion of Lignocellulose into Bio-Ethanol

Pretreatment is the first step to treat lignocellulosic biomass. It involves acid hydrolysis, alkali swelling, and steam explosion (Venkata Mohan et al. 2008). The rationale behind pretreatment is to make lignocellulosic waste more available to cellulases. In the second step (i.e., saccharification), cellulose and hemicelluloses components are converted into fermentable sugars that can be fermented into ethanol and other value-added products.

Saccharification of lignocellulosic biomass is the most critical step for bio-ethanol production. The optimum temperature and pH for cellulases for the conversion of lignocellulose into fermentable sugars have been reported to be 40°C–50°C and 4–5, respectively. Enzymatic hydrolysis of lignocellulosic waste is considered as cheap, nontoxic, and corrosion free as compared to chemical hydrolysis (Sun and Cheng 2002; Talebnia et al. 2010). In addition, no inhibitory end product is created during enzymatic hydrolysis of lignocellulose. However, enzymatic hydrolysis critically depends on some factors like concentration of the substrates, enzyme loading, time, and operating temperature of saccharification (Tucker et al. 2003). Incorporation of surfactants during enzymatic hydrolysis enhances the conversion of cellulose into fermentable sugars (Eriksson et al. 2002). Tween 20 has been reported to be the most effective for improving enzyme hydrolysis.

## 10.11 PROBLEMS ASSOCIATED WITH THE USE OF CELLULASES

Progress in biotechnology of cellulases and related enzymes is truly remarkable and attracting worldwide attention. Lignocellulose serves as the potential source for biofuels, biofertilizers, animal feed, and chemicals, besides being the raw material for the paper industry. Exploitation of renewable resources requires either chemical or biological treatment of the material, and in the latter context, cellulases have gained wide popularity over the past several decades. Extensive research has uncovered the mechanisms of microbial cellulase production and has led to the development of technologies for production and applications of cellulose-degrading enzymes. However, there is no single process that is cost-effective and efficient in the conversion of the natural lignocellulosic materials. The use of the current commercial preparations of cellulase for the bioconversion of lignocellulosic waste is economically not unfeasible. The reduction in the cost of production of cellulases by optimizing growth conditions, processes, and improvement in their performance by protein engineering and microbial genetics to improve the properties of cellulases will be major goals for future cellulase research. Optimization of growth conditions and processes has been studied extensively in order to improve cellulase production. For instance, empirical optimization of process variables to improve productivity has been the focus of earlier research using fermentation for cellulase production.

Several commercial production technologies employ SmF and use hypersecretary mutants. Despite several attempts directed toward generation of hyperproducers by mutation, the cost of enzymes could not be reduced. SSF utilizing lignocellulosic biomass, particularly by using host-/substrate-specific microbes, is another alternative strategy of cellulase production. There are several reports on filamentous fungi used in the production of enzyme complex for the degradation of lignocellulose. Cellobiose is an extremely potent inhibitor of the cellobiohydrolases (CBH) and endoglucanase (EG) full-form biosynthesis. *Trichoderma* and other cellulase-producing microbes produce little β-glucosidase as compared to other cellulolytic enzymes. The lower production of β-glucosidase reduces the capacity to hydrolyze the cellobiose to glucose, thereby leading to feedback inhibition of cellulase production. This issue has been addressed by various means, such as the addition of exogenous β-glucosidases to eliminate the cellobiose and incorporation of β-glucosidase genes into the organism for overproduction of enzymes.

## 10.12 CONCLUSION AND FUTURE PROSPECTS

In recent years, one of the most important biotechnological applications of microbial enzymes has been the conversion of agricultural wastes and all lignocellulosics into products of commercial interest, such as ethanol, glucose, and single-cell products. The key element in the bioconversion process of lignocellulosics to these useful products are the hydrolytic enzymes, mainly cellulases. The bioconversions of cellulosic materials are now an issue of extensive research as a contribution to the development of a large-scale conversion process beneficial to humankind. Such a process would help in alleviating insufficient food and animal feeds, solving the waste disposal problem, and diminishing man's reliance on fossil fuels by offering a convenient and renewable source of energy in the form of glucose. A wide spectrum of cellulolytic microorganisms, mainly fungi and bacteria, have been isolated and identified in the past few years, and efforts to apply their qualities are being made. Fungi are the main cellulase-producing microorganisms and *Aspergillus* and *Trichoderma* are the main fungal genera that are used for commercial production of cellulases.

In the future, developments in process design and medium formulations will no longer be sufficient, and controlled genetic manipulations into the physiology of cellulase producers toward getting improved cellulase production will be key to making the cellulase production process more cost-effective. Major tasks ahead include overriding the feedback control by glucose and development of integrated bioprocesses for the production of cellulases. Improvements in activities of cellulase or incorporation of desired features to enzymes by protein engineering are other areas where cellulase research has to be advanced. Active site modifications can be exercised through site-directed mutagenesis, and the mutant proteins could be utilized for understanding the mechanisms of action as well as for altering the substrate specificities or improving their activities.

## ACKNOWLEDGMENTS

The authors are thankful to Head, Department of Environmental Science, Babasaheb Bhimrao Ambedkar University (A Central University), Lucknow, India. Young Scientist Fellowship (Award No: CST/242/, dated 12/05/2015, Council of Science and Technology Uttar Pradesh, Lucknow, India) to first author Dr. Shiv Shankar is gratefully acknowledged.

## REFERENCES

Abdel-Fattah, A.F., Osman, M.Y., and Abdel-Naby, M.A. 1997. Production and immobilization of cellobiose from *Aspergillus niger* A20. *Chemical Engineering Journal* 68(2–3): 189–196. DOI: http://dx.doi.org/10.1016/S1385-8947(97)00082-X.

Acharya, S., and Chaudhary, A. 2012. Bioprospecting thermophiles for cellulase production: A review. *Brazilian Journal of Microbiology* 43(3): 844–856. DOI: http://dx.doi.org/10.1590/S1517-83822012000300001.

Adsul, M.G., Ghule, J.E., Singh, R., Shaikh, H., Bastawde, K.B., Gokhale, D.V., and Varma, A.J. 2004. Polysaccharides from bagasse: Applications in cellulase and xylanase production. *Carbohydrate Polymers* 57(1): 67–72. DOI: http://dx.doi.org/10.1016/j.carbpol.2004.04.001.

Ang, S.K., Shaza, E.M., Adibah, Y., Suraini, A.A., and Madihah, M.S. 2013. Production of cellulases and xylanase by *Aspergillus fumigatus* SK1 using untreated oil palm trunk through solid state fermentation. *Process Biochemistry* 48(9): 1293–1302. DOI: http://dx.doi.org/10.1016/j.procbio.2013.06.019.

Aro, N., Pakula, T., and Penttilä, M. 2005. Transcriptional regulation of plant cell wall degradation by filamentous fungi. *FEMS Microbiology Reviews* 29(4): 719–739. DOI: http://dx.doi.org/10.1016/j.femsre.2004.11.006.

Asztalos, A., Daniels, M., Sethi, A., Shen, T., Langan, P., Redondo, A., and Gnanakaran, S. 2012. A coarse-grained model for synergistic action of multiple enzymes on cellulose. *Biotechnology for Biofuels* 5(1): 55. DOI: http://dx.doi.org/10.1186/1754-6834-5-55.

Bajpai, P. 2012. *Biotechnology for pulp and paper processing*. Springer, New York.

Bansal, N., Tewari, R., Soni, R., and Soni, S.K. 2012. Production of cellulases from *Aspergillus niger* NS-2 in solid state fermentation on agricultural and kitchen waste residues. *Waste Management* 32(7): 1341–1346. DOI: http://dx.doi.org/10.1016/j.wasman.2012.03.006.

Barr, B.K., Hsieh, Y.L., Ganem, B., and Wilson, D.B. 1996. Identification of two functionally different classes of exocellulases. *Biochemistry* 35(2): 586–592. DOI: http://dx.doi.org/10.1021/bi9520388.

Bayer, E.A., Morag, E., and Lamed, R. 1994. The cellulosome—A treasure-trove for biotechnology. *Trends in Biotechnology* 12(9): 379–386. DOI: http://dx.doi.org/10.1016/0167-7799(94)90039-6.

Belghith, H., Ellouz-Chaabouni, S., and Gargouri, A. 2001. Biostoning of denims by *Penicillium occitanis* (Pol6) cellulases. *Journal of Biotechnology* 89(2–3): 257–262.

Brijwani, K., Oberoi, H.S., and Vadlani, P.V. 2010. Production of a cellulolytic enzyme system in mixed-culture solid-state fermentation of soybean hulls supplemented with wheat bran. *Process Biochemistry* 45(1): 120–128. DOI: http://dx.doi.org/10.1016/j.procbio.2009.08.015.

Bronnenmeier, K., Kern, A., Liebl, W., and Staudenbauer, W.L. 1995. Purification of Thermotoga maritima enzymes for the degradation of cellulosic materials. *Applied and Environmental Microbiology* 61(4): 1399–1407.

Buaban, B., Inoue, H., Yano, S., Tanapongpipat, S., Ruanglek, V., Champreda, V., Pichyangkura, R., Rengpipat, S., and Eurwilaichitr, L. 2010. Bioethanol production from ball milled bagasse using an on-site produced fungal enzyme cocktail and xylose-fermenting *Pichia stipitis*. *Journal of Bioscience and Bioengineering* 110(1): 18–25. DOI: http://dx.doi.org/10.1016/j.jbiosc.2009.12.003.

Castro, A.M., Castilho, L.R., and Freire, D.M.G. 2015. Performance of a fixed-bed solid-state fermentation bioreactor with forced aeration for the production of hydrolases by *Aspergillus awamori*. *Biochemical Engineering Journal* 93: 303–308. DOI: http://dx.doi.org/10.1016/j.bej.2014.10.016.

Chandel, A.K., Gonçalves, B.C.M., Strap, J.L., and da Silva, S.S. 2015. Biodelignification of lignocellulose substrates: An intrinsic and sustainable pretreatment strategy for clean energy production. *Critical Reviews in Biotechnology* 35(3): 281–293. DOI: http://dx.doi.org/10.3109/07388551.2013.841638.

Chaudhary, P., Kumar, N.N., and Deobagkar, D.N. 1997. The glucanases of cellulomonas. *Biotechnology Advances* 15(2): 315–331. DOI: http://dx.doi.org/10.1016/S0734-9750(97)00010-4.

Cherry, J.R., and Fidantsef, A.L. 2003. Directed evolution of industrial enzymes: An update. *Current Opinion in Biotechnology* 14(4): 438–443.

Das, A., Paul, T., Halder, S.K., Jana, A., Maity, C., Mohapatra, P.K.D., Pati, B.R., and Mondal, K.C. 2013. Production of cellulolytic enzymes by *Aspergillus fumigatus* ABK9 in wheat bran-rice straw mixed substrate and use of cocktail enzymes for deinking of waste office paper pulp. *Bioresource Technology* 128: 290–296. DOI: http://dx.doi.org/10.1016/j.biortech.2012.10.080.

Davies, G., and Henrissat, B. 1995. Structures and mechanisms of glycosyl hydrolases. *Structure* 3(9): 853–859. DOI: http://dx.doi.org/10.1016/S0969-2126(01)00220-9.

Delabona Pda, S., Farinas, C.S., da Silva, M.R., Azzoni, S.F., and Pradella, J.G. 2012. Use of a new *Trichoderma harzianum* strain isolated from the Amazon rainforest with pretreated sugar cane bagasse for on-site cellulase production. *Bioresource Technology* 107: 517–521. DOI: http://dx.doi.org/10.1016/j.biortech.2011.12.048.

Deswal, D., Khasa, Y.P., and Kuhad, R.C. 2011. Optimization of cellulase production by a brown rot fungus *Fomitopsis* Sp. RCK2010 under solid state fermentation. *Bioresource Technology* 102(10): 6065–6072. DOI: http://dx.doi.org/10.1016/j.biortech.2011.03.032.

Dienes, D., Egyházi, A., and Réczey, K. 2004. Treatment of recycled fiber with *Trichoderma* cellulases. *Industrial Crops and Products* 20(1): 11–21. DOI: http://dx.doi.org/10.1016/j.indcrop.2003.12.009.

Elisashvili, V., Kachlishvili, E., Tsiklauri, N., Metreveli, E., Khardziani, T., and Agathos, S.N. 2009. Lignocellulose-degrading enzyme production by white-rot basidiomycetes isolated from the forests of Georgia. *World Journal of Microbiology and Biotechnology* 25(2): 331–339. DOI: http://dx.doi.org/10.1007/s11274-008-9897-x.

Eriksson, K.-E.L., Blanchette, R.A., and Ander, P. 1990. *Microbial and enzymatic degradation of wood and wood components*. Springer Series in Wood Science. Springer, Berlin, Heidelberg.

Eriksson, T., Börjesson, J., and Tjerneld, F. 2002. Mechanism of surfactant effect in enzymatic hydrolysis of lignocellulose. *Enzyme and Microbial Technology* 31(3): 353–364. DOI: http://dx.doi.org/10.1016/S0141-0229(02)00134-5.

Falkoski, D.L., Guimarães, V.M., de Almeida, M.N., Alfenas, A.C., Colodette, J.L., and de Rezende, S.T. 2013. *Chrysoporthe cubensis*: A new source of cellulases and hemicellulases to application in biomass saccharification processes. *Bioresource Technology* 130: 296–305. DOI: http://dx.doi.org/10.1016/j.biortech.2012.11.140.

Foust, T.D., Aden, A., Dutta, A., and Phillips, S. 2009. An economic and environmental comparison of a biochemical and a thermochemical lignocellulosic ethanol conversion processes. *Cellulose* 16(4): 547–565. DOI: http://dx.doi.org/10.1007/s10570-009-9317-x.

Fujii, T., Xu, F., Inoue, H., Murakami, K., and Sawayama, S. 2009. Enzymatic hydrolyzing performance of *Acremonium cellulolyticus* and *Trichoderma reesei* against three lignocellulosic materials. *Biotechnol. Biofuels* 2(1): 2–24.

Gautam, S.P., Bundela, P.S., Pandey, A.K., Khan, J., Awasthi, M.K., and Sarsaiya, S. 2011. Optimization for the production of cellulase enzyme from municipal solid waste residue by two novel cellulolytic fungi. *Biotechnology Research International* 2011: 1–8. DOI: http://dx.doi.org/10.4061/2011/810425.

Gil, N., Gil, C., Amaral, M.E., Costa, A.P., and Duarte, A.P. 2009. Use of enzymes to improve the refining of a bleached eucalyptus globulus kraft pulp. *Biochemical Engineering Journal* 46(2): 89–95. DOI: http://dx.doi.org/10.1016/j.bej.2009.04.011.

Gunny, A.A., Arbain, D., Jamal, P., and Gumba, R.E. 2015. Improvement of halophilic cellulase production from locally isolated fungal strain. *Saudi Journal of Biological Sciences* 22(4): 476–483. DOI: http://dx.doi.org/10.1016/j.sjbs.2014.11.021.

Gusakov, A.V., Salanovich, T.N., Antonov, A.I., Ustinov, B.B., Okunev, O.N., Burlingame, R., Emalfarb, M., Baez, M., and Sinitsyn, A.P. 2007. Design of highly efficient cellulase mixtures for enzymatic hydrolysis of cellulose. *Biotechnology and Bioengineering* 97(5): 1028–1038. DOI: http://dx.doi.org/10.1002/bit.21329.

Heck, J.X., Hertz. P.F., and Ayub, M.A.Z. 2002. Cellulase and xylanase productions by isolated Amazon bacillus strains using soybean industrial residue based solid-state cultivation. *Brazilian Journal of Microbiology* 33(3): 213–218. DOI: http://dx.doi.org/10.1590/S1517-83822002000300005.

Henrissat, B. 1991. A classification of glycosyl hydrolases based on amino acid sequence similarities. *The Biochemical Journal* 280(Pt 2): 309–316.

Henrissat, B., Teeri, T.T., and Warren, R.A.J. 1998. A scheme for designating enzymes that hydrolyse the polysaccharides in the cell walls of plants. *FEBS Letters* 425(2): 352–354. DOI: http://dx.doi.org/10.1016/S0014-5793(98)00265-8.

Jindou, S., Borovok, I., Rincon, M.T., Flint, H.J., Antonopoulos, D.A., Berg, M.E., White, B.A., Bayer, E.A., and Lamed, R. 2006. Conservation and divergence in cellulosome architecture between two strains of *Ruminococcus flavefaciens*. *Journal of Bacteriology* 188(22): 7971–7976. DOI: http://dx.doi.org/10.1128/JB.00973-06.

Ju, L.K., and Afolabi, O.A. 1999. Wastepaper hydrolysate as soluble inducing substrate for cellulase production in continuous culture of *Trichoderma reesei*. *Biotechnology Progress* 15(1): 91–97. DOI: http://dx.doi.org/10.1021/bp980116n.

Kalogeris, E., Fountoukides, G., Kekos, D., and Macris, B.J. 1999. Design of a solid-state bioreactor for thermophilic microorganisms. *Bioresource Technology* 67(3): 313–315. DOI: http://dx.doi.org/10.1016/S0960-8524(98)00124-2.

Kleywegt, G.J., Zou, J.Y., Divne, C., Davies, G.J., Sinning, I., Stâhlberg, J., Reinikainen, T., Srisodsuk, M., Teeri, T.T., and Jones, T.A. 1997. The crystal structure of the catalytic core domain of endoglucanase I from *Trichoderma reesei* at 3.6 A resolution, and a comparison with related enzymes. *Journal of Molecular Biology* 272(3): 383–397. DOI: http://dx.doi.org/10.1006/jmbi.1997.1243.

Krishna, C. 1999. Production of bacterial cellulases by solid state bioprocessing of banana wastes. *Bioresource Technology* 69(3): 231–239. DOI: http://dx.doi.org/10.1016/S0960-8524(98)00193-X.

Kuhad, R.C., Gupta, R., and Singh, A. 2011. Microbial cellulases and their industrial applications. *Enzyme Research* 2011: 1–10. DOI: http://dx.doi.org/10.4061/2011/280696.

Kumar, D., and Murthy, G.S. 2013. Stochastic molecular model of enzymatic hydrolysis of cellulose for ethanol production. *Biotechnology for Biofuels* 6(1): 63. DOI: http://dx.doi.org/10.1186/1754-6834-6-63.

Kumar, R., Singh, S., and Singh, O.V. 2008. Bioconversion of lignocellulosic biomass: Biochemical and molecular perspectives. *Journal of Industrial Microbiology & Biotechnology* 35(5): 377–391. DOI: http://dx.doi.org/10.1007/s10295-008-0327-8.

Lecourt, M., Sigoillot, J.-C., and Petit-Conil, M. 2010. Cellulase-assisted refining of chemical pulps: Impact of enzymatic charge and refining intensity on energy consumption and pulp quality. *Process Biochemistry* 45(8): 1274–1278. DOI: http://dx.doi.org/10.1016/j.procbio.2010.04.019.

Lee, C.K., Darah, I., and Ibrahim, C.O. 2007. Enzymatic deinking of laser printed office waste papers: Some governing parameters on deinking efficiency. *Bioresource Technology* 98(8): 1684–1689. DOI: http://dx.doi.org/10.1016/j.biortech.2006.05.052.

Leschine, S.B. 1995. Cellulose degradation in anaerobic environments. *Annual Review of Microbiology* 49(1): 399–426. DOI: http://dx.doi.org/10.1146/annurev.mi.49.100195.002151.

Li, D.-C., Lu, M., Li, Y.L., and Lu, J. 2003. Purification and characterization of an endocellulase from the thermophilic fungus *Chaetomium thermophilum* CT2. *Enzyme and Microbial Technology* 33(7): 932–937. DOI: http://dx.doi.org/10.1016/S0141-0229(03)00245-X.

Lin, Y.-S., Lee, W.-C., Duan, K.-J., and Lin, Y.-H. 2013. Ethanol production by simultaneous saccharification and fermentation in rotary drum reactor using thermotolerant *Kluyveromyces marxianus*. *Applied Energy* 105: 389–394. DOI: http://dx.doi.org/10.1016/j.apenergy.2012.12.020.

Liu, Y.-T., Luo, Z.-Y., Long, C.-N., Wang, H.-D., Long, M.-N., and Hu, Z. 2011. Cellulase production in a new mutant strain of *Penicillium decumbens* ML-017 by solid state fermentation with rice bran. *New Biotechnology* 28(6): 733–737. DOI: http://dx.doi.org/10.1016/j.nbt.2010.12.003.

Lynd, L.R., van Zyl, W.H., McBride, J.E., and Laser, M. 2005. Consolidated bioprocessing of cellulosic biomass: an update. *Curr. Opin. Biotechnol.* 16(5): 577–583.

Lynd, L.R., Weimer, P.J., van Zyl, W.H., and Pretorius, I.S. 2002. Microbial cellulose utilization: Fundamentals and biotechnology. *Microbiology and Molecular Biology Reviews: MMBR* 66(3): 506–577.

Matkar, K., Chapla, D., Divecha, J., Nighojkar, A., and Madamwar, D. 2013. Production of cellulase by a newly isolated strain of *Aspergillus sydowii* and its optimization under submerged fermentation. *International Biodeterioration & Biodegradation* 78: 24–33. DOI: http://dx.doi.org/10.1016/j.ibiod.2012.12.002.

Medve, J., Karlsson, J., Lee, D., and Tjerneld, F. 1998. Hydrolysis of microcrystalline cellulose by cellobiohydrolase I and endoglucanase II from *Trichoderma reesei*: Adsorption, sugar production pattern, and synergism of the enzymes. *Biotechnology and Bioengineering* 59 (5): 621–634. DOI: http://dx.doi.org/10.1002/(SICI)1097-0290(19980905)59:5<621::AID-BIT13>3.0.CO;2-C.

Medve, J., Ståhlberg, J., and Tjerneld, F. 1994. Adsorption and synergism of cellobiohydrolase I and II of *Trichoderma reesei* during hydrolysis of microcrystalline cellulose. *Biotechnology and Bioengineering* 44(9): 1064–1073. DOI: http://dx.doi.org/10.1002/bit.260440907.

Narra, M., Dixit, G., Divecha, J., Madamwar, D., and Shah, A.R. 2012. Production of cellulases by solid state fermentation with *Aspergillus terreus* and enzymatic hydrolysis of mild alkali-treated rice straw. *Bioresource Technology* 121: 355–361. DOI: http://dx.doi.org/10.1016/j.biortech.2012.05.140.

Ncube, T., Howard, R.L., Abotsi, E.K., Jansen van Rensburg, E.L., and Ncube, I. 2012. Jatropha curcas seed cake as substrate for production of xylanase and cellulase by *Aspergillus niger* FGSCA733 in solid-state fermentation. *Industrial Crops and Products* 37(1): 118–123. DOI: http://dx.doi.org/10.1016/j.indcrop.2011.11.024.

Nidetzky, B., Steiner, W., Hayn, M., and Claeyssens, M. 1994. Cellulose hydrolysis by the cellulases from *Trichoderma reesei*: A new model for synergistic interaction. *Biochemical Journal* 298(3): 705–710. DOI: http://dx.doi.org/10.1042/bj2980705.

Nigam, P., and Singh, D. 1996. Processing of Agricultural wastes in solid state fermentation for microbial protein production. *J. Sci. Ind. Res.* 55: 373–380.

Oksanen, A., Edelmann, K., Kataja-aho, J., and Suurnäkki, A. 2011. Enhancing dewatering of thermo-mechanical pulp (TMP) based papermaking through enzymatic treatment. *Holzforschung* 65(6): 787–797. DOI: http://dx.doi.org/10.1515/HF.2011.083.

Oksanen, T., Pere, J., Paavilainen, L., Buchert, J., and Viikari, L. 2000. Treatment of recycled kraft pulps with *Trichoderma reesei* hemicellulases and cellulases. *Journal of Biotechnology* 78(1): 39–48.

Pandey, S., Singh, S., Yadav, A.N., Nain, L., and Saxena, A.K. 2013. Phylogenetic diversity and characterization of novel and efficient cellulase producing bacterial isolates from various extreme environments. *Bioscience, Biotechnology, and Biochemistry* 77(7): 1474–1480. DOI: http://dx.doi.org/10.1271/bbb.130121.

Paudel, Y.P., and Qin, W. 2015. Characterization of novel cellulase-producing bacteria isolated from rotting wood samples. *Applied Biochemistry and Biotechnology* 177(5): 1186–1198. DOI: http://dx.doi.org/10.1007/s12010-015-1806-9.

Pothiraj, C., and Eyini, M. 2007. Enzyme activities and substrate degradation by fungal isolates on cassava waste during solid state fermentation. *Mycobiology* 35(4): 196. DOI: http://dx.doi.org/10.4489/MYCO.2007.35.4.196.

Prasad, P., Singh, T., and Bedi, S. 2013. Characterization of the cellulolytic enzyme produced by *Streptomyces griseorubens* (Accession No. AB184139) isolated from Indian soil. *Journal of King Saud University—Science* 25(3): 245–250. DOI: http://dx.doi.org/10.1016/j.jksus.2013.03.003.

Rahkamo, L., Siika-Aho, M., Vehviläinen, M., Dolk, M., Viikari, L., Nousiainen, P., and Buchert, J. 1996. Modification of hardwood dissolving pulp with purified *Trichoderma reesei* cellulases. *Cellulose* 3(1): 153–163. DOI: http://dx.doi.org/10.1007/BF02228798.

Reczey, K., Szengyel, Zs., Eklund, R., and Zacchi, G. 1996. Cellulase production by *T. reesei*. *Bioresource Technology* 57(1): 25–30. DOI: http://dx.doi.org/10.1016/0960-8524(96)00038-7.

Romero, M.D., Aguado, J., González, L., and Laderom, M. 1999. Cellulase production by *Neurospora crassa* on wheat straw. *Enzyme and Microbial Technology* 25(3–5): 244–250. DOI: http://dx.doi.org/10.1016/S0141-0229(99)00035-6.

Rouvinen, J., Bergfors, T., Teeri, T., Knowles, J.K., and Jones, T.K. 1990. Three-dimensional structure of cellobiohydrolase II from *Trichoderma reesei*. *Science* 249(4967): 380–386.

Saha, B.C. 2004. Production, purification and properties of endoglucanase from a newly isolated strain of *Mucor circinelloides*. *Process Biochemistry* 39(12): 1871–1876. DOI: http://dx.doi.org/10.1016/j.procbio.2003.09.013.

Schafner, D.W., and Toledo, R.T. 1992. Cellulase production in continuous culture by *Trichoderma reesei* on xylose-based media. *Biotechnology and Bioengineering* 39(8): 865–869. DOI: http://dx.doi.org/10.1002/bit.260390808.

Schwarz, W.H. 2001. The cellulosome and cellulose by anaerobic bacteria. *Applied Microbiology and Biotechnology* 56(5–6): 634–649. DOI: http://dx.doi.org/10.1007/s002530100710.

Shafique, S. 2011. Strain improvement in *Trichoderma viride* through mutation for overexpression of cellulase and characterization of mutants using random amplified polymorphic DNA (RAPD). *African Journal of Biotechnology* 10(84): 19590–19597. DOI: http://dx.doi.org/10.5897/AJB10.2464.

Shahzadi, T., Mehmood, S., Irshad, M., Anwar, Z., Afroz, A., Zeeshan, N., Rashid, U., and Sughra, K. 2014. Advances in lignocellulosic biotechnology: A brief review on lignocellulosic biomass and cellulases. *Advances in Bioscience and Biotechnology* 5(3): 246–251. DOI: http://dx.doi.org/10.4236/abb.2014.53031.

Singh, R., Kumar, R., Bishnoi, K., and Bishnoi, N.R. 2009. Optimization of synergistic parameters for thermostable cellulase activity of *Aspergillus heteromorphus* using response surface methodology. *Biochemical Engineering Journal* 48(1): 28–35. DOI: http://dx.doi.org/10.1016/j.bej.2009.08.001.

Singhania, R.R., Patel, A.K., Soccol, C.R., and Pandey, A. 2009. Recent advances in solid-state fermentation. *Biochemical Engineering Journal* 44(1): 13–18. DOI: http://dx.doi.org/10.1016/j.bej.2008.10.019.

Singhania, R.R., Sukumaran, R.K., and Pandey, A. 2007. Improved cellulase production by *Trichoderma reesei* RUT C30 under SSF through process optimization. *Applied Biochemistry and Biotechnology* 142(1): 60–70. DOI: http://dx.doi.org/10.1007/s12010-007-0019-2.

Sun, Y., and Cheng, J. 2002. Hydrolysis of lignocellulosic materials for ethanol production: A review. *Bioresource Technology* 83(1): 1–11. DOI: http://dx.doi.org/10.1016/S0960-8524(01)00212-7.

Szabó, I.J., Johansson, G., and Pettersson, G. 1996. Optimized cellulase production by *Phanerochaete chrysosporium*: Control of catabolite repression by fed-batch cultivation. *Journal of Biotechnology* 48(3): 221–230. DOI: http://dx.doi.org/10.1016/0168-1656(96)01512-X.

Taechapoempol, K., Sreethawong, T., Rangsunvigit, P., Namprohm, W., Thamprajamchit, B., Rengpipat, S., and Chavadej, S. 2011. Cellulase-producing bacteria from Thai higher termites, *Microcerotermes* sp.: Enzymatic activities and ionic liquid tolerance. *Applied Biochemistry and Biotechnology* 164(2): 204–219. DOI: http://dx.doi.org/10.1007/s12010-010-9128-4.

Takashima, S., Nakamura, A., Hidaka, M., Masaki, H., and Uozumi, T. 1999. Molecular cloning and expression of the novel fungal beta-glucosidase genes from *Humicola grisea* and *Trichoderma reesei*. *Journal of Biochemistry* 125(4): 728–736.

Talebnia, F., Karakashev, D., and Angelidaki, I. 2010. Production of bioethanol from wheat straw: An overview on pretreatment, hydrolysis and fermentation. *Bioresource Technology* 101(13): 4744–4753. DOI: http://dx.doi.org/10.1016/j.biortech.2009.11.080.

Tanskul, S., Amornthatree, K., and Jaturonlak, N. 2013. A new cellulose-producing bacterium, *Rhodococcus* sp. MI 2: Screening and optimization of culture conditions. *Carbohydrate Polymers* 92(1): 421–428. DOI: http://dx.doi.org/10.1016/j.carbpol.2012.09.017.

Teeri, T.T. 1997. Crystalline cellulose degradation: New insight into the function of cellobiohydrolases. *Trends in Biotechnology* 15(5): 160–167. DOI: http://dx.doi.org/10.1016/S0167-7799(97)01032-9.

Teeri, T.T., Koivula, A., Linderm, M., Wohlfahrt, G., Divne, C., and Jones, T.A. 1998. *Trichoderma Reesei* cellobiohydrolases: Why so efficient on crystalline cellulose? *Biochemical Society Transactions* 26(2): 173–177. DOI: http://dx.doi.org/10.1042/bst0260173.

Tirado, M.C., Clarke, R., Jaykus, L.A., McQuatters-Gollop, A., and Frank, J.M. 2010. Climate change and food safety: A review. *Food Research International* 43(7): 1745–1765. DOI: http://dx.doi.org/10.1016/j.foodres.2010.07.003.

Tomme, P., Warren, R.A.J., and Gilkes, N.R. 1995. Cellulose hydrolysis by bacteria and fungi. *Advances in Microbial Physiology* 37: 1–81.

Tucker, M.P., Kim, K.H., Newman, M.M., and Nguyen, Q.A. 2003. Effects of temperature and moisture on dilute-acid steam explosion pretreatment of corn stover and cellulase enzyme digestibility. *Applied Biochemistry and Biotechnology* 105: 165–177.

Usami, S., Kirimura, K., Imura, M., and Morikawa, S. 1990. Cellular localization of the constitutive β-Glucosidase in *Trichoderma viride*. *Journal of Fermentation and Bioengineering* 70(3): 185–187. DOI: http://dx.doi.org/10.1016/0922-338X(90)90183-W.

Venkata Mohan, S., Lalit Babu, V., and Sarma, P.N. 2008. Effect of various pretreatment methods on anaerobic mixed microflora to enhance biohydrogen production utilizing dairy wastewater as substrate. *Bioresource Technology* 99(1): 59–67. DOI: http://dx.doi.org/10.1016/j.biortech.2006.12.004.

Vladut-Talor, M., Kauri, T., and Kushner, D.J. 1986. Effects of cellulose on growth, enzyme production, and ultrastructure of a *Cellulomonas* species. *Archives of Microbiology* 144(3): 191–195. DOI: http://dx.doi.org/10.1007/BF00410945.

Vu, V.H., Pham, T.A., and Kim, K. 2011. Improvement of fungal cellulase production by mutation and optimization of solid state fermentation. *Mycobiology* 39(1): 20. DOI: http://dx.doi.org/10.4489/MYCO.2011.39.1.020.

Vyas, S., and Lachke, A. 2003. Biodeinking of mixed office waste paper by alkaline active cellulases from alkalotolerant *Fusarium* sp. *Enzyme and Microbial Technology* 32(2): 236–245. DOI: http://dx.doi.org/10.1016/S0141-0229(02)00273-9.

Watanabe, H., and Tokuda, G. 2010. Cellulolytic systems in insects. *Annual Review of Entomology* 55(1): 609–632. DOI: http://dx.doi.org/10.1146/annurev-ento-112408-085319.

Wen, Z., Liao, W., and Chen, S. 2005. Production of cellulase/β-glucosidase by the mixed fungi culture *Trichoderma reesei* and *Aspergillus phoenicis* on dairy manure. *Process Biochemistry* 40(9): 3087–3094. DOI: http://dx.doi.org/10.1016/j.procbio.2005.03.044.

White, A.R. 1982. Visualization of cellulases and cellulose degradation. In *Cellulose and other natural polymer systems,* Malcolm Brown, R. (ed.), Springer, Boston, MA, pp. 489–509.

Wilson, D.B. 2011. Microbial diversity of cellulose hydrolysis. *Current Opinion in Microbiology* 14(3): 259–263. DOI: http://dx.doi.org/10.1016/j.mib.2011.04.004.

Wood, T.M. 1992. Fungal cellulases. *Biochemical Society Transactions* 20(1): 46–53. DOI: http://dx.doi.org/10.1042/bst0200046.

Wu, X., McLaren, J., Madl, R., and Wang, D. 2010. Biofuels from lignocellulosic biomass. In *Sustainable biotechnology*, Singh, O.V., and Harvey, S.P. (eds.), Springer, Dordrecht, The Netherlands, pp. 19–41.

Xia, L., and Cen, P. 1999. Cellulase production by solid state fermentation on lignocellulosic waste from the xylose industry. *Process Biochemistry* 34(9): 909–912. DOI: http://dx.doi.org/10.1016/S0032-9592(99)00015-1.

Yamada, M., Amano, Y., Horikawa, E., Nozaki, K., and Kanda, T. 2005. Mode of action of cellulases on dyed cotton with a reactive dye. *Bioscience, Biotechnology, and Biochemistry* 69(1): 45–50. DOI: http://dx.doi.org/10.1271/bbb.69.45.

Yang, X.Q., Zhao, X.X., Liu, C.Y., Zheng, Y., and Qian, S.J. 2009. Decolorization of azo, triphenylmethane and anthraquinone dyes by a newly isolated *Trametes* sp. SQ01 and its laccase. *Process Biochemistry* 44(10): 1185–1189. DOI: http://dx.doi.org/10.1016/j.procbio.2009.06.015.

Yuan, L., Wang, W., Pei, Y., and Lu, F. 2012. Screening and identification of cellulase-producing strain of *Fusarium oxysporum*. *Procedia Environmental Sciences* 12: 1213–1219. DOI: http://dx.doi.org/10.1016/j.proenv.2012.01.410.

Zhang, Y.-H.P., and Lynd, L.R. 2004. Toward an aggregated understanding of enzymatic hydrolysis of cellulose: Noncomplexed cellulase systems. *Biotechnology and Bioengineering* 88(7): 797–824. DOI: http://dx.doi.org/10.1002/bit.20282.

Zhu, J.Y., Pan, X.J., Wang, G.S., and Gleisner, R. 2009. Sulfite pretreatment (SPORL) for robust enzymatic saccharification of spruce and red pine. *Bioresource Technology* 100(8): 2411–2418. DOI: http://dx.doi.org/10.1016/j.biortech.2008.10.057.

# 11 Phytoremediation
## *A Green Technology for Remediation of Metal-Contaminated Sites*

*Manoj Kumar and Rajesh Singh*

## CONTENTS

11.1 Introduction....306
11.2 Sources of Heavy Metals Pollution and Toxicity in the Environment....308
 11.2.1 Sources of Heavy Metals Pollution in the Environment....308
  11.2.1.1 Natural Sources....308
  11.2.1.2 Anthropogenic Sources....309
 11.2.2 Toxic Effects of Heavy Metals in the Environment....309
  11.2.2.1 Cadmium and Its Toxic Effects....310
  11.2.2.2 Mercury and Its Toxic Effects....310
  11.2.2.3 Lead and Its Toxic Effects....311
  11.2.2.4 Arsenic and Its Toxic Effects....311
  11.2.2.5 Chromium and Its Toxic Effects....312
11.3 Types of Phytoremediation....313
 11.3.1 Phytoextraction....315
 11.3.2 Rhizofiltration....316
 11.3.3 Phytostabilization....317
 11.3.4 Phytovolatilization....317
 11.3.5 Phytodegradation....317
 11.3.6 Rhizodegradation....318
11.4 Mechanism of Heavy Metal Accumulation in Plants....318
 11.4.1 Metal Solubilization from Soil Matrix....319
 11.4.2 Metal Uptake into the Plant Root....320
 11.4.3 Accumulation of Heavy Metals Root to Shoot....320
 11.4.4 Metal Detoxification and Chelation....320
 11.4.5 Metal Sequestration and Its Volatilization....321
11.5 Plants Used for Phytoremediation of Heavy Metals....321
11.6 Phytoremediation Technology: Task of Plant Growth-Promoting Rhizobacteria....322
11.7 Genetically Modified Plants Approach for Phytoremediation....323
11.8 Advantages and Disadvantages of Phytoremediation Technology....325

11.9 Prospects and Future Challenges ........ 325
11.10 Conclusion ........ 327
Acknowledgment ........ 328
References ........ 328

## 11.1 INTRODUCTION

Contamination of the environment with inorganic pollutants like toxic metals has become a major environmental problem. The unsafe disposal of industrial wastes is the prime source of introduction of toxic metals into the environment. Unlike organic pollutants, metals are nonbiodegradable and, therefore, persist for a long time in the environment. These are highly toxic in nature and cause serious environmental problems and health hazards in humans, animals, and plants. Therefore, the search for an eco-friendly remediation approach for the cleaning of metal-contaminated sites and safety of human health is the immediate requirement (Abou-Shanab et al. 2006).

In the past 10 years, phytoremediation approaches have meticulously been scrutinized as the able remediation technology to remove heavy metals, organic contamination, many hazardous environmental contaminants like landfill leachates, pesticides, polycyclic aromatic hydrocarbons (PAHs), radionuclides, petroleum, crude oil, chlorinated solvents, polychlorinated biphenyls (PCBs), explosives, ammunitions, and even poisonous gases (Pilon-Smits 2005).

Since antiquity human beings have been interfering and interacting with the natural environment for their survival and well-being. Inorganic environmental contaminants occur as natural elements in the soil, including plant macronutrients viz. nitrates and phosphates; micronutrients viz. Ni, Mo, Mn Zn, Cr, Fe, and Cu; nonessential elements viz. Pb, As Cd, Co, Hg, and W; and radionuclide's viz. $^{238}U$, $^{137}Cs$. Organic pollutants include oil, gasoline, benzene, toluene, PCBs, and PAHs. Environmental contaminants, whether inorganic or organic, affect all surrounding environments including aquatic life and human health (Bridge 2004). Due to acute toxicity associated with inorganic contaminants, like heavy metals, these are considered as environmental priority pollutants and are targeted for cleanup processes (Verma and Suthar 2015).

The health hazards and toxic effects to the environment associated with heavy metal soil contamination with trace elements together with higher cost of removal and replacement of polluted soil have prompted the development of alternative and cost-effective technologies to recover and protect the degraded land. As a competing technology, phytoremediation offers significant benefits over conventional tools for the remediation of heavy metal-contaminated land sites due to it being cost-effective and eco-friendly (Sekhar et al. 2005; Fischerova et al. 2006). Phytoremediation, plant-based technologies are applicable for detoxifying heavy metals from areas of low concentrations with shallow soils and water, although longer treatment times may be required (Adriano 2009). Due to their sedentary nature, plants have involved a deferent capability for dealing with toxic compounds in their environment (Abhilash et al. 2009). Phytoremediation called green technology is a plant-based bioremediation technology using plants with their interaction rhizospheric microbial process for the onsite remediation of organic and inorganic pollutants (Chandra et al. 2015). Phytoremediation is the only cost-effective alternative technology to developing countries (Chandra et al. 2015). Phytoextraction and rhizofiltration are used as bio-ore to extract metal, called phytomining (Angle et al. 2001), and energy can be produced through biomass burning (Li et al. 2003). Environmental sustainable research efforts are currently underway into the economic viability of phytoremediation technologies (Ghosh and Singh 2005) with several plant species like *Helianthus annuus* and *Arabidopsis thaliana* genetically modified species now recognized as suitable for phytoremediation.

In plant-based phytoextraction technology, the plants used to absorb toxic metals and help to purify (clean up) them from soils or wastewaters are technically termed as *hyperaccumulators*. Metal-accumulating plants or metal hyperaccumulators are resistant to the high metal concentrations and are capable of accumulating them into their roots, shoots, and leaves (Chandra et al. 2015). In addition, hyperaccumulators may differ from normal plants by their higher capacity to accumulate metal ions from the soil (Ho-Man et al. 2013). However, the inhibition of plant growth and biomass in metal-contaminated sites may limit the efficiency of phytoremediation processes (Kumar et al. 1995; Burd et al. 2000). Therefore, the possibilities of enhancing plant growth and biomass of metal accumulators using plant growth-promoting rhizobacteria (PGPR) have been suggested (Sheng and Xia 2006). PGPR can produce indole acetic acid (IAA), siderophores, and 1-aminocyclopropane-1-carboxylate (ACC). Deaminase is the ability to stimulate plant growth, lowering the level of ethylene by consuming ACC, the immediate precursor of ethylene in plants growing in the presence of heavy metals, and helping plants to acquire sufficient iron for optimal growth (Glick et al. 1995; Wang et al. 2000; Rajkumar et al. 2006).

Most heavy metals have low mobility in soil (Garbisu and Alkorta 2001) and are not easily absorbed by plant roots. The plant–microbe interaction can enhance the heavy metal bioavailability in the rhizosphere (Yang et al. 2005; Saravanan et al. 2007). In addition, the phytoremediation efficiency can also be promoted by the addition of accelerants or chelators and soil amendments (N, P, K fertilizers) to soil that increases the bioavailability of heavy metals (Blaylock and Huang 2000; Buchell-Witschel and Egli 2001; Wei et al. 2008; Karami and Shamsuddin 2010; Garba et al. 2012). A direct method for promoting the effectiveness of phytoremediation is to overexpress in genetically modified plants the genes involved in metabolism, uptake, or transport of specific pollutants, transported to the root to shoot (Shiota et al. 1994; Rugh 2004; Cherian and Oliveira 2005; Kramer 2005; Eapen et al. 2007; Aken 2008; Doty 2008; Macek et al. 2008).

This chapter provides an extensive review on several phytoremediation technologies, including phytoextraction, phytodegradation, rhizofiltration, phytostabilization, and phytovolatilization with their advantages and disadvantages. In addition, the basic heavy metals phytoremediation mechanism has been described. Appropriate PGPR is one of the most useful and eco-friendly tools that may enhance the efficiency of phytoremediation. The applications of genetically modified plants may be very efficient in remediating toxic metals contaminated sites. Thus, phytoremediation can be a cost-effective and sustainable approach for the restoration of toxic heavy metal-contaminated sites.

## 11.2 SOURCES OF HEAVY METALS POLLUTION AND TOXICITY IN THE ENVIRONMENT

### 11.2.1 Sources of Heavy Metals Pollution in the Environment

Heavy metals pollution is a hazardous environmental pollution concern throughout the world because of their highly toxic and nonbiodegradable nature (Adriano 2001). The major heavy metals contaminants include cadmium (Cd), chromium (Cr), mercury (Hg), lead (Pb), nickel (Ni), and arsenic (As) (USEPA 1997; Lasat 2002).

Heavy metals are present naturally in the soil at various levels and other sources are anthropogenic activity (Angelone and Bini 1992). Generally, most of the plant requires essential metals for cells (e.g., Mn, Ni, Zn Cu, Fe), and all heavy metals are toxic in nature at higher concentrations (Marschner 1995). All heavy metals may become highly toxic because they may cause oxidative stress (Pilon-Smits and Pilon 2002). The natural geological processes such as volcanic eruptions, continental dusts, as well as anthropogenic activities like mining and smelting, combustion of fossil fuels, phosphate fertilizer, military operations, discharge of wastewaters from different industries like leather industries, pulp and paper industries, distillery industries, and small-scale industries like battery production industries, etc., lead to the emission of heavy metals in the environment (Hooda 2007).

The area of the detrimental effects of heavy metals is dependent not only on the total metals concentration but also on the bioavailability of heavy metals (Kim and Owens 2010). Heavy metals enter the food chain by bioaccumulation and biomagnification and cause a variety of toxic effects in living organisms and the ecosystem. It is now well documented that several human diseases or dysfunctions have resulted from chronic exposure to heavy metals. Exposure to Cd (itai-itai disease), Hg (Minamata disease), As (Arsenicosis disease), and Pb (Pb poisoning), as well as exposure to acute Cd and Zn concentrations often results in gastrointestinal and respiratory, heart, brain, and kidney damage (Vassilev et al. 2004).

More than 50,000 metal-contaminated sites await remediation in the United States alone (Ensley 2000). Approximately 80% of U.S. Superfund sites (designated by the U.S. Environmental Protection Agency as priority sites for cleanup) contain heavy metals, often mixed with organic pollutants (Ensley 2000). Conventional *in situ* and *ex situ* remediation methods for organic and inorganic environment contaminants include soil washing, excavation and reburial of soil, and pump and treat systems for water (Glass 1999). The current costs concerned with U.S. metal contamination site remediation are $7 to $8 billion per year, of which ~35% involves metals remediation (Glass 1999, 2000). The use of green plants for heavy metal soil contamination remediation in a sustainable manner offers an attractive alternative, because it is solar driven and can be carried out *in situ*, minimizing cost and human exposure (Salt et al. 1998).

In the environment, heavy metals enter naturally from natural as well as anthropogenic sources, which are discussed below.

#### 11.2.1.1 Natural Sources

The most important natural source of heavy metals contamination in the environment is the geological parent material or rock outcroppings. The composition and concentration of heavy metals largely depend on the rock type and environmental conditions activating the weathering process. The geological plant materials generally have high concentrations of Cr, Cd, Cu, Zn, Ni, Pb, Hg, Co, Mn, and Sn.

However, the class-wise heavy metals concentrations vary within the rocks. Sedimentary rocks are mainly responsible for soil formation and contribute only a minute quantity of toxic metals to the environment, since these are not generally or easily weathered (Nagajyoti et al. 2010). Within the class of sedimentary rocks, shale has the highest concentrations of Cr, Mn, Co, Ni, Cu, Zn, Cd, Sn, Hg, and Pb followed by limestone and sand stone (Nagajyoti et al. 2010). But, many igneous rocks such as olivine, augite, and hornblende are reported to contribute a considerable amount of Mn, Co, Ni, Cu, and Zn into the soil (Nagajyoti et al. 2010).

#### 11.2.1.2 Anthropogenic Sources

The anthropogenic input of various toxic metals into the environment has increased drastically since the industrial revolution because different metals are widely used in various industrial processes. Various industries such as metallurgical, chemical, refractory brick, leather, wood preservation, pigments, and dyes are the major sources of toxic metals contamination in the environment. In this way, millions of people worldwide working in various industries such as pigment production, chrome plating, stainless steel welding, and leather tanning are being exposed to various toxic metals leading to the development of various types of serious environmental problems and health hazards.

### 11.2.2 Toxic Effects of Heavy Metals in the Environment

Industrialization and extraction of natural resources have resulted in large-scale environmental contamination and severe health hazards to animals and human beings. The contamination of soils, groundwater, sediments, surface water, and air with toxic metals and chemicals is one of the major problems facing the world today. The need to remediate these natural resources (soil, water, and air) has led to the development of new technologies that emphasized mainly the destruction of pollutants rather than the conventional approaches of their safe disposal into the environment because of their potential to enter the food chain.

Anthropogenic activities like metalliferous mining and smelting, agricultural, waste disposal, or industry discharge add a significant amount of different toxic metals such as Cu, Cd, Cr, Zn, Ni, Hg, Pb, Co, Pd, and As into the environment, and if these metals enter the food chain cause severe health hazards. However, some of these metals like arsenic, copper, iron, and nickel, are also required in the body in trace amounts for various metabolic activities, but their high concentration causes various cytotoxic, carcinogenic, mutagenic, and various metallic disorders in living organisms as shown in Table 11.1. Plants also require some metals in very small amounts for their proper growth and optimum performance, but the increasing concentration of several metals in soil and water ecosystems due to various anthropogenic activities has created an alarming situation for all life forms.

**TABLE 11.1**
**Heavy Metals, Sources, and Their Toxicity**

| Heavy Metal | Sources of Contamination | Toxicity |
|---|---|---|
| Cadmium (Cd) | Industrial wastewater and dusts, fumes, and polluted food | Glucosuria, osteomalacia, aminoaciduria, and emphysemia |
| Chromium (Cr) | Industrial wastewater like lather industries, industrial dusts, and polluted food | Ulcer and respiratory disease, cancer |
| Lead (Pb) | Industrial wastewater like sewage wastewater and dusts, fumes, and polluted food | Anemia, peripheral neuropathy |
| Arsenic (As) | Industrial wastewater and dusts; production pesticides, herbicides, and insecticides | Perforation of nasal septum, respiratory cancer, peripheral neuropathy, dermatoses, skin cancer |
| Mercury (Hg) | Industrial wastewater and fumes, vapor, and polluted food | Chronic rhinitis and sinusitis, respiratory cancer, dermatitis |

*Source:* Bharti, R., *Int. J. Sci. Res. Publ.*, 4, 1–5, 2014.

### 11.2.2.1 Cadmium and Its Toxic Effects

Cadmium occurs naturally in ores together with zinc, lead, and copper. Cadmium compounds are used as stabilizers in PVC products, color pigments, several alloys, and most commonly in rechargeable nickel and cadmium batteries. Metallic cadmium has been mostly used as an anti-corrosing agent (cadmiation). Cadmium is also present as a pollutant in phosphate fertilizers. Cadmium-containing products are rarely recycled but are frequently dumped together with household wastes, thereby contaminating the environment, especially if the waste is incinerated.

The inhalation of cadmium fumes or particles can be life threatening. Although acute pulmonary effects and deaths are uncommon, sporadic cases are still occurring (Seidal et al. 1993; Barbee and Prince 1999). Cadmium exposure may also cause kidney damage. The first sign of the renal lesion is tubular dysfunction, which is evidenced by an increased excretion of low molecular weight proteins such as β2-microglobulin and α1-microglobulin (protein HC) or enzymes such as *N*-acetyl-β-D-glucosaminidase (NAG) (WHO 1992; Jarup et al. 1998). It has been suggested that tubular damage is reversible, but there is overwhelming evidence that the cadmium-induced tubular damage is indeed irreversible (Jarup et al. 1998).

According to WHO, a urinary excretion of 10 mol/mmol of creatinine (corresponding to 200 mg of Cd/kg kidney cortex) would constitute a "critical limit" below which the kidney damage would not occur (WHO 1992). Several reports have shown that lower cadmium levels may also cause kidney damage and/or bone effects (Jarup 2003).

### 11.2.2.2 Mercury and Its Toxic Effects

Mercury is a chemical element with symbol Hg and atomic number 80. Metallic mercury is used in thermometers, barometers, and instruments used to measure blood pressure. Mercury is largely used in the electrochemical process of chlorine manufacturing, where mercury is used as an electrode in the chlor-alkali industry.

Acute mercury exposure may give rise to lung damage. Chronic poisoning is characterized by neurological and psychological symptoms such as tremor, changes in personality, restlessness, anxiety, sleep disturbance, and depression (Jarup 2003). The symptoms are reversible after cessation of exposure. Because of the blood–brain barrier, there is no central nervous system involvement related to inorganic mercury exposure. Metallic mercury may cause kidney damage, which is reversible if exposure has stopped. It has been also possible to detect proteinuria at relatively low levels of occupational exposure.

#### 11.2.2.3 Lead and Its Toxic Effects

Lead is a chemical element in the carbon group with symbol Pb, atomic number 82, atomic mass 207.2, density 11.4 g $cm^{-3}$, melting point 327.4°C, and boiling point 1750°C. It is a naturally occurring, bluish gray metal usually found in the form of minerals combined with other elements such as sulfur (i.e., PbS, $PbSO_4$) or oxygen ($PbCO_3$), and its concentration ranges from 10 to 30 mg $kg^{-1}$ in the earth's crust. Lead is a soft, malleable, and heavy post-transition metal. Metallic lead has a bluish white color after being freshly cut, but it soon turns into a dull grayish color when exposed to air. The general population gets exposed to lead from air and food (Jarup 2003). However, occupational exposure to inorganic lead mainly occurs in mines and smelters as well as welding of lead-painted metal, and in battery plants; however, low or moderate exposure may take place in the glass industry. High levels of air emissions may pollute areas near lead mines and smelters. Airborne lead can be deposited on soil and water and, thus, finally reaches into the human or animal body.

The symptoms of acute lead poisoning are headache, irritability, abdominal pain, and various disorders related to the nervous system (Steenland and Boffetta 2000). Lead encephalopathy is characterized by sleeplessness and restlessness. Children may be affected by behavioral disturbances and learning and concentration difficulties. In severe cases of lead encephalopathy, the affected person may suffer from acute psychosis, confusion, and reduced consciousness. People who have been exposed to lead for a long period of time may suffer from memory deterioration, prolonged reaction time, and reduced ability to understand. Individuals with average blood lead levels under 3 µmol/L may show signs of peripheral nervous symptoms with reduced nerve conduction velocity and reduced dermal sensibility (Jarup 2003).

#### 11.2.2.4 Arsenic and Its Toxic Effects

Arsenic is a widely distributed metalloid that occurs in rock, soil, water, and air. Inorganic arsenic is a metalloid present in group VA and period 4 of the periodic table that occurs in a wide variety of mineral ores as $AS_2O_3$ and can be recovered by processing ores containing mostly Cu, Pb, Zn, Ag, and Au. It is also present in ashes from coal combustion. Arsenic has the following properties: atomic number 33, atomic mass 75, density 5.72 g $cm^{-3}$, melting point 817°C, and boiling point 613°C, and exhibits fairly complex chemistry and can also be present in several oxidation states (−III, 0, III, V). In aerobic environments, As(V) is dominant, usually in arsenate ($AsO_4^{3-}$) forms in various protonation states: $H_3AsO_4$, $H_2AsO_4^-$, $HAsO_4^{2-}$, and $AsO_4^{3-}$. Arsenate and other anionic forms of arsenic behave as chelator and can precipitate in the presence of metal cations.

Since arsenic is often present in anionic form, it does not form complexes with simple anions such as $Cl^-$ and $SO_4^{2-}$. Arsenic speciation also includes organ metallic forms such as methyl arsenic acid $(CH_3)AsO_2H_2$ and dimethyl arsenic acid $(CH_3)_2AsO_2H$. Many As compounds adsorb strongly to soils and are therefore transported only over short distances in groundwater and surface water. Arsenic is reported to cause skin damage, increased risk of cancer, and problems with the circulatory system.

The inorganic form of arsenic is acutely toxic, and intake in large quantities may lead to gastrointestinal symptoms, severe disturbances in cardiovascular and central nervous systems, and even death. Arsenic exposure through drinking water is reported to cause skin, lung, kidney, and bladder cancers (WHO 2001). In an affected person, the skin cancer is preceded by directly observable precancerous lesions. Uncertainties in the estimation of past exposures are important while assessing the exposure-response relationships, but the arsenic contamination in drinking water at the level of 100 μg/L leads to the development of cancer, whereas the concentration of arsenic from 50 to 100 μg/L is found to be associated with the precursors of skin cancer (Jarup 2003). The relationships between arsenic exposure and health effects are less clear. There is strong evidence for hypertension and cardiovascular disease, but the evidence is only suggestive for diabetes and reproductive effects and weak for cerebrovascular disease, long-term neurological effects, and cancer at sites other than lung, bladder, kidney, and skin (WHO 2001).

#### 11.2.2.5 Chromium and Its Toxic Effects

Chromium is a *d*-block transition metal placed in group VIB in the periodic table and has atomic number 24, atomic mass 52, density 7.19 g $cm^{-3}$, melting point 1875°C, and boiling point 2665°C. It is one of the less common elements and does not occur naturally in elemental form but only in compounds. Chromium is mined as a primary ore product in the form of mineral chromate ($FeCr_2O_4$). The major sources of Cr contamination in the environment include wastewater discharge from the tannery industry (Saxena and Bharagava 2015; Saxena et al. 2016). In nature, chromium mainly exits in trivalent and hexavalent forms. Chromium is a naturally occurring heavy metal that exists in air, water, soil, and food. It is now considered as one of the major environmental pollutants due to its ecological, nutritional, and environmental toxicity.

Cr(VI) can enter the body when people breathe air, eat food, or drink water contaminated with it. Cr(VI) is also found in house dust and soil, which can be ingested or inhaled. Of the various forms of chromium, Cr(VI) is the most common form and is toxic in nature. Many Cr(VI) compounds have been found to be carcinogenic in nature, but the evidence to date indicates that the carcinogenicity is site specific and limited to the lung and sino-nasal cavity and dependent on exposure intensity (Salem et al. 2000; Saxena et al. 2016). Inhaling a relatively high concentration of Cr(VI) can cause a runny nose, sneezing, itching, nosebleeds, ulcers, and holes in nasal septum. Short-term high-level inhalational exposure can cause adverse effects at the contact site, including ulcers, irritation of the nasal mucosa, and holes in nasal septum. Ingestion of very high Cr(VI) doses can cause kidney and liver damage, nausea, irritation of the gastrointestinal tract, stomach ulcers, convulsions, and death. While dermal exposures may cause skin ulcers or allergic reactions. Cr(VI) is one of the most highly allergenic metals, second to nickel, and studies on mice given high doses of Cr(VI) have shown reproductive abnormalities including reduced litter size and decreased fetal weight (ATSDR 2000).

## 11.3 TYPES OF PHYTOREMEDIATION

Phytoremediation as a discipline in environment sciences was established in the late 1970s following the discovery of a series of hyperaccumulators (Brooke et al. 1977). Root exudation is the main mechanism providing the carbon source in a plant. In this mechanism, they stimulate the degradation of organic chemicals in the rhizospheric region by the release of root exudates and enzymes and build up the organic carbon in the soil (Mukhopadhyay and Maiti 2010).

The term "phytoremediation" consists of the Greek prefix *phyto* meant for plant, attached to the Latin root *remedium* meant to cure or remedy. The concept of phytoremediation was emerging in the 1980s from the ability some plant species displayed in accumulating high toxic heavy metals concentrations in their tissues or organs. Along the years, a number of related technologies were developed that enabled the practical application of higher plants to decontaminate soil and water, and then "phytoremediation" began to be used in the scientific literature in 1993. The definition later evolved into "phytotechnologies" (ITRC 2001), meaning different phytoremediation technologies, which can be applied to remediate organic and inorganic pollutants like heavy metal soil contamination through (1) stabilization; (2) volatilization; (3) metabolism, including rhizospheric degradation with microbial interaction, where rhizospheric degradation and root exudates provide the carbon source; and (4) accumulation and sequestration (McCutcheon and Schnoor 2003). Xenobiotic metabolism in humans, animals, and higher plants usually occurs through three main biochemical processes: conversion or transformation (phase I), conjugation (phase II), and compartmentalization (phase III) (Schmidt et al. 2006). In humans and animals, the liver is the main site of xenobiotic metabolism, where the normally nonpolar and lipophilic xenobiotics are metabolized into the more soluble forms, which are typically excreted in urine (Sandermann 1992).

Phytoremediation is an eco-friendly, noninvasive, plant-based green technology for *in situ* treatment of pollutants, accomplished by use of plants with their associated microbial interaction for uptake, sequestration, or volatilization of pollutants from contaminated soils, wastewater, and sediments (Chandra et al. 2015). This phytoremediation technology can be used for both organic and inorganic pollutants present in soil (solid substrate), wastewater (liquid substrate), or the air (Salt et al. 1998). The application of phytoremediation technology has been shown by many workers (Table 11.2). With phytoremediation, it is possible to grow plants in a polluted matrix, either artificially (constructed wetlands) or naturally, for a required growth period; to remove pollutants from the matrix, degradation (detoxification) of the pollutants, such as organic and inorganic pollutants, inorganic-polluted heavy metals, and organic pollutant PHA compounds, occurs (Chandra et al. 2015).

**TABLE 11.2**
**Different Plant-Based Technologies for Heavy Metal Remediation**

| Mechanism | Pollutant | Media | Plant | Status | References |
|---|---|---|---|---|---|
| Phytoextraction | Zn, C, $As^{137}$, and Cr | Soil | *Datura stramonium* and *Chenopodium murale* *Catharanthus roseus, Anogeissus latifolia* | Applied | Varun et al. (2012), Fulekar et al. (2010), Mathur et al. (2010) |
| Phytodegradation | As, Zn and Cd, Pb, U | Soil | *Cassia fistula, Vetiveria, Sesbania, Viola, Sedum, Rumex* *Jatropha curcas* L., *Brassica juncea* | Applied Field demo | Preeti et al. (2011), Mukhopadhyay and Maiti (2010), Mangkoedihardjo and Surahmaida (2008), Huhle et al. (2008) |
| Phytostabilization | Mn | Soil | *Chondrilla juncea* and *Chenopodium botrys* | Soil | Cheraghi et al. (2011) |
| Phytostabilization | Cd | Soil | Sunflower | Applied | Zadeh et al. (2008) |

*Sources*: Wani et al., *Journal of the Science of Food and Agriculture* 92(5): 1054–1063, 2012; Chandra, R., Saxena, G., and Kumar V., Phytoremediation of environmental pollutants: An eco-sustainable green technology to environmental management, in *Advances in Biodegradation and Bioremediation of Industrial Waste*, Ram Chandra (Ed.), CRC Press, Boca Raton, FL, 1–30, 2015.

Phytoremediation technology has the following characteristic features:

1. Phytoremediation is based on the phyto-techniques that make use of green plants to achieve environmental goals. These phyto-techniques use plants to extract, degrade, contain, or immobilize pollutants from contaminated soil, groundwater, surface water, and other polluted media. Phytoremediation remediates a broad range of pollutants by different mechanisms depending on the application, but all mechanisms are not applicable to all pollutants (McCutcheon and Schnoor 2003).
2. It cleans up moderate to low levels of selected elemental and organic pollutants over large areas.
3. It maintains sites by treating residual pollution after achieving cleanup.
4. It acts as a buffer against potential waste releases.
5. It aids voluntary cleanup efforts.
6. It facilitates nonpoint source pollution control.
7. It offers an effective form of monitored natural attenuation.

Phytoremediation efficiencies and uptake mechanisms of pollutants (either organic or inorganic) are greatly influenced by plant species and their characteristics, root–zone interactions, properties of medium, chemical properties of contaminants, bioavailability of contaminants, effects of added chelating agents, environmental conditions, etc. (Cunningham and Ow 1996; Tangahu et al. 2011). Hence, depending on the detoxification process, applicability, medium, type, and extent of pollution, phytoremediation processes can be classified as phytoextraction, rhizofiltration, phytostabilization (for inorganic contaminants), phytotransformation, rhizodegradation, and phytovolatilization (for organic contaminants) (Raskin and Ensley 2000; Sureshvarr et al. 2010; Ali et al. 2013).

### 11.3.1 Phytoextraction

Phytoextraction involves the removal of organic and inorganic contaminants like heavy metals translocated from root to shoot or leaves and subsequent storage of contaminants by the plant (Abhilash et al. 2009). Phytoextraction detoxifies soil heavy metal concentrations, higher to lower concentrations with a high capacity for metal accumulation in root to shoots or leaves (Barcelo and Poschenrieder 2003). Phytoextraction is root to shoots or leaves translocation of heavy metals. It is also known as phytoaccumulation, phytoabsorption, and phytosequestration. These green plants remove environmental pollutants like heavy metals or organics from soil by concentrating them in harvestable plant parts (Chandra et al. 2015). This process involves the following:

1. Cultivation of the suitable plant, more biomass/crop species on the polluted site
2. Removal of harvestable plant parts like whole plant containing heavy metal from site
3. Postharvest treatments (include composting, compacting, thermal treatments) to reduce the biomass volume and/or weight for disposal as a hazardous waste or for its recycling to recover valuable metals

Two types of phytoextraction processes have been suggested, continuous or natural phytoextraction and inoculation, enhanced or chemically assisted phytoextraction (Lombi et al. 2001). Continuous phytoextraction is the use of plants, usually hyperaccumulators that accumulate particularly high levels of the toxic heavy metals pollutants throughout their lifetime. The induced phytoextraction enhances toxin accumulation at a single time point by the addition of accelerants or chelators to soil and use of microbial consortium–microbial rhizospheric interaction.

Metals such as Cu, Ni, and Zn are best suitable for phytoremediation techniques for phytoextraction, as approximately 400 known plants that absorb unusually large amounts of metals have high affinity for accumulating these metals (Blaylock and Huang et al. 2000). The main factors limiting phytoextraction efficiency are as follows:

1. Soil heavy metals concentration phytoavailability
2. Metals translocation to aboveground plant parts, root to shoot or leaves

There are two important factors that should be considered when evaluating the potential of a plant as phytoextractor: bioconcentration and biomass production. The former is defined as the ratio between the concentration of the pollutant in the shoot and in the soil. It serves as an indicator of the capacity of a plant to accumulate toxic compounds. Biomass production is also critical in order for phytoextraction to be commercially viable, since it decreases the number of crops required to complete remediation of a given site (McGrath and Zhao 2003).

### 11.3.2 Rhizofiltration

Rhizofiltration technology is generally used for wastewater treatment by using aquatic plants, floating plants, submersed plants, and terrestrial plants. This is an innovative green and eco-friendly technology. It is used for removing organic and inorganic pollutants, like heavy metal–polluted surface water or wastewaters such as domestic wastewater, industrial discharge, agricultural runoff, or acid mine drainage by adsorption or precipitation of metals onto roots or absorption by roots or other submerged organs of metal-tolerant aquatic plants. Wastewater heavy metals contamination translocates from root to shoot or leaves (Jadia and Fulekar 2008). Rhizofiltration technology is similar to phytoextraction, but is mainly related to the remediation of contaminated water rather than polluted soils, groundwater, surface water, and industrial wastewater. In the process of attempting to enhance rhizofiltration, it was discovered that young plant seedlings grown in aerated water (aqua cultured) are often more effective than roots in removing heavy metals from water (Salt et al. 1997). The advantages of rhizofiltration are as follows:

1. The ability to use both terrestrial and aquatic plants is subdivided into floating and submerged plants for either *in situ* or *ex situ* applications.
2. Pollutants do not have to be translocated to the shoots, but inorganic contaminants like heavy metal contamination can be translocated root to shoot or leaves.

Some terrestrial plants like Indian mustard are more efficient in removing Cd, Cr, Cu, Ni, Pb, and Zn, whereas sunflower absorbs Pb and U (Dushenkov et al. 1997) from hydroponic solutions. Indian mustard could effectively remove a wide range (4–500 mg/L) of Pb concentration (Raskin and Ensley 2000).

### 11.3.3 Phytostabilization

Phytostabilization, the plant-based technology, uses metal-tolerant plants to stop the mobility of heavy metals, thus minimizing the risk of further environmental contamination by leaching into groundwater (Salt et al. 1995). Phytostabilization processes inactivate the soil contaminants through plant roots and are primarily used for the remediation of soil contaminants, sediment, and sludge (USEPA 2000). In phytostabilization technology, plants minimize soil erosion, greatly minimizing surface water runoff and sediment available to receiving streams. Phytostabilization technology has advantages including that disposal of hazardous material/biomass is not required (USEPA 2000) and it is very effective when rapid immobilization is needed to prevent contamination of ground surface waters and soil. However, the major disadvantage is that the contaminants always remain in soil and therefore require regular monitoring.

### 11.3.4 Phytovolatilization

Phytovolatilization technology involves translocation of soil organic contaminants from root to shoots then to leaf. Phytovolatilization is a process in which plants uptake organic and inorganic contaminants from soil and release them in volatile form into the atmosphere through the transpiration mechanism. As water travels from the roots to the shoot or leaves along the vascular system of the plants, it is changed and modified along the way. Then, some of the contaminants move through the plants to shoot or leaves and evaporate into the atmosphere (Sakakibara et al. 2007). The phytovolatilization mechanism has been observed for both organic and inorganic contaminants including trichloroethylene (TCE), which has been observed in the off-gas from plant leaves in laboratory and field experiments (Compton et al. 1998) as well as in the production of volatile elemental mercury by genetically engineered *Arabidopsis thaliana* grown in the presence of ionic mercury (Rugh et al. 1996; Bizily et al. 1999).

Phytoextraction and phytovolatilization of metals by hyperaccumulating plants offer a viable remediation on commercial projects (Sakakibara et al. 2007). It has been primarily used for mercury removal where the mercuric ion is converted into less toxic elemental Hg (Ghosh and Singh 2005). It is the most controversial of all phytoremediation technologies since some metals, like Se, Hg, and As, may exist in gaseous state in the environment. It is also reported that some naturally occurring or genetically modified plants, like *Chara canescens* (musk grass), *Brassica juncea* (Indian mustard), and *Arabidopsis thaliana,* possess the capability to absorb heavy metals and convert them to gaseous state within the plant and subsequently release them into the atmosphere (Ghosh and Singh 2005).

### 11.3.5 Phytodegradation

Phytodegradation is a plant-based technology for the transformation of one form to another form, higher concentration to lower concentration forms via rhizospheric microorganisms through the action of enzymatic reaction, secreted within plant tissue and plant root exudates (Suresh and Ravishankar 2004). The metabolic processes involved in phytodegradation have strong similarities to those used by animals for modification and degradation of drugs and other toxins (Abhilash et al. 2009). This has given rise to a conceptual model for phytodegradation known as the "green liver" model (Sanderman 1994). Plants are able to produce enzymes that catalyze and enhance the organic compound degradation. Plant exudates and plant tissue produce some enzymes, such as dehalogenase and oxygenase, which help in degradation of organic pollutant. It is independent of enzymatic activity of rhizospheric microorganisms. Some plant enzymes have been identified, which are involved in conversion of ammunition wastes, chlorinated solvents such as TCE (trichloroethylene), and others, which degrade organic herbicides (Newman et al. 1997). Plant enzymes that metabolize contaminants may be released into the vicinity of rhizosphere, where they may participate in pollutant transformation of one form to another from higher concentration to lower concentration. Enzymes such as nitro-reductase, dehalogenase, peroxidase, nitrogenase, and laccase have been discovered in plant sediments and soils (Suresh and Ravishankar 2004).

### 11.3.6 Rhizodegradation

Rhizodegradation is an indirect mechanism that involves the conversion of one form of pollutant to another form by microbes present in the plant rhizospheric region. Microbial interaction in plant rhizospheric region showed more microbial diversity (Abhilash et al. 2009). Rhizodegradation is a process by which plant root exudates supplied carbon source substrates, stimulate microbial diversity in plant root zones (rhizospheric region) for removal of undesirable organic and inorganic contaminants in soil. Rhizodegradation, also known as phytostimulation, is the degradation of organic and inorganic contaminants in the rhizosphere (area of soil surrounding the roots of the plants) by means of microbial activity, which is promoted by the presence of plant roots. Phytodegradation is a rhizospheric degradation that involves the enzymatic conversion of organic compounds from higher concentration to lower concentration, through rhizospheric microbial enzymatic activity (Jones et al. 2004; Kirk et al. 2005).

## 11.4 MECHANISM OF HEAVY METAL ACCUMULATION IN PLANTS

Good long-term planning for promoting phytoextraction is to understand and exploit the biological processes involved in metal acquisition, transport, and roots to shoot or leaves accumulation (Raskin et al. 1997). New advances in plant biotechnology provide the means to rapidly subsidize the mechanical possibilities of phytoextraction. Until now, we have had less awareness about the role of the microbial mechanism in phytoremediation. Green plants are inherently adapted to hyperaccumulate heavy metals from soil-contaminated sites. The process of metal uptake/hyperaccumulation by a heavy metal-accumulating plant has several steps by which the metals transport from the

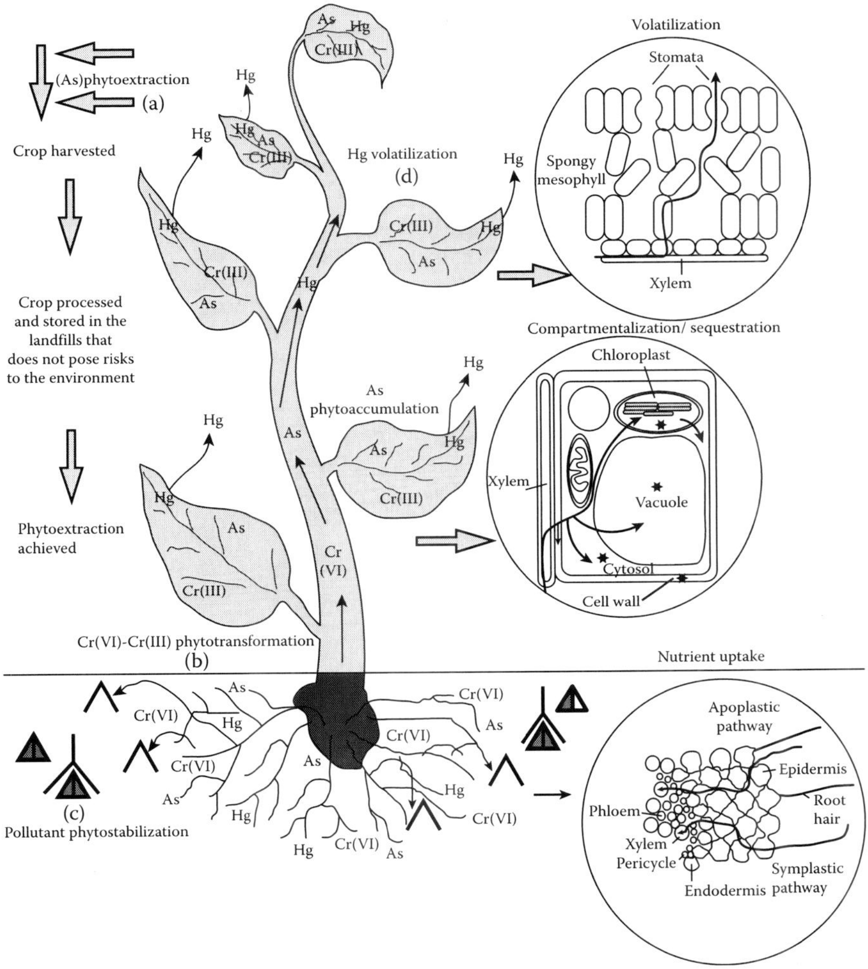


**FIGURE 11.1** Different mechanisms of phytoremediation: (a) phytoextraction of As from the soil to aerial parts of the plant (leaves and stems), (b) phytotransformation of Cr(VI) from the soil to Cr(III) in the aerial parts of the plant, (c) phytostabilization of metal contaminants in soil, and (d) phytovolatization of Hg from the soil.

soil into the plant roots to shoot or leaves and are finally stored in cell vacuoles (Ho-Man et al. 2013). The phytoextraction mechanism used by plant-based technology for the accumulation of heavy metals from contaminated sites is accomplished in many steps (Mukhopadhyay and Maiti 2010) as shown in Figure 11.1.

### 11.4.1 Metal Solubilization from Soil Matrix

Heavy metal soil contaminants are a major worldwide environmental problem with implications for human health. The presence of metals in soil has two main origins:

anthropogenic activity and natural activity, the major cause of high levels of metals in mining operations by anthropogenic activity (Hutton and Symon 1986; Battarbee et al. 1998; Nriagu 1989). Generally in soil the metals occur in insoluble form, and therefore plants are not able to uptake these from soil. Plant roots interaction with microbial diversity can significantly alter the physicochemical characteristics of the rhizosphere by their metabolic activities, for example, metabolite excretion (Whitelaw 2000). The rhizospheric fungi can alter the composition and properties of soil as a result of metabolic processes. The bioavailability of toxic heavy metals in soil can be enhanced to a considerable extent by the fungi in soil (Ho-Man et al. 2013). However, two main actions performed by green plants to solubilize the metals from soil include rhizospheric acidification due to acts of plasma membrane proton transport and secretion of metal chelators (Lasat 2000; Mukhopadhyay and Maiti 2010).

### 11.4.2 Metal Uptake into the Plant Root

Plant roots account for ~20%–50% of plant biomass (Raskin et al. 1997). Environmentally hazardous heavy metals are too insoluble to move freely in the plant vascular system. To absorb the soil-bound heavy metals, plants have to mobilize them into the soil solution and translocate them from roots to shoot or leaves. Generally, soluble metals penetrate the plant root via two pathways: symplast and apoplast. Xylem is the most important vascular tissue that helps in the movement of solutes upward by passage of the plasma membrane of root endodermal cells through the apoplastic way. Once the metal is in the xylem, the stream of the xylem sap shoots it to the cells of the leaves (Peer et al. 2005). However, the excluder plants survive by using the strategy to exclude or pump the toxic metals outside of the plant (Hall 2002).

### 11.4.3 Accumulation of Heavy Metals Root to Shoot

Heavy metals are translocated within the plant mainly via the xylem of the plant by the action of a membrane pump or channel. After the loading of solutes into the xylem, the metals are transported to the cells of leaves by crossing the membrane through the movement of xylem sap. However, the types of cells wherein the metals are loaded vary among the hyperaccumulator plants (Oropeza-Garcia et al. 2014). The hyperaccumulators show root proliferation in localities with high metal concentrations, increased release of metals in insoluble forms, increased release of metals desorbed to soil particles, and increased flux into the symplasm, but decreased flux into the root vacuole and/or increased flow from root symplasm to xylem, and uptake from xylem to leaf symplasm and/or increased transfer into the leaf vacuoles (Macnair 2003). The toxic metals accumulation ability of plants may be promoted throughout the use of express metal-chelating agents by enhancing the obtainment and transport of metals from roots to shoot or leaves.

### 11.4.4 Metal Detoxification and Chelation

Plants possess a variety of processes for the transformation of highly toxic metals by complications into less toxic forms and from higher concentrations to lower concentrations.

Plants have different uptake, transport, and sequestration mechanisms for different oxidation states of toxic metals. Within the plant cells, metals with various oxygen states can undergo transformation from one form to another form, from higher concentration to lower concentration or by complexation (Peer et al. 2005). However, the effects of toxins chelation by endogenous plant compounds can have similar effects on these properties. Many chelators use thio group as ligands, and sulfur (S) biosynthetic pathways are critical for metal hyperaccumulators and their ensuing phytoremediation (Van Huysen et al. 2004).

### 11.4.5 Metal Sequestration and Its Volatilization

Metal sequestration is the last step followed by volatilization. It mainly takes place in the vacuole present in a plant, wherein metal-ligand complexes are transported across the vacuolar membrane. For example, metal compartmentalization may enable *Pteris vittata* (an As hyperaccumulator) to tolerate exposure to As and to maintain normal growth after having absorbed a relatively large amount of As (Chen et al. 2005). The possible reason is that the vacuole may suffer the effects of maximum concentrations of As on cell metabolism in order to assure the normal growth and development of the plants (Ho-Man et al. 2013). However, the heavy metals may also remain in the cell wall rather than cross the plasma membrane into the cell as the negatively charged sites on the cell walls may interact with polyvalent cations (Wang and Evangelou 1994).

## 11.5 Plants Used for Phytoremediation of Heavy Metals

A number of green plants including herbs, shrubs, and trees (both terrestrial and aquatic) have been reported to be endowed with magnificent abilities for restoration and reclamation of contaminated environments (Sinha et al. 2007). The selection of plants is a main factor in the supernova of remediation of heavy metal-contaminated places. Plants possess remarkable potential for the accumulation of high concentrations of higher toxic metals from a contaminated place. Plants accumulating appetency concentration of heavy metals in the shoot or leaves (i.e., 100- to 1000-fold without showing any visible symptoms are known as hyperaccumulators (Baker and Whiting 2002).

However, the choice of indigenous plant species for phytoremediation is a tough assignment. The required criteria for plants to be suitable for phytoremediation should include the capability to hyperaccumulate heavy metals, ability to withstand high salt concentration and pH, fast growth rate, high biomass, be easily harvestable, and must uptake and translocate metals to roots to shoot or leaves (Sharma 2011). But, insufficient biomass and growth rate are major problems associated with hyperaccumulators (Kramer and Chardonnens 2001). This problem can be overcome by transferring desired traits of hyperaccumulators into high biomass plants by the application of genetic engineering (Kramer and Chardonnens 2001). Baker et al. (2000) reported different species as hyperaccumulators based on their capacity to tolerate toxic metals concentration (see Table 11.3). In addition, the search for suitable plants for

**TABLE 11.3**
**Types of Metal Hyperaccumulator Plant Species**

| Species | Metal | References |
|---|---|---|
| *Biden spilosa* | Cd | Sun et al. (2007) |
| *Solanum nigrum* L. | Cd | Sun et al. (2008) |
| *Sedum alferedii* | Cd | Sun et al. (2009) |
| *Brassica juncea* | Ni and Cr | Saraswat and Rai (2009) |
| *Thlaspi caerulescens* | Cd, Zn, and Pb | Banasova et al. (2008) |
| *Helicotylenchus indicus* | Pb | Sekara et al. (2005) |
| *Alyssum lesbiacum* | Ni | Cluis (2004) |
| *Pistia stratiotes* | Zn, Pb, Ni, Hg, Cu, Cd, and Cr | Odjegba and Fasidi (2004) |
| *Pityrogramma calomelanos* | As | Dembitsky and Rezanka (2003) |
| *Thordisa villosa*, *Croton bonplandianus* | Cu | Rajakaruna and Böhm (2002) |

*Sources*: Ahmadpour, P., et al., *African Journal of Biotechnology*, 11(76), 14036–14043, 2012; Chandra, R., et al., Phytoremediation of environmental pollutants: An eco-sustainable green technology to environmental management, in *Advances in Biodegradation and Bioremediation of Industrial Waste*, Ram Chandra (Ed.), CRC Press, Boca Raton, FL, 1–30, 2015.

phytoremediation of toxic metals is required and should be tested in the field (Kramer 2003). In order to distinguish the hyperaccumulators from typical plants or accumulators, a set of threshold values of elemental concentrations in plant biomass (dry weight) are used to define the hyperaccumulation: Mn and Zn hyperaccumulating plants contain greater than 10,000 mg/g (Reeves and Baker 2000), hyperaccumulators of As, Co, Cu, Ni, Se, and Pb have greater than 1000 mg/g (Reeves and Baker 2000), whereas the hyperaccumulators of Cd have greater than 100 mg/g (Reeves and Baker 2000). Selection has distinguished Indian mustard seedlings as particularly productive in sorbing the divalent cations of toxic metals (Salt et al. 1997). Vegetation-based remediation has an evident potential for the accumulation and immobilization of persistent contaminants in land (Ho-Man et al. 2013).

## 11.6 PHYTOREMEDIATION TECHNOLOGY: TASK OF PLANT GROWTH-PROMOTING RHIZOBACTERIA

Most of the microbial diversity remains in the rhizospheric region. Rhizospheric microorganisms, which are present in the rhizospheric region of plants, are known as rhizobacteria. Many microbial species of microbes like *Pseudomonas, Arthrobacter, Burkholderia, Bacillus, Serratia*, *Azospirillum, Azotobacter, Klebsiella, Enterobacter,* and *Alcaligenes* have been reported to promote the plant's growth and biomass (Kloepper et al. 1989; Glick 1995; Joseph et al. 2007).

Plant growth-promoting rhizobacteria (PGPR) have been reported to play an important role in soil contamination remediation, particularly in overcoming plant stress under flooded, high-temperature, and acidic conditions (Lucy et al. 2004). PGPR are being used in the agricultural field to increase plant food production as well as growth and protection of plants from pathogens causing disease. The harmful effects of heavy metals in aquatic life and soil from the environment on plants can be reduced with the use of PGPR (Joseph et al. 2007).

The PGPR enhance the phytoremediation of the heavy metals contaminated sites. Contaminated nutrient-poor places can be improved by actionable metal-tolerant microbes that provide key needed plant nutrients. Actionable metal-tolerant microbes therefore may be dynamic in promoting the detoxification of heavy metals contaminated soils (Glick 2003). These plant–microbe interactions can act in transforming organic pollutants from one form to another by improving phytoremediation. The rhizospheric microbes play an important role in maintaining the soil structure and fertility and play a role in biocontrol (control of plant pathogens) and in remediating contaminated environmental soils. The rhizospheric microbes, PGPR, phosphate-solubilizing bacteria (PSB), mycorrhizal-helping bacteria (MHB), and arbuscular mycorrhizal fungi (AMF), in the rhizospheric region of plants growing on metal-contaminated soils play a main role in the phytoremediation of toxic metals (Kloepper et al. 2004).

PGPR involve a multifarious group of free-living rhizospheric soil bacteria (around the plant root region) that can promote host plant growth and play a main role in mitigating the toxic effects of heavy metals on the plants (Belimov et al. 2004). PGPR microbial iron- siderophore complexes can be taken up by the plants, and thereby serve as an iron source for plants (Wang et al. 1993). Some PGPR can increase the growth and biomass of plants in the presence of heavy metals contaminated soil including nickel, lead, cadmium, and zinc (Burd et al. 2000), thus allowing plants to develop longer plant roots and become better established during early stages of growth (Glick et al. 1998). The major benefit in application of PGPR is their tolerance to the heavy metal concentrated soil. The effects of PGPR on plant growth and biomass and/or heavy metal-contaminated polluted soil are given in Table 11.4.

## 11.7 GENETICALLY MODIFIED PLANTS APPROACH FOR PHYTOREMEDIATION

The potential of genetic engineering to enhancing the bioremediation of xenobiotics has been recognized since the early 1980s with initial attempts being focused on microorganisms (Rugh et al. 1998; Rosser et al. 2001; Sung et al. 2003; Rugh 2004; Singleton 2007; Doty 2008). To overcome the possible drawbacks of phytoremediation technology, the genetic engineering may be a suitable approach. The efficiency of plants for phytoremediation of toxic metals can be improved by using genetic engineering technologies. Transgenic plants for phytoremediation were first developed for the remediation of heavy metals contaminated sites, for example, *Nicotiana tabacum*, expressing a yeast metallothione gene for higher tolerance to cadmium, or *Arabidopsis thaliana* overexpressing a mercuric ion reductase gene for higher tolerance to mercury (Misra and Gedama 1989; Rugh et al. 1996).

Genetic engineering can integrate genes from other organisms to improve phytoremediation ability in plants (Ruiz and Daniell 2009). Genetic engineering may help

**TABLE 11.4**
**Major Role of Plant Growth-Promoting Rhizobacteria (PGPR) in the Plants for Heavy Metal Contamination Remediation**

| PGPR | Plant | Heavy Metal | Effect (s) | References |
|---|---|---|---|---|
| *Bradyrhizobium* sp., *Pseudomonas* sp., *Ochrobactrum cytisi* | *Lupinus luteus* | Pb, Cu, Cd | Decreased metal accumulation; however, plant biomass increased | Dary et al. (2010) |
| *Bacillus subtilis*, *Bacillus cereus*, *Pseudomonas aeruginosa*, *Flavobacterium* sp. | *Orychophragmus violaceus* | Zn | Increased shoot biomass and Zn accumulation | He et al. (2010) |
| *Pseudomonas aeruginosa*, *Ralstonia metalidurans* | Maize | Cr, Pb<br>Cr | Increased the uptake by shoot by a factor of 5.4 and 3.4, respectively<br>Increased accumulation of Cr in shoots by a factor of 5.2 | Braud et al. (2009) |
| *Achromobacter xylosoxidans* strain A×10 | *Brassica juncea* | Cu | Increased the length of root and shoot, fresh and dry weight significantly, and extensively improved the Cu uptake of *B. juncea* plants as compared to the control | Ma et al. (2009) |
| *Microbacterium* sp. G16, *Pseudomonas fluorescens* G10 | Rape | Pb | Increased root elongation of inoculated rape seedlings and total Pb accumulation as compared to the control plants | Sheng et al. (2008) |
| *Burkholderia* sp. J62 | Maize and tomato | Cd, Pb | Increased the biomass of maize and tomato plant significantly; the increased Pb and Cd content in tissue varied from 38% to 192% and from 5% to 191%, respectively | Jiang et al. (2008) |
| *Bradyrhizoium* sp. RM8 | Green gram var. K851 | Ni, Zn | Increased plant growth and decreased uptake of heavy metals by plant | Wani et al. (2007) |

*Source*: Karami, A., and Shamsuddin, Z.H., *African Journal of Biotechnology*, 9(25), 3689–3698, 2010.

the plants to develop tolerance against atmospheric factors as well as high concentration of toxic metals. New scholars are focused on understanding the significance of plant–microbe interactions in the transformation of organic pollutants (Doty 2008). However, the application of metal hyperaccumulators in constructed wetlands to treat municipal and industrial wastewater has been increasing in the recent years. Genetically modified plants reported for the detoxified heavy metals are presented in Table 11.5. A promising biotechnology approach for promoting phytoremediation is to overexpress genes, whose protein substrates are involved in heavy metals being uprooted, transported, and sequestered, transforming organic pollutants from one form to another. It is clear that plants can be transgenic to improve their tolerance, uprooted, translocated, and have accumulation and vaporization ability for heavy metals phytoremediation.

## 11.8 ADVANTAGES AND DISADVANTAGES OF PHYTOREMEDIATION TECHNOLOGY

Plants may accumulate higher concentrations of hazardous wastes than many microorganisms used for bioremediation (Abhilash et al. 2009). The major advantage of phytoremediation is that it is a cost-effective technology that is plant based or green technology. It can be a sustainable and promising, less expensive technology for use in remediating heavy metals contaminants in soil (both organic and inorganic) that might otherwise be leached out from soil through rainwater and contaminate the nearby watercourses. This technology is favorable for large spaces in which other approaches would be expensive and ineffective (Vidali 2001). The estimated cost for phytoremediation of metal-contaminated soil is in the range of US$25–$100 per ton (Glass 2000) in contrast to approximately US$150–$350 per ton for conventional excavation landfilling approaches (Ensley 2000). In addition, phytoremediation is beneficial, including carbon sequestration, soil stabilization, and residual biomass, as a possible source of biofuel production. However, the major barrier to the implementation of phytoremediation is that it is a new technology, which is at a nascent stage and is not yet fully developed (Vidali 2001; Rajakaruna et al. 2006). The main advantages and disadvantages of phytoremediation technology are briefed in Table 11.6.

## 11.9 PROSPECTS AND FUTURE CHALLENGES

Sustainable research is required for higher biomass and growth, enhancing heavy metals rooted up, translocation, root to shoot or leaves, and tolerance via genetically modified plants for effective phytoremediation. Phytoremediation, a green technology, is a less expensive, eco-friendly, and sustainable approach. It is a plant-based technology that uses terrestrial hyperaccumulators or aquatic plants (floating and submerged) to remove organic and inorganic pollutants, like heavy metals, by accumulating in their above-ground plant parts. In the commercialization of phytoremediation, there are many obstacles that need to be addressed, and the genetic capability of plants needs to be optimized to make this technology economically viable.

**TABLE 11.5**
**Genetically Engineered Plants for Phytoremediation with Other Additional Benefits**

| Transgenic Plant | Targeted Pollutant | Proposed Additional Benefits | References |
|---|---|---|---|
| *Brassica juncea* with ATP sulfurylase from *Arabidopsis thaliana* and SeCys methyltransferase (*SMT*) from *Astragalus bisulcatus* | Enhanced Se accumulation, tolerance, and volatilization | Biodiesel production, carbon sequestration | Dhankher et al. (2012) |
| Hybrid poplar (*Populus* seiboldii × *Populus grandidentata)* with manganese perodixase (*MnP*) gene from *Trametes versicolor* | Increased degradation of bisphenol A | Biomass for bioenergy, pulp, charcoal, carbon sequestration | Iimura et al. (2007) |
| *Populus deltoides* with bacterial mercuric ion reductase (*merA*) gene | Enhanced mercuric ion reduction and resistance | Biomass for bioenergy, pulp, charcoal, carbon sequestration | Che et al. (2003) |
| *Populus canescens* overexpressing γ-glutamylcysteine synthetase | Tolerance to Zn stress | Biomass for bioenergy, pulp, charcoal, carbon sequestration | Bittsanszkya et al. (2005) |
| Hybrid aspen (*Populus tremula* × *Populus tremuloides*) expressing bacterial nitroreductase (*pnrA*) | Enhanced bioremediation of TNT | Biomass for bioenergy, pulp, charcoal, carbon sequestration | Van Dillewijin et al. (2008) |
| Hybrid poplar (*Populus tremula* × *Populus alba*) | Removal of TCE, vinyl chloride, $CCl_4$, benzene, and chloroform | Biomass for bioenergy, pulp, charcoal, carbon sequestration | Doty et al. (2007) |
| *Populus trichocarpa* overexpressing γ-glutamylcysteine synthetase from poplar | Increased tolerance to chloroacetanilide herbicides | Biomass for bioenergy, pulp, charcoal, carbon sequestration | Gullner et al. (2001) |

*Source*: Chandra, R., et al., Phytoremediation of environmental pollutants: An eco-sustainable green technology to environmental management, *Advances in Biodegradation and Bioremediation of Industrial Waste*, Ram Chandra (Ed.), CRC Press, Boca Raton, FL, 1–30, 2015.

The complete mechanism of metal accumulation and technology for the safe disposal of plant biomass is required (Jadia and Fulekar 2009). The development of transgenic plants through the application of genetic engineering with better phytoremediation potential is still required. The transgenic plants must be regularly tested in the field for better acceptance worldwide (Suresh and Ravishankar 2004).

**TABLE 11.6**
**Advantages and Disadvantages of Phytoremediation Technology**

| Advantages | Limitations |
|---|---|
| Low cost and esthetically pleasing (no excavation required) | The plant must be able to grow in the polluted media. |
| Soil stabilization and reduced water leaching and transport of inorganic materials in the soil | The plant can accumulate inorganic materials that it can reach through root growth and are soluble in soil. |
| Generation of a recyclable metal-rich plant residue | It is a time-consuming process that can take years for pollutant concentrations to reach regulatory levels (long-term commitment). |
| Applicability to a wide range of toxic metals and radionuclides | The pollutant must be within or drawn toward the root zones of plants that are actively growing. |
| Minimal environmental disturbance as compared to conventional remedial methods | It must not pose harm to human health or become a further environmental problem. |
| Removal of secondary air or water-borne wastes | Climatic conditions are the limiting factor. |
| Enhanced regulatory and public acceptance | The introduction of exotic plant species may affect biodiversity. |

*Source*: Chandra, R., et al., Phytoremediation of environmental pollutants: An eco-sustainable green technology to environmental management, in *Advances in Biodegradation and Bioremediation of Industrial Waste*, Ram Chandra (Ed.), CRC Press, Boca Raton, FL, 1–30, 2015.

Moreover, the search for suitable metal hyperaccumulators continues to be on the environmental agenda.

## 11.10 CONCLUSION

In the effort to limit metal contamination, more efficient strategies for remediation have been sought, and a new technology known as phytoremediation has been advocated. Phytoremediation is a plant-based green technology, is less expensive, and is an esthetically pleasing and environmentally safe technology. Phytoremediation technology is actionable to a range of pollutants, including organic and inorganic contamination and radionuclides. In phytoremediation, plants accumulate the desired concentration of toxic metals from contaminated sites in their harvestable plant parts, and accumulated metals can be recovered through the burning of plant biomass. It can be a green solution for toxic metal-contaminated sites, thereby enhancing the future development of with less environmental ramifications. Therefore, it helps in the conservation of natural resources polluted with toxic metals. It will be the most favorable alternative for developing countries like India. Further, the search/development of transgenic plants, which can tolerate high metal concentrations, is required for making this technology acceptable worldwide.

## ACKNOWLEDGMENT

The author (MK) is highly thankful to the University Grant Commission (UGC), Government of India (GOI), New Delhi, for fellowship support.

## REFERENCES

Abhilash, P.C., Jamil, S., and Singh, N. 2009. Transgenic plants for enhanced biodegradation and phytoremediation of organic xenobiotics. *Biotechnology Advances* 27: 474–488.

Abou-Shanab, R.A, Angle, J.S., and Chaney, R.L. 2006. Bacterial inoculants affecting nickel uptake by *Alyssum murale* from low, moderate and high Ni soils. *Soil Biology and Biochemistry* 38: 2882–2889.

Adriano, D.C. 2001. Trace elements in the terrestrial environments: Biogeochemistry, bioavailability, and risks of metals. *Springer Science & Business Media*, New York, pp. 1–867.

Adriano, D.C. 2009. Bioremediation of contaminated soils. In *Agronomy monographs*. Soil Science Society of America, Madison, WI, pp. 457–508.

Ahmadpour, P., Ahmadpour, F., Mahmud, T.M.M., Arifin Abdu, H., Soleimani, M., and Hosseini Tayefeh, F. 2012. Phytoremediation of heavy metals: A green technology. *African Journal of Biotechnology* 11(76): 14036–14043.

Aken, B.V. 2008. Transgenic plants for phytoremediation: Helping nature to clean up environmental pollution. *Trends in Biotechnology* 26: 225–227.

Ali, H., Khan, E., and Sajad, M.A. 2013. Phytoremediation of heavy metals: Concepts and applications. *Chemosphere* 91: 869–881.

Angelone, M., and Bini, C. 1992. Trace element concentrations in soils and plants of Western Europe. In *Biogeochemistry of trace metals,* Adriano, D.C. (ed.), Lewis, Boca Raton, FL, pp. 19–60.

Angle, J.S., Chaney, R.L., Baker A.J.M. et al. 2001. Developing commercial phytoextraction technologies: Practical consideration. *South African Journal of Science* 97: 619–623.

ATSDR. 2000. *Toxicological Profile for Chromium.* U.S. Department of Health and Human Services Public Health Service. Atlanta, GA. https://www.atsdr.cdc.gov.

Baker, A.J.M., McGrath, S.P., Reeves, R.D., and Smith, J.A.C. 2000. Metal hyperaccumulator plants: A review of the ecology and physiology of a biological resource for phytoremediation of metal-polluted soil. In *Phytoremediation of Contaminated Soil and Water*, Terry, T. and Banuelos, G. (eds.), Lewis Publishers, CRC Press, Boca Raton, FL, pp. 85–108.

Baker, A.J.M., and Whiting, S.M. 2002. In Search for the Holy Grail-another step in understanding metal hyper accumulation. *New Phytologist* 155: 1–7.

Banasova, V., Horak, O., Nadubinska, M., and Ciamporova, M. 2008. Heavy metal content in *Thlaspi caerulescens* J. et C. Presl growing on metalliferous and non-metalliferous soils in Central Slovakia. *International Journal of Environmental Pollution* 33(2): 133–145.

Barbee, J.Y., Jr, and Prince, T.S. 1999. Acute respiratory distress syndrome in a welder exposed to metal fumes. *Southern Medical Association* 92: 510–512

Barcelo, J. and Poschenrieder, C. 2003 Phytoremediation principles and perspectives. contributions to science. *Open Journal of Ecology* 2: 333–334.

Battarbee, R.W., Anderson, N.J., Appleby, P.G., Flower, R.J., Fritz, S.C., Haworth, E.Y., Higgit, S., et al. 1998. *Lake acidification in the United Kingdom 1800–1986*. ENSIS, London, pp. 66.

Belimov, A.A., Kunakova, A.M., Safronova, V.I., Stepanok, V.V., Yudkin, L.Y., Alekseev, Y.V., and Kozhemyakov, A.P. 2004. Employment of rhizobacteria for the inoculation of barley plants cultivated in soil contaminated with lead and cadmium. *Microbiology (Moscow)* 73(1): 99–106.

Bharti, R. 2014. Phytoremediation of heavy metal toxicity and role of soil and rhizobacteria. *International Journal Scientific and Research Publication* 4:1–5.

Bittsanszkya, A., Kömives, T., Gullner, G., Gyulai, G., Kiss, J., Heszky, L., Radimszky, L., and Rennenberg, H. 2005. Ability of transgenic poplers with elevated glutathione content to tolerate Zinc ($2^+$) stress. *Environmental International* 31: 251–254.

Bizily, S., Rugh, C.L., Summer, A.O., and Meagher, R.B. 1999. Phytoremediation of methylmercury pollution: merB expression in *Arabidopsis thaliana* confers resistance to organomercurials. *Proceeding of Natural Academy of Science of the USA* 96: 6808–6813.

Braud, A., Jezequel, K., Bazot, S., and Lebeau, T. 2009. Enhanced phytoextraction of an agricultural Cr and Pb contaminated soil by bioaugmentation with siderophore producing bacteria. *Chemosphere* 74: 280–286.

Bridge, G. 2004. Contested terrain: Mining and the environment. *Annual Review of Environment and Resources* 29: 205–259.

Brooke, R.R., Lee, J., Reeves, R.D., and Jaffre, V.1977. Detection of nickel ferrous rocks by analysis of herbarium specimens of indicator plants. *Journal of Geochemical Exploration* 7: 49–57.

Buchell-Wtschel, M., and Egli, T. 2001. Environmental fate and microbial degradation of aminopolycarboxylic acids. *FEMS Microbiology Review* 25: 69–106.

Burd, G.I., Dixon, D.G., and Glick, B.R. 2000. Plant growth promoting bacteria that decrease heavy metal toxicity in plants. *Canadian Journal of Microbiology* 46(3): 237–245.

Chandra, R., Saxena, G., and Kumar, V. 2015. Phytoremediation of environmental pollutants: An eco-sustainable green technology to environmental management. In *Advances in biodegradation and bioremediation of industrial aste,* Chandra, R. (ed.), CRC Press, Boca Raton, FL, pp. 1–30.

Che, D., Meagher, R.B., Heaton, A.C., Lima, A., Rugh, C.L., and Merkle, S.A. 2003. Expression of mercuric ion reductase in eastern cottonwood (*Populus deltoides*) confers mercuric iron reduction and resistance. *Plant Biotechnology Journal* 1: 311–319.

Chen, X., Wu, C.H., Tang, J.J., and Hu, S.J. 2005. Arbuscular mycorrhizae enhance metal lead uptake and growth of host plants under a sand culture experiment. *Chemosphere* 60: 665–671.

Cheraghi, M., Lorestani, B., Khorasani, N., Yousefi, N., and Karami, M. 2011. Findings on the phytoextraction and phytostabilization of soils contaminated with heavy metals. *Biology of Trace Elements and Research* 144: 1133–1141.

Cherian, S., and Oliveira, M.M. 2005. Transgenic plants in phytoremediation: Recent advances and new possibilities. *Environmental Science Technology* 39: 9377–9390.

Cluis, C. 2004. Junk-greedy greens: Phytoremediation as a new option for soil decontamination. *Biotechnology Journal* 2: 61–67.

Compton, H.R., Haroski, D.M., Hirsch, S.R., and Wrobel, J.G. 1998. Pilot-scale use of trees to address voc contamination. *Proceedings of the lnternational Conference on Remediation of Chlorinated and Recalcitrant Compounds* 4: 245–250.

Cunningham, S.D., and Ow, D.W. 1996. Promises and prospect of phytoremediation. *Plant Physiology* 110: 715–719.

Dary, M., Chamber-Perez, M.A., Palomares, A.J., and Pajuelo, E. 2010. In situ phytostabilisation of heavy metal polluted soils using *Lupinus luteus* inoculated with metal resistant plant-growth promoting rhizobacteria. *Journal of Hazardous Material* 177: 323–330.

Dembitsky, V.M., and Rezanka, T. 2003. Natural occurrence of arseno-compounds in plants, lichens, fungi, algal species, and microorganisms. *Plant Science* 165(6): 1177–1192.

Dhankher, O.P., Pilon-Smits, E.A., Meagher, R.B., and Doty, S. 2012. Biotechnological approaches for phytoremediation. In *Plant biotechnology and agriculture,* Arie Altman, A., and Hasegawa, P. M., Academic Press, Oxford, UK, pp. 309–328.

Doty, S.L. 2008. Enhancing phytoremediation through the use of transgenic plants and entophytes. *New Phytology* 179: 318–33.

Doty, S.L., James, C.A., Moore, A.L., Vajzovic, A., Singleton, G.L., Ma, C., Khan, Z. et al. 2007. Enhanced phytoremediation of volatile environmental pollutants with transgenic trees. *Proceedings of the National Academy of Sciences of the USA* 104(43): 16816–16821.

Dushenkov, S., Vasudev, D., Kapulnik, Y., Gleba, D., Fleisher, D., Ting K.C., and Ensley. B. 1997. Removal of uranium from water using terrestrial plants. *Environmental Science and Technology* 31(12): 3468–3474.

Eapen, S., Singh, S., and D'Souza, S.F. 2007. Advances in development of transgenic plants for remediation of xenobiotic pollutants. *Biotechnology Advances* 25:442–451.

Ensley, B.D. 2000. Rationale for use of phytoremediation. In *Phytoremediation toxic metals: Using plants to clean up environment,* Raskin, I., and Ensley, B. D. (eds.), Wiley, New York, pp. 1.

Fischerova, Z., Tlustos, P., Szakova, J., and Sichorova, K. 2006. A comparison of phytoremediation capability of selected plant species for given trace elements. *Environmental Pollution* 144: 93–100.

Fulekar, M.H., Singh, A., Thorat, V., Kaushik, C.P., and Eapen, S. 2010. Phytoremediation of 137 Cs from low level nuclear waste using *Catharanthus roseus. Indian Journal of Pure and Applied Physics* 48: 516–519.

Garba, S.T., Osemeahon, A.S., Maina, H.M., and Barminas, J.T. 2012. Ethylene diamine tetraacetate (EDTA)-assisted phytoremediation of heavy metal contaminated soil by *Eleusine indica* L. Gearth. *Journal of Environmental and Chemical Ecotoxicology* 4(5): 103–109.

Garbisu, C., and Alkorta, I. 2001. Phytoextraction: A cost effective plant based technology for the removal of metals from the environments. *Bioresource Technology* 77: 229–236.

Ghosh, M., and Singh, S.P. 2005. A review on phytoremediation of heavy metals and utilization of its byproducts. *Applied Ecology and Environmental Research* 3: 1–18.

Glass, D.J. 1999. Current market trends in phytoremediation. *International Journal of Phytoremediation* 1(1): 1–8.

Glass, D.J. 2000. Economic potential of phytoremediation. In *Phytoremediation of toxic metals-using plants to clean up the environment,* Raskin, I., and Ensley, B. D. (eds.), Wiley, New York, pp. 15–33.

Glick, B., Karaturovic, D., and Newell, P. 1995. A novel procedure for rapid isolation of plant growth promoting *Pseudomonas. Canadian Journal of Microbiology* 41: 533–536.

Glick, B.R. 2003. Phytoremediation: synergistic use of plants and bacteria to clean up the environment. *Biotechnology Advances* 21(5): 383–393.

Glick, B.R., Penrose, D.M., and Li, J.P. 1998. A model for the lowering of plant ethylene concentrations by plant growth-promoting bacteria. *Journal of Theoretical Biology* 190(1): 63–68.

Gullner, G., Kömives, T., and Rennenberg, H. 2001. Enhanced tolerance of popler plants overexpressing gamma-glutamylcysteine synthetase towards chloroacetanilide herbicides. *Journal Experimental Botany* 52: 971–979.

Hall, J.L. 2002. Cellular mechanism for heavy metal detoxification and tolerance. *Journal of Experimental Botany* 53: 1–11.

He, C.Q., Tan, G.E., Liang, X., Du, W., Chen., Y. L., Zhi, G.Y., and Zhu, Y. 2010. Effect of Zn-tolerant bacterial strains on growth and Zn accumulation in *Orychophragmus violaceus. Applied Soil Ecology* 44: 1–5.

Ho-Man, L., Zhen-Wen, W., Zhi-Hong, Y., Kin-Lam, Y., Xiao-Ling, P., and Kwai-Chung, C. 2013. Interactions between arbuscular mycorrhizae and plants in phytoremediation of metal-contaminated soils: A review. *Pedosphere* 23(5): 549–563.

Hooda, V. 2007. Phytoremediation of toxic metals from soil and waste water. *Journal of Environmental Biology* 28: 367–371.

Huhle, B., Heilmeier, H., and Merkel, B. 2008. Potential of *Brassica juncea* and *Helianthus annuus* in phytoremediation for uranium. In *Uranium Mine Hydrogeology,* Springer - Verlag Berlin Heidelberg, pp. 307–318.

Hutton, M., and Symon, C. 1986. The quantities of cadmium, lead, mercury and arsenic entering the U.K. environment from human activities. *Science of Total Environment* 57: 129–150.

Iimura, Y., Yoshizumi, M., Sonoki, T., Uesugi, M., Tatsumi, K., Horiuchi, K., Kajita, S., and Katayama, S. 2007. Hybrid aspen with a transgene for fungal manganese peroxidase is a potential contributor to phytoremediation of the environment contaminated with bisphenol. *Journal of the Indian Academy of Wood Science* 53: 541–544.

ITRC. 2001. *Phytotechnology technical and regulatory guidance document.* Interstate Technology and Regulatory Cooperation Work Group, Phytotechnologies Work Team, Suite, Washington, DC.

Jadia, C.D., and Fulekar, M.H. 2008. Phytotoxicity and remediation of heavy metals by fibrous root grass (sorghum). *Journal of Applied Biosciences* 10: 491–499.

Jadia, C.D., and Fulekar, M.H., 2009. Phytoremediation of heavy metals: recent techniques. *African Journal of Biotechnology* 8: 921–928.

Jarup, L. 2003. Hazards of heavy metal contamination. *British Medical Bulletin* 68: 167–182.

Jarup, L. Berglund, M., Elinder, C., Nordberg, G., and Vahter, M. 1998. Health effects of cadmium exposure – A review of literature and a risk estimate. *Scandinavian Journal of Work, Environment and Health* 24: 1–52.

Jiang, C.Y., Sheng, X.F., Qian, M., and Wang, Q.Y. 2008. Isolation and characterization of a heavy metal-resistant *Burkholderia* sp. from heavy metal-contaminated paddy field soil and its potential in promoting plant growth and heavy metal accumulation in metal polluted soil. *Chemosphere* 72: 157–164.

Jones, R., Sun, W., Tang, C.S., and Robert, F.M. 2004. Phytoremediation of petroleum hydrocarbons in tropical coastal soils. II. Microbial response to plant roots and contaminant. *Environmental Science and Pollution Research* 11: 340–346.

Joseph, B., Patra, R.R., and Lawrence R. 2007. Characterization of plant growth promoting rhizobacteria associated with chickpea (*Cicer arietinum* L). *International Journal of Plant Production* 1(2): 141–152.

Karami, A., and Shamsuddin, Z.H. 2010. Phytoremediation of heavy metals with several efficiency enhancer methods. *African Journal of Biotechnology* 9(25): 3689–3698.

Kim, K.R., and Owens, G. 2010. Potential for enhanced phytoremediation of landfills using biosolids – a review. *Journal of Environmental Management* 91: 791–797.

Kirk, J., Klironomos, J., Lee, H., and Trevors, J.T. 2005. The effects of perennial ryegrass and alfalfa on microbial abundance and diversity in petroleum contaminated soil. *Environmental Pollution* 133: 455–465.

Kloepper, J.W., Lifshitz, R., and Zablotowicz, R. 1989. Free living bacterial inocula for enhancing crop productivity. *Trends in Biotechnology* 7(2): 39–43.

Kloepper, J.W., Ryu, C.M., and Zhang, S.A. 2004. Induced systemic resistance and promotion of plant growth by *Bacillus* spp. *Phytopathology* 94: 1259–1266.

Kramer, U. 2003. Phytoremediation to phytochelatin-plant trace metal homeostasis. *New Phytologist* 158: 4–6.

Kramer, U. 2005. Phytoremediation: Novel approaches to cleaning up polluted soils. *Current Opinion in Biotechnology* 16: 133–41.

Kramer, U., and Chardonnens, A.N. 2001. The use of transgenic plants in the bioremediation of soils contaminated with trace elements. *Applied Microbiology and Biotechnology* 55: 661–72.

Kumar, P.B.A.N., Dushenkov, V., Motto, H., and Raskin, I. 1995. Phytoextraction: The use of plants to remove heavy metals. *Environmental Science Technology* 29: 1232–1238.

Lasat, M.M. 2000. Phytoextraction of metals from contaminated soil: A review of plant/soil/ metal interaction and assessment of pertinent agronomic issues. *Journal of Hazardous Substance Research* 2: 1–25.

Lasat, M.M. 2002. Phytoextraction of toxic metals: a review of biological mechanisms. *Journal of Environmental Quality* 31: 109–120.

Li, Y.M., Chaney, R., Brewer, E., Rosenberg, R., Angle, S.J., Baker, A.J.M., Reeves, R.D., and Nelkin, J. 2003. Development of technology for commercial phytoextraction of nickel: Economic and technical consideration. *Plant and Soil* 249: 107–115.

Lombi, E., Zhao, F., McGrath, S., Young, S., and Sacchi, G. 2001. Physiological evidence for a high-affinity cadmium transporter highly expressed in Thlaspi caerulescens ecotype. *New Phytologist* 149: 53–60.

Lucy, M., Reed, E., and Glick, B.R. 2004. Applications of free living plant growth-promoting rhizobacteria. *Antonie Van Leeuwenhoek* 86: 1–25.

Ma, Y., Rajkumar, M., and Freitas, H. 2009. Inoculation of plant growth promoting bacterium *Achromobacter xylosoxidans* strain Ax10 for the improvement of copper phytoextraction by *Brassica juncea*. *Journal Environment Management* 90: 831–837.

Macek, T., Kotrba, P., Svatos, A., Novakova, M., Demnerova, K., and Mackova, M. 2008. Novel roles for genetically modified plants in environmental protection. *Trends in Biotechnology* 26: 146–52.

Macnair, M.R. 2003. The hyperaccumulation of metals by plants. *Advances in Botanical Research* 40: 63–105.

Mangkoedihardjo, S., and Surahmaida. 2008. *Jatropha curcas* L. for phytoremediation of lead and cadmium polluted soil. *World Applied Science Journal* 4: 519–522.

Marschner, H. 1995. *Mineral nutrition of higher plants,* (2nd edn.), Academic press, London, pp. 889.

Mathur, N., Singh, J., Bohra, S., Bohra, A., Mehboob Vyas, M., and Vyas, A. 2010. Phytoremediation potential of some multipurpose tree species of Indian Thar desert in oil contaminated soil. *Advances in Environmental Biology* 4(2): 131–137.

McCutcheon, S.C., and Schnoor, J.L. 2003. *Phytoremediation: Transformation and control of contaminants*, Wiley, Hoboken, NJ, pp. 987.

McGrath, S.P., and Zhao, F.J. 2003. Phytoextraction of metals and metalloids from contaminated soils. *Current Opinion in Biotechnology* 14: 277–282.

Misra, S., and Gedamu, L. 1989. Heavy metal tolerant transgenic *Brassica napus* L. and *Nicotiana tabacum* L. plants. *Theoretical Applied Genetics* 78: 161–8.

Mukhopadhyay, S., and Maiti, S.K. 2010. Phytoremediation of metal mine waste. *Applied Ecology and Environmental Research* 8(3): 207–222.

Nagajyoti K.D. 2010 Heavy metals, occurrence and toxicity for plants: A review *Environmental Chemistry Letters* 8: 199–216.

Newman, L.A., Strand, S.E., Choe, N., Duffy, J., Ekuan, G., Ruszaj, M., Shurtleff, B.B., Wilmoth, J., Heilman, P., and Gordon, M.P. 1997. Uptake and biotransformation of trichloroethylene by hybrid poplars. *Environmental Science Technology* 31: 1062–1067.

Nriagu, J.O 1989. A global assessment of natural sources of atmospheric trace metals. *Nature* 338: 47–49.

Odjegba, V.J., and Fasidi, I.O. 2004. Accumulation of trace elements by *Pistia stratiotes*: Implications for phytoremediation. *Ecotoxicology* 13(7): 637–646.

Oropeza-Garcia, N., Hausler, R., Glaus, M., Vega-Azamar, R., and Romero-Lopez. 2014. Transport of heavy metals in materials with diameter analogous to xylem vessels. *International Journal of Environmental Research* 8(1): 123–132.

Peer, W.A., Baxter, I.R., Richards E.L., Freeman, J.L., and Murphy, A.S. 2005. Phytoremediation and hyperaccumulator plants. In *Molecular biology of metal homeostasis and detoxification,* Tamas, M. J., and Martinoia, E. (eds.), Springer-Verlag, Berlin, Heidelberg, pp. 299–340.

Pilon-Smits, E. 2005. Phytoremediation. *Annual Review in Plant Biology* 56: 15–39.

Pilon-Smits, E., and Pilon, M. 2002. Phytoremediation of metals using transgenic plants. *Critical Review in Plant Science* 21: 439–456.

Preeti, P.P., Tripathi, A.K., and Shikha, G. 2011. Phytoremediation of arsenic using *Cassia fistula* Linn. seedling. *International Journal of Research and Chemistry of Environment* 1: 24–28.

Rajakaruna, N., and Böhm, B.A. 2002. Serpentine and its vegetation: A preliminary study from Sri Lanka. *Journal of Applied Botany-Angewandte Botanik* 76: 20-28.

Rajakaruna, N., Tompkins, K.M., and Pavicevic, P.G. 2006. Phytoremediation: An affordable green technology for the clean-up of metal contaminated sites in Sri Lanka. *Ceylon Journal of Science (Biological Sciences).* 35: 25-39.

Rajkumar, M., Nagendran, R., Lee, K.J., Lee, W.H., and Kim, S.Z. 2006. Influence of plant growth promoting bacteria and $Cr^{6+}$ on the growth of Indian mustard. *Chemosphere* 62: 741–748.

Raskin, I., and Ensley, B.D. 2000. Recent developments for in situ treatment of metal contaminated soils. In *Phytoremediation of toxic metals: Using Plants to Clean Up the Environment,* Wiley, New York.

Raskin, L., Smith, R.D., and Salt, D.E. 1997. Phytoremediation of metals: Using plants to remove pollutants from the environment. *Current Opinion in Biotechnology* 8221–226.

Reeves, R.D., and Baker, A.J.M. 2000. Metal-accumulating plants. In *Phytoremediation of toxic metals: Using plants to clean up environment*, Raskin, I., and Ensley, B. D. (Eds.), Wiley, New York pp. 193–229.

Rosser, S.J., French, C.E., and Bruce, N.C. 2001. Engineering plants for the phytoremediation of explosives. *In Vitro Cell Development Biology of Plant* 37: 330–333.

Rugh, C.L. 2004. Genetically engineered phytoremediation: one man's trash is another man's transgene. *Trends in Biotechnology* 22: 496–8.

Rugh, C.L., Senecoff, J.F., Meagher, R.B, and Merkle, S.A. 1998. Development of transgenic yellow poplar for mercury phytoremediation. *Nature Biotechnology* 16: 925–958.

Rugh, C.L., Wilde, D., Stack, N.M., Thompson, D.M., Summer, A.O., and Meagher, R.B. 1996. Mercuric ion reduction and resistance in transgenic *Arabidopsis thaliana* plants expressing a modified bacterial merA gene. *Proceeding of Natural Academy of Science of the USA* 93: 3182–3187.

Ruiz, O.N., and Daniell, H. 2009. Genetic engineering to enhance mercury phytoremediation. *Current Opinion in Biotechnology* 20(2): 213–9.

Sakakibara, M., Aya, W., Masahiro, I., Sakae, S., and Toshikazu, K. 2007. Phytoextraction and phytovolatilization of arsenic from As-contaminated soil by *Pteris vittata*. *Proceedings of the Annual International Conference on Soils, Sediments, Water and Energy* 12: 26.

Salem, H.M., Eweida, E.A., and Farag, A. 2000. *Heavy metals in drinking water and their environmental impact on human health*. Cairo University, Giza, Egypt, pp. 542–556.

Salt, D.E., Blaylock, M., Kumar, P.B.A.N., Dushenkov, V., Ensley, B.D., Chet, L., and Raskin, L. 1995. Phytoremediation: A novel strategy for the removal of toxic metals from the environment using plants. *Biotechnology* 13(2): 468–474.

Salt, D.E, Pickering, I.J, Prince, R.C., Gleba, D., Dushkenov, S., Smith, R.D., and Raskin, I. 1997. Metal accumulation by aquacultured seeds of Indian mustard. *Environmental Science and Technology* 31: 1636–1644.

Salt, D.E., Smith, R.D., and Raskin, I. 1998. Phytoremediation. *Annual Reviews of Plant Physiology, Plant Molecular Biology* 49: 643–668.

Sandermann, H. 1992. *Plant metabolism of xenobiotics*. Dewey, T.G. (ed.), Plenum Press Cambridge, MA, pp. 20, 294.

Sanderman, H., Jr. 1994. Higher plant metabolism of xenobiotics: The green liver concept. *Pharmacogenetics* 4: 225–41.

Saraswat, S., and Rai, J.P.N. 2009. Phytoextraction potential of six plant species grown in multi-metal contaminated soil. *Chemical Ecology* 25(1): 1–11.

Saravanan, V.S., Madhaiyan, M., and Thangaraju, M. 2007. Solubilization of zinc compounds by the diazotrophic, plant growth promoting bacterium *Gluconacetobacter diazotrophicus*. *Chemosphere* 66: 1794–1798.

Saxena, G., and Bharagava, R.N. 2015. Persistent organic pollutants and bacterial communities present during the treatment of tannery wastewater. In *Environmental waste management*, Ram Chandra (Ed.), CRC Press, Boca Raton, FL, pp. 217–247.

Saxena, G., Chandra, R., and Bharagava, R.N. 2016. Environmental pollution, toxicity profile and treatment approaches for tannery wastewater and its chemical pollutants. *Reviews of Environmental Contamination and Toxicology* 240: 31–69.

Schmidt, B., Joussen, N., Bode, M., and Schuphan, I. 2006. Oxidative metabolic profiling of xenobiotics by human P450s expressed in tobacco cell suspension cultures. *Biochemical Society Transactions* 34: 1241–1245.

Sekhar, K.C., Kamala, C.T., Chary, N.S., Balaram, V., and Garcia, G. 2005. Potential of Hemidesmus indicus for phytoextraction of lead from industrially contaminated soils. *Chemosphere* 58: 507–514.

Seidal, K., Jorgensen, N., Elinder, C.G., Sjogren, B., and Vahter, M. 1993. Fatal cadmium-induced pneumonitis. *Scandinavian Journal of Work, Environment & Health* 19: 429–431.

Sekara, A., Poniedzialeek, M., Ciura, J., and Jedrszczyk, E. 2005. Cadmium and lead accumulation and distribution in the organs of nine crops: Implications for phytoremediation. *Polish Journal of Environmental Studies* 14(4): 509–516.

Sharma, H. 2011. Metal hyperaccumulation in plants: A review focusing on phytoremediation technology. *Journal of Environmental Science and Technology* 4: 118–138.

Sheng, X., and Xia, J. 2006. Improvement of rape (*Brassica napus*) plant growth and cadmium uptake by cadmium-resistant bacteria. *Chemosphere* 64: 1036–1042.

Sheng, X.F., Xia, J.J., Jiang, C.Y., He, L.Y., and Qian, M. 2008. Characterization of heavy metal-resistant endophytic bacteria from rape (*Brassica napus*) roots and their potential in promoting the growth and lead accumulation of rape. *Environment Pollution* 156: 1164–1170.

Shiota, N., Nagasawa, A., Sakakai, T., Yabusaki, Y., and Ohkawa, H. 1994. Herbicide-resistant tobacco plants expressing the fused enzyme between rat cytochrome P4501A1 (CYP1A1) and yeast NADPH-cytochrome P450 oxidoreductase. *Plant Physiology* 106: 17.

Singleton, G.L. 2007. *Genetic analysis of transgenic plant for enhanced phytoremediation*. University of Washington, Seattle, WA.

Sinha, R.K., Heart, S., and Tandon, P.K. 2007. Phytoremediation: Role of plants in contaminated site management. In *Environmental bioremediation technologies,* Singh, S. N., and Tripathi, R. D. (Eds.), Springer-Verlag, Berlin, Heidelberg, pp. 315–330.

Steenland K, and Boffetta P. 2000. Lead and cancer in humans. *American Journal of Industrial Medicine* 38: 295–9.

Sun, Q., Ye, Z.H., Wang, X.R., and Wong, M.H. 2007. Cadmium hyperaccumulation leads to an increase of glutathione rather than phytochelatins in the cadmium hyperaccumulator *Sedum alfredii*. *Journal of Plant Physiology* 164(11): 1489–1498.

Sun, Y., Zhou, Q., and Diao, C. 2008. Effects of cadmium and arsenic on growth and metal accumulation of Cd-hyperaccumulator *Solanum nigrum L. Bioresource Technology* 99(5): 1103–1110.

Sun, Y., Zhou, Q., Wang, L., and Liu, W. 2009. Cadmium tolerance and accumulation characteristics of *Biden spilosa* L. as a potential Cd-hyperaccumulator. *Journal of Hazardous Material* 161(2–3): 808–814.

Sung, K., Munster, C.L., Rhykerd, R., Drew, M.C., and Corapcioglu, M.Y. 2003. The use of vegetation to remediate soil freshly contaminated by recalcitrant contaminants. *Water Research* 37: 2408–18.

Suresh, B., and Ravishankar, G. 2004. Phytoremediation—A novel and promising approach for environmental clean-up. *Critical Reviews in Biotechnology* 24: 97–124.

Sureshvarr, K., Bharathiraja, B., Jayakumar, M., Jayamuthunagai, J., and Balaji, L. 2010. Removal of azo dye compounds from paper industries wastes using phytoremediation methodology. *International Journal of Chemical Science* 8(1): 687–700.

Tangahu, B.V., Abdullah, S.R.S., Basri, H., Idris, M., Anuar, N., and Mukhlisin, M., 2011. A review on heavy metals (As, Pb and Hg) uptake by plants through phytoremediation. *International Journal of Chemical Engineering* 201: 1–31.

United States Environmental Protection Agency (USEPA). 1997. *Cleaning up the nation's waste sites: Markets and technology trends*. EPA/542/R-96/005. Office of Solid Waste and Emergency Response, Washington, DC.

United States Environmental Protection Agency (USEPA). 2000. *Introduction to phytoremediation*. EPA 600/R-99/107.U.S. Environmental Protection Agency, Office of Research and Development, Cincinnati, OH.

Van Dillewijin, P., Couselo, J.L., Corredoira, E., Delgado, A., Wittich, R.-M., Ballester, A., and Ramos, J.L. 2008. Bioremediation of 2, 4, 6-trinitrotoluene by bacterial nitroreductase expressing transgenic aspen. *Environmental Science and Technology* 42: 7405–7410.

Van Huysen, T., Terry, N., and Pilon-Smits, E.A.H. 2004. Exploring the selenium phytoremediation potential of transgenic Indian mustard overexpressing ATP sulfurylase or cystathionine-y-synthase. *International Journal of Phytoremediation* 6: 1–8.

Varun, M.D., Souza, R., Pratas, J., and Paul, M.S. 2012. Metal contamination of soils and plants associated with the glass industry in North Central India: Prospects of phytoremediation. *Environmental Science and Pollution Research* 19: 269–281.

Vassilev, A., Lindon, F.C., Ramaho, J.C., Do Ceu Matos, M., and Bareiro, M.G. 2004. Shoot cadmium accumulation and photosynthetic performance of barley plants exposed to high cadmium treatments. *Journal of Plant Nutrition* 27(3): 775–795.

Verma, R., and Suthar, S. 2015. Lead and cadmium removal from water using duckweed—*Lemna gibba L.*: Impact of pH and initial metal load. *Alexandria Engineering Journal* 4(54): 1297–1304.

Vidali, M. 2001. Bioremediation. An overview. *Pure Applied Chemistry* 73(7): 1163–1172.

Wang, C., Knill E., Glick B.R., and Défago G. 2000. Effect of transferring 1-aminocyclopropane-1-carboxylic acid (ACC) deaminase genes into *Pseudomonas fluorescens* strain CHA0 and its gacA derivative CHA96 on their growth-promoting and disease-suppressive capacities. *Canadian Journal of Microbiology* 46: 898–907.

Wang, J., and Evangelou, V.P. 1994. Metal tolerance aspects of plant cell walls and vacuoles. In *Handbook of plant and crop physiology,* Pessaraki, M. (ed.), Marcel Dekker, Inc, New York, pp. 695–717.

Wang, Y., Brown, H.N., Crowley, D.E., and Szaniszlo, P.J.1993. Evidence for direct utilization of a siderophore, ferroxamine B, in axenically grown cucumber. *Plant Cell Environment* 16(5): 579–585.

Wani, P.A., Khan, M.S., and Zaidi, A. 2007. Effect of metal tolerant plant growth promoting *Bradyrhizobium* sp. (vigna) on growth, symbiosis, seed yield and metal uptake by green gram plants. *Chemosphere* 70: 36–45.

Wani, S.P., Dixin,Y., Li, Z., Dar, W.D., and Chander, G. 2012. Enhancing agricultural productivity and rural incomes through sustainable use of natural resources in the semi-arid tropics. *Journal of the Science of Food and Agriculture* 92(5): 1054–1063.

Wei, S.H., Teixeira da Silva, J.A., and Zhou, Q.X. 2008. Agro-improving method of phytoextracting heavy metal contaminated soil. *Journal of Hazardous Materials* 150: 662–668.

Whitelaw, M.A. 2000. Growth promotion of plants inoculated with phosphate-solubilizing fungi. *Advances in Agronomy* 69: 99–151.

World Health Organization.1992. *Cadmium. Environmental health criteria.* WHO, Geneva.

World Health Organization. 2001. *Arsenic in drinking water: WHO Information Fact Sheet No 210.* WHO, Geneva.

Yang, X.O., Feng, Y., He, Z.L., and Stoffella, P.J. 2005. Molecular mechanisms of heavy metal hyperaccumulation and phytoremediation. *Journal of Trace Elements in Medicine and Biology* 18: 339–353.

Zadeh, B.M., Savaghebi-Firozabadi, G.R., Alikhani, H.A, and Hosseini, H.M. 2008. Effect of sunflower and amaranthus culture and application of inoculants on phytoremediation of the soils contaminated with cadmium. *American-Eurasian Journal of Agriculture and Environment Science* 4: 93–103.

# 12 Solid Waste
## *Environmental Threats and Management*

*V. Sudharsan Varma and Ajay S. Kalamdhad*

**CONTENTS**

12.1 Introduction ........ 338
12.2 Types of Municipal Solid Waste ........ 339
12.2.1 Type and Quantity of Waste ........ 340
12.3 Sources of Solid Waste Generation and Characteristics ........ 341
12.4 Environmental Threats and Pollution from Solid Waste ........ 343
12.4.1 Impacts of Solid Waste on Environment ........ 343
12.4.1.1 Groundwater Contamination ........ 343
12.4.1.2 Greenhouse Gas Emissions from Landfills ........ 344
12.4.1.3 Materials Recovery Facilities ........ 347
12.4.2 Impacts of Solid Waste on Health ........ 348
12.4.2.1 Environmental Pollution and Impacts of Exposure ........ 349
12.4.2.2 Toxicity ........ 350
12.5 Key Elements and Issues in Solid Waste Management ........ 351
12.6 Strategies for Integrated Solid Waste Management ........ 352
12.6.1 Waste Prevention ........ 353
12.6.2 Recycling and Composting ........ 354
12.6.3 Disposal (Landfilling and Combustion) ........ 354
12.6.4 Developing a Plan for Integrated Solid Waste Management ........ 354
12.7 Solid Wastes Processing and Management ........ 356
12.7.1 Composting ........ 356
12.7.2 Bioenergy Production ........ 358
12.7.3 Incineration ........ 361
12.7.4 Recycling ........ 362
12.7.5 Landfilling ........ 363
12.8 Prospects and Future Challenges in Solid Waste Management ........ 364
12.8.1 Recommendations ........ 365
12.9 Conclusion ........ 365
References ........ 365

## 12.1 INTRODUCTION

Solid waste management and disposal practices have become major public health and environmental concerns in the urban areas of many developing countries due to improper management practices, unsafe practices of open dumping, and disorganized administration. Waste generation has witnessed an increasing trend parallel to the development of industrialization, urbanization, and rapid growth of the population. The problem has become one of the primary urban environmental issues. The management of urban solid waste is generally governed by the municipal body of the city, which is termed as the *formal sector*. The formal sector is responsible for waste collection, transportation, processing, and disposal. This also includes other private organizations involved in converting the waste to marketable products such as compost, biogas, and RDF pellets. The other sector that contributes greatly to managing the waste is the informal sector in most developing countries, which never accounts in the authorized system. The informal sector consists of many groups, such as waste-pickers (rag-pickers), itinerant buyers, small scrap dealers, and wholesalers, who, together, recycle about 20% of the waste. A typical solid waste management system in a developing country displays an array of problems, including low collection coverage and irregular collection services, crude open dumping and burning without air and water pollution control, the breeding of flies and vermin, and the handling and control of informal waste picking or scavenging activities. These public health, environmental, and management problems are caused by various factors that constrain the development of effective solid waste management systems. These can be categorized into technical, financial, institutional, economic, and social constraints.

Solid waste management divisions in developing countries are allocated with huge amounts of money for cleaning and disposal activities, but still there needs to be huge improvement. The improper management can be pointed out, not as a technical issue by the governing bodies, but also due to the influence of political, legal, cultural, and environmental factors involved in the system. These are the main factors interlinked inside the system that make the waste system a complex story in most developing countries (Kumar et al. 2005). One major influencing factor in the waste management system can be observed as the solid waste generation data, which are highly unreliable. It is reported that urban areas of Asia spend about US$25 billion on solid waste management every year, and this is expected to increase to at least US$50 billion in 2025. Most of the solid waste in Japan is incinerated, which is costing 10 times more than the collection cost. However, in developing countries, 80% of the money is spent for collection, and the waste is landfilled which is the cheapest and practical option.

Municipal governments are usually the responsible agency for solid waste collection and disposal, but the magnitude of the problem is well beyond the ability of any municipal government, and thus, they need help. In addition to other levels of government, businesses and the general community need to be more involved in waste management. Generally, solid waste planners place too much emphasis on residential waste; this waste represents only about 30% of the overall municipal waste stream but often receives the lion's share of attention. The waste components requiring priority attention in Asia are organics and paper. Indonesia and the Philippines as well as parts of China and India are the Asian countries facing the greatest waste management challenge, based on projected waste generation rates and relative affluence to deal with the problem. In terms of waste management trends, no region of the world faces a greater need to break the inextricable link between waste generation rates and affluence than Asia. For example, if Asia follows lifestyle trends of the United States and Canada (as Hong Kong already seems to be doing) versus the more typical European urban resident, the world would need to supply about 500 million tons more resources in 2025. Asia should pursue regional approaches to many solid waste management problems (e.g., packaging regulations and import/export rules). Urban residents generate two to three times more solid waste than their fellow rural citizens. Municipalities should charge for waste disposal and possibly collection, based on generation rates. Industrialized countries contain 16% of the world's population but use about 75% of the world's paper supply. Residents of India, Indonesia, and China, for example, are aspiring to be as affluent as more industrialized nations. This would require a doubling of the world's current level of paper production.

Developing countries face serious problems in managing their solid waste, and their waste generation is increasing proportionally with the rise in population, urbanization, and other issues related to waste management practices. The solid waste generated and the problems faced in managing such waste in these developing countries are similar, however with variances between regions and locations based on geographic, sociocultural, industrial, infrastructure, legal, and environmental factors. Open dumping and landfilling are the common practices practiced by most of the Asian countries and other developing nations. These practices have crucial impact on environmental hazards causing ecological imbalances with respect to land, water, and air pollution. In this chapter, we briefly described the solid waste management issues, sources of solid wastes generation and characteristics, and environmental threats and pollution from solid wastes. Finally, the best strategies for integrated solid waste management, processing, and treatment options are discussed as well.

## 12.2 TYPES OF MUNICIPAL SOLID WASTE

The term *municipal solid waste* (MSW) is generally used to describe most of the nonhazardous solid waste from a city, town, or village, which requires routine collection and transport to a processing or disposal site. Sources of MSW include private homes, commercial establishments and institutions, as well as industrial facilities. MSW refers to any nonliquid waste, which is generated by an individual person, household, small business, or institution, such as a school or hospital. This type of waste is commonly called trash or garbage and includes everyday items, things that are broken, food that has spoiled, or simply any item a person no longer needs or wants. The most common items disposed of as MSW are paper, yard trimmings, food, plastics, metals, rubbers, and textiles. In recent years, the amount of electronic waste, also known as e-waste, has increased drastically as people become more reliant on electronics, such as computers and cell phones that are replaced and disposed of frequently.

The composition of MSW differs for different countries and regions. Moreover, a major portion of MSW generated in most developing Asian countries was dominated by biodegradable organic fractions composed of food wastes, yard wastes, and mixed paper. Food wastes dominate the major portion of the waste generated in most

**TABLE 12.1**
**Types of Municipal Solid Waste**

| Source | Waste Generators | Types of Solid Wastes |
|---|---|---|
| Residential | Single and multifamily dwellings | Food wastes, paper, cardboard, plastics, textiles, leather, yard wastes, wood, glass, metals, ashes, special wastes (e.g., bulky items, consumer electronics, white goods, batteries, oil, tires), and household hazardous wastes |
| Commercial | Stores, hotels, restaurants, markets, office buildings, etc. | Paper, cardboard, plastics, wood, food wastes, glass, metals, special wastes, hazardous wastes |
| Institutional | Schools, hospitals, prisons, government centers | Same as commercial |
| Construction and demolition | New construction sites, road repair, renovation sites, demolition of buildings | Wood, steel, concrete, dirt, etc. |
| Municipal services | Street cleaning, landscaping, parks, beaches, other recreational areas, water and wastewater treatment plants | Street sweepings; landscape and tree trimmings; general wastes from parks, beaches, and other recreational areas; sludge |

*Source*: Haug, R.T., *The Practical Handbook of Compost Engineering*, CRC Press, Boca Raton, FL, 1993.

developing countries in Asia like China, India, Sri Lanka, and Thailand. In this regard, waste can be characterized as highly biodegradable with high moisture content in which the disposal management should be considered. However, solid wastes can be classified into different types depending on their source (Table 12.1).

### 12.2.1 Type and Quantity of Waste

The type and quantity of waste generated vary greatly, and the main factors affecting these are as follows:

- The geographical region (developed or less-developed country or region)
- Sociocultural practices and material levels among affected population
- Seasonal variations (affecting types of food available)
- The stage of emergency (volume and composition of waste may change over time)
- The packaging of food rations

In general, the volume of waste generated is likely to be small and largely degradable where the population is of rural origin and the food rations supplied are unpackaged dry foodstuffs. Displaced urban populations are more likely to generate larger volumes of nondegradable waste, especially where packaged food rations are provided.

The solid waste generated from human settlements varies independently based on the size of the population. However, the waste generated can be categorized into two subdivisions, which originate classes of sources: residential (dwellings) and non-residential (commercial, institutional/services, construction/demolition, and special). Each class produces different solid waste, which classifies accordingly with the generating source, as follows (Buenrostro et al. 2001):

1. *Residential waste*: The waste generated in dwellings, either houses or apartments.
2. *Commercial waste*: The waste generated in commercial facilities, department stores, supermarkets, restaurants, marketplaces, and ambulant markets.
3. *Institutional and service waste*: The waste generated in governmental and private offices, education centers, museums, libraries, archaeological zones, and recreation centers, such as movie theaters and stadiums.
4. *Construction and demolition waste*: The waste generated in construction and demolition sites.
5. *Special waste*: The waste, which needs special techniques for control, either because of being relatively hazardous, because of its condition or state, or because control is enforced by the standing environmental regulations. This waste is generated in sectors such as scientific research, health, industrial and automobile maintenance shops, human and veterinarian drugstores, airports and terrestrial transportation terminals, among others.
6. *Industrial waste*: The waste generated in any process of extraction, benefit, transformation, and production of goods.
7. *Agricultural and animal husbandry waste*: The waste generated in agricultural and animal husbandry activities.

## 12.3 SOURCES OF SOLID WASTE GENERATION AND CHARACTERISTICS

Municipal solid waste (MSW) or urban solid waste is normally composed of food wastes, rubbish, demolition and construction wastes, street sweepings, garden wastes, abandoned vehicles and appliances, and treatment plant residues (Table 12.2). Quantity and composition of MSW vary greatly for different municipalities and time of the year. Factors influencing the characteristics of MSW are climate, social customs, per capita income, and degree of urbanization and industrialization. The composition of MSW as collected may vary greatly depending on geographical region and season. The typical moisture content of MSW may vary from 15% to 40% depending on the composition of the waste and the climatic conditions. The density of MSW depends upon the composition and degree of compaction. The uncompacted density of MSW is around 150 kg/m$^3$ (250 lb/yd$^3$). The density of collected solid waste is 235–350 kg/m$^3$. The energy content of MSW as collected is 9890 kJ/kg (4260 BTU/lb). Information of chemical composition of the organic portion of MSW is important for many processes such as incineration, composting, biodegradability, leachate generation, and others. The ultimate analysis of the organic fraction of MSW is in terms of the constituent's carbon, hydrogen, oxygen, nitrogen, sulfur, and ash (Singh et al. 2014).

**TABLE 12.2**
**Descriptions of the Main Collection Systems**

| System | Description | Advantages | Disadvantages |
|---|---|---|---|
| **SHARED: Residents can bring out waste at any time** | | | |
| Dumping at designated location | Residents and other generators are required to dump their waste at a specified location or in a masonry enclosure. | Low capital costs | Loading the waste into trucks is slow and unhygienic. Waste is scattered around the collection point. Adjacent residents and shopkeepers protest about the smell and appearance. |
| Shared container | Residents and other generators put their waste inside a container, which is emptied or removed. | Low operating costs | If containers are not maintained, they quickly corrode or are damaged. Adjacent residents complain about the smell and appearance. |
| **INDIVIDUAL: The generators need a suitable container and must store the waste on their property until it is collected** | | | |
| Block collection | Collector sounds horn or rings bell and waits at specified locations for residents to bring waste to the collection vehicle. | Economical; less waste on streets; no permanent container or storage to cause complaints | If all family members are out when collector comes, waste must be left outside for collection. It may be scattered by wind, animals, and waste pickers. |
| Curbside collection | Waste is left outside property in a container and picked up by passing vehicle, or swept up and collected by sweeper. | Convenient; no permanent public storage | Waste that is left out may be scattered by wind, animals, children, or waste pickers. If collection service is delayed, waste may not be collected for some time, causing considerable nuisance. |
| Door-to-door collection | Waste collector knocks on each door or rings doorbell and waits for waste to be brought out by resident. | Convenient for resident; little waste on street | Residents must be available to hand waste over. Not suitable for apartment buildings because of the amount of walking required. |
| Yard collection | Collection laborer enters property to remove waste. | Very convenient for residents; no waste in street | The most expensive system, because of the walking involved. Cultural beliefs, security considerations, or architectural styles may prevent laborers from entering property. |

*Source*: Hann, H.C., et al., Involving micro- and small enterprises in Municipal solid waste management: Guidelines for municipal managers, International Training Center of the ILP, SKAT, WASTE, Turin, Italy, pp. 94–95, 1998.

## 12.4 ENVIRONMENTAL THREATS AND POLLUTION FROM SOLID WASTE

Municipal solid waste is generally composed of biodegradable and non-biodegradable materials. Normally the waste is dumped on the outskirts of the urban areas, turning into the child sources of contamination due to the incubation and proliferation of flies, mosquitoes, and rodents. These activities in turn become the disease transmitters that affect population's health, which has its organic defenses in a formative and creative state. Since there is no segregation for MSW before disposal, it is leading to more emission of greenhouse gases and leachate production due to large fractions of organic matter (fruit and vegetable peels, food waste) (Suthar et al. 2005). The leachate from these wastes majorly contaminates the groundwater (Pokhrel and Viraraghavan 2005). In addition, these illegally dumped wastes have adverse effects on human health and the environment (Achankeng 2003). Moreover, these emissions are mainly due to the result of landfilling and other lifecycle activities. With its high biodegradability nature, the organic waste of the vegetable market is causing much nuisance after reaching the landfill (Bouallagui et al. 2004). However, incineration of solid waste generally results in the production of more polluting gases and other toxic solid residues in the land.

### 12.4.1 Impacts of Solid Waste on Environment

#### 12.4.1.1 Groundwater Contamination

Landfilling is the simplest, cheapest, and most cost-effective method of disposing waste in both developed and developing nations of the world (Barrett and Lawlor 1995). The impacts of solid waste pollution during its processing and management activities have adverse effects on the environment (i.e., ground and surface water and greenhouse gas emissions). Groundwater is a precious resource of drinking activities and should be treated as such, but due to the waste deposited in open dumps and landfills and the by-products produced due to the biological activities, it becomes a part of the prevailing hydrological system. Landfills have been identified as one of the major threats to groundwater resources (Fatta et al. 1999). The leachate generated during the biological decomposition of organic waste through a series of complex biochemical reactions during degradation percolate through the deposit and mobilize different components within the waste. In addition, the water derived from the rainfall, snowmelt, and groundwater along with the biological fluids from the dumps containing high organic carbon content in the form of fulvic acids migrate downward and contaminate the groundwater (Ugwu and Nwosu 2009).

The leachates generated from municipal landfills are highly concentrated with dissolved organic matters, inorganic compounds such as Ammonium, Ca, Mg, Na, K, Fe, sulfate, chlorides, and many other heavy metals (Lee and Jones-Lee 1993; Christensen et al. 1998). Areas near landfills have a greater possibility of groundwater contamination because of the potential pollution source of leachate originating from the nearby site. Such contamination of the groundwater resource poses a substantial risk to the local resource user and to the natural environment. Leachate migration into the soil is greatly affected by the type of waste deposited. The permeability of the deposited waste can be largely reduced by the compaction activity of the waste before dumping. Moreover, application of a topsoil cover between the loadings of the waste to landfills induces layering. These practices are reported to bring about preferential flow paths through landfills (Anilkumar et al. 2015).

The pollution effects on water by solid waste dumping around municipal landfills have been reported by many researchers (Longe and Balogun 2010; Akinbile and Yusoff 2011; Kumar et al. 2012). The reports highlighted the microbiological identification and physicochemical analysis of leachate and groundwater. The impacts of landfills on groundwater quality, quantitative analysis of level of water contamination, and identification of possible threats to the local environments and residents have been well documented. Many other approaches have also been reported to assess the contamination of groundwater either by the experimental determination of the impurities or their estimation through mathematical modeling.

The organic fractions of MSW after dumping are attacked by microorganisms, mainly bacteria, thereby changing the oxidation–reduction potential of the liquid present and probably influencing the quality of different constituents. The biochemical reactions include the dissolution, hydrolysis, oxidization, and reduction processes by these bacteria converting the complex substances to simpler monomers with by-products of leachate, $CO_2$, $CH_4$, $NH_3$, and sulfide gas. Landfills are built to concentrate the waste in compacted layers to reduce volume and monitored to control the liquid and gaseous effluent in order to protect the environment and human health. However, in open landfills nearby or above an aquifer, water percolates through these landfills containing the partially decomposed waste and reaches the groundwater table (WHO 2006). The downward flow of leachate from the landfills threatens underlying groundwater resources, whereas outward flow may result in leachate springs yielding water of a poor, often dangerous quality at the periphery of the waste deposit. Leachate springs represent a major risk to public health; therefore, their detection in situation assessment is vital so as to prevent access to such springs (Syukri et al. 2013).

#### 12.4.1.2 Greenhouse Gas Emissions from Landfills

Global climate change is mainly contributed during the disposal and processing of waste that results in emissions of several greenhouse gases (GHGs). Methane is considered the most significant gas released during the breakdown of organic matter in landfills contributing to the GHGs. Other such activities in landfill during the disposal and processing of waste also produce GHGs, but these are mainly in the form of carbon dioxide (a less powerful GHG). Hegde et al. (2003) have reported that 60% methane ($CH_4$) and 40% carbon dioxide ($CO_2$) together with other trace gases are released during the decomposition of MSW in landfills. In addition to the emission of GHGs during biological mechanisms, even the recycling of waste produces some emissions. The emission of GHGs varies depending on the type of waste composition and the rate of degradation occurring in the landfill with respect to the age of waste.

Since landfilling is the most common waste management practice, the methane released is 20 times more potent as a GHG than carbon dioxide. Instead of releasing the methane into the atmosphere, alternatively the methane can be captured and used as a source of energy. Since the waste is partially decomposed, remaining carbon can be sequestered instead of being released into the atmosphere. The other alternative includes reducing disposal of organic materials in the dumpsite, and instead composting or combustion would be practiced to reduce the amount of GHGs emissions.

In most of the developing countries, there is no adequate and estimated data on the emission of $CH_4$ from landfill. The major reason for the lack of adequate data is due to the various types of processing of MSW such as sorting of recyclable and compostable materials before being sent to the landfills. These activities drastically change the characteristics of waste materials reaching the landfills sites, which directly influence the GHG emissions. It can be well predicted and expected that the complexity of the cause will grow in the future, and the emissions of GHGs in developing countries like India would increase further, if such practices are continued. The Intergovernmental Panel on Climate Change (IPCC) uses a theoretical approach for estimating methane emission through site-specific models that consider local conditions, soil type, climate, and methane oxidation rates to calculate overall methane emissions.

The key factors that affect the rate of decomposition of organic materials to release methane in landfills, include pH, moisture content, temperature, microorganisms, composition, density, and particle size. Bogner et al. (2008) had reported that the landfill methane production might range between 0.003 and 3000 $g/m^2/day$. It was further reported that not all landfill methane is released into the atmosphere; however, part of the methane is utilized during the biochemical reactions occurring inside the landfill and oxidized into $CO_2$.

Fractionates of carbon from the landfilled organic waste can be sequestered indefinitely in landfills depending on local conditions as suggested by performing many laboratory experiments. However, there are limited cases in which the GHG emissions can be minimized from landfills through waste management practices. Compared to landfilling, composting emits lower GHGs, and for yard waste GHGs, emissions are roughly comparable from landfilling. Moreover, landfilling of paper waste causes higher GHG emissions as compared to recycling and incineration.

Jha et al. (2008) reported the field measurement of $CH_4$ emission in Chennai to identify key variables that influence GHG emissions as well as to compare the resulting inventory estimates with that of IPCC recommended methodologies. $N_2O$ and $CO_2$ were also measured to assess the nature of emission. The methane emission was reported to be 0.12 Gg/year, whereas $N_2O$ emission is about 1 ton/year in Chennai. Lower emission of methane was reported due to the lower height of MSW deposits in the landfill area, uncontrolled leaching of organic matter, open burning of MSW in landfill, and climatic conditions. Since most of the waste is degraded aerobically, the $CO_2$ emission rate was found to be 1.16 Gg/year.

In Chennai, $CH_4$ emission was reported to be about 0.12 Gg/year, whereas $N_2O$ emission is about 1 ton/year. Since the organic wastes are degraded aerobically, and the $CH_4$ emissions are less while the $CO_2$ emission was found to be 1.16 Gg/year of $CO_2$. Lower emission of $CH_4$ is also due to lower height of MSW deposits in the landfill area, uncontrolled leaching of organic matter, open burning of MSW in landfill, and climatic conditions. The difference between $CH_4$ emission estimates of measurement and IPCC methodologies as well as uncertainties is mainly due to lack of certain site- and region-specific data as well as model equation assumptions. It is also important to study

MSW reaching to landfills along with the generation and composition determination at source for $CH_4$ emission inventory as intermediate stages of waste handling also influence its quantity.

Significant processes in a landfill include microbial and other oxidation of $CH_4$ within the landfill's surface layer, which has been estimated to be between 10% and 20% (Onk 1996). Open burning practiced by rag pickers to get recyclable materials like metals leads to burning of rags, textiles, wood, decomposable matter, leather, and rubber. This practice may burn up to 75% of combustible materials (Sinha 1997). Burning of MSW by rag pickers at landfill sites was common in Chennai. In such cases, $CH_4$ emission estimates for Kodungaiyur (KDG) is 0.59 Gg, which may be further reduced to 0.53 Gg if we were to consider 10% $CH_4$ oxidation in the topmost layer. Similarly, the $CH_4$ estimate for the Perungudi (PGD) dumping ground would be 0.72 and 0.64 Gg, respectively, for the year 2000. Inventory estimates based on field measurements were lower than calculated estimates using Tier I and Tier II of the IPCC recommended methodologies.

In Canada, landfill sites produce about 27 million metric tons of carbon dioxide equivalent in methane emissions annually. Sixty-four facilities across Canada recover landfill gas, reducing GHG emissions from landfills by 6.9 million metric tons of carbon dioxide equivalent each year. The recovery of landfill gas has increased by 40% since 1990. Internationally, Canada is working in partnership with Colombia, Chile, and Mexico on various projects that offer a suite of options to direct organic matter from landfills and reduce methane emissions.

Current mitigation technologies include landfill gas recovery, improved landfill practices, and engineered wastewater management. In addition, significant GHG generation is avoided through controlled composting, state-of-the-art incineration, and expanded sanitation coverage. Reduced waste generation and the exploitation of energy from waste (landfill gas, incineration, and anaerobic digester biogas) produce an indirect reduction of GHG emissions through the conservation of raw materials, improved energy and resource efficiency, and fossil fuel avoidance. Flexible strategies and financial incentives can expand waste management options to achieve GHG mitigation goals; local technology decisions are influenced by a variety of factors such as waste quantity and characteristics, cost and financing issues, infrastructure requirements including available land area, collection and transport considerations, and regulatory constraints. Existing studies on mitigation potentials and costs for the waste sector tend to focus on landfill $CH_4$ as the baseline. The commercial recovery of landfill $CH_4$ as a source of renewable energy has been practiced at full scale since 1975 and currently exceeds 105 Mt $CO_2$-eq per year. Although landfill $CH_4$ emissions from developed countries have been largely stabilized, emissions from developing countries are increasing as more controlled (anaerobic) landfilling practices are implemented; these emissions could be reduced by accelerating the introduction of engineered gas recovery, increasing rates of waste minimization and recycling, and implementing alternative waste management strategies provided they are affordable, effective, and sustainable.

The emissions from landfill sites are not measured but are modeled as detailed in the IPCC guidelines on waste. Methane is emitted from landfill sites as fugitive emissions. The models used take into account the local weather conditions (humidity and climate), which have a high impact on the overall process. A simple model may divide the landfill process into three steps:

1. During the initial active phase, the degradable portion of the landfilled waste undergoes both aerobic and anaerobic degradation; as the landfill is not sealed during this phase, both methane and biogenic carbon dioxide are emitted.
2. Once the landfill has been sealed, the "methanogenic" phase takes place, and the landfill gas contains greater quantities of methane than in the first phase.
3. Finally, as different types of carbon degrade at different rates (sugars and fat degrade relatively quickly, while cellulose degrades more slowly) and there is very little degradation of lignin products (contained in wood) in landfill, these materials may not degrade appreciably even over much longer time periods.

Stopping landfill today, therefore, would not put an end to methane emissions immediately and as such, but a slow decline in methane emissions from old landfill sites would become apparent. A key measure to mitigate methane emissions is the capture of the methane, especially during the second phase, and its subsequent combustion to produce energy. Methane is combusted to carbon dioxide, which is not entered in the inventory (organic carbon). In the UN Framework Convention on Climate Change (UNFCCC) reporting system, the methane emitted to the atmosphere is entered in the "accounts."

#### 12.4.1.3 Materials Recovery Facilities

There are a variety of types and purposes of materials recovery facilities (also known as materials recycling facilities or MRFs), which may range from bulking up of source-separated fractions of the waste stream with limited sorting and processing taking place, to advanced mechanical or manual separation processes of mixed recyclables. MRFs may be attached to, or incorporated within, transfer stations or other waste facilities or may be separate dedicated facilities dealing purely with the recyclable fraction of collected municipal waste. Some MRFs incorporate a degree of materials processing to enhance the quality or value of the materials extracted from the municipal stream (e.g., by granulating plastics on site). Moreover, the materials recovery facilities in the United Kingdom do not handle green or putrescible elements of household wastes and are considered to process "dry" recyclables only (e.g., paper, plastics, glass, metals, textiles, etc.); these are also known as "clean" MRFs. Due to market acceptability of recyclate and operational experience over the last 10 years in the United Kingdom, it is unlikely that MRFs processing organic wastes or "dirty" MRFs will find any significant future application in the United Kingdom, and this study does not address the health or emissions impacts of this type of facility.

Emissions from the plant are likely to be primarily fugitive emissions to the air from waste handling and sorting. The only residue to land is likely to be the reject fraction from the process, which would go to landfill or incineration. There are no leachate and limited water emission impacts from clean MRFs unless further processing of materials takes place on the same site. The key aspect of controlling emissions from MRFs relates to the management of incoming wastes to ensure that wastes are suitable for handing at a MRF and that the residence time of wastes in a MRF is minimized. Assuming the materials are representative of MSW, then abnormal emissions could potentially arise if waste remains in the facility for an extended period of time. This could give rise to increased emissions of odor, and possibly microorganisms.

By their nature, MRFs provide an opportunity for materials in the waste stream to be recycled, reducing the need to use raw materials. Reprocessing materials in this way could result in increase or decrease in emissions and effects at locations remote from the MRF itself. Reprocessing recycled materials often has associated environmental emissions: for example, if more heat is needed for reprocessing compared to that which would otherwise be required, or if long transportation distances are required. In other cases (such as the use of recycled glass cullet), recycling can require less energy or result in lower emissions than required for manufacture using new materials. In addition, if recycling results in a reduction in demand for raw materials, this would result in lower emissions associated with the extraction of raw materials. These complex issues lie outside the scope of this report, which focuses on the potential emissions and effects associated with the facility itself. Emissions from MRFs are fugitive in nature and arise during the tipping and sorting process. The processing of recyclables in an MRF occurs in a housed facility, which may operate under negative pressure to allow the cleaning of the air emitted from the facility through the use of a biofilter or other control mechanism.

The only direct emission from MRFs to land is through the landfilling of the reject fraction of the plant. This fraction will vary depending on the type of collection system in place, which feeds into the facility, the numbers of different materials being collected, and the effectiveness of separation by the householder and the efficiency and configuration of the plant operation itself. Typically, reject fractions are in the order of 5%–15% of the input material (however due to the variables noted above may be considerably at variance with this range) and comprise fines and contraries (e.g., bottle tops, contaminated recyclables, or products where no market may be available for a particular facility, e.g., carrier bags, yogurt containers).

### 12.4.2 Impacts of Solid Waste on Health

Indiscriminate disposal of solid waste in dumpsites located within urban areas has proved to be a problem to nearby residents in most developing cities of the world, and free town is no exception. Open dumps have environmental safeguards; they can pose major public health threats and environmental effects in urban cities.

The group at risk from the unscientific disposal of solid waste includes the population in areas where there is no proper waste disposal method, especially the preschool children; waste workers; and workers in facilities producing toxic and infectious material. Other high-risk groups include populations living close to a waste dump and those whose water supply has become contaminated either due to waste dumping or leakage from landfill sites. Uncollected solid waste also increases risk of injury and infection. In particular, organic domestic waste poses a serious threat, since the waste ferments, creating conditions favorable to the survival and growth of microbial pathogens. Direct handling of solid waste can result in various types of infectious and chronic diseases with the waste workers and the rag pickers being the most vulnerable. The following list the characteristics of different types of solid waste:

- *Corrosive*: These are wastes that include acids or bases that are capable of corroding mental containers (e.g., tanks).
- *Ignitable*: These are wastes that can create fires under certain conditions (e.g., waste oils and solvents).
- *Reactive*: These are unstable in nature; they cause explosions and toxic fumes when heated.
- *Toxic*: These are wastes that are harmful or fatal when ingested or absorbed.

#### 12.4.2.1 Environmental Pollution and Impacts of Exposure

Heavy metals are metallic elements that are present in both natural and contaminated environments. In natural environments, they occur at low concentrations. However, at high concentrations as is the case in contaminated environments, they result in public health impacts. The elements that are of concern include lead, mercury, cadmium, arsenic, chromium, zinc, nickel, and copper. Heavy metals may be released into the environment from metal smelting and refining industries, scrap metal, plastic and rubber industries, and various consumer products and from burning of waste containing these elements. On release to the air, the elements travel for large distances and are deposited onto the soil, vegetation, and water depending on their density. Once deposited, these metals are not degraded and persist in the environment for many years, poisoning humans through inhalation, ingestion, and skin absorption. Acute exposure leads to nausea, anorexia, vomiting, gastrointestinal abnormalities, and dermatitis. (See also Table 12.3.)

**TABLE 12.3**
**Sources, Risk Levels, and Health Effects from Exposure to Heavy Metals**

| Heavy Metal | Sources of Environmental Exposure | Minimum Risk Level | Chronic Exposure Toxicity Effects |
|---|---|---|---|
| Lead (Pb) | Industrial, vehicular emissions, paints and burning of plastics, etc. | Blood lead levels below 10 ug/dL of blood | Impairment of neurological development, suppression of the hematological system, and kidney failure |
| Mercury (Hg) | Electronics, plastic waste, pesticides, pharmaceutical and dental waste | Below 10 ug/dL of blood; oral exposure of 4 mg/day | Gastrointestinal disorders, respiratory irritation, renal failure, and neurotoxicity |
| Cadmium | Electronics, plastics, batteries, and contaminated water | Below 1 ug/dL of blood | Irritation of the lungs and gastrointestinal tract, kidney damage, abnormalities of the skeletal system, and cancer of the lungs and prostrate |

μg/dL, micrograms per deciliter of blood; mg/kg, milligrams per kilogram.

Persistent organic pollutants are long-lasting, nonbiodegradable organic compounds that accumulate in the food chain, especially fish and livestock, and pose serious health risks to humans. These dissolve poorly in water and get readily stored in fatty tissue; hence, they may be passed to infants through breast milk. These chemicals include aldrin, dieldrin, dichlorodiphenyl trichloroethane (DDT), endrin, heptachlor, toxaphene, chlordane, hexachlorobenzene, mirex, pesticides, and polychlorinated m-biphenyls (PCBs) all of which are to be phased out and/or eliminated under the international environmental agreements.

Gouveia and Prado (2009) highlighted that in a number of health surveys a wide range of health problems, including respiratory system problems, irritation of the skin, eyes and nose, gastrointestinal problems, psychological disorders, and allergies, have been discovered. In addition, dumpsites closer to residential areas are always feeding places for dogs and cats. These pets, together with rodents, carry diseases with them to nearby homesteads.

According to Medina (2002), pollution, a major environmental effect of dumpsites, is not directly transferred from land to people, except in the case of dusts and direct contact with toxic materials. Pollutants deposited on land usually enter the human body through the medium of contaminated crops, animals, food products, or water. In addition, the dumpsite has smelly and unsightly conditions. These conditions are worse in the summer because of extreme temperatures, which speed up the rate of bacterial action on biodegradable organic material.

#### 12.4.2.2 Toxicity

An individual might be exposed to a particular substance by one or more routes, for example, by breathing in air containing the substance, by consuming food or drink containing the substance, or by contact of the substance on the skin. Any substance to which we are exposed has the potential to cause harmful effects. The harm that might be caused by exposure to a particular substance is determined by dose that is the amount of substance experienced by an individual. In general, the higher the dose, the higher the risk of adverse effects, and the more severe any effects would be expected to be.

This means that at a high enough dose, even an innocuous substance such as water can be lethal. Conversely, at low enough doses, no substance will be toxic (with the exception of a few cancer-causing chemicals and even for these substances, exposure at very low concentrations will have a vanishingly small likelihood of any significant effect on health). Even substances that are essential to our bodies, such as iron, can be toxic at high doses.

The substances emitted from management of MSWs have a range of potential health effects, dependent on the dose received. The epidemiological research indicates that adverse health effects are in general not observed in populations living close to MSW management facilities. This means that exposure to the substances emitted from these facilities is too low for significant adverse health effects to be observed in the field. However, an indication of the type of health issues, which might in principle be of concern to researchers investigating the potential health effects of MSW management, can be gained from considering the health effects that could arise from a sufficiently high dose of the substances emitted from these facilities:

- Eye irritation: volatile organic compounds
- Bronchitis: particulate matter, sulfur dioxide
- Increased susceptibility to respiratory infection: sulfur dioxide
- Asthma attacks: nitrogen dioxide
- Reduction in oxygen-carrying capacity of blood: carbon monoxide
- Effects on the central nervous system: lead, manganese, carbon monoxide
- Effects on the immune system: lead, dioxins, mercury, polycyclic aromatic hydrocarbons, benzene, polychlorinated biphenyls, organochlorine compounds including vinyl chloride, nickel, chromium, toluene
- Reproductive effects: arsenic, benzene, cadmium, chlorinated compounds, lead, mercury, polycyclic aromatic hydrocarbons, polychlorinated biphenyls
- Cancer: polycyclic aromatic hydrocarbons, arsenic, nickel, chromium, vinyl chloride, benzene
- Effects on the liver: arsenic, polychlorinated biphenyls, chloroform, vinyl chloride
- Effects on the kidney: mercury, cadmium, chromium, arsenic, lead, halogenated hydrocarbons, organic solvents, and pesticides

The majority of these health effects are characterized by a threshold of effects that is a level of exposure below which no adverse health effects would be expected. The severity of any effect is likely to increase as the dose increases. This is sometimes referred to as a "nonstochastic" effect. However, in the case of some carcinogens, there is in principle no reason to expect there to be a threshold of effect, which is referred as a "stochastic" effect. In this case, the severity of the effect is not related to the dose, but the likelihood of the effect occurring is related to the dose.

## 12.5 KEY ELEMENTS AND ISSUES IN SOLID WASTE MANAGEMENT

The problem associated with the management of solid wastes in today's society are complex because of the quantity and diverse nature of the wastes, the development of sprawling urban areas, the funding limitations for public services in many large cities, the impacts of technology, and the emerging limitations in both energy and raw materials. As a consequence, if solid waste management is to be accomplished in an efficient and orderly manner, the fundamental aspects and relationships involved must be identified, adjusted for uniformity of data, and understood clearly. Esthetics, land use, health, water pollution, air pollution, and economic considerations make proper solid waste storage, collection, and disposal of solid wastes municipal and individual functions that must be taken seriously. Indiscriminate dumping of solid waste and failure of the collection system in a populated community would soon cause many health problems. Odors, flies, rats, roaches, crickets, wandering dogs and cats, and fires would dispel any remaining doubts of the importance of proper solid waste storage, collection, and disposal. In this lecture note, the activities associated

**TABLE 12.4**
**Functional Elements of Solid Waste Management**

| Functional Element | Description |
|---|---|
| Waste generation | Those activities in which materials are identified as no longer being of value and are either thrown away or gathered together for disposal |
| On-site handling, storage, and processing | Those activities associated with the handling, storage, and processing of solid wastes at or near the point of generation |
| Collection | Those activities associated with the gathering of solid wastes and the hauling of wastes after collection to the location where the collection vehicle is emptied |
| Transfer and transport | Those activates associated with (1) the transfer of wastes from the smaller collection vehicle to the larger transport equipment and (2) the subsequent transport of the wastes, usually over long distance to the disposal site |
| Processing and recovery | Those techniques, equipment, and facilities used both to improve the efficiency of the other functional elements and to recover usable materials, conversion products, or energy from solid wastes |
| Disposal | Those activities associated with ultimate disposal of solid wastes including those wastes collected and transported directly to a landfill site, semisolid wastes (sludge) from wastewater treatment plants, incinerator residue, compost, or other substances from the solid waste processing plants that are of no further use. |

with the management of solid wastes from the point of generation to final disposal have been grouped into six functional elements (Table 12.4):

1. Waste generation
2. On-site handling, storage, and processing
3. Collection
4. Transfer and transport
5. Processing and recovery
6. Disposal

## 12.6 STRATEGIES FOR INTEGRATED SOLID WASTE MANAGEMENT

However, waste management remains one of the costliest public services as conventional waste management systems are not well suited to deal with increased waste generation rates and new and special waste streams. In most cases, the revenue from waste management activities is not large enough to compensate for the expenditures.

The World Bank estimates that in developing countries, it is common for municipalities to spend 20%–50% of their available budget on solid waste management. Services cover, however, only about 40%–70% of all urban solid wastes, with the remainder being uncollected and less than 50% of the population being served.

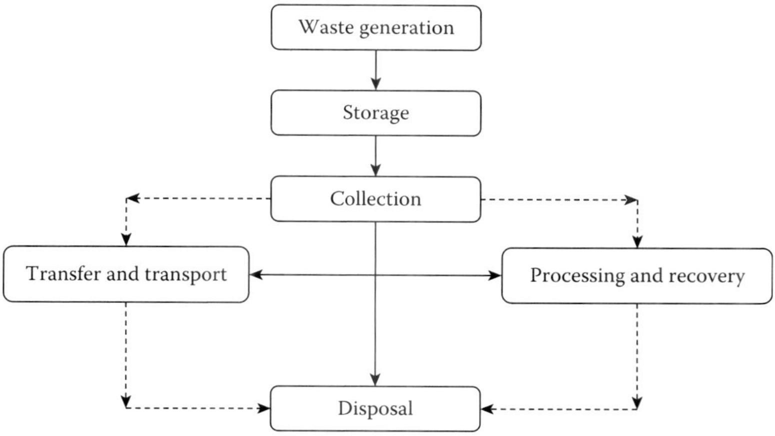


**FIGURE 12.1** Key elements in a municipal solid waste management system.

The approach to waste management in many developing countries has been rather piecemeal—concentrating on certain aspects of waste management (e.g., collection or disposal). However, in many developing countries, authorities have realized that waste contains valuable components that can be recovered as materials for recycling and as a resource to generate energy and thus as a substitute for fossil fuels. Waste is being looked at as a resource.

There is a clear need for strategies to redesign conventional waste generation systems in such a way that they can effectively and efficiently handle growing amounts of waste with diversified waste streams. To respond to this need, the work plan for the focal area on integrated solid waste management (ISWM) proposes to promote an integrated approach to solid waste management, which will enable local/national authorities to reduce the overall amount of waste generated and to recover valuable materials for recycling and for the generation of energy. This has the potential to augment the revenue of waste management activities, which will in turn help to compensate the expenditures for solid waste management.

ISWM is a comprehensive waste prevention, recycling, composting, and disposal program. An effective ISWM system considers how to prevent, recycle, and manage solid waste in ways that most effectively protect human health and the environment. ISWM involves evaluating local needs and conditions, and then selecting and combining the most appropriate waste management activities for those conditions. The major ISWM activities are waste prevention, recycling and composting, and combustion and disposal in properly designed, constructed, and managed landfills (see Figure 12.1). Each of these activities requires careful planning, financing, collection, and transport, all of which are discussed.

### 12.6.1 Waste Prevention

Waste prevention—also called "source reduction"—seeks to prevent waste from being generated. Waste prevention strategies include using less packaging, designing products to last longer, and reusing products and materials. Waste prevention helps reduce handling, treatment, and disposal costs and ultimately reduces the generation of methane.

### 12.6.2 Recycling and Composting

Recycling is a process that involves collecting, reprocessing, and/or recovering certain waste materials (e.g., glass, metal, plastics, paper) to make new materials or products. Some recycled organic materials are rich in nutrients and can be used to improve soils. The conversion of waste materials into soil additives is called *composting*. Recycling and composting generate many environmental and economic benefits. For example, they create jobs and income, supply valuable raw materials to industry, produce soil-enhancing compost, and reduce GHG emissions and the number of landfills and combustion facilities.

### 12.6.3 Disposal (Landfilling and Combustion)

These activities are used to manage waste that cannot be prevented or recycled. One way to dispose of waste is to place it in properly designed, constructed, and managed landfills, where it is safely contained. Another way to handle this waste is through combustion. Combustion is the controlled burning of waste, which helps reduce its volume. If the technology is available, properly designed, constructed, and managed landfills can be used to generate energy by recovering methane. Similarly, combustion facilities produce steam and water as a by-product that can be used to generate energy.

### 12.6.4 Developing a Plan for Integrated Solid Waste Management

Planning is the first step in designing or improving a waste management system. Waste management planners should, for example, take into consideration institutional, social, financial, economic, technical, and environmental factors. These factors vary from place to place. Based on these factors, each community has the challenge of selecting the combination of waste management activities that best suits its needs. Because ISWM involves both short- and long-term choices, it is critical to set achievable goals (Table 12.5). While developing your ISWM plan, you should identify goals or objectives (e.g., protect human health, protect water supplies, eliminate open dumping, increase recycling or composting). The ISWM plan will help guide you through the implementation process. Do not neglect to ask for the community's input in developing your plan, so as to ensure an informed public and to increase public acceptance.

Government plays an important role in developing and enforcing waste management standards, providing funding, and managing day-to-day operations of solid waste management activities. Each level of government may have responsibility in your ISWM plan: national governments typically set standards for solid waste management; the state, provincial, or regional governments may help monitor and enforce these standards; and local governments often play the primary role of managing solid waste activities on a daily basis. All levels may also provide funding for solid waste

**TABLE 12.5**
**Important Questions to Consider and Steps to Take When Developing an Integrated Solid Waste Management Plan**

| Factors | Questions to Consider | Steps to Take |
|---|---|---|
| Institutional (laws and processes) | Are existing laws and policies adequate to allow the government to properly implement ISWM? | • Establish a national policy and pass laws on solid waste.<br>• Manage standards and practices.<br>• Identify the roles and responsibilities of each level of government.<br>• Ensure the local government has the authority and resources to implement an ISWM plan. |
| Social (local customs and religious practices, public education) | What types of waste does your community generate and how it is managed? | • Encourage citizen participation in all phases of waste management planning to help gain community awareness, input, and acceptance. |
| Financial (funding) | Where will you go to get funds for creating a solid waste management system? | • Identify sources that can provide funding for solid waste management, including general revenues or user fees, the private sector, and government or international agency grants and loans. |
| Economic (costs and job creation) | What will it cost to implement various waste management activities? | • Calculate the initial capital investment requirements and long-term operating and maintenance costs associated with the various waste management activities.<br>• Evaluate the public's ability and willingness to pay.<br>• Evaluate activities based on effectiveness in handling waste and potential for job creation. |
| Technical (location and equipment) | Where will you build collection and disposal facilities and what equipment will you need? | • Include geological factors, transport distances, and projected waste generation in siting and design considerations.<br>• Determine what equipment and training will be necessary to perform the waste management tasks. (See How to Establish Recycling and Composting Programs? What Are the Components of Waste Collection and Transport? and What Are the Options for Waste Disposal?) |
| Environmental (natural resources and human health) | Will solid waste management activities (e.g., landfilling or combustion) affect the environment? | • Establish procedures to verify the protection of groundwater and drinking water.<br>• Monitor compliance with the national standards to ensure human health risks are minimized. |

management activities. Two primary costs must be considered in any waste management system: initial capital costs (to purchase equipment or construct new facilities) and ongoing operations and maintenance costs. These costs can be funded in a number of ways including private equity, government loans, local taxes, or user fees.

## 12.7 SOLID WASTES PROCESSING AND MANAGEMENT

### 12.7.1 Composting

Composting is a microbiological conversion of organic residues of plant and animal origin to manure rich in humus and nutrients by various microorganisms including bacteria, fungi, and actinomycetes in the presence of oxygen (Figure 12.2). During the process, it releases by-products such as carbon dioxide, water, and heat (Bharadwaj 1995; Abbasi and Ramasamy 1999; Varma and Kalamdhad 2014a).

The phases in the composting processes can be distinguished according to temperature patterns as shown in Figure 12.3. In the mesophilic phase, the microorganisms acclimatize and colonize in the new environment in the compost heap. The growth

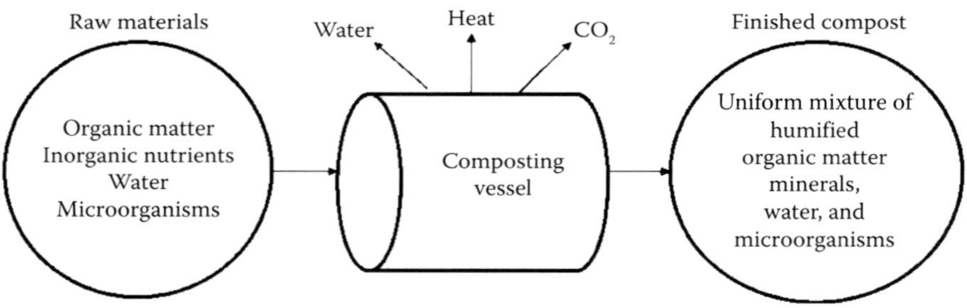


**FIGURE 12.2** Composting process. (After Haug, R.T., *The Practical Handbook of Compost Engineering*, CRC Press, Boca Raton, FL, 1993.)

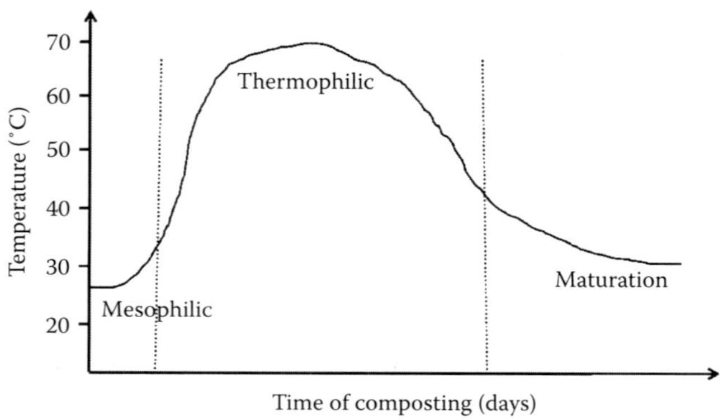


**FIGURE 12.3** Pattern of temperature during composting.

phase is characterized by the rise of biologically produced temperature to the mesophilic level. In the thermophilic phase, the temperature rises to the highest level with stabilization of waste and pathogen destruction, which are more effective. During the maturation phase, the temperature decreases to mesophilic and consequently ambient levels (Figure 12.3). In addition, humification takes place in which some of the complex organics are converted into humic colloids that are closely associated with minerals (iron, calcium, nitrogen, etc.) and finally to humus. Oxidation of ammonia to nitrite ($NO_2^-$) and finally nitrate ($NO_3^-$) also takes place.

During composting, about 50% of added organic matter is completely mineralized due to the degradation of easily degradable compounds such as proteins, cellulose, and hemicellulose by microorganisms. The final residual organic matter consists of humic-like substances, which are highly nonbiodegradable and also the most stable fraction of mature compost (Chefetz et al. 1996; Varma et al. 2015). In composting, the organic matter degradation is carried out by a diversity of microorganisms including mesophilic bacteria, spore-forming bacteria, fungi, and actinomycetes to transform them into stable humic components. However, the degradation pattern and humification during composting are considered to follow different patterns depending on the raw materials used for composting.

The organic waste composition of MSW has been estimated to be in the range of 51%–57% in most of the cities of India (Annepu 2012; Rawat et al. 2013). The higher biodegradable wastes are mainly due to the organic fraction of vegetables, fruits, yard, and other agricultural wastes. Landfilling and illegal open dumping of MSW remain the primary waste disposal strategies in India. When these organic fractions are landfilled, it results in more leachate production causing major environmental pollution due to their higher moisture content and biodegradable nature. In addition, landfilling of these biodegradable wastes is leading to emission of GHGs. The best alternative for the issue is the source segregation of wet and dry waste at the generation point and opting suitable treatment process. Many researchers and government policies have recommended composting instead of landfilling, incineration, and pyrolysis for the processing of these organic wastes as the sustainable method (Kulcu and Yaldiz 2004; JnNURM 2012; MSWR 2012; UNEP 2014). The active microbial metabolism is highly dependent on the type of waste materials used for degradation and the proper aeration rate. Since the thermophilic stage is directly related to the degradation rate (Kalamdhad et al. 2009), proper mixing and controlled aeration rate are necessary for maintaining the elevated temperature in the composting system, otherwise materials will be cooled due to increased heat transfer (Haug 1993). Diaz et al. (2002) had reported aeration as the most important factor during the composting process and the aeration to the composting system can be provided by three major ways (1) manual turning of the composting materials, (2) passive aeration, and (3) forced aeration by using blowers (Kulcu and Yaldiz 2004).

Decentralized composting of organic waste has several advantages in reducing the waste quantities that are transported and also the transportation costs. Moreover, application of these types of systems has a positive effect on overall MSW management costs (Kalamdhad et al. 2009; Varma and Kalamdhad 2014b). One such technique includes the use of a rotary drum composter. Composting time can be drastically reduced to 15–20 days with higher degradation of organic matter without any leachate production. Moreover, the final compost had been completely stabilized with lower carbon dioxide ($CO_2$) evolution and oxygen uptake rate (OUR). Organic fractions of MSW, vegetable waste, water hyacinth, and many other wastes had been successfully composted using rotary drum composters (Tolvanen et al. 2005). In addition, Gajalakshmi and Abbasi (2008) have reported the possible advantages of composting and the use of rotary drum for an effective degradation process. The conversion of municipal organic waste to biogas has become increasingly popular in recent years as a sustainable technology producing green energy. Various high-tech designs of different scales have been evolved, but significant potential for biogas use still exists. For instance, the European energy production from biogas reached 6 million tons of oil equivalents with a yearly increase of more than 20%. On the contrary, large proportions of people in tropical areas have no access to a secure source of energy and rely on traditional biomass, such as wood and charcoal.

The chemical and biological processes occurring during composting are complex and closely related to the initial substrate mix and the combination of waste materials added. The best way of achieving compost highly depends on the initial combination of waste materials. Such optimum combination of waste materials can be obtained by making multiple numbers of experimental trials with different proportions of waste material to achieve quality compost. Here, it can be said that huge amounts of resources are needed to run the process for each variable to the very range. There are many literature reports available on the composting of different organic waste with varying combinations of waste materials such as cow dung, sawdust, dry leaves, and rice straw for effective composting (Liao et al. 1996; Nayak et al. 2013; An et al. 2014; Sarika et al. 2014). The process by its very nature with action of indigenous microorganisms on the substrate, eludes the classical modeling (Wang et al. 2016).

### 12.7.2 Bioenergy Production

Biogas production depends on different factors. One of the factors is the volatile solid (VS) content of the feedstock and the level of biological activity in the digester. According to Hartmann et al. (2004), the type of solid waste and a lignin material in the feedstock influence the biological activity and degradability of the substrate. The biogas production rates are also influenced by the balance of carbon and nitrogen in feeding material. The C/N ratio should be in the range between 25:1 and 30:1.

When the anaerobic digester works properly, the conversion of the intermediate products (i.e., the products of the first three steps) is virtually complete, so that the concentrations of these are low at any time. In the hydrolysis process, the macromolecules like proteins, polysaccharides, and fats that compose the cellular mass of the microorganisms are converted into smaller molecules that are soluble in water: peptides, saccharides, and fatty acids. The hydrolysis or solubilization process is done by exo-enzymes excreted by fermentative bacteria. Hydrolysis is a relatively slow process and generally, it limits the rate of the overall AD process. Polymers are transformed into soluble monomers through enzymatic hydrolysis:

$$n\ C_6H_{10}O_5 + n\ H_2O \xrightarrow{Hydrolysis} n\ C_6H_{12}O_6 \tag{12.1}$$

Reaction 12.1 is catalyzed by extracellular microbial enzymes known as hydrolyses or lyses. Depending on the type of the reaction they catalyze, these hydrolyses can be esterase, glycosidase, or peptidase. For example, lipases hydrolyze the ester bonds of lipids to produce fatty acids and glycerol. Lyses, on the other side, catalyze the nonhydrolytic removal of groups from substrates. The major class of anaerobic bacteria degrading cellulose includes *Bacteroides succinogenes, Clostridium lochhadii, Clostridium cellobioporus, Ruminococcus flavefaciens, Ruminococcus albus, Butyrivibrio fibrosolvens, Clostridium thermocellum, Clostridium stercorarium,* and *Micromonospora bispora.* The dungs of various animals such as cow, pig, etc., have been used as inoculum in anaerobic digestion of food waste. The anaerobes present in the dungs belong to the digestive system of the species. The predominant bacteria found to degrade the hemicelluloses in the rumen are *Bacteroides ruminicola, B. fibrisolvens, R. flavefaciens,* and *R. albus*. The second step of the AD process is acidogenesis or acidification as given in Reaction 12.2, a process that results in the conversion of the hydrolyzed products into simple molecules with a low molecular weight, like VFA (e.g., acetic, propionic, and butyric acid), alcohols, aldehydes, and gases like $CO_2$, $H_2$, and $NH_3$. The acidification is affected by a very diverse group of bacteria, the majority of which are strictly anaerobic (i.e., the presence of oxidants like oxygen or nitrate is toxic). The presence of oxygen-utilizing bacteria is important to remove all oxygen that might be introduced into the system. The acidogenic bacteria are able to metabolize organic material down to a very low pH of around 4.

The monomer results of the first reaction become substrates for the microorganisms in the second stage where they are converted into organic acids by a group of bacteria. In the third step, acetogenesis, the products of the acidification are converted into acetic acids, hydrogen, and $CO_2$ by acetogenic bacteria. The first three steps of AD are often grouped together as acid fermentation. It is important to note that in the acid fermentation, no organic material is removed from the liquid phase: it is transformed into a form suitable as substrate for the subsequent process of methanogenesis:

$$\text{n } C_6H_{12}O_6 \xrightarrow{\textit{Acid form bacteria}} 3\text{ n } CH_3CO\ OH \tag{12.2}$$

In the final step of the AD process as in reaction 12.3, the products of the acid fermentation (mainly acetic acid) are converted into $CO_2$ and $CH_4$ by acetoclastic methanogens. Only then will organic material be removed as the produced methane gas will largely desorb from the liquid phase. In each of the four sequential steps, the catabolic reactions described above develop together with anabolic activity. The free energy exhausted from the reactions is partially utilized for synthesis of the anaerobes populations. Therefore, a large fraction of the digested organic matter is converted into biogas. These organic acids, primarily acetic acid, form the substrate for the third stage:

$$CH_3CO\ OH \xrightarrow{\textit{Methane-forming bacteria}} CH_4 + CO_2 \tag{12.3}$$

In the third step, $CH_4$ can be generated in two routes, fermenting acetic acid to $CH_4$ and $CO_2$ by acetoclastic methanogens and using $CO_2$ as a source of carbon and hydrogen as a reducing agent by hydrogenotrophic methanogens or formate generated by other bacterial species as given in Reactions 12.4 through 12.6. The most commonly found methanogens genera in the biogas reactors are *Methanobacterium, Methanothermobacter* (formerly *Methanobacterium*), *Methanobrevibacter, Methanosarcina,* and *Methanosaeta* (formerly *Methanotrix*):

$$CO_2 + 4H_2 \xrightarrow{Reduction} CH_4 + 3H_2O \tag{12.4}$$

Similarly, $CO_2$ can be hydrolyzed to carbonic acid and to methane:

$$CO_2 + H_2O \xrightarrow{Hydrolysis} H_3CO_3 \tag{12.5}$$

$$4H_2 + H_2CO_3 \xrightarrow{Reduction} CH_4 + 3H_2O \tag{12.6}$$

$CO_2$ in the biogas is undesirable. It is removed for optimum performance of biogas as fuel. $CO_2$ is removed by passing the gas into lime water, which turns milky due to formation of calcium carbonate. $H_2S$ is also another undesirable in the biogas due to the presence of sulfate in the substrate. However, the percentage of $H_2S$ present in biogas varies according to sulfate percentage.

In the anaerobic digestion process, different types of bacteria degrade the organic matter successively in a multistep process and parallel reactions. The anaerobic digestion process of complex organic polymers is commonly divided into three interrelated steps: hydrolysis, fermentation (also known as acidogenesis), oxidation (acetogenesis), and methanogenesis.

In the first step, complex organic polymers such as polysaccharides, proteins, and lipids (fat and grease) are hydrolyzed by extracellular enzymes into soluble products. The size of these soluble products must be small enough to allow their transport across the cell membrane of bacteria. Hydrolysis is a rather slow and energy-consuming process and is normally considered as the overall rate-limiting step for the complete anaerobic digestion of complex polymers.

The monomers produced from the hydrolysis process are then degraded by a large diversity of facultative anaerobes through many fermentative pathways. The degradation of these compounds results in the production of carbon dioxide, hydrogen gas, alcohols, organic acids, some organic nitrogen compounds, and some organic sulfur compounds. The most important of the organic acids is acetate, since it can be used directly as a substrate by methanogenic bacteria.

Acetate can be produced not only through the fermentation of soluble organic compounds but also through acetogenesis. In this step, low molecular weight volatile fatty acids are converted into acetate, hydrogen gas, and carbon dioxide by acetogenic bacteria. This conversion process can only be thermodynamically favored if the partial hydrogen pressure is kept low. Thus, efficient removal of the produced hydrogen gas is necessary.

Finally, methane gas is produced by methane-producing bacteria. Methane is formed around 66% from acetate by means of acetate decarboxylation proceeded by acetoclastic methanogenic sp. and 34% from carbon dioxide reduction by hydrogen, catalyzed by hydrogen-utilizing (hydrogenophilic) methanogenic bacteria.

In particular, hydrogen-utilizing methanogenic bacteria may be responsible for the low partial pressure of hydrogen gas in anaerobic reactors. Thus, they create optimal conditions for acetogenic bacteria to break down the hydrolyzed organic compounds other than $CO_2$, $H_2$, and acetate into substrates for methanogenic bacteria. Alternatively, sulfate-reducing bacteria or autotrophic acetogenic bacteria may also use hydrogen for sulfate reduction or acetate production from $CO_2 + H_2$ and thus decrease the hydrogen partial pressure.

### 12.7.3 Incineration

Basically, waste incineration is the oxidation of combustible materials contained in the waste. Waste is generally a highly heterogeneous material, consisting essentially of organic substances, minerals, metals, and water. During incineration, flue-gases are created that will contain the majority of available fuel energy as heat. The organic substances in waste will burn when they have reached the necessary ignition temperature and come into contact with oxygen. The actual combustion process takes place in the gas phase in fractions of seconds and simultaneously releases energy. Where the calorific value of the waste and oxygen supply is sufficient, this can lead to a thermal chain reaction and self-supporting combustion (i.e., there is no need for the addition of other fuels).

Although approaches vary greatly, the incineration sector may approximately be divided into the following main subsectors:

1. Mixed municipal waste incineration: this treats typically mixed and largely untreated household and domestic wastes, but may sometimes include certain industrial and commercial wastes (industrial and commercial wastes are also separately incinerated in dedicated industrial or commercial nonhazardous waste incinerators).
2. Pretreated municipal or other pretreated waste incineration: these are installations that treat wastes that have been selectively collected, pretreated, or prepared in some way, such that the characteristics of the waste differ from mixed waste. Specifically prepared refuse-derived fuel incinerators fall into this subsector.
3. Hazardous waste incineration: this includes incineration on industrial sites and incineration at merchant plants (that usually receive a very wide variety of wastes).
4. Sewage sludge incineration: in some locations sewage sludge is incinerated separately from other wastes in dedicated installations; in others such waste is combined with other wastes (e.g., municipal wastes) for its incineration.
5. Clinical waste incineration: there are dedicated installations for the treatment of clinical wastes, typically those arising at hospitals and other healthcare institutions. They exist as centralized facilities or onsite at the individual hospital, etc. In some cases, certain clinical wastes are treated in other installations, for example, with mixed municipal or hazardous wastes.

Incineration is a thermal treatment technology used to reduce the volume of waste requiring final disposal. Incineration can typically reduce the waste volume by over

90%, and it is one of the widely used technologies for treating MSW prior to disposal at landfills. Most modern incineration plants incorporate heat recovery as well as power generation facilities to recover the heat energy in the waste.

To ensure that the gas emissions meet the stringent standards imposed by regulatory bodies (e.g., EU Waste Incineration Directive) for public health and environmental protection, modern incineration plants adopt a number of advanced design and process controls as well as exhaust gas-cleaning measures as follows:

- *Combustion*: Waste is continuously fed into the furnace by an overhead crane. The waste is combusted in the specially designed furnace at a high temperature of >850°C for more than 2 seconds with a sufficient supply of air so as to ensure complete burning of the waste and to prevent the formation of dioxins and carbon monoxide.
- *Boiler/steam turbine*: Heat from the combustion is used to generate steam in the boiler. The steam then drives the turbine, which is coupled to the electricity generator. The excess heat generated can also be used for other purposes (e.g., heat for swimming pool).
- *Exhaust gas cleaning*: Exhaust gas from the boiler is typically cleaned by the following advanced pollution control systems to ensure compliance with stringent environmental standards:
    - Dry or wet scrubbers: Spray lime powder or fine atomized slurry into the hot exhaust gas to neutralize and remove polluted acidic gases (sulfur oxides, hydrogen chloride).
    - Activated carbon injection: Carbon injected to adsorb and remove any heavy metal and organic pollutants (e.g., dioxins) in the exhaust gas.
- *Bag house filter*: To filter and remove dust and fine particulates by selective noncatalytic reduction to remove nitrogen oxides, which is a cause of urban smog, by reacting them with ammonia or urea.
- *Ash residues handling*: The ash residues from incineration generally include bottom ash from the furnace and fly ash from the exhaust gas-cleaning units. The bottom ash is either reused as construction material or disposed of at landfills. Fly ash is typically stabilized and solidified by reagents (e.g., cement) and disposed of at dedicated landfill with continuous environmental monitoring. Ash melting that uses the heat energy in the incinerator to melt the ash residues at a high temperature is a technology used in some places. The melted products are inert and contain no hazardous materials so that they may be reused (e.g., as construction material). Comparatively, ash melting is more expensive but it has the advantages of further volume reduction and fixation of any hazardous materials.

### 12.7.4 Recycling

The reuse and recycling of solid wastes can also serve as a means for conservation of energy. In light of this, there are four principles guiding the management of solid wastes. Recognizing that our resources are finite and continued pollution will be difficult to rectify in coming times, the management of solid wastes has become important.

Before disposal, a waste should be considered for the following possibilities:

- Reduction in raw materials and solid waste quantities
- Reuse of waste materials
- Materials recovery
- Energy recovery

A number of materials present in municipal and industrial waste are suitable for recovery and recycling. About 10%–15% of solid wastes are recoverable. Most suitable candidates are the wastes generated by paper, cardboard, glass, ferrous metals, metals (mostly aluminum), and rubber. On the contrary, plastics, leather, textile, and food wastes are unsuitable candidates for materials recovery.

The first benefit of waste recycling is that it creates job opportunities for many people involved in the waste management process, such as waste collectors. In addition, people who sort and process recycled items also earn a living from waste recycling. This implies that more money in the community will be spent on local stores and dispensaries. This has various benefits such as more taxes and revenues to the city or state governments, production of affordable and durable goods, clean and healthy environment, clean water, and fresh air to breathe. To summarize, waste recycling leads to a healthier economy for everyone concerned (Moore 2008).

The second benefit of waste recycling is fiscal efficiency as reported by various municipalities that point out fiscal benefits accrued from the adoption of recycling initiatives, mostly in the form of reduced landfill costs. Landfills pose a substantial danger to humans and animals; hence, they are not the best method of waste management. In addition, biodegradable wastes like vegetables and eggshells at home can be recycled and utilized as manure; this helps in saving money that would have been used to buy fertilizer and instead channels it to other beneficial uses. Furthermore, recycled materials are less expensive than the original products and also friendly to the environment.

The third advantage of recycling is that it saves energy; this is because less energy is required to process recycled materials than to manufacture virgin materials. The energy saving attribute of waste recycling culminates to a number of merits such as decreasing pollution and reducing costs of production, which creates less anxiety on human health and the natural environment. Recycling reduces the need for raw materials; as a result, forests can be preserved and conserved leading to a clean environment and rich and fertile soils, and enriching the esthetic value of planet Earth.

### 12.7.5 Landfilling

Landfilling of the solid waste can be considered as the least preferable option as per the waste management hierarchy and should be limited. The option of landfilling is to majorly reduce the negative effects of the solid waste.

The objective of the Landfill Directive is to prevent or reduce as far as possible negative effects on the environment, in particular on surface water, groundwater, soil, air, and on human health from the landfilling of waste by introducing stringent technical requirements for waste and landfills.

The directive defines the different categories of waste (municipal waste, hazardous waste, nonhazardous waste, and inert waste) and applies to all landfills, defined as waste disposal sites for the deposit of waste onto or into land. Landfills are divided into three classes:

- Landfills for hazardous waste
- Landfills for nonhazardous waste
- Landfills for inert waste

The directive does not apply to:

- The spreading on the soil of sludges (including sewage sludges and sludges resulting from dredging operations).
- The use in landfills of inert waste for redevelopment or restoration work.
- The deposit of unpolluted soil or of nonhazardous inert waste resulting from prospecting and extraction, treatment, and storage of mineral resources as well as from the operation of quarries.
- The deposit of nonhazardous dredging sludges alongside small waterways from which they have been dredged and of nonhazardous sludges in surface water, including the bed and its subsoil.

## 12.8 PROSPECTS AND FUTURE CHALLENGES IN SOLID WASTE MANAGEMENT

Daily-generated solid waste contains several recyclable materials and energy, which can be successfully processed and used again as resources. In fact, the production of waste could be a sign of ineffective processing and recycling activity of waste generated. Hence, the governing bodies and the policies implemented consequently play an important role in the reduction of waste dumping and use of the resources on the environment. There is a huge responsibility on us to do the necessary actions and follow such practices to produce less waste, recycle, and use the natural resources by reducing the pollution load on environment.

One of the most important contributing factors for the successful solid waste management issues is the systematic, synergetic, and dynamic linkage of the solid waste governing authority and the public participation. It is very important to have the involvement and participation of opinion leaders, decision-makers, environmental service providers, and government agencies.

Hence, it can be emphasized that reducing wastes and maximizing reuse facilities should be of higher priority in the waste management agenda. The environmental, social, and economic benefits of ISWM should be an emerging agenda worldwide for effective management practices. The following methods could be effectively followed for the proper treatment and disposal of solid waste (i.e., sanitary landfilling, composting, biomethanization/pyrolysis/gasification, palletization, and incineration).

### 12.8.1 Recommendations

1. Increased synergy between the formal and informal sector should be promoted. UN organizations (UNOs) should provide assistance to the municipal authorities to bridge the gap between the two sectors. Further, they should integrate the potential of the informal sector.
2. An awareness program should be promoted regarding the environmental and health risks of poor solid waste management.
3. Indigenous technologies for sustainable solid waste management should be promoted and developed.
4. Participation and involvement of nongovernmental organizations, the private sector, academia, and research institutions should be increased.

## 12.9 CONCLUSION

Finally, it can be concluded that the accumulation of solid waste in open lands and improper disposing may appear as another threat to the environment and also decrease clean lands. The result of improper disposal and, therefore, negative impact of solid waste on soil and surrounding nature is that it is impossible to use the lands where the waste was disposed for any practical purposes, such as agriculture, or building some residential zones, for instance. Hence, the suitable option is the segregation of waste at the generation point and adoption of suitable treatment options. By this practice, the solid waste generated would be effectively managed. Thus, taking into account all the above-mentioned points, it is possible to conclude that solid waste needs to be disposed of properly in order to minimize its negative impact on human health and the environment.

## REFERENCES

Abbasi, S.A., and Ramasamy, E.V. 1999. *Biotechnological methods of pollution control*. Orient Longman (Universities Press India Ltd), Hyderabad, India.

Achankeng, E. 2003. Globalization, urbanization and municipal solid waste management in Africa. *Proceedings of the African Studies Association of Australasia and Pacific, Africa on a Global Stage.*

Akinbile, C.O., and Yusoff, M.S. 2011. Environmental impact of leachate pollution on groundwater supplies in Akure, Nigeria. *International Journal of Environmental Science and Development* 2: 81–86.

An, Y., Li, G., Wu, W., Huang, J., He, W., and Zhu, H. 2014. Generation, collection and transportation, disposal and recycling of kitchen waste: A case study in Shanghai. *Waste Management & Research* 32(3): 245–248.

Anil Kumar, A., Sukumaran, D., and Vincent, S.G.T. 2015. Effect of municipal solid waste leachate on ground water quality of Thiruvananthapuram district, Kerala, India. *Applied Ecology and Environmental Sciences* 3(5): 151–157.

Annepu, R.K. 2012. *Sustainable solid waste management in India.* Earth Engineering Center, Columbia University, New York, NY.

Barrett, A., and Lawlor, J. 1995. *The economics of waste management in Ireland.* Economic and Social Research Institute, Dublin, pp. 1–129.

Bharadwaj, K.K.R. 1995. Improvements in microbial compost technology: A special reference to microbiology of composting. In *Wealth from waste,* Khanna, S., and Mohan, K. (eds.), Tata Energy Research Institute, New Delhi, pp. 115–135.

Bogner, J., Pipatti, R., Hashimoto, S., Diaz, C., Mareckova, K., Diaz, L., Kjeldsen, P., et al. 2008. Mitigation of global green house gas emissions from waste: Conclusions and strategies from the Intergovernmental Panel on Climate Change (IPCC) Fourth Assessment Report. Working Group III (Mitigation). *Waste Management & Research* 26(1): 11–32.

Bouallagui, H., Torrijos, M., Godon, J.J., Moletta, R., Cheikh, R.B., Touhami, Y., Delgenes, J.P., and Hamdi, M. 2004. Two-phase anaerobic digestion of fruit and vegetable wastes: Bioreactors performance. *Biochemical Engineering Journal* 21(2): 193–197.

Buenrostro, O., Bocco, G., and Cram, S. 2001. Classification of sources of municipal solid wastes in developing countries. *Resources, Conservation and Recycling* 32(1): 29–41.

Chefetz, B., Hatcher, P.G., Hadar, Y., and Chen, Y. 1996. Chemical and biological characterization of organic matter during composting of municipal solid waste. *Journal of Environmental Quality* 25(4): 776–785.

Christensen, J.B., Jensen, D.L., Gron, C., Filip, Z., and Christensen, T.H. 1998. Characterization of the dissolved organic carbon in landfill leachate-polluted groundwater. *Water Research* 32: 125–135.

Diaz, M.J., Madejon, E., Lopez, F., Lopez, R., and Cabrera, F. 2002. Optimization of the rate vinasse/grape marc for co-composting process. *Process Biochemistry* 37(10): 1143–1150.

Fatta, D., Papadopoulos, A., and Loizidou, M. 1999. A study on the landfill leachate and its impact on the groundwater quality of the greater area. *Environmental Geochemistry and Health* 21: 175–190.

Gajalakshmi, S., and Abbasi, S.A. 2008. Solid waste management by composting: State of the art. *Critical Reviews in Environmental Science and Technology* 38: 311–400.

Gouveia, N., and Prado, R.R. 2009. Health risks in areas close to urban solid waste landfill sites. *Revista de Saúde Pública* 44(5): 1–8.

Hann, H.C., Coad, A., and Lardinois, I. 1998. *Involving micro- and small enterprises in Municipal solid waste management: Guidelines for municipal managers*. International Training Center of the ILP, SKAT, WASTE, Turin, Italy, pp. 94–95.

Hartmann, H., Møller, H.B., and Ahring, B.K. 2004. Efficiency of anaerobic treatment of the organic fraction of MSW: Collection and pre-treatment. *Waste Management and Research* 22(1): 35–41.

Haug, R.T. 1993. *The practical handbook of compost engineering*. Lewis, Boca Raton, FL.

Hegde, U., Chang, T.-C., and Yang, S.-S. 2003. Methane and carbon dioxide emissions from Shan-ch-ku landfill site in northern Taiwan. *Chemosphere* 52: 1275–1285.

Jha, A.K., Sharma, C., Singh, N., Ramesh, R., Purvaja, R., and Gupta, P.K. 2008. Greenhouse gas emissions from municipal solid waste management in Indian mega-cities: A case study of Chennai landfill sites. *Chemosphere* 71(4): 750–758.

JnNURM (Jawaharlal Nehru National Urban Renewal Mission). 2012. *Toolkit for solid waste management.* Ministry of Urban Development, Govt. of India, New Delhi, India.

Kalamdhad, A.S., Singh, Y.K., Ali, M., Khwairakpam, M., and Kazmi, A.A. 2009. Rotary drum composting of vegetable waste and tree leaves. *Bioresource Technology* 100: 6442–6450.

Kulcu, R., and Yaldiz, O. 2004. Determination of aeration rate and kinetics of composting some agricultural wastes. *Bioresource Technology* 93(1): 49–57.

Kumar, S., Tripathi, V.R., and Garg, S.K. 2012. Physicochemical and microbiological assessment of recreational and drinking waters. *Environmental Monitoring and Assessment* 184: 2691–2698.

Kumar, V., Sharp, A., and Harnpornchai, N. 2005. Improving the solid waste management in phnom Pench city: a strategic approach. *Improving the solid waste management in phnom Pench city: a strategic approach* 25(1): 101–109.

Lee, G.F., and Jones-Lee, A. 1993. *Groundwater quality protection: A suggested approach for water utilities*. Report to the CA/NV AWWA Section Source Water Quality Committee, August 1993, p. 8. http://www.gfredlee.com/Groundwater_Quality_Publications.html.

Liao, P.H., Jones, L., Lau, A.K., Walkemeyer, S., Egan, B., and Holbek, N. 1997. Composting of fish wastes in a full-scale in-vessel system. *Bioresource Technology* 59(2): 163–168.

Longe, E.O., and Balogun, M.R. 2010. Groundwater quality assessment near a municipal landfill, Lagos, Nigeria. *Research Journal of Applied Sciences, Engineering and Technology* 2: 39–44.

Medina, M. 2002. *Globalization, development and municipal solid waste management in Third World Countries*. E1 Colegio de la Frontera, Tijuana, Mexico. http://www.gdnet.org/pdf/2002AwardsMedalsWinners/outstandingResearchDevelopment/martindinaMaetinezpaper.pdf.

Moore, C. 2008. Synthetic polymers in the marine environment: A rapidly increasing, long-term threat. *Environmental Research* 108(2): 131–139.

Municipal Solid Waste Management Rule (MSWR). 2012. Ministry of Urban Development, Govt. of India, New Delhi, India.

Nayak, A.K., Varma, V.S., and Kalamdhad, A.S. 2013. Effects of various C/N ratios during vermicomposting of sewage sludge using Eisenia Fetida. *Journal of Environmental Science and Technology* 6(2): 63–78.

Onk, H. 1996. Dutch abiogenic emissions of methane and nitrous oxide and options for emission reduction. *Energy Conversion Management* 37(6–8): 985–989.

Pokhrel, D., and Viraraghavan, T. 2005. Municipal solid waste management in Nepal: Practices and challenges. *Waste Management* 5(5): 555–562.

Rawat, M., Ramanathan, A.L., and Kuriakose, T. 2013. Characterisation of municipal solid waste compost (MSWC) from selected Indian cities—A case study for its sustainable utilisation. *Journal of Environmental Protection and Ecology* 4: 163–171.

Sarika, D., Singh, J., Prasad, R., Vishan, I., Varma, V.S., and Kalamdhad, A.S. 2014. Study of physico-chemical and biochemical parameters during rotary drum composting of water hyacinth. *International Journal of Recycling of Organic Waste in Agriculture* 3(3): 1–10.

Singh, G.K., Gupta, K., and Chaudhary, S. 2014. Solid waste management: Its sources, collection, transportation and recycling. *International Journal of Environmental Science and Development* 5(4): 347–351.

Sinha, C. 1997. Open burning of municipal solid waste: State level analysis. *TERI Monitoring on Environmental Science* 2(2): 71–75.

Suthar, S.S., Watts, J., Sandhu, M., Rana, S., Kanwal, A., Gupta, D., and Meena, M.S. 2005. Vermicomposting of kitchen waste by using *Eisenia fetida* (Savigny). *Asian Journal of Microbiology, Biotechnology & Environmental Science* 7(3): 541–544.

Syukri, M., Saad, R., and Marwan. 2013. Leachate migration delineation using 2-D electrical resistivity imaging (2-DERI) at Gampong Jawa, Banda Aceh. *EJGE* 18: 1505–1510.

Tolvanen, O., Nykanen, J., Nivukoski, U., Himanen, M., Veijanen, A., and Hanninen, K. 2005. Occupational hygiene in a finish drum composting plants. *Waste Management* 25(4): 427–433.

Ugwu, S.A., and Nwosu, J.I. 2009. Effect of waste dumps on groundwater in Choba using geophysical method. *Journal of Applied Sciences and Environmental Management* 13(1): 85–89.

UNEP. 2014. http://www.unep.or.jp /ietc/estdir/pub/msw/sp/sp4/sp4_1.asp (accessed July 1, 2016).

Varma, V.S., and Kalamdhad, A.S. 2014a. Evolution of chemical and biological characterization during thermophilic composting of vegetable waste using rotary drum composter. *International Journal of Environmental Science and Technology* 12(6): 2015–2024.

Varma, V.S., and Kalamdhad, A.S. 2014b. Stability and microbial community analysis during rotary drum composting of vegetable waste. *International Journal of Recycling of Organic Waste in Agriculture* 3: 52.

Varma, V.S., Ramu, K., and Kalamdhad, A.S. 2015. Carbon decomposition by inoculating Phanerochaete chrysosporium during drum composting of agricultural waste. *Environmental Science and Pollution Research* 22(10): 7851–7858.

Wang, H., Wang, S.Y., Li, H.Y., Wang, B., Zhou, Q.S., Zhang, X.M., Li, J., and Zhang, Z.J. 2016. Decomposition and humification of dissolved organic matter in swine manure during housefly larvae composting. *Waste Management & Research* 34(5): 465–473.

WHO. 2006. *Protecting groundwater for health: Managing the quality of drinking-water sources.* Schmoll, O., Howard, G., Chilton, J., and Chorus, I. (eds.), IWA Publishing, London, UK. ISBN: 1843390795, pp. 1–175.

# 13 Pollutants in Tannery Wastewater

## *Pharmacological Effects, and Bioremediation Approaches for Human Health Protection and Environmental Safety*

*Swetlana Gautam, Gaurav Kaithwas, Ram Naresh Bharagava, and Gaurav Saxena*

**CONTENTS**

13.1 Introduction ........ 370
13.2 Major Pollutants in Tannery Wastewater ........ 373
  13.2.1 Solids ........ 374
  13.2.2 Nitrogen ........ 374
  13.2.3 Chromium Compounds ........ 374
  13.2.4 Phenolic and Chlorophenolic Compounds ........ 375
  13.2.5 Phthalates ........ 375
  13.2.6 Dyes ........ 375
  13.2.7 Other Heavy Metals ........ 376
    13.2.7.1 Aluminum ........ 376
    13.2.7.2 Zinc ........ 377
    13.2.7.3 Cadmium ........ 377
    13.2.7.4 Lead ........ 377
  13.2.8 Sulfide ($S^{2-}$) ........ 378
  13.2.9 Natural Salts (Chloride [$Cl^-$]) ........ 378
  13.2.10 Oils and Grease ........ 378
13.3 Pharmacological Aspects of Tannery Wastewater ........ 378
  13.3.1 Respiratory System ........ 378
    13.3.1.1 Sulfide Toxicity ........ 378

13.3.2 Urinary System ....................................................................... 379
13.3.2.1 Chronic Cadmium Toxicity ........................................... 379
13.3.2.2 Aluminum Toxicity........................................................ 379
13.3.3 Hypertension and Cardiovascular System ..................................... 379
13.3.3.1 Cadmium Toxicity ........................................................ 379
13.3.3.2 Lead Toxicity ................................................................ 380
13.3.3.3 Sulfide Toxicity.............................................................. 380
13.3.3.4 Chloride Toxicity .......................................................... 381
13.3.4 Nervous System ......................................................................... 381
13.3.4.1 Aluminum Toxicity........................................................ 381
13.3.4.2 Sulfide Toxicity.............................................................. 382
13.3.5 Digestive System ........................................................................ 382
13.3.5.1 Zinc Toxicity.................................................................. 382
13.3.6 Skeletal Effects ........................................................................... 383
13.3.6.1 Acute Cadmium Toxicity............................................... 383
13.3.7 Exposure during Pregnancy ........................................................ 383
13.3.7.1 Lead Toxicity ................................................................ 383
13.3.8 Carcinogenic Property ................................................................ 383
13.3.8.1 Chromium Toxicity........................................................ 383
13.3.8.2 Cadmium Toxicity ......................................................... 384
13.3.8.3 Lead Toxicity ................................................................ 385
13.4 Bioremediation Approaches for Tannery Wastewater Pollutants................ 385
13.4.1 Bioremediation............................................................................ 385
13.4.2 Phytoremediation ........................................................................ 388
13.4.3 Constructed Wetlands ................................................................. 388
13.5 Conclusion ......................................................................................... 389
Acknowledgments....................................................................................... 389
References .................................................................................................. 390

## 13.1 INTRODUCTION

Leather processing has materialized as an important economical and industrial basis for the growth of many developing countries. The annual capacity for the production of leather is more than 10 million worldwide. Major problems have arisen from decades of production from the leather industry leading to poor environmental quality issues (Bosnic et al. 2000; Shakir et al. 2012). The tannery effluent from the leather industry generates indiscriminate removal of organic and inorganic substances along with the discharge of suspended or gas-solids oil and grease, nitrogen-containing compounds, and heavy metals either alone or in their reduced salt form (Song et al. 2004).

In some instances, the effluent is discharged into the sewage system, which is an indirect discharge. In this indirect discharge, tannery wastewater undergoes full-scale treatment before being returned to the environment. But, the effluent being discharged directly into the streams and river has to be treated effectively through bioremediation (Gowd and Govil 2008).

The wastewater generated is characterized by a high chemical oxygen demand (COD), biological oxygen demand (BOD), total dissolved solids (TDSs), total suspended solids (TSSs), chromium (III), and phenolics with high pH, strong odor, and dark brown color. It contains a variety of organic and inorganic pollutants with various nutrients such as nitrogen and phosphorus that can lead to eutrophication of water bodies (Durai and Rajasimmam 2011; Saxena et al. 2016) (Table 13.1). In addition, the dark brown color of wastewater hinders the photosynthesis process by blocking sunlight penetration, and it is therefore deleterious to aquatic life (Durai and Rajasimmam 2011; Saxena et al. 2016). However, the major pollutants present in tannery wastewater include chromium, tannins or syntans (STs), phenolics, phthalates, and azo dyes (Kumar et al. 2008; Lofrano et al. 2013; Dixit et al. 2015; Saxena et al. 2016). Further, the presence of pathogens in water and wastewater has been reviewed by many workers (Bharagava et al. 2014; Saxena et al. 2015). Tannery wastewaters are also rich in organic and inorganic constituents and thus may provide a chance for a variety of pathogenic bacteria to flourish and contaminate the receiving water bodies, as these constituents may act as a source of nutrients (Verma et al. 2008; Bharagava et al. 2014). Many reports are available on the presence of a variety of pathogenic bacteria remaining in tannery wastewater even after the secondary treatment process (Verma et al. 2008; Ramteke et al. 2010; Bharagava et al. 2014).

The composition of the effluent waste includes sulfides; natural salts such as sulfates and chlorides; and heavy metals compounds including chromium, copper, arsenic, lead, zinc, and cadmium. From all the heavy metals present in the tannery waste, chromium is the major constituent present in a high concentration, which is being used in tannery processing (Akan et al. 2009; Singh et al. 2010; Singare et al. 2011). Declination in the amount of oxygen and unacceptable changes in the pH ranges result in the susceptible loss of aquatic and plant life. Evidenced literature have emerged concerning the presence of heavy metal ions in the sensitive fish exposed to tannery effluent. Concentration and distribution of the chromium, copper, lead, and zinc vary in different tissues of the aquatic fish, of which liver is most susceptible to the metallic deposition and toxicity followed by the gills and gut. The relatively lower amount of accumulation was found in muscle tissues of the fish. The toxicity of these metallic ions has affected the life and growth of the aquatic animals (Gbem et al. 2001; Aich et al. 2011). Tannery effluents have also been demonstrated as an endocrine-disrupting compound in aquatic animals. The ranges of the concentration sometimes are high enough to result in deleterious reproduction consequences (Schwarzenbach et al. 2010).

The toxicity of tannery effluent is sometimes uncertain following the dilution of its components on the surface water. The nature of the individual components and their degradation rates, both in suspension and as part of sediment, vary on the dilution on its discharge on water. A wide variety of aquatic life and their responses to the tannery effluent may also vary, thus varying the type of toxicity caused by the tannery water waste components. Lack of specific data regarding the individual component of toxicity needs to be addressed. Hence, it is not possible to demonstrate the effects of complex or multiple components of tannery effluent on aquatic life including fish, invertebrates, algae, etc., and on human life.

The types of responses to an equal exposure of tannery components differ, and the degrees of toxicity standards are used for the protection of aquatic and human life.

**TABLE 13.1**
**Applications, Toxicity, and $LD_{50}$ for Chemicals Used during Leather Production in the Leather Industry**

| Name of Chemical | Applications | $LD_{50}$ in Rats (oral mg/kg) | Target Organs |
|---|---|---|---|
| Pentachlorophenol (PCP) (a carcinogen) | Applied as a biocide in preservative for raw hides/skins | 2000 | Eyes, nose, skin, respiratory tract, blood, kidney, liver, immune system, and reproductive system |
| Di-butyl phthalate (DBP) (an endocrine disrupting chemical) | Applied as a plasticizer in artificial leather manufacturing | 7499 | Eyes, lungs, gastrointestinal (GI) tract, and testes |
| Benzyl butyl phthalate (BBP) (an endocrine disrupting chemical) | Applied in preparation of microporous artificial leather coating/water vapor-permeable sheet materials | 2330 | Eyes, lungs, liver, and reproductive system |
| Bis(2-ethylhexyl) phthalate (DEHP) (an endocrine disrupting chemical) | Applied as a plasticizer in artificial leather manufacturing | 30,000 | Liver and testes |
| Short-chain, chlorinated paraffins | Additive for leather treatment (gives smoothness), leather clothing and belts, and as oiling agent | 3090 | Liver, kidney, and thyroid |
| Anthracene (a carcinogen) | Additive during tanning | 16,000 | Kidneys and liver |
| Nonyl phenol (an endocrine disrupting chemical and xenoestrogen) | Applied during finishing | 1475 | Blood, lungs, eyes, skin, central nervous system (CNS), kidneys, and testes |
| N-methyl pyrrolidone | Applied as a coalescene, plasticizers, and wetting agents | 3914 | Eyes, kidneys, lymphatic system, liver, lung, and testes |
| Methylisothiazolinone (a carcinogen) | Applied as a biocide | 1800 | Skin and eyes |
| Organotin compounds (Dibutyl tin) (a carcinogen) | Applied as a catalyst | 175 | GI tract and liver |
| Azo dyes (Orange II) (a carcinogen) | Applied as a dyeing agent | 3418 | Blood, liver, and testes |
| Hexachlorobenzene (a carcinogen) | Applied for raw hide/skins preservation | 10,000 | Reproductive system |
| Chromium (a carcinogen) | Applied as a tanning agent | 3250 | Kidneys, CNS, and hematopoietic system |

*(Continued)*

**TABLE 13.1** *(CONTINUED)*
**Applications, Toxicity, and $LD_{50}$ for Chemicals Used during Leather Production in the Leather Industry**

| Name of Chemical | Applications | $LD_{50}$ in Rats (oral mg/kg) | Target Organs |
|---|---|---|---|
| Formaldehyde (a carcinogen) | Applied in finishing of leather | 100 | Eyes and lungs |
| Arsenic (a carcinogen) | Applied in finishing of leather | 763 | Liver, kidneys, skin, lungs, and lymphatic system |
| Sodium dichromate | Applied in preparation of chrome-tanning salts | NA[a] | Blood, kidneys, heart, lungs, and eyes |
| Cobalt dichloride | Applied in dyeing and finishing | 80 | Skin, lungs, liver, kidney, and heart |
| Cadmium sulfate (pigment) | Applied as fastening agents and used in marking and surfacing of material | 280 | Lungs, liver, tissues, and reproductive system |
| Lead chromate (pigment) | Applied as fastening agents and used in marking and surfacing of material | 1000 | Lungs, liver, tissues, and reproductive system |

*Source*: Dixit, S., et al., *J. Clean. Prod.*, 87, 39–49, 2015; Saxena, G., et al., *Rev. Environ. Contam. Toxicol.*, 240, 31–69, 2016.
[a] NA, not available.

Environmental quality standards (EQS) set the limits of the components of tannery waste discharged into the streams and rivers after the treatment of the effluent. In the maintenance of aquatic life, EQS sets the limit of components tannery water waste, which can influence the life and growth of organisms. The heavy metals present in the tannery effluent are classified as "High/Moderate-Acute" and "Chronic toxicity" (Bosnic et al. 2000).

Wastewater irrigation led to the accumulation of heavy metals in agricultural soil as well as in crops and plants. The plants uptake the tannery components and they get accumulated in the parts of the agricultural plants. From the evidenced literature, the components of the tannery effluent have deteriorated the functioning and morphology of crops/plants in the soil. The major effects were seen in the growth of the plants and their seed germination property (Singh et al. 2010). Heavy metals and other pollutants of tannery components are taken up and accumulated by plants. Eventually, this influences the ecosystem; human health risks are in turn exposed to plants; and animals are majorly susceptible to their growth and quality of life.

## 13.2 MAJOR POLLUTANTS IN TANNERY WASTEWATER

Industrial wastewater is considered as the major source of environmental pollution (Saxena and Bharagava 2016; Saxena et al. 2016). Tannery wastewater is a high-strength wastewater and contains a variety of organic and inorganic pollutants

(Saxena and Bharagava 2015; Saxena et al. 2016). Organic pollutants mainly contain phenols, tannins, chlorophenols, recalcitrant dyes, and phthalates; whereas inorganic pollutants contain a variety of toxic metals such as cadmium, chromium, lead, etc. These pollutants together constitute toxicity of tannery wastewater in the environment upon disposal after treatment at the wastewater treatment plant at tannery industries or combined effluent treatment plant (CETP). A brief description of pollutants present in tannery wastewater is given below.

### 13.2.1 Solids

Solids are the insoluble matter contained in the tannery wastewater. These solids fall in the distinct categories named suspended and gross solids. These solids are generated from all the stages of the tannery process. Either they are composed of fine leather particles or residue from reagents and chemicals used in processing (Bosnic et al. 2000; Buljan et al. 2011).

Suspended solids can be seen on shaking the effluent and may settle within a few minutes of shaking the wastewater. Fine particles of suspended solids may take a considerable amount of time to settle down. They can be easily filtered through the filter papers. These particles produce sludge problems due to their large amounts, which can easily block the sludge pumps and pipes. On discharge into the streams and river, they form a thick layer on the water surface because of their suspension nature and, hence, alter the quality of water and, consequently, plant and aquatic life. Before discharge of the water waste into the streams, the solids need to be removed and dewatered (Bosnic et al. 2000). Gross solids are larger particles than suspended particles and can be seen clearly. They comprise the majority of leather particles, hair debris, and fleshing residues. They are the major threat to the water bodies as their accumulation leads to blockage and stagnation. This solid also settles and blocks the pipes and pumps on its discharge and damages the structure of pipe work and motors (Skrypski-Mantele and Bridle 1995).

### 13.2.2 Nitrogen

Nitrogen is present in several components of tannery waste as part of its chemical structure. The most common components are ammonia and the proteinaceous material containing nitrogen. These are the major threats to the environment as they deprive water bodies of oxygen and result in toxic anaerobic conditions. The problems occur due to a large amount of nitrogen present in the wastewater. Bacteria accomplish the breaking and decomposition of ammonia compounds, and this process needs a large amount of oxygen. The breakdown product is nitrogen gas, which is nontoxic. However, the need for a large amount of oxygen raises the high oxygen demand for the survival of the bacteria and eventually degrades the quality of water (Bosnic et al. 2000; Sumathi et al. 2014).

### 13.2.3 Chromium Compounds

Chromium is used in leather processing as a chrome-tanning process, which is a commonly employed method for permanent preservation. Chromium is heavily discharged in an effluent waste in the chrome-tanning process, since only up to 70% of the basic chromium sulfate is taken up by leather; hence, the major presence of chromium in tannery effluent raises ecological concerns (Leghouchi et al. 2009; Aich et al. 2011).

A soluble form of chromium ion (Cr-IV) has been used for industrial purposes for centuries. The major stable oxidative states of Cr element are hexavalent (Cr-VI) and trivalent (Cr-III), and chromium metal as well (Kolomaznik et al. 2008). Chromium has always been recognized for its irritant and allergic nature, and the experience from excessive exposure shows it is a human carcinogen. Cancer has been established from the chronic exposure of most toxic hexavalent forms, and it has been accepted widely as causing toxic responses in humans. Trivalent chromium is recognized as a principal irritant in toxic responses, but not enough studies have been conducted to determine their carcinogenicity (Chandra and Kulshreshtha 2004; Saranraj and Sujitha 2013).

### 13.2.4 Phenolic and Chlorophenolic Compounds

Tannery wastewater contains a considerable amount of phenolic compounds such as pentachlorophenol (PCP). The exposure to chlorinated phenols is possible particularly to PCP, which is highly carcinogenic, teratogenic, and mutagenic in nature and causes toxicity to living beings by inhibiting oxidative phosphorylation, inactivating the respiratory enzymes, and damaging the mitochondrial structure (Jain et al. 2005; Tripathi et al. 2011; Verma and Maurya 2013). The high concentration of PCP can also cause obstruction in the circulatory system, lungs, heart failure, and damage to the central nervous system (USDHHS 2001; Tewari et al. 2011; Dixit et al. 2015).

### 13.2.5 Phthalates

There has been an increasing concern regarding the release of many endocrine disrupting compounds (EDCs) along with tannery wastewater in the environment. EDCs disturb the delicate hormonal balance, compromise the reproductive fitness of animals, and ultimately may lead to carcinogenesis (Dixit et al. 2015). Kumar et al. (2008) have detected many EDCs like nonylphenol (NP), 4-aminobiphenyl, hexachlorobenzene, and benzidine in tannery wastewater collected from the northern region of India and tested their toxicity on the reproductive system of male rats. However, the presence of phthalates (EDCs) such as bis(2-ethylhexyl) phthalate (DEHP), dibutyl phthalate (DBP), bis(2-methoxyethyl) phthalate in tannery wastewater has also been reported (Alam et al. 2009, 2010).

### 13.2.6 Dyes

Tannery wastewater also contains azo dyes that are highly persistent in nature due to their complex chemical structure and xenobiotic nature leading to environmental pollution (Nachiyar and Rajkumar 2003; Gurulakshmi et al. 2008; Baccar et al. 2011; Mahmood et al. 2013; Dixit et al. 2015). Thus, the removal of azo dyes from tannery wastewater is essential because of their high mutagenicity, carcinogenicity, and intense coloration problems of contaminated aquatic resources (Osugi et al. 2009; Saratale et al. 2010). The discharge of azo dyes into the surface water also leads to esthetic problems and obstructs light penetration and oxygen transport into the water bodies, finally affecting aquatic life (Khalid et al. 2008; Chen et al. 2011). Moreover, these dyestuffs have also been reported to cause some other serious problems such as dermatitis, skin and eye irritation, and respiratory problems in human beings (Keharia and Madamwar 2003).

### 13.2.7 Other Heavy Metals

Heavy metals are trace elements and some have bio-importance. They have relatively high density and are biotoxic or poisonous at high concentrations. However, the toxicity of heavy metals concerns more of its chemical property rather its density. The major heavy metals present in tannery effluent and disrupting the quality of organism life are chromium, lead, cadmium, zinc, and aluminum. The concentration of the heavy metals varies on dilution on the surface water. Some heavy metals (iron, zinc, and magnesium) have been reported as bio-important to humans, and limits for the recommended dietary allowance have been set under EQS (Fraga 2005). Some of the heavy metals have no bio-importance; some that are present in living systems have not been demonstrated for their biological functions, and their role in biochemistry and physiology has not yet been recognized. Metals like lead, cadmium, mercury, and arsenic have no bio-importance and are toxic to human life and to other living organisms even at very low concentrations (Smith and Huyck 1999).

Epidemiological studies along with experimental studies have demonstrated the bio-disorders in humans on exposure to heavy metals above the permissible limits. Cytotoxicity of the metals has also been demonstrated as a potential mutagen in experimental animals. Metals may react with the proteins and enzymes to form toxic stable compounds hindering their biological functions. Some heavy metals are found to damage the structure of the cell walls or their organelles, even destroying the DNA molecules (Duruibe et al. 2007).

#### 13.2.7.1 Aluminum

Aluminum is a serious environmental toxicant and has been implicated in the pathogenesis of a number of disorders in humans including Parkinson's, osteomalacia, and Alzheimer's. Aluminum is an abundant metal, and it comprises about 8% of the Earth's surface and is the most common element in the Earth's crust. Physiologically, aluminum has no biological function in the human body, but it is added to pharmaceutical and food products including drinking water (Scheuhammer 1987). In the whole of the total aluminum concentration/amount, trivalent species $Al^{3+}$ only react to human physiology and are critically considered for toxicity. This form of species is highly aqueous, and the water waste from the tannery industry is found to comprise a low level of aluminum, but not enough for manifesting human toxicity. The pH of the solution may have a profound effect on the $Al^{3+}$ concentration. Under alkaline conditions, the concentration of $Al^{3+}$ switches to the aluminates species $Al(OH)_4(aq)$. At physiological pH, the soluble form of $Al(OH)_3$ evolves from the aluminum species, and solubility increases more toward the change of pH toward acidic media (Exley and Birchall 1992). No assigned biological function of aluminum is found. A higher affinity for the binding of aluminum to oxygen and nitrogen-containing compounds, particularly to organic compounds, and high accumulation of aluminum in organs and tissues leads to dysfunction and toxicity, leading to serious disorders. After exposure to aluminum through food and drinking water contamination, the main site of aluminum absorption is the gastrointestinal tract. The aluminum is also absorbed through the skin, olfactory, and oral epithelia. Once absorbed, the aluminum in the intestines could be completely excreted through urine from the body. Under higher dietary levels of aluminum and/or kidney dysfunctioning, the significant deposition of aluminum occurs in tissue and organs. Tissue aluminum concentration increases with the human age. Once the aluminum reaches the bloodstream, it binds to the transferrin and citrate, and like $Fe^{3+}$ its uptake is mainly regulated via the binding of the aluminum transferrin complex to the transferrin receptor. Aluminum crosses the blood–brain barrier and gets accumulated in neurons and glial cells.

#### 13.2.7.2 Zinc

Zinc (Zn) is the biological molecule present ubiquitously in the environment. It is imperative as a trace element for the growth and development of organisms including microorganisms and plant and animal life (Rout and Das 2009). Zn is the only metal that is present as cofactor in more than 300 enzymes, such as RNA polymerase, superoxide dismutase, and angiotensin I converting enzymes. Zn is appreciated for its major role in physiological functioning in the human body, of which the deficiency plays an important role in several diseases (Fraga 2005). It plays a key role in human physiology; zinc works as a cofactor in many enzymes, and as part of proteins, which provides structural functions in cells. Zn also acts as a signaling molecule in several physiological pathways and is necessary for maintenance of the cytoskeletal structure (Oteiza and Mackenzie 2005).

#### 13.2.7.3 Cadmium

The major source of cadmium (Cd) for exposure in a general population is food or the plant readily takes up Cd through soil irrigated from wastewater from the tannery industry containing a large amount of Cd or from fertilizers containing Cd. In mammals, the intestinal ion of Cd is dose dependent; however, it is toxic at very low levels (Scheuhammer 1987).

#### 13.2.7.4 Lead

Lead (Pb) is the most significant toxin of the heavy metals and is ubiquitously detectable in all phases of the inert environment and in all biological systems. The lead had no biological function. It is mainly hazardous if ingested in food, processed food, drinking water, and other products, and majorly from industrial exposure. This heavy metal is found in tannery wastewater in moderate to lower concentrations yet is toxic. Pb is toxic at even low concentrations of chronic exposure. The general population is exposed to the inorganic form of Pb mainly through ingestion by food and water, and by inhalation.

### 13.2.8 Sulfide ($S^{2-}$)

Sulfide can be an environmental hazard in tannery effluent resulting from the use of sodium sulfide and sodium hydrosulfide in a tannery process. In the de-hairing process of animal skin, the breakdown of the hair also results in generation of sulfide contents. This sulfide remains largely in the tannery solution unless there is a change in pH. The amount of change in pH of the effluent toward alkaline involves the formation of hydrogen sulfide. The lower the pH (below 9.5), the higher is the formation of sulfide.

### 13.2.9 Natural Salts (Chloride [$Cl^-$])

Sodium chloride (NaCl), commonly referred to as common salt, is used in large quantities in hide and skin preservation or the pickling process. Sodium chloride is highly stable and soluble in water, and hence, is unaffected in the environment in large amounts. A considerable quantity of sodium chloride rises up to maximum levels in drinking water, which is unacceptable in EQS. Unacceptable deposition of NaCl in drinking water and underground water, especially in the industrial area, is the major problem the environment is facing (Meneelys et al. 1953; Bosnic et al. 2000).

### 13.2.10 Oils and Grease

Oils in the form of fatty acids and grease are released and discharged into the wastewater generated from the leather manufacturing process. These oils are released either from the animal skin structure or from inter-reactions between the fatty components of tannery waste to produce oily components and grease. They block the pipework and treatment system on agglomeration of the grease and oils. On discharge of the wastewater into rivers and other water bodies, they float on the water surface and form a blanket or mats.

## 13.3 PHARMACOLOGICAL ASPECTS OF TANNERY WASTEWATER

### 13.3.1 Respiratory System

#### 13.3.1.1 Sulfide Toxicity

Hydrogen sulfide is a colorless gas heavier than air. It is a well-known toxic gas. In more than five decades, many findings have been generated and much more has been learned about sulfide toxicity. Hydrogen sulfide toxicity is still a problem because of its occupational exposure from industrial toxicity. Hydrogen sulfide is fairly soluble in water and is a physiologically important biomolecule. It has been recognized as a signaling molecule and cytoprotectant as well. Hydrogen sulfide is synthesized in humans by cysteine enzymes along with the major enzymes including cystathione beta-synthase, cystathione gamma-lyase, and 3-mercaptopyruvate sulfur transferase (Guidotti 2010; Kimura 2011).

Pulmonary edema is a well-recognized injury caused by acute effects of hydrogen sulfide. It irritates the epithelia of the deep lungs. Hydrogen sulfide deeply penetrates into the respiratory tract leading to alveolar injury due to their relatively low solubility. Acute episodes of exposure of hydrogen sulfide could be recovered easily as it does not disrupt the basement of the cells of the alveolar endothelia and is moderately cytotoxic. The irritant property of hydrogen sulfide also results in superficial irritation of the cornea and conjunctiva.

Distinguishably, hydrogen sulfide also causes olfactory paralysis on exposure to its high concentration. Its highly odorous property (smells like a rotten egg) causes severe odor problems. Olfactory paralysis on exposure of hydrogen sulfide is cause by two distinctive mechanisms. Prolonged and intense odor may cause olfactory fatigue at lower-level exposure, and people may become accustomed to the odor problem. In contrast, neurotoxic effects of hydrogen sulfide can lead to olfactory paralysis. Prolonged exposure may damage olfactory bulb and fibers, which may result in permanent damage to the sense of smell.

### 13.3.2 Urinary System

#### 13.3.2.1 Chronic Cadmium Toxicity

The critical organ affected by chronic Cd toxicity is generally considered to be the kidney. Long-term exposure results in renal dysfunction characterized by tubular proteinuria. Cd is toxic to renal tubular cells and glomeruli leading to impaired dysfunctioning of renal tubular and glomerular function. In the initial stage, tubular cell degeneration occurs in progression to fibrosis and inflammatory reactions.

In chronic renal toxicity, the cadmium–metallothionein (Cd–MT) complex plays a major role affecting physiological function over a longer duration of accumulation. MT is a low molecular weight binding protein ubiquitous in most organs, but it exists in higher concentrations in liver and kidney. The major role that MT plays is detoxifying the cadmium bound to it, but due to the long biological half-life of the MT complex, it tends to accumulate with time.

#### 13.3.2.2 Aluminum Toxicity

Chronic aluminum exposure has been implicated in renal dysfunction and further complications of osteomalacia on regular hemodialysis. Osteomalacia is a condition of bone softening through deficiency of calcium and vitamin D. However, aluminum-induced osteomalacia typically differs from the classical vitamin D deficient pathology. In such disease, aluminum deposition and concentration is found in the bones and serum, and the cure could not be possible through classical methods of vitamin D dietary supplementation despite osteomalacia having typical biochemical features (Boyce et al. 1982; Abreo and Glass 1993).

### 13.3.3 Hypertension and Cardiovascular System

#### 13.3.3.1 Cadmium Toxicity

Many epidemiological studies suggest that cadmium is an etiological agent for essential hypertension. Animal studies have concluded that cadmium is toxic for myocardial functions. Moreover, epidemiologically, cadmium toxicity increases systolic

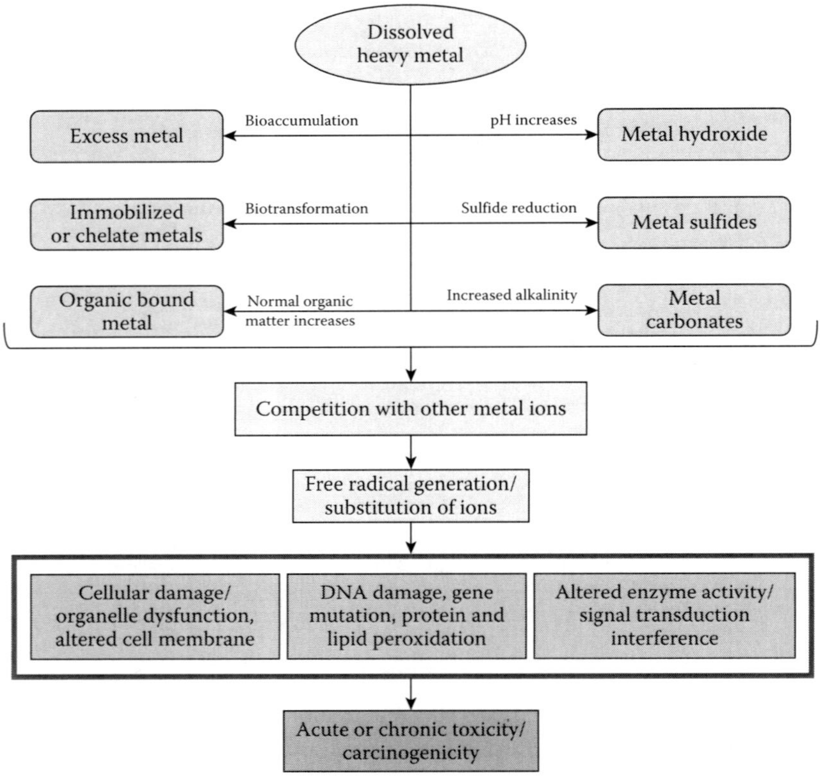


**FIGURE 13.1** Mechanism of carcinogenesis from metals present in effluent from the tanning industry.

blood pressure but has no effect on diastolic blood pressure. Cadmium toxicity on exposure via drinking water develops electrocardiographic and biochemical changes and impairment of the function of the myocardium. These abnormalities may be related to decreased myocardial contractility and decreased high-energy phosphate stored in the myocardium. Epidemiological studies have shown the implication of cadmium toxicity in carcinogenesis, particularly prostate cancer and lung cancer, when exposed through inhalation chronically (Figure 13.1).

#### 13.3.3.2 Lead Toxicity

Many of the studies also concluded that cases of elevated lead compounds levels correlated with increased hypertension and associated cardiovascular diseases. Studies that contradict the finding of increase in blood pressure upon elevation of lead compound accumulation are also present, so this finding is not supported.

#### 13.3.3.3 Sulfide Toxicity

Hydrogen sulfide is a neurotransmitter molecule that facilitates the induction of hippocampal long-term potentiation (LTP) by enhancing the activity of NMDA receptor in neurons, which alone does not induce LTP. Hydrogen sulfide is also a smooth muscle relaxant. Its production and release from the thoracic aorta, portal vein, and the ileum through cystathione-gamma-lyase enzymes causes smooth muscle relaxations. Literature suggests that it is physiologically a smooth muscle relaxant and may regulate blood pressure. Relaxation mostly occurs by opening ATP-dependent potassium channels in the smooth muscles. Along with these physiological activities, hydrogen sulfide also protects cardiac muscles from ischemic injury. It also attenuates the isoproterenol-induced decrease in myocardial contractility and decreases the size of infarction induced by left coronary artery ligation. Acute toxicities of hydrogen sulfide exposure in high concentrations include pulmonary edema or irritant toxicity, irrigative conjunctivitis and inflammation of cornea, olfactory dysfunctioning, and gastrointestinal toxicity including jaundice and diarrhea (Beauchamp et al. 1984).

#### 13.3.3.4 Chloride Toxicity

It is established that dietary sodium is important or necessary for the development for certain types of human hypertension. Experimental findings support the fact that only dietary and body sodium alone plays an important role in the development of human essential hypertension, and it does not increase blood pressure in normal human beings. It is either sodium load or chloride rising to unacceptable levels that is a serious threat to human health (Meneely 1953; Dahl and Love 1954; Kawasaki et al. 1978).

### 13.3.4 Nervous System

#### 13.3.4.1 Aluminum Toxicity

Aluminum is considered a pro-oxidant and in high concentration causes oxidative stress and formation of free radicals. Imbalance between the oxidant and antioxidant activity of aluminum results in favoring oxidative stress and disrupting redox signaling and molecular level damage. This oxidative stress status has been found in association with neurodegeneration disorders, following aluminum accumulation in brain cells. Aluminum changes the levels of redox enzymes such as superoxide dismutase and catalase, and causes considerable changes in biomarkers of the peroxidation. Evidenced literature reported that the formation of free radicals and disruption of molecular signaling result in neurodegeneration disorders such as Parkinsonism, Alzheimer's disease, dementia, and toxicity of other organs including hepatic damage. Apart from the disrupting oxidative stress, aluminum also influences neurotransmission negatively. Recent advances in findings state that aluminum may disrupt the physical property of the synaptic membrane that uptakes the neurotransmitters. Aluminum also caused dysfunctioning of the neuronal enzymes by inhibiting or disrupting the structural formation of enzymes responsible for the synthesis of neurotransmitters. The pathophysiology of Parkinsonism disorders characterizes dopamine deficiency in the nigrostriatal region of the brain. Experimental findings reported that high levels of aluminum in neurons inhibit the enzymes responsible for dopamine formation, decrease the dopamine level in the striatum by 40%, and alter the metabolism of dopamine. Aluminum also inhibits the conversion of dopamine to norepinephrine through dopamine β-hydroxylase and promotes alpha synuclein aggregation of fibrillation, which again enhances neuronal degeneration. Aluminum toxication also imbalances and decreases the dopamine receptors in striatum and brain cortex leading to more pronounced neurodegenerative disorders. Interrupting the key events of the neurotransmission and coagulation of neurodegenerative disorder, aluminum toxicity studies need more insight into the etiology of brain dysfunction (Krewski et al. 2007; Mohammadirad and Abdollahi 2011).

Aluminum toxication also affects the cholinergic and GABAergic metabolism. In animal models, severe cognitive functioning impairments were reported on chronic aluminum exposure associated with a decrease in acetylcholine synthesis and its metabolism. Aluminum exposure increases the levels of excitatory neurotransmitter including glutamine and glutamate, and inhibits the GABA neurotransmitter suggesting excitotoxic neuronal damage disorders. Aluminum also decreases the levels of serotonin, which is the key neurotransmitter responsible for maintaining the "biological clock" for sleeping, eating, and neuro-endocrinal function (Verstraeten et al. 2008).

#### 13.3.4.2 Sulfide Toxicity

Complication of hydrogen sulfide toxicity, as with all direct toxic effects, leads to more prevalent secondary effects. Even at low concentrations, exposure to the gas induces headache and nausea and short-term cognitive changes. Nonspecifically, short-term memory loss and posttraumatic stress are common. Seizures disorders have also been reported in a few cases, but their clinical descriptions were not validated in most cases. The neurotoxicity of hydrogen sulfide is not simple. Accumulated reports and findings suggest that chronic brain injury and chronic hydrogen sulfide exposure predominately results in anoxic brain damage (Reiffenstein et al. 1992).

Besides the physiological importance of hydrogen sulfide in human organs, the sulfide concentration present in tissue following its toxicity still unknown. Concentration and duration of hydrogen sulfide exposure are much more important for its toxicity. Surprisingly, human brain concentration of hydrogen sulfide is 1.5 μg/g of brain tissues. For poisoning from hydrogen sulfide, the concentration has to cross 3.1 μg/g of brain tissue; this indistinguishable increase in hydrogen sulfide concentration follows widespread environmental pollution, of which tannery waste alone is enough.

### 13.3.5 Digestive System

#### 13.3.5.1 Zinc Toxicity

Critically, zinc is recognized for its role in immune functions, maintenance of homeostasis, role in oxidative stress, etc. In addition to the importance of zinc in vital biological functions, its deficiency also may complicate some diseases, such as autoimmune and neurological diseases. Zinc deficiency has been associated with malnutrition in several countries (Aggett and Harries 1979; Chasapis et al. 2012).

Although it is vital for human growth and development, a higher concentration of zinc exposure always leads to toxicity and poisoning. Excess zinc ingestion in the human body can result in various biological dysfunctioning, including oxidative stress, aberrations in cell functioning, and survival. Acute toxicity of zinc on exposure of higher concentration symptoms includes anorexia, vomiting, nausea, dizziness, lethargy, diarrhea, and bleeding gastric erosion.

### 13.3.6 Skeletal Effects

#### 13.3.6.1 Acute Cadmium Toxicity

Acute toxicity of cadmium results from ingestion of relatively high concentrations in contaminated food and drinking water. Depending on the severity of exposure, the symptoms include nausea, vomiting, abdominal cramps, muscular weakness, and dyspnea (Goyer and Clarkson 1996). Cadmium toxicity is fairly accumulated in a large amount in the organs (such as the kidney and liver). Gastrointestinal absorption of cadmium occurs by dietary deficiency of calcium and iron. Calcium binding protein increases in states of calcium deficiency, and this binding protein enhances the uptake of cadmium in organs. Cadmium exposure also affects the bone tissues, but the mechanism by which this occurs is not yet clear. In calcium deficiency, the accumulation of cadmium along with vitamin D deficiency results in severe bone degeneration and diseases like osteomalacia, osteoporosis, and spontaneous fractures (Goyer and Clarkson 1996).

### 13.3.7 Exposure during Pregnancy

#### 13.3.7.1 Lead Toxicity

A notably serious effect of lead toxicity has been associated with neonatal death in humans on long-term exposure. Clinically, lead toxicity is critically recognized as teratogenic. In one study, workers exposed to lead were found with low sperm count and abnormal sperm motility, which have been associated with sterility in humans. Chronic exposure to lead was found to decrease the testicular endocrine functions in workers. An effect of chronic exposure of lead in females results in abortion or preterm delivery. In this aspect, epidemiological studies have been conducted to determine the relationship of prenatal lead exposure and birth outcome. Chronic lead exposure has increased the risk of preterm delivery. Moreover, gestational duration was reduced, and decreased birth weight was reported (Goyer and Clarkson 1996).

### 13.3.8 Carcinogenic Property

#### 13.3.8.1 Chromium Toxicity

Under physiological pH conditions, the hexavalent chromium compound exists as anoxion $(CrO_4)^{-2}$, which is not commonly found. Anoxion has a close resemblance to sulfates and phosphates. Due to their mimic property, anoxion is taken by the cells via the anionic transport system and gets accumulated in the cell. The toxicity of this metal varies with the physiological pH. Chromate ions $(CrO_4)^{-2}$ predominated at the above range of physiological pH where dichromate exists at pH in a range of 2 to 6 (Messer et al. 1999). Chromate accumulation in the cells in physiological pH sometimes gets much higher than the extracellular levels. After chromate accumulation, hexavalent spontaneously reduced to trivalent (Cr-III) compounds by various intracellular reductants, including ascorbic acid and glutathione. Once the trivalent is formed, the formed compound is bound to cellular constituents. Various experimental studies for the use of trivalent (Cr-III) compounds as nutrition supplements are ongoing. Evidence of experimental data on the beneficial effects of Cr-III in human has been established. Trivalent Cr-III helps in insulin-dependent glucose regulation. However, the trivalent compounds have shown some of its irritant and allergen activity in humans, but somehow it lacks the carcinogenic property, and thus, this finding makes it necessary to continue research into its essential role in nutritional supplementation (Vincent 2011).

Ascorbic acid is the major reducer for hexavalent chromium to trivalent chromium in the cell. Hexavalents are either converted to trivalents directly or through various intermediate Cr-IV and Cr-V depending on the nature of the cell and the type of reducer. Hexavalent chromium is a human carcinogen of which lung cancers have been reported in the literature. Other than lung cancers, hexavalent chromium has been established and recognized for its role in other cancer such as gastrointestinal systems. From various cases and epidemiological studies reported, chromium was recognized for lung cancers on exposure to smokers or nonsmokers. In the past, workers in the chromic pigments industrial area were diagnosed with lung cancer (Dayan and Paine 2001).

Hexavalent chromium is a major cause of cumulative exposure and the study firmly concluded hexavalent carcinogenicity. The type of toxicity differs with the route of exposure of chromium compounds. Hexavalent chromium on inhalation irritates the respiratory tract and causes ulceration of the nasal septum. Rhinitis, pneumonia, and bronchospasm have also been reported for hexavalent toxicity via inhalation (Costa and Klein 2006). The chronic accumulation of chromates in cells results in cytotoxicity via disrupting the molecular functioning of the cells and causing damage. On attack at molecular levels of the cells, chromium can damage DNA in several ways. Chromates, in particular, sodium chromates and potassium chromates, dichromates, and strontium chromates induce gene mutation. They also damage DNA via chromosomal aberration, sister chromatid exchange, cell transformation, and mutation in numbers of DNA targets. This eventually results in alteration of DNA replication and transcription (Dayan and Paine 2001; Matsumoto et al. 2006).

#### 13.3.8.2 Cadmium Toxicity

Gastrointestinal absorption of cadmium is low. Cadmium accumulates through time bound to a low molecular weight sulfhydryl-rich protein named metallothione (MT). The synthesis of MT increases on human exposure to heavy metals like zinc, copper, and mercury. The role of MT for biological function needs to be fully defined, but MT typically functions as storage and metabolism of zinc and copper in human organs. It also detoxifies heavy metals like cadmium. However, this MT bound to cadmium has a biological half-life of nearly 20 years, and deposition of cadmium in organs is a time-dependent process that increases with age. Cadmium metallothione (Cd–MT) complex catabolism is slow, and chronic accumulation may result in renal toxicity and its dysfunctioning. For this chronic renal toxicity, great concern has been devoted to levels of cadmium intake for the general population (Shaikh et al. 1999).

#### 13.3.8.3 Lead Toxicity

Lead has also been recognized for renal toxicity as one of the oldest health effects due to chronic exposure. In the experimental animal model, lead compounds have also been found to be a renal carcinogen on chronic exposure, where the acute toxicity of lead holds for nephrotoxicity, which is reversible. Nephrotoxicity is limited to functional and morphological changes to tubular cells. These pathological changes are reversible by treatment. In patients with renal toxicity due to lead exposure, the incidences of gouty deposition of uric acid are found in some cases. However, the relationship between elevation of blood uric acid and lead poisoning is still not clear (Goyer and Clarkson 1996).

A number of epidemiological studies confirm that chronic occupational lead exposure can cause increased incidences of cancers. The most common tumors were found in respiratory and digestive systems. Since then hundreds of case reports have been studied and suggested higher incidences of lung cancers in workers and the general population in industrial areas. Lead compounds have also been shown to stimulate cell proliferation and increase survival of tumor cells. As a result, lead is classed as a 2B carcinogen.

## 13.4 BIOREMEDIATION APPROACHES FOR TANNERY WASTEWATER POLLUTANTS

### 13.4.1 Bioremediation

Tannery wastewater is a high strength wastewater that is characterized by very high BOD, COD, and TDS, and contains a variety of organic and inorganic pollutants such as chromium and sulfide with its low biodegradability that is of serious environmental concern (Saxena and Bharagava 2015; Saxena et al. 2016). The physicochemical treatment methods are not efficient and eco-friendly because these methods use a large amount of hazardous chemicals and generate a large quantity of chromium-containing sludge in the environment. Therefore, bioremediation is an eco-friendly way to reduce pollutant load from tannery wastewater prior to its safe disposal into the environment.

Bioremediation is the eco-friendly method for the degradation and detoxification of industrial wastewaters wherein microbes such as algae, fungi, and bacteria are involved to degrade and detoxify the pollutants, whether organic and inorganic in nature, present in industrial wastewaters. Saxena et al. (2016) have reviewed various microorganisms involved in the bioremediation of tannery wastewater (Table 13.2). Megharaj et al. (2003) isolated chromium-resistant bacterial strain *Bacillus cereus* S-6 from effluents from a tannery, which was used for the reduction of toxic hexavalent chromium into less toxic trivalent chromium. At an initial hexavalent chromium concentration of 100 μg/mL, the cytosol and membrane preparation of the strain were able to reduce almost 67% and 43% of hexavalent chromium within a 24-hour incubation period while the heat killed cytosol and membrane preparation reduced 24% and 18% within the same time period. They reported that tannery effluent may change the characteristics of soil and interfere with the intake of water by plants. The presence of sulfide and chromium in tannery effluent affects plant life and soil productivity.

**TABLE 13.2**
**Microorganisms Reported in the Bioremediation of Tannery Wastewater**

| References | Microorganisms | Chemical Oxygen Demand Removal (%) | Biological Oxygen Demand Removal (%) | Cr Removal (%) |
|---|---|---|---|---|
| Kim et al. (2014) | *Brachymonas denitrificans* | 98.3 | — | 88.5 |
| Noorjahan (2014) | *Escherichia coli* | 90 | 90 | 63.8 |
| | *Bacillus* sp. | 95.4 | 95.4 | 73.5 |
| Elmagd and Mahmoud (2014) | Mixed culture | 98.3 | 98.4 | 98.3 |
| Sharma and Malviya (2014) | *Fusarium chlamydosporium* SPFS2-g | 71.80 | — | — |
| Yusuf et al. (2013) | *Bacillus subtilis* | 87.6 | — | — |
| | *Pseudomonas fragi* | 85.2 | | |
| El-Bestawy et al. (2013) | *Providencia vermicola* W9B-11, *E. coli* O7:K1 CE10, *Bacillus* sp. 58, *Bacillus amyloliquefaciens* T004, *Pseudomonas stutzeri* M15-10-3, *Bacillus* sp. PL47 | 79.16 | 94.14 | 93.66 |
| Mandal et al. (2010) | *Thiobacillus ferrooxidans* | 69 | 72 | 5 |
| Nanda et al. (2010) | *Nostoc* sp. | 37.8 | 48.6 | — |
| Ramteke et al. (2010) | *E. coli* | 98.46 | 90 | — |
| | *Vibrio* sp. | 87.5 | | |
| | *Pseudomonas* sp. | 96.15 | | |
| Sivaprakasam et al. (2008) | *Pseudomonas aeruginosa, Bacillus flexus, Exiguobacterium homiense, Staphylococcus aureus* | 80 | — | — |
| Vankar and Bajpai (2008) | *Trichoderma* sp. | — | — | 97.93 |
| Onyancha et al. (2008) | *Spirogyra condensate* *Rhizocolonium hieroglyphicum* | — | — | >75 |
| Srivastava et al. (2007) | *Acinetobacter* sp. | — | — | 90 |
| Rajasimman et al. (2007) | Mixed culture | 46–85 | 65–93 | — |
| Wang et al. (2007) | *Acidithiobacillus thiooxidans* | — | — | 99.7 |
| Srivastava and Thakur (2006) | *Aspergillus* sp. | — | — | — |
| | *Hirsutella* sp. | | | 70 |
| Lefebvre et al. (2005) | Halophiles | 95 | — | — |
| Thanigavel (2004) | Mixed culture | 89.5 | — | — |
| Shakoori et al. (2000) | Bacterial strain | — | — | 87 |

Several reports have indicated the biological reduction of hexavalent chromium by both aerobic and anaerobic microorganisms. Biological reduction of hexavalent chromium usually occurs at a neutral pH range and generates an insignificant quantity of chemical sludge as well as offers a potential cost-effective remediation strategy (Donati et al. 2003). Subsequent studies have shown that the capability of hexavalent chromium reduction is widespread and is shown in organisms such as *Bacillus subtilis, Pseudomonas aeruginosa, Achromobacter, Eurydice, Micrococcus roseus,* and *Escherichia coli* (Flores and Perez 1999) as well as *Pseudomonas ambigua* (Guha et al. 2001), *Pseudomonas fluorescens* (Jonnalagadda et al. 2007), *Enterobacter cloacae* (Komori et al. 1990), *Streptomyces* sp. (Romanenko and Korenkov 1977), *Pseudomonas putida* (Schulin 2007), *Desulfovibrio desulfuricans, Desulfovibrio vulgaris* (Lovley and Phillips 1994), and *Pseudomonas liquefaciens* (Sultan and Hasnain 2005).

Srinath et al. (2002) studied that *Bacillus circulans* and *Bacillus megaterium* are able to bioaccumulate 34.5 and 32 mg chromium/g dry weight, respectively, and brought the residual concentration of hexavalent chromium to the permissible limit in 24 hours when the initial concentration was 50 mg hexavalent chromium/L. They stated that biosorption of hexavalent chromium was shown by *Bacillus megaterium* and another strain *Bacillus coagulans*. Living and dead cells of *Bacillus coagulans* biosorbed 23.8 and 39.9 mg chromium/g dry weight, respectively, whereas, 15.7 and 30.7 mg chromium/g dry weight were biosorbed by living and dead cells of *Bacillus megaterium,* respectively.

Muhammed Faisal and Shahida Hasnain (2004) demonstrated that the ability of *Brevibacterium* cells to accumulate toxic hexavalent chromium at different chromate concentrations (100, 500, and 1000 μg/mL) in different time intervals (15 minutes, 2 hours, and 4 hours). They showed that the *Arthrobacter oxydans* does a complete uptake of hexavalent chromium concentration (35 mg/mL) in about 10 days.

Igwe and Abia (2006) found that the chromium was bioremoved from tannery wastewater by *Aspergillus oryzae. Aspergillus oryzae* can grow in different concentrations of chromium (120–1080 mg/L). They observed that maximum biomass growth and chromium removal rate at pH 3.3, trivalent chromium concentration equal to 240 mg/L, and inoculum size equal to 0.12% (dry weight) were 0.25 (dry weight) and 94.2%, respectively. Shaili Srivastava and Indu Shekhar Thakur (2006) studied the relationship between the hexavalent chromium resistance of a culturable microbial community and hexavalent chromium resistance and reducing ability of the population. Srivastava and Thakur (2007) isolated chromium-resistant bacterial strain *Bacillus cereus* S-6 from effluents of tannery and used it for the reduction of toxic hexavalent chromium into less toxic trivalent chromium. At an initial hexavalent chromium concentration of 100 μg/mL, the cytosol and membrane preparation of the strain were able to reduce almost 67% and 43% of hexavalent chromium within a 24-hour incubation period while the heat-killed cytosol and membrane preparation reduced 24% and 18% within the same time period.

Saranraj et al. (2010) also isolated a bacterial strain from tannery effluent and identified it as *Enterococcus casseliflavus.* It showed a high-level resistance of 800 μg/mL chromium. The minimal inhibitory concentration of chromium was found to be 512 μg/mL of potassium dichromate in nutrient broth medium. The chromium adsorption was more significant by the live cells than killed cells at different time intervals. It was observed that the inoculation of *Enterococcus casseliflavus* reduced the BOD and COD values of tannery effluent. The maximum adsorption of chromium occurred at a temperature between 35°C and 45°C and at a pH of 7–7.5.

### 13.4.2 Phytoremediation

Phytoremediation is an eco-sustainable technology used for the remediation of industrial wastewaters or contaminated soil and uses plants composed of two components, one by the root colonizing microbes and the other by the plants themselves, which accumulate metabolites and convert the toxic compounds into nontoxic products. Phytoremediation is an emerging green technology used to clean up the contaminated sites, which combines the disciplines of plant physiology, soil chemistry, and soil microbiology. It is cost-effective and has esthetic advantages and long-term applicability. It is best applied at sites with shallow contamination of organic, nutrient, or metal pollutants that are amenable to one of the five applications: phytotransformation, rhizoremediation, phytostabilization, phytoextraction, and rhizofiltration. After sufficient plant growth and metal accumulation, the above-ground parts of the plant are harvested and removed, resulting in permanent removal of metals from the contaminated site. A lot of literature can be found in the public domain on the phytoremediation of chromium from contaminated sites (Shahandeh and Hossner 2000; Mohanty and Patra 2012).

### 13.4.3 Constructed Wetlands

Constructed wetlands (CWs) are man-engineered, eco-friendly systems designed to remove the pollutants from highly polluted industrial and municipal wastewaters. The use of CWs for the treatment of industrial wastewater has developed rapidly in current years and is successfully employed to remove a diverse group of pollutants from industrial wastewaters. The proper functioning of a wetland system depends on the complex relationship between the plants, microorganisms, soil, wastewater characteristics, and operational parameters (Aguilar et al. 2008). In this regard, several efforts have been made to select a suitable plant species capable of tolerating and removing pollutants from tannery wastewater (Mant et al. 2004; Calheiros et al. 2007, 2008, 2012), to select suitable supporting media/substrate for proper growth and development of wetland plants (Calheiros et al. 2008), as well as to study the bacterial community dynamics in CWs (Aguilar et al. 2008; Calheiros et al. 2009a, 2009b). The plant roots and rhizomes are the major sites of microbial degradation/transformation of pollutants, and subsequently the purification of wastewater because microbes form a biofilm on root surface and substrates (Stottmeister et al. 2003; Gagnon et al. 2007; Munch et al. 2007). However, the availability of nutrients or other environmental parameters affects the biofilm formation (Kierek-Pearson and Karatan 2005). Therefore, detailed profiling of complex microbial populations is required to understand the proper functioning of CWs and phytoremediation processes (Chandra et al. 2015). Culture-dependent techniques are known to be insufficient to study the microbial community structure because numerous microorganisms are unculturable in lab conditions (Ward et al. 1990).

Hence, the molecular techniques such as random amplified polymorphic DNA (RAPD), polymerase chain reaction (PCR), and denaturation gradient gel electrophoresis (DGGE) are used for the study of microbial community structure, composition, and diversity in CW system (Calheiros et al. 2009a, 2012).

Mant et al. (2004) have studied the phytoremediation potential of *Pennisetum purpureum*, *Brachiaria decumbens,* and *Phragmites australis* in CWs for the removal of chromium (ranging from 10 to 20 mg Cr $dm^{-3}$) from tannery wastewater. In addition, the potentials of *Canna indica*, *Typha latifolia, P. australis, Stenotaphrum secundatum,* and *Iris pseudacorus* in CWs for the treatment of tannery wastewater under two different hydraulic loading rates at 3 and 6 cm/day have been studied, and it was found that only *P. australis* and *T. latifolia* were able to establish successfully (Calheiros et al. 2007). Further, these authors also evaluated *Arundo donax* and *Sarcocornia fruticosa* in two series of horizontal subsurface flow CWs used to treat tannery wastewater received from a conventional biological treatment plant and reported the removal of COD (51% and 80%) and $BOD_5$ (53% and 90%) for COD inlet: 68–425 mg/L and for $BOD_5$ inlet: 16–220 mg/L (Calheiros et al. 2012). In addition, the use of tannery wastewater as a growth medium for Arthrospira (*Spirulina*) has been recently suggested (Dunn et al. 2013). However, the chromium salt can be retained in wetlands with nonspecialized supporting media (Dotro et al. 2012).

## 13.5 CONCLUSION

Tannery wastewater is ranked as one of the major environmental pollutants among all the industrial wastewaters due to the presence of a variety of toxic and hazardous chemicals such as chromium, chlorophenols, oils, resins, biocides, and phthalates, which create a negative image of LIs (Lofrano et al. 2013; Dixit et al. 2015; Saxena et al. 2016). These hazardous compounds have various pharmacological effects in humans and animals, especially endocrine disrupting chemicals such as phthalates that cause disruption of the endocrine system in males as evidenced from a report available on the male rat (Kumar et al. 2008). Therefore, the proper treatment of tannery wastewater before disposal into the environment is required for human health protection and environmental safety. Further, physicochemical methods are generally applied in developing countries for the treatment of tannery wastewater, which are not effective as these require toxic chemicals for the treatment and generate a large amount of sludge as a secondary pollutant in the environment. Hence, biological approaches such as bioremediation can be effective tools for the effective degradation and detoxification of tannery wastewater due to their eco-friendly and cost-effective nature.

## ACKNOWLEDGMENTS

The authors are extremely grateful to the "Science and Engineering Research Board" (SERB), Department of Science and Technology (DST), Government of India (GOI), New Delhi, India, for financial support as "Major Research Project" (Grant No.: SB/EMEQ-357/2013) for this work. The financial support as Rajeev Gandhi National Fellowship (RGNF) to Swetlana Gautam and University Grant Commission (UGC) Fellowship to Mr. Gaurav Saxena are also duly acknowledged.

## REFERENCES

Abreo, K., and Glass, J. 1993. Cellular, biochemical, and molecular mechanisms of aluminum toxicity. *Nephrology and Dialysis Transplantation* 8 (Suppl 1): 5–11.

Aggett, P.J., and Harries, J.T. 1979. Current status of zinc in health and disease states. *Archives of Disease in Childhood* 54(12): 909.

Aguilar, J.R.P., Cabriales, J.J.P., and Vega, M.M. 2008. Identification and characterization of sulfur oxidizing bacteria in an artificial wetland that treats wastewater from a tannery. *Int. J. Phytoremediation* 10(5): 359–370.

Aich, A., Chattopadhyay, B., Datta, S., and Mukhopadhyay, S.K. 2011. Impact of composite tannery effluent on the amino-transferase activities in a fish biosystem, using Guppy fish (*Poecilia reticulata*) as an experimental model. *Toxicological and Environmental Chemistry* 93(1): 85–91.

Akan, J.C., Abdulrahman, F.I., Ayodele, J.T., and Ogugbuaja, V.O. 2009. Impact of tannery and textile effluent on the chemical characteristics of Challawa River, Kano State, Nigeria. *Australian Journal of Basic and Applied Sciences* 3(3): 1933–1947.

Alam, M.Z., Ahmad, S., and Malik, A. 2009. Genotoxic and mutagenic potential of agricultural soil irrigated with tannery effluents at Jajmau (Kanpur), India. *Archieves of Environmental Contamination and Toxicology* 57(3): 463–476.

Alam, M.Z., Ahmad, S., Malik, A., and Ahmad, M. 2010. Mutagenicity and genotoxicity of tannery effluents used for irrigation at Kanpur, India. *Ecotoxicology and Environmental Safety* 73(5): 1620–1628.

Baccar, R., Blanquez, P., Bouzid, J., Feki, M., Attiya, H., and Sarra, M. 2011. Decolorization of a tannery dye: From fungal screening to bioreactor application. *Biochemical Engineering Journal* 56(3): 184–189.

Beauchamp, R.O., Bus, J.S., Popp, J.A., Boreiko, C.J., Andjelkovich, D.A., and Leber, P. 1984. A critical review of the literature on hydrogen sulfide toxicity. *CRC Critical Reviews in Toxicology* 13(1): 25–97.

Bharagava, R.N., Yadav, S., and Chandra, R. 2014. Antibiotic and heavy metal resistance properties of bacteria isolated from the aeration lagoons of common effluent treatment plant (CETP) of tannery industries (Unnao, India). *Indian Journal of Biotechnology* 13(4): 514–519.

Bosnic, M., Buljan, J., and Daniels, R.P. 2000. *Pollutants in tannery effluents*. UNIDO, Vienna. Rev.

Boyce, B.F., Elder, H.Y., Elliot, H.L., Fogelman, I., Fell, G.S., Junor, B.J., Beastall, G., and Boyle, I.T. 1982. Hypercalcaemic osteomalacia due to aluminium toxicity. *The Lancet* 320(8306): 1009–1013.

Buljan, J., Kral, I., Clonfero, G., Bosnic, G., and Schmel, F. 2011. *Introduction to treatment of tannery effluents*. United nations industrial development organization (UNIDO), Vienna.

Calheiros, C.S.C., Duque, A.F., Moura, A., Henriques, I.S., Correia, A., Rangel, A.O.S.S., and Castro, P.M.L. 2009a. Changes in the bacterial community structure in two-stage constructed wetlands with different plants for industrial wastewater treatment. *Bioresour. Technol.* 100(13): 3228–3235.

Calheiros, C.S.C., Quiterio, P.V.B., Silva, G., Crispim, L.F.C., Brix, H., Moura, S.C., and Castro, P.M.L. 2012. Use of constructed wetland systems with Arundo and Sarcocornia for polishing high salinity tannery wastewater. *J. Environ Manage.* 95(1): 66–71.

Calheiros, C.S.C., Rangel, A.O.S.S., and Castro, P.M.L. 2007. Constructed wetland systems vegetated with different plants applied to the treatment of tannery wastewater. *Water Res.* 41(8): 1790–1798.

Calheiros, C.S.C., Rangel, A.O.S.S., and Castro, P.M.L. 2008. Evaluation of different substrates to support the growth of Typha latifolia in constructed wetlands treating tannery wastewater over long-term operation. *Bioresour. Technol.* 99(15): 6866–6877.

Calheiros, C.S.C.., Rangel, A.O.S.S., and Castro, P.M.L. 2009b. Treatment of industrial wastewater with two-stage constructed wetlands planted with Typha latifolia and Phragmites australis. *Bioresour. Technol.* 100(13): 3205–3213.

Chandra, P., and Kulshreshtha, K. 2004. Chromium accumulation and toxicity in aquatic vascular plants. *The Botanical Review* 70(3): 313–327.

Chandra, R., Saxena, G., and Kumar, V. 2015. Phytoremediation of environmental pollutants: An eco-sustainable green technology to environmental management. In *Advances in biodegradation and bioremediation of industrial waste*, Chandra, R. (ed.), CRC Press, Boca Raton, FL, pp. 1–30.

Chasapis, C.T., Loutsidou, A.C., Spiliopoulou, C.A., and Stefanidou, M.E. 2012. Zinc and human health: An update. *Archives of Toxicology* 86(4): 521–534.

Chen, G., Huang, M.H., Chen, L., and Chen, D.H. 2011. A batch decolorization and kinetic study of Reactive Black 5 by a bacterial strain *Enterobacter* sp. GY-1. *International Biodeterioration and Biodegradation* 65(6): 790–796.

Costa, M., and Klein, C.B. 2006. Toxicity and carcinogenicity of chromium compounds in humans. *Critical Reviews in Toxicology* 36(2): 155–163.

Dahl, L.K., and Love, R.A. 1954. Evidence for relationship between sodium (chloride) intake and human essential hypertension. *AMA Archives of Internal Medicine* 94(4): 525–531.

Dayan, A.D., and Paine, A.J. 2001. Mechanisms of chromium toxicity, carcinogenicity and allergenicity: Review of the literature from 1985 to 2000. *Human and Experimental Toxicology* 20(9): 439–451.

Dixit, S., Yadav, A., Dwivedi, P.D., and Das, M. 2015. Toxic hazards of leather industry and technologies to combat threat: A review. *Journal of Cleaner Production* 87: 39–49.

Donati, E., Oliver, C., and Curutchet, G. 2003. Reduction of chromium (VI) by the indirect action of *Thiobacillus thioparus*. *Brazilian Journal of Chemical Engineering* 20(1): 1999–2005.

Dotro, G., Castro, S., Tujchneider, O., Piovano, N., Paris, M., Faggi, A., Palazolo, P., Larsen, D., and Fitch, M. 2012. Performance of pilot-scale constructed wetlands for secondary treatment of chromium bearing tannery wastewaters. *J. Hazard Mater* 239–240: 142–151.

Dunn, K., Maart, B., and Rose, P. 2013. *Arthrospira* (Spirulina) in tannery wastewaters. Part 2: Evaluation of tannery wastewater as production media for the mass culture of *Arthrospira* biomass. *Water SA* 59(2): 279–284.

Durai, G., and Rajasimmam, M. 2011. Biological treatment of tannery wastewater—A review. *Journal of Environmental Science and Technology* 4: 1–17.

Duruibe, J.O., Ogwuegbu, M.O., and Egwurugwu, J.N. 2007. Heavy metal pollution and human biotoxic effects. *International Journal of Physical Sciences* 2(5): 112–118.

El-Bestawy, E., Al-Fassi, F., Amer, R., and Aburokba, R. 2013. Biological treatment of leather-tanning industrial wastewater using free living bacteria. *Adv. Life Sci. Technol.* 12: 46–65.

Elmagd, A.M., and Mahmoud, M.S. 2014. Tannery wastewater treatment using activated sludge process system (lab scale modeling). *International Journal of Engineering, Technology and Research* 2(5): 21–28.

Exley, C., and Birchall, J.D. 1992. The cellular toxicity of aluminium. *Journal of Theoretical Biology* 159(1): 83–98.

Faisal, M., and Hasnain, S. 2004. Microbial conversion of Cr VI to Cr III in industrial effluent. *African Journal of Biotechnology* 3: 610–617.

Flores, A., and Perez, J.M. 1999. Cytotoxicity, apoptosis and in vitro DNA damage induced by potassium chromate. *Toxicology and Applied Pharmacology* 161: 75–81.

Fraga, C.G. 2005. Relevance, essentiality and toxicity of trace elements in human health. *Molecular Aspects of Medicine* 26(4): 235–244.

Gagnon, V., Chazarenc, F., Comeau, Y., and Brisson, J. 2007. Influence of macrophyte species on microbial density and activity in constructed wetlands. *Water Sci. Technol.* 56(3): 249–254.

Gbem, T.T., Balogun, J.K., Lawal, F.A., and Annune, P.A. 2001. Trace metal accumulation in *Clarias gariepinus* (Teugels) exposed to sublethal levels of tannery effluent. *Science of the Total Environment* 271(1): 1–9.

Gowd, S.S., and Govil, P.K. 2008. Distribution of heavy metals in surface water of Ranipet industrial area in Tamil Nadu, India. *Environmental Monitoring and Assessment* 136 (1–3): 197–207.

Goyer, R.A., and Clarkson, T.W. 1996. Toxic effects of metals. Casarett & Doull's Toxicology. In *The basic science of poisons*, 5th Edition, Klaassen, C.D. (ed.), McGraw-Hill Health Professions Division, New York.

Guha, H., Jayachandran, K., and Maurrasse, F. 2001. Kinetics of chromium (VI) reduction by a type strain *Shewanella alga* under different growth conditions. *Environmental Pollution* 115: 209–218.

Guidotti, T.L. 2010. Hydrogen sulfide advances in understanding human toxicity. *International Journal of Toxicology* 29(6): 569–581.

Gurulakshmi, M., Sudarmani, D.N.P., and Venba, R. 2008. Biodegradation of leather acid dye by *Bacillus subtilis*. *Advances in Biotechnology* 7: 12–19.

Igwe, J.C., and Abia, A.A. 2006. A bioseparation process for removing heavy metals from waste water using biosorbents. *African Journal of Biotechnology* 5(12): 1167–1179.

Jain, R.K., Kapur, M., Labana, S., Lal, B., Sarma, P.M., Bhattacharya, D., and Thakur, I.S. 2005. Microbial diversity: Application of microorganisms for the biodegradation of xenobiotics. *Current Science* 89(1): 101–112.

Jonnalagadda, R.R., Rathinam, A., Kalarical, J.S. and Balachandran, U.N. 2007. Biological removal of carcinogenic chromium (VI) using mixed *Pseudomonas* strains. *Journal of General and Applied Microbiology* 53(2): 71–91.

Kawasaki, T., Delea, C.S., Bartter, F.C., and Smith, H. 1978. The effect of high-sodium and low-sodium intakes on blood pressure and other related variables in human subjects with idiopathic hypertension. *The American Journal of Medicine* 64(2): 193–198.

Keharia, H., and Madamwar, D. 2003. Bioremediation concepts for treatment of dye containing wastewater: A review. *Indian Journal of Experimental Biology* 41(9): 1068–1075.

Khalid, A., Arshad, M., and Crowly, D.E. 2008. Accelerated dechlorination of structurally different azo dyes by newly isolated bacterial strains. *Applied Microbiology and Biotechnology* 78(2): 361–369.

Kierek-Pearson, K., and Karatan, E. 2005. Biofilm development in bacteria. *Adv. Appl. Microbiol.* 57: 79.

Kim, I.-S., Ekpeghere, K.I., Ha, S.-Y., Kim, B.-S., Song, B., Kim, J.-T., Kim, H.-G., and Koh, S.-C. 2014. Full-scale biological treatment of tannery wastewater using the novel microbial consortium BM-S-1. *Journal of Environmental Science and Health Part A* 49(3): 355–364.

Kimura, H. 2011. Hydrogen sulfide: Its production, release and functions. *Amino Acids* 41(1): 113–121.

Kolomaznik, K., Adamek, M., Andel, I., and Uhlirova, M. 2008. Leather waste-potential threat to human health, and a new technology of its treatment. *Journal of Hazardous Materials* 160(2): 514–520.

Komori, K.A., Rivas, A., Toda, K., and Ohtake, H. 1990. A method for removal of toxic chromium using dialysis-sac cultures of a chromate-reducing strain of *Enterobacter cloacae*. *Applied Microbiology and Biotechnology* 33: 117–119.

Krewski, D., Yokel, R.A., Nieboer, E., Borchelt, D., Cohen, J., Harry, J., Kacew, S., Lindsay, J., Mahfouz, A.M., and Rondeau, V. 2007. Human health risk assessment for aluminium, aluminium oxide, and aluminium hydroxide. *Journal of Toxicology and Environmental Health Part B Critical Review* 10 Suppl 1: 1–269.

Kumar, V., Majumdar, C., and Roy, P. 2008. Effects of endocrine disrupting chemicals from leather industry effluents on male reproductive system. *Journal of Steroid Biochemistry and Molecular Biology* 111(3–5): 208–216.

Lefebvre, O.N., Vasudevan, N., Torrijos, M., Thanasekaran, K., and Moletta, R. 2005. Halophilic biological treatment of tannery soaks liquor in a sequencing batch reactor. *Water Research* 39(8): 1471–1480.

Leghouchi, E., Laib, E., and Guerbet, M. 2009. Evaluation of chromium contamination in water, sediment and vegetation caused by the tannery of Jijel (Algeria): A case study. *Environmental Monitoring and Assessment* 153(1–4): 111–117.

Lofrano, G., Meric, S., Zengin, G.E., and Orhon, D. 2013. Chemical and biological treatment technologies for leather tannery chemicals and wastewaters: A review. *Science of the Total Environment* 461–462: 265–281.

Lovley, D.R., and Phillips, E.J.P. 1994. Reduction of chromate by *Desulfovibrio vulgaris* and its C3 cytochrome. *Applied and Environmental Microbiology* 60: 726–728.

Mahmood, S., Khalid, A., Mahmood, T., Arshad, M., and Ahamad, R. 2013. Potential of newly isolated bacterial strains for simultaneous removal of hexavalent chromium and reactive black-5 azo dye from tannery effluent. *Journal of Chemical Technology and Biotechnology* 88(8): 1506–1513.

Mandal, T., Dasgupta, D., Mandal, S., and Datta, S. 2010. Treatment of leather industry by aerobic biological fenton oxidation process. *Journal of Hazardous Material* 180(1–3): 204–211.

Mant, C., Costa, S., Williams, J., and Tambourgi, E. 2004. Phytoremediation of chromium by model constructed wetland. *Bioresour. Technol.* 97(15): 1767–1772.

Matsumoto, S.T., Mantovani, M.S., Malaguttii, M.I., Dias, A.L., Fonseca, I.C., and Marin-Morales, M.A. 2006. Genotoxicity and mutagenicity of water contaminated with tannery effluents, as evaluated by the micronucleus test and comet assay using the fish *Oreochromis niloticus* and chromosome aberrations in onion root-tips. *Genetics and Molecular Biology* 29(1): 148–158.

Megharaj, M., Avudainayagam, S., and Naidu, R. 2003. Toxicity of hexavalent chromium and reduction by bacteria isolated from soil and contaminated with tannery waste. *Current Microbiology* 47: 51–54.

Meneely, G.R., Tucker, R.G., Darby, W.J., and Auerbach, S.H. 1953. Chronic sodium chloride toxicity in the albino rat II. Occurrence of hypertension and of a syndrome of edema and renal failure. *The Journal of Experimental Medicine* 98(1): 71–80.

Messer, R.L., Bishop, S., and Lucas, L.C. 1999. Effects of metallic ion toxicity on human gingival fibroblasts morphology. *Biomaterials* 20(18): 1647–1657.

Mohammadirad, A., and Abdollahi, M. 2011. A systematic review on oxidant/antioxidant imbalance in aluminium toxicity. *International Journal of Pharmacology* 7(1): 12–21.

Mohanty, M., and Patra, H.K. 2012. Phytoremediation potential of paragrass—An in situ approach for chromium contaminated soil. *International Journal of Phytoremediation* 14(8): 796–805.

Munch, C.H., Neu, T., Kuschk, P., and Roske, I. 2007. The root surface as the definitive detail for microbial transformation processes in constructed wetlands-a biofilm characteristic. *Water. Sci. Technol.* 56(3): 271–276.

Nachiyar, C.V., and Rajkumar, G.S. 2003. Degradation of a tannery and textile dye, Navitan Fast Blue S5R by *Pseudomonas aeruginosa*. *World Journal of Microbiology and Biotechnology* 19(6): 609–614.

Nanda, S., Sarangi, P.K., and Abraham, J. 2010. Cyanobacterial remediation of industrial effluents I. Tannery effluents. *New York Science Journal* 3(12): 32–36.

Noorjahan, C.M. 2014. Physicochemical characteristics, identification of bacteria and biodegradation of industrial effluent. *Journal of Bioremediation and Biodegradation* 5: 229.

Onyancha, D., Mavura, W., Ngila, J., Ongoma, P., and Chacha, J. 2008. Studies of chromium removal from tannery wastewaters by algae biosorbents, *Spirogyra condensate* and *Rhizoclonium hieroglyphicum*. *Journal of Hazardous Material* 158(2–3): 605–614.

Osugi, M.E., Rajeshwar, K., Ferraz, E.R.A., de Oliveira, D.P., Araujo, A.R., and Zanoni, M.V.W. 2009. Comparision of oxidation efficiency of disperse dyes by chemical and photoelectrocatalytic chlorination and removal of mutagenic activity. *Electrochimica Acta* 54(7): 2086–2093.

Oteiza, P.I., and Mackenzie, G.G. 2005. Zinc, oxidant-triggered cell signaling, and human health. *Molecular Aspects of Medicine* 26(4): 245–255.

Rajasimman, M., Jayakumar, M., Ravindranath, E., and Chitra, K. 2007. Treatment of solid and liquid wastes from tanneries in an UASB reactor. *Proceedings of 60th Annual Session of Indian Institute of Chemical Engineers*, CHEMCON-2007, Kolkata, India.

Ramteke, P.W., Awasthi, S., Srinath, T., and Joseph, B. 2010. Efficiency assessment of common effluent treatment plant (CETP) treating tannery effluents. *Environmental Monitoring and Assessment* 169(1–4): 125–131.

Reiffenstein, R.J., Hulbert, W.C., and Roth, S.H. 1992. Toxicology of hydrogen sulfide. *Annual Review of Pharmacology and Toxicology* 32(1): 109–134.

Romanenko, V.I., and Korenkov, V.N. 1977. A pure culture of bacterial cells assimilating chromates and bichromates as hydrogen acceptors when grown under anaerobic conditions. *Mikrobiologiya* 46: 414–417.

Rout, G.R., and Das, P. 2009. Effect of metal toxicity on plant growth and metabolism: I. Zinc. In *Sustainable agriculture*, Eric Lichtfouse, Mireille Navarrete, Philippe Debaeke, Souchere Véronique, Caroline Alberola, (eds.), Springer, The Netherlands, pp. 873–884.

Saranraj, P., and Sujitha, D. 2013. Microbial bioremediation of chromium in tannery effluent: A review. *International Journal of Microbiological Research* 4(3): 305–320.

Saranraj, P., Stella, D., Reetha, D., and Mythili, K. 2010. Bioadsorption of chromium resistant *Enterococcus casseliflavus* isolated from tannery effluent. *Journal of Ecobiotechnology* 2(7): 17–22.

Saratale, R.G., Saratale, G.D., Chang, J.S., and Govindwar, S.P. 2010. Decolorization and biodegradation of reactive dyes and dye wastewater by a developed bacterial consortium. *Biodegradation* 21(6): 999–1015.

Saxena, G., and Bharagava, R.N. 2015. Persistent organic pollutants and bacterial communities present during the treatment of tannery wastewater. In *Environmental waste management*, Chandra, R. (eds.), CRC Press, Boca Raton, FL, pp. 217–247.

Saxena, G., and Bharagava, R.N. 2016. Ram Chandra: Advances in biodegradation and bioremediation of industrial waste. *Clean Technology and Environmental Policy* 18: 979–980. doi: http://dx.doi.org/10.1007/s10098-015-1084-9.

Saxena, G., Bharagava, R.N., Kaithwas, G., and Raj, A. 2015. Microbial indicators, pathogens and methods for their monitoring in water environment. *Journal of Water and Health* 13: 319–339. doi: http://dx.doi.org/10.2166/wh.2014.275.

Saxena, G., Chandra, R., and Bharagava, R.N. 2016. Environmental pollution, toxicity profile and treatment approaches for tannery wastewater and its chemical pollutants. *Reviews of Environmental Contamination and Toxicology* 240: 31–69. doi: http://dx.doi.org/10.1007/398_2015_5009.

Scheuhammer, A.M. 1987. The chronic toxicity of aluminium, cadmium, mercury, and lead in birds: A review. *Environmental Pollution* 46(4): 263–295.

Schulin, R. 2007. Heavy metal contamination along a soil transect in the vicinity of the iron smelter of Kremikovtzi (Bulgaria). *Geoderma* 140: 52–61.

Schwarzenbach, R.P., Egli, T., Hofstetter, T.B., Von Gunten, U., and Wehrli, B. 2010. Global water pollution and human health. *Annual Review of Environment and Resources* 35: 109–136.

Shahandeh, H., and Hossner, L.R. 2000. Plant screening for chromium phytoremediation. *International Journal of Phytoremediation* 2(1): 31–51.

Shaikh, Z.A., Vu, T.T., and Zaman, K. 1999. Oxidative stress as a mechanism of chronic cadmium-induced hepatotoxicity and renal toxicity and protection by antioxidants. *Toxicology and Applied Pharmacology* 154(3): 256–263.

Shakir, L., Ejaz, S., Ashraf, M., Qureshi, N.A., Anjum, A.A., Iltaf, I., and Javeed, A. 2012. Ecotoxicological risks associated with tannery effluent wastewater. *Environmental Toxicology and Pharmacology* 34(2): 180–191.

Shakoori, A.R., Makhdoom, M., and Haq, R.U. 2000. Hexavalent chromium reduction by a dichromate resistant gram-positive bacterium isolated from effluents of tanneries. *Applied Microbiology and Biotechnology* 53(3): 348–351.

Sharma, S., and Malaviya, P. 2014. Bioremediation of tannery wastewater by *Aspergillus niger* SPFSL2-a isolated from tannery sludge. *International Journal of Basic and Applied Sciences* 2(3): 88–93.

Singare, P.U., Jagtap, A.G., and Lokhande, R.S. 2011. Water pollution by discharge effluents from Gove Industrial Area of Maharashtra, India: Dispersion of heavy metals and their toxic effects. *International Journal of Global Environmental Issues* 11(1): 28–36.

Singh, A., Sharma, R.K., Agrawal, M., and Marshall, F.M. 2010. Risk assessment of heavy metal toxicity through contaminated vegetables from waste water irrigated area of Varanasi, India. *Tropical Ecology* 51(2): 375–387.

Sivaprakasam, S., Mahadevan, S., Sekar, S., and Rajakumar, S. 2008. Biological treatment of tannery wastewater by using salt-tolerant bacterial strains. *Microbial Cell Factories* 7: 15.

Skrypski-Mäntele, S., and Bridle, T.R. 1995. Environmentally sound disposal of tannery sludge. *Water Research* 29(4): 1033–1039.

Smith, K.S., and Huyck, H.L. 1999. An overview of the abundance, relative mobility, bioavailability, and human toxicity of metals. *The Environmental Geochemistry of Mineral Deposits* 6: 29–70.

Song, Z., Williams, C.J., and Edyvean, R.G. 2004. Treatment of tannery wastewater by chemical coagulation. *Desalination* 164(3): 249–259.

Srinath, T., Verma, T., Ramteke, P.W., and Garg, S.K. 2002. Chromium (VI) biosorption and bioaccumulation by chromate resistant bacteria. *Tannery Technology* 48(4): 427–435.

Srivastava, S., and Thakur, S.K. 2006. Biosorption potency of *Aspergillus niger* for removal of chromium (VI). *Journal of Biomedical Life Science* 53(3): 232–237.

Srivastava, S., and Thakur, S.K. 2007. Evaluation of biosorption potency of *Acinetobacter* sp. for removal of hexavalent chromium from tannery effluent. *Journal of Earth and Environmental Science* 18(5): 637–646.

Srivastava, S., Ahmad, A.H., and Thakur, I.S. 2007. Removal of chromium and pentachlorophenol from tannery wastewaters. *Bioresource Technology* 98(5): 1128–1132.

Stottmeister, U., Wiener, A., Kuschk, P., Kappelmeyer, U., Kastner, M., Bederski, O., Muller, RA., and Moormann, H. 2003. Effects of plants and microorganisms in constructed wetlands for wastewater treatment. *Biotechnol. Adv.* 22(1–2): 93–117.

Sultan, S., and Hasnain, S. 2005. Chromate reduction capability of a Gram positive bacterium isolated from effluent of dying industry. *Environmental Contamination and Toxicology* 75: 699–706.

Sumathi, C., MohanaPriya, D., Swarnalatha, S., Dinesh, M.G., and Sekaran, G. 2014. Production of prodigiosin using tannery fleshing and evaluating its pharmacological effects. *The Scientific World Journal* 2014: 290327.

Tewari, C.P., Shukla, S., and Pandey, P. 2011 Biodegradation of pentachlorophenol (PCP) by consortium of *Flavobacterium* sp. in tannery effluent. *Journal of Environmental Research and Development* 7(2A): 876–882.

Thanigavel, M. 2004. Biodegradation of tannery effluent in fluidized bed bioreactor with low density biomass support. M.Tech Thesis. Annamalai University, Tamilnadu, India.

Tripathi, M., Vikram, S., Jain, R.K., and Garg, S.K. 2011. Isolation and growth characteristics of chromium (VI) and pentachlorophenol tolerant bacterial isolate from treated tannery effluent for its possible use in simultaneous bioremediation. *Indian Journal of Microbiology* 51(1): 61–69.

USDHHS. 2001. United States Department of Health and Human Services (USDHHS): Public Health Statement, In *Toxicological profile for pentachlorophenol*, Prepared by Syracuse Research corporation, Washington, D.C., pp. 1–11.

Vankar, P.S., and Bajpai, D. 2008. Phytoremediation of chrome-VI of tannery effluent by *Trichoderma* species. *Desalination* 222(1–3): 255–262.

Verma, T., and Maurya, A. 2013. Isolation of potential bacteria from tannery effluent capable to simultaneously tolerate hexavalent chromium and pentachlorophenol and its possible use in effluent bioremediation. *International Journal of Engineering and Science* 2: 64–69.

Verma, T., Ramteke, P.W., and Garg, S.K. 2008. Quality assessment of treated tannery wastewater with special emphasis on pathogenic *E. coli* detection through serotyping. *Environmental Monitoring and Assessment* 145(1–3): 243–249.

Verstraeten, S.V., Aimo, L., and Oteiza, P.I. 2008. Aluminium and lead: Molecular mechanisms of brain toxicity. *Archives of Toxicology* 82(11): 789–802.

Vincent, J. 2011. *The nutritional biochemistry of chromium (III)*. Elsevier, Amsterdam, The Netherlands.

Wang, K., Li, W., Gong, X., Li, X., Liu, W., He, C., Wang, Z., Minh, Q.N., Chen, C.-L., and Wang, J.-Y. 2014. Biological pretreatment of tannery wastewater using a full-scale hydrolysis acidification system. *International Biodeterioration and Biodegradation* 95(Part A): 41–45. doi: http://dx.doi.org/10.1016/j.ibiod.2014.05.019.

Wang, Y.S., Pan, Z.Y., Lang, J.M., Xu, J.M., and Zheng, Y.G. 2007. Bioleaching of chromium from tannery sludge by indigenous, *Acidithiobacillus thiooxidans*. *J. Hazard Mater* 147 (1–2): 319–334.

Ward, D.M., Weller, R., and Bateson, M.M. 1990. 16S rRNA sequences reveal numerous uncultured microorganisms in a natural community. *Nature* 345(6270): 63–65.

Yusuf, R.O., Noor, Z.Z., Abu Hassan, M.A., Agarry, S.E., and Solomon, B.O. 2013. A comparison of the efficacy of two strains of *Bacillus subtilis* and *Pseudomonas fragi* in the treatment of tannery wastewater. *Desalination and Water Treatment* 51(16–18): 3189–3195.

# 14 Constructed Wetlands

## *An Emerging Phytotechnology for Degradation and Detoxification of Industrial Wastewaters*

*Ram Naresh Bharagava, Gaurav Saxena, and Pankaj Chowdhary*

**CONTENTS**

14.1 Introduction....398
14.2 Wetland: Types and Characteristics....399
  14.2.1 Natural Wetlands....399
  14.2.2 Constructed Wetlands....400
    14.2.2.1 Surface Flow Constructed Wetland....400
    14.2.2.2 Subsurface Flow Constructed Wetland....402
    14.2.2.3 Hybrid Constructed Wetland....402
14.3 The Role of Plants and Microbes in Degradation and Detoxification of Industrial Wastewaters in a Constructed Wetland....403
  14.3.1 The Role of Plants in Constructed Wetlands for Wastewater Treatment....403
  14.3.2 The Role of Microbes in Constructed Wetlands for Wastewater Treatment....405
14.4 Mechanism of Pollutants Removal from Wastewaters in Constructed Wetland....406
  14.4.1 Mechanisms for Nitrogen Removal from Wastewaters....406
  14.4.2 Mechanisms for Phosphorous Removal from Wastewaters....406
  14.4.3 Mechanisms for Sulfur Removal from Wastewaters....408
14.5 Techniques to Study the Microbial Community Structure and Metabolic Potential in Constructed Wetlands....408
14.6 Applications of Constructed Wetlands in the Treatment of Industrial Wastewaters....415
  14.6.1 Constructed Wetlands for Tannery Wastewater....415
  14.6.2 Constructed Wetlands for Winery and Distillery Wastewater....416

14.6.3 Constructed Wetlands for Pulp and Paper Mill Wastewater ..... 417
14.6.4 Constructed Wetlands for Petrochemical and Oil Refinery Wastewater ..... 417
14.6.5 Constructed Wetlands for Textile Wastewater ..... 418
14.7 Monitoring and Management of Constructed Wetlands ..... 419
14.8 Merits and Demerits of Constructed Wetland Technology ..... 419
14.9 Conclusion ..... 420
Acknowledgments ..... 420
References ..... 420

## 14.1 INTRODUCTION

Currently, the whole world is facing an environmental crisis due to the contamination of natural resources. Industries are the major contributors in the national economy of many developing countries; however, these are also the major polluters worldwide due to their potentially toxic wastewater discharged into the environment. The wastewaters discharged from various industries are mainly characterized by the high biochemical oxygen demand (BOD), chemical oxygen demand (COD), total suspended solids (TSSs), total dissolved solids (TDSs), total nitrogen (TN), total phosphorus (TP), phenols, and a variety of organic and inorganic pollutants such as cadmium (Cd), chromium (Cr), arsenic (As), mercury (Hg), lead (Pb), etc., as well as total fecal coliforms (Saxena et al. 2015a; Saxena et al. 2016a). The discharge of untreated industrial wastewater causes soil and water pollution and also makes water unsuitable for drinking, land irrigation, fish production, or recreation purposes (Kivaisi et al. 2001; Vymazal 2010). This problem is expected to worsen in the near future, unless measures are taken to control and treat the wastewaters. In most of the developing countries, a centralized (conventional) system is used for the treatment and management of wastewater. However, the operation and management costs of a centralized wastewater treatment system are often high due to its technologically complex nature, and it is therefore not affordable to many developing countries (Massoud et al. 2009; Libralato et al. 2012). Hence, there is a need for a low-cost alternative for the adequate treatment of wastewater prior to its reuse and disposal into the environment.

Constructed wetlands (CWs) are considered as a low-cost alternative for wastewater treatment, especially in the case of developing countries, where advance treatment technologies are neither available nor affordable. A CW is used for the treatment of various types of wastewater such as municipal, agricultural, mining, industrial wastewaters, etc. (Kadlec et al. 2000; Khan et al. 2009). In recent years, the use of constructed wetland technology (CWT) has now been recognized as a reliable biological wastewater treatment technology worldwide including in America, Europe, Czech Republic, China, Greece, Argentina, and the Netherlands (IWA 2000; Whitney et al. 2003; Chen et al. 2006; Lesage et al. 2007; Vymazal et al. 2007; Khan et al. 2009). CWs contain natural processes of aquatic macrophytes that not only accumulate pollutants from wastewater directly into their tissues but also act as catalysts for purification reactions that usually occur in the rhizospheres of the plants (Chandra et al. 2015; Jenssen et al. 1993; Khan et al. 2009). However, various physicochemical and biological processes take place in the rhizosphere region due to the interaction between plants, microorganisms, and soil/sediments and are meant for the removal of various pollutants from wastewater prior to its disposal into the environment (Stottmeister et al. 2003; Kosolapov et al. 2004; Faulwetter et al. 2009; Khan et al. 2009; Truu et al. 2009).

The treatment efficiency of CW systems mainly depends on the wetland design, hydraulic loading rate (HLR), type of wastewater to be treated, microbial interactions, and climatic factors; however, a low HLR and a long hydraulic retention time are required for best treatment efficiency (Vymazal et al. 2010). Further, the industrial sector has also focused a great deal of attention on CWs for the removal of heavy metals (HMs) from wastewaters (Hadad et al. 2006; Maine et al. 2006; Jayaweera et al. 2008; Khan et al. 2009). A range of factors including pH of water and sediment, mobilization and uptake from the soil, compartmentalization and sequestration within the root, efficiency of xylem loading and transport (transfer factors), distribution between metal sinks in the aerial parts, sequestration and storage in leaf cells, and plant growing and transpiration rates can also effect the remediation processes of the industrial wastewaters in CWs (Hadad et al. 2006; Khan et al. 2006; Khan et al. 2009). CWs are the sustainable wastewater treatment options and are associated with low energy consumption, low capital cost, with low mechanical technology requirements. Therefore, the treatment of industrial wastewaters using CW systems could be an efficient alternative to conventional treatment systems, especially for small communities, typically rural or suburban areas, due to low treatment and maintenance costs (Soukup et al. 1994; Solano et al. 2003; Babatunde et al. 2008).

This chapter provides detailed knowledge on the wetlands, types, characteristics, and design and operational parameters with their merits and demerits in view to their role in the treatment and management of hazardous industrial wastewaters for environmental safety. In addition, the mechanisms of pollutants removal from wastewaters during treatment in CWs have been discussed. Moreover, the roles of plants and microbes in the degradation and detoxification of pollutants from various types of industrial wastewaters in CWs is also discussed. Further, the current and emerging molecular approaches for characterizing microbial community structure and activity in CWs have also been discussed.

## 14.2 WETLAND: TYPES AND CHARACTERISTICS

A wetland is a land area that is saturated with water, either permanently or seasonally having the vegetation of aquatic plants, which are adapted to the hydric soil and play a major role in environmental management such as in water purification, flood control, provision of shoreline stability, and role as a carbon sink; hence, they are regarded as the most diverse of all the ecosystems that provide many important ecosystem services to human society and serve as home to a variety of plants and wildlife (Sundaravadivel and Vigneswaran 2001; Truu et al. 2009; Vymazal 2011). Wetlands are mainly classified into two types: natural and constructed.

### 14.2.1 Natural Wetlands

Natural wetlands are the transition areas between terrestrial and aquatic systems and are often called the "kidneys of the landscape" due to their ability to transform and store organic matter and nutrients and also play a major role in environmental management (Sundaravadivel and Vigneswaran 2001; Truu et al. 2009; Vymazal 2011). Natural wetlands are principally dominated by woody aquatic plants and are generally termed as swamps; those with soft stemmed plants as marshes; and those with mosses as bogs; however, swamps and marshes can be of either freshwater or saltwater (often regarded as mangroves) (Kivaisi 2001; Sundaravadivel and Vigneswaran 2001). Some important functions of natural wetlands include the following: support a rich diversity of wildlife and fisheries, act as the "natural purifiers of water," and reduce or remove contaminants including organic matter, inorganics, trace organics, and pathogens from the water (Kivaisi 2001). All these functions provide a convincing argument for wetland preservation. Moreover, wetlands have also shown great potential for wastewater treatment (Khan et al. 2009). Therefore, efforts have been made to use the CW systems for wastewater treatment.

### 14.2.2 Constructed Wetlands

CWs are the man-engineered systems that have been designed and constructed for the removal of a variety of organic and inorganic pollutants as well as microbial pathogens from various types of wastewaters (Ghermandi et al. 2007; Vymazal 2007; Snow et al. 2008; Cooper 2009; Kadlec and Wallace 2009; Chen 2011; Choudhary et al. 2011; Vymazal 2011). CWs are similar to natural wetlands in many cases; however, they also enhance certain services and cause a decline in other services as compared to natural wetlands and are intentionally designed to provide higher direct use value, for example, food production, recreation, and wastewater treatment (Foley et al. 2005; Yang et al. 2008). CWs may be categorized on the basis of various design parameters, but the three most important criteria are hydrology (open water-surface flow and subsurface flow [SSF]), type of macrophyte growth (emergent, submerged, free-floating), and flow path (horizontal and vertical). A hybrid wetland system is also constructed by combining different types of CWs with each other to utilize the specific advantages of the different CWs (Vymazal 2005, 2008; Marchand et al. 2010; Vymazal 2011) (Figure 14.1). A brief on surface flow (SF), SSF, and hybrid CWs is presented below.

#### 14.2.2.1 Surface Flow Constructed Wetland

A wetland system with aboveground water flow is referred to as SF-CW and typically consists of a sequence of treatment cells with water control structures that regulate flow and hydro-period to optimize the maintenance and functioning of vegetation communities (Chen 2011). These systems are similar to natural wetlands and are often vegetated by a variety of densely rooted aquatic plant species such as *Typha* species and *Phragmites australis* and have a target water depth of 20–40 cm (Chen 2011). In SF-CWs, the near-surface layer is aerobic while the deeper waters and substrate are usually anaerobic and used to treat the acid mine drainage and agricultural runoff (Chaudhary et al. 2011). SF-CWs exhibit a broad spectrum of biological characteristics like natural marshes, which are helpful in removing the various pollutants from water/wastewaters and therefore improve the water quality (Ghermandi et al. 2007; Chen 2011). These wetland systems can be further classified on the basis of dominant

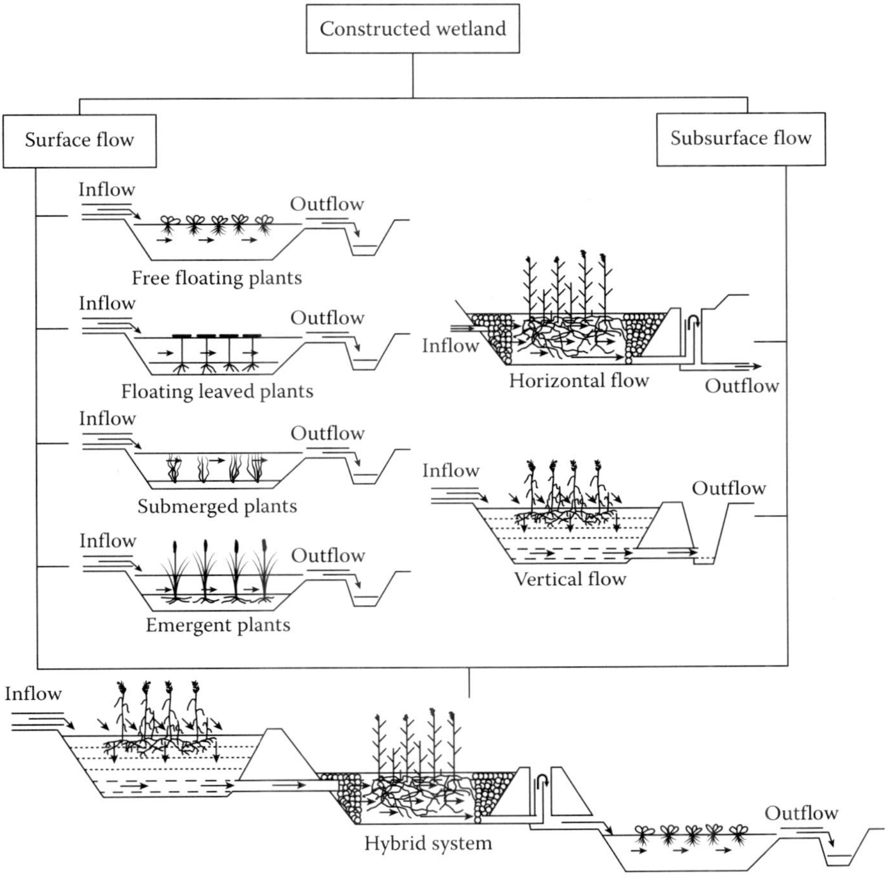


**FIGURE 14.1** Classification of constructed wetlands for wastewater treatment. (Adapted from Hiraishi, T. et al., Intergovernmental panel on climate change, IPCC, Switzerland, 2014.)

vegetation community into the following (Brix 1994; Vymazal 2007; Chen 2011): free-floating aquatic vegetation (FFAV), emergent aquatic vegetation (EAV), or submerged aquatic vegetation (SAV) dominated systems. The use of these wetland systems is generally popular in North American countries, and many are in operation in Australia, Asian, and European countries (Ghermandi et al. 2007; Kadlec and Wallace 2009; Chen 2011). The major advantages associated with SF-CWs are the low capital and operating costs; however, these wetland systems generally have lower contaminant removal efficiency as compare to SSF systems (Lee et al. 2009; Chaudhary et al. 2011).

The removal efficiencies more than 70% can be achieved in SF-CWs for TSSs, COD, BOD, and pathogens, primarily bacteria and viruses, whereas 40%–50% and 40%–90% for N and P, respectively (Andersson et al. 2005; Vymazal 2007; Kadlec and Wallace 2009; Chen 2011). However, in SF-CWs, the removal efficiencies for toxic metals vary from metal to metal (Ghermandi et al. 2007; Arroyo et al. 2010). Further, the pollutant removal efficiency in SF-CWs depends on a number of factors including pollutant loading, hydrologic regime, and vegetation type (Kadlec and Wallace 2009; Chen 2011). Moreover, hydraulic and hydrologic conditions also strongly affect the biotic community composition, biogeochemical processes, and the fate of pollutants in SF-CWs (Reddy et al. 1999; Mitsch and Gosselink 2007; Kadlec and Wallace 2009; Chen 2011).

#### 14.2.2.2 Subsurface Flow Constructed Wetland

A wetland system with belowground water flow is referred to as a SSF-CW that mostly employs gravel as the main media to support the growth of plants, and wastewater flows through the substrate where it comes in contact with microorganisms living on the surfaces of plant roots and substrate allowing pollutants removal from wastewaters (Cooper et al. 1996; Kadlec and Knight 1996; Chaudhary et al. 2011; Saeed et al. 2012). SSF-CWs are further divided on the basis of wastewater flow into the two following groups (Saeed et al. 2012): vertical flow (VF: wastewater flow is vertical) and horizontal flow (HF: wastewater flow is horizontal) wetland systems. HF CWs can effectively remove the organic pollutants (TSS, $BOD_5$, and COD) from the wastewater (UN-HABITAT 2008). The major removal mechanism for nitrogen in HF CWs is denitrification; however, the removal of ammonia is limited due to lack of oxygen in the filtration bed as a consequence of permanent waterlogged conditions (Vymazal 2007, 2010). Phosphorus is removed primarily by ligand exchange reactions, where phosphate displaces water or hydroxyls from the surface of iron and aluminum hydrous oxides; however, removal of P is usually low in HF CWs unless special materials are used (Vymazal 2007, 2010). HF CWs have always been used to treat domestic and municipal wastewaters around the world; additionally, HF CWs are used to treat many other types of wastewaters including industrial and agricultural, landfill leachate, and runoff waters (Vymazal 2010). In contrast to HF CWs, the VF CWs are considerably smaller than HF systems and hence require less land, have much greater oxygen transfer capacity that results in the good nitrification, and can efficiently remove $BOD_5$, COD, and pathogens (UN-HABITAT 2008). VF CWs are far more aerobic than HF CWs and, hence, provide suitable conditions for nitrification and are also very effective in the removal of organics and suspended solids (Vymazal 2010). VF CWs are often used to treat domestic and municipal wastewater; however, there are many reports available on the use of VF CWs for various types of wastewater, such as refinery effluent, composting leachate, airport runoff, and dairy or cheese production wastewater (Vymazal 2010).

#### 14.2.2.3 Hybrid Constructed Wetland

HF wetland is approved to remove $BOD_5$ and TSS for secondary wastewater treatment but not for nitrification due to the limited oxygen transfer capacity (UN-HABITAT 2008). As a result, there has been a growing interest in VF wetlands because they have a much greater oxygen transfer capacity and considerably less area requirement than HF. But VF wetlands also have some limitations, including that they are less efficient in solids removal and can become clogged if the media selection is not correct (UN-HABITAT 2008). Due to these reasons, there has been a growing interest in combined or hybrid constructed wetlands (HCWs). In these systems, the advantages and disadvantages of the HF and VF can be combined to complement each other (UN-HABITAT 2008). Depending on the purpose, HCWs could be either HF wetland followed by VF wetland or VF wetland followed by HF wetland to achieve high wastewater treatment efficiency (UN-HABITAT 2008; Vymazal 2010). HCWs are especially used for the removal of ammonia-N and total-N from wastewater and to treat a variety of wastewaters such as from landfill leachate, compost leaching, slaughterhouse, shrimp and fish aquaculture, or wineries (Vymazal 2010).

## 14.3 THE ROLE OF PLANTS AND MICROBES IN DEGRADATION AND DETOXIFICATION OF INDUSTRIAL WASTEWATERS IN A CONSTRUCTED WETLAND

### 14.3.1 The Role of Plants in Constructed Wetlands for Wastewater Treatment

Wetland plants (aquatic macrophytes) play an important role in pollutant removal from wastewaters through direct uptake, by mechanical resistance to flow that promotes settling of particles, and by providing a substrate for the attachment of algae and other microorganisms, which sequester or break down pollutants (Brix 1994; Wetzel 2001; Chen et al. 2011). Wetland plants are often viewed as transient nutrient storage compartments because they absorb nutrients during the growing season and release nutrients at senescence (Chen et al. 2011). The major roles played by aquatic plants in CWs during wastewater treatment are presented in Table 14.1. However, some of the important functions of aquatic plants in CWs are discussed below (Shelef et al. 2013):

The physical effects of plant roots help in the filtration, flow velocity reduction, improved sedimentation, decreased resuspension, and even the distribution of water and prevention of clogging:

1. The plant rhizosphere helps in improving conditions for various microorganisms, which are considered as the key drivers during wastewater treatment in CWs.
2. The release of exudates and gas ($O_2$) by plant roots affects various mechanisms during wastewater treatment in CWs including redox reactions, which determine the nitrogen fate, oxidation of some phytotoxins, and enhancement of microbial activity. In addition, oxygen leakage increases soil redox potential and aerobic niches, which in turn improves degradation, supports heavy-metal sedimentation, and increases nitrification, whereas the excretion of carbon from plant roots causes an increase in denitrification.
3. Aquatic plants play an important role in the phytoremediation of toxic metals by accumulating them from wastewaters during treatment in CWs.
4. Plants have a critical role in determining the dynamics of water loss, mainly by dictating water loss through evaporation and plant transpiration, that is, evapotranspiration (ET). Water loss from the CWs through ET slows flow velocities, which induces longer retention times, and increased pollutant and salt concentrations in the water. The transfer of water to the atmosphere is sometimes an advantage, mainly in humid environments. But in arid regions, where treated wastewater is destined for reuse, water loss may be disadvantageous.

**TABLE 14.1**
**Role of Wetland Plants Constructed Wetlands for Wastewater Treatment**

| Property of the Macrophytes | Role in the Treatment Wetlands |
|---|---|
| Physical effects of root structure | Filtration (filter out large debris) |
| | Reduction in velocity, promotion of sedimentation, and decreased resuspension |
| | Prevention of medium clogging |
| | Improved hydraulic conductivity |
| Roots act as a base for microorganisms | Provide surface for microbial attachment |
| | Release of gases and exudates for microbial activity |
| | Oxygen leakage helps to provide the aerobic niches |
| | Oxygen leakage enhances the aerobic degradation of organic pollutants |
| | Oxygen leakage supports the precipitation of heavy metal pollutants |
| | Oxygen leakage increases the nitrification |
| | Excretion of carbon increases the denitrification |
| | Release of antibiotics, phytometallophores, and phytochelatins |
| | Root exudates promote metal chelation to prevent metal toxicity |
| Plant nutrient uptake | Storage and uptake of nutrients |
| | Metal phytoremediation |
| | Salt phytoremediation |
| Evapotranspiration | Increased loss of water |
| Microclimatic conditions | Light attenuation (reduced photosynthesis) reduces algal growth |
| | Insulation from frost in the winter |
| | Insulation from radiation in the spring |
| | Reduced wind velocity |
| | Stabilization of the sediment surface (less soil erosion) |
| Other functions | Pathogens removal |
| | Insect and odor control |
| | Wastewater gardens |
| | Increased wildlife diversity |
| | Esthetic appearance of the treatment wetland |
| | Bioindicators |
| Plant production | Ornamental plant production |
| | Production of fiber for construction material |
| | Bioenergy crops |
| | Animal feed |

*Source*: Modified from Brix, *Wat. Sci. Tech.* 35: 11–17, 1997; Sundaravadivel, M., and Vigneswaran, S., *Crit. Rev. Environ. Sci. Technol.*, 31(4), 351–409, 2001; Vymazal, J., *Environ. Sci. Technol.*, 45, 61–69, 2011; Shelef, O., et al., *Water*, 5, 405–419, 2013.

5. The physical structure of plants growing in the CW medium affects the microclimatic conditions in the CW, which includes shade that prevents algal growth, and insulation from radiation in the spring and frost in the winter. Moreover, the reduced wind velocity by the plants' upper parts may also stabilize the sediment surface.
6. CW plants can be used for the production of marketable goods, plants may be used as bioenergy crops, water lilies may be used for esthetics, and plants can be used as animal feed.
7. Plants can be used as bioindicators of CW performance in the planning phase.
8. Salt-tolerant plants (halophyte) such as *Bassia indica* can be used for salt phytoremediation.
9. Other functions of plants in the CWs include esthetic appearance and the elimination of pathogens, insects, and offensive odors.

### 14.3.2 The Role of Microbes in Constructed Wetlands for Wastewater Treatment

Constructed wetlands provide an ideal environment to support the growth of organisms, which break down the pollutants present in wastewater by biodegradation. Bacteria, fungi, and algae are the common organisms in wetlands. Among microorganisms, the presence of bacteria is extremely important, which often act as catalysts for pollutants removal in SSF wetlands (Saeed and Sun 2012). In constructed wetlands, the main role in the transformation and mineralization of nutrients and organic pollutants is played not by plants but by microorganisms. Depending on the oxygen input by helophytes and the availability of other electron acceptors, the contaminants in the wastewater are metabolized in various ways. In SSF systems, aerobic processes only predominate near roots and on the rhizoplane (the surface of the roots), whereas in the zones that are largely free of oxygen, anaerobic processes such as denitrification, sulfate reduction, and/or methanogenesis take place (Stottmeister et al. 2003). Bacteria, which are unicellular prokaryotic organisms, can be classified according to their metabolic requirements (Bitton 2005; Kadlec and Wallace 2009; Saeed and Sun 2012). Heterotrophic bacteria utilize organic carbon for the formation of cell tissue, while autotrophic bacteria use carbon from $CO_2$. Organisms that use light as their energy source are referred to as phototrophic bacteria and can be classified as either photoheterotrophic (sulfur bacteria) or photo-autotrophic (photosynthetic bacteria) (Saeed and Sun 2012). Organisms deriving energy from chemical reactions are known as chemotrophs, and they can be classified as either chemo-heterotrophic (most of the bacteria) or chemo-autotrophic (*Nitrosomonas* and *Nitrobacter*) (Saeed and Sun 2012). The respiration process of bacteria is carried out in the presence or absence of oxygen. Aerobic respiration is the process that utilizes oxygen as the terminal electron acceptor, and carbohydrates are decomposed to $CO_2$, water, and energy (38 ATP molecules are produced for each fully oxidized glucose molecule) (Saeed and Sun 2012). Anaerobic respiration occurs in the absence of oxygen by a wider range of anaerobic bacteria that exist in treatment wetlands, promoting different biodegradation routes, such as *Pseudomonas* and *Bacillus* employ $NO_3$-N as final electron acceptor, producing $NO_2$-N, $N_2O$, or nitrogen ($N_2$) gas; *Desulfovibrio* bacteria utilize sulfate as the final electron acceptor and produce $H_2S$; whereas *Methanobacterium* use carbonate to form methane (Saeed and Sun 2012).

## 14.4 MECHANISM OF POLLUTANTS REMOVAL FROM WASTEWATERS IN CONSTRUCTED WETLAND

CWs mainly employ two main mechanisms (that is, liquid/solid separations and transformations) for the degradation and detoxification of organic and inorganic pollutants from wastewaters (EPA 2000; Chaudhary et al. 2011). Separations typically include gravity separation filtration, absorption, adsorption, ion exchange, stripping, and leaching, whereas transformations may be chemical, including oxidation/reduction reactions, flocculation, acid/base reactions, precipitation, and biochemical reactions occurring under aerobic or anaerobic conditions facilitated by root zone environment (Chaudhary et al. 2011). The overall processes taking place in CWs for the removal of pollutants are divided into three categories (that is, physical, chemical, and biological), which are presented in Table 14.2. Further, the mechanisms responsible for the removal of nitrogen, phosphorous, and sulfur are described below.

### 14.4.1 Mechanisms for Nitrogen Removal from Wastewaters

Nitrogen (N) is one of the major pollutants, existing in wastewater in both organic and inorganic forms in many industrial wastewaters that can cause eutrophication, affect dissolved oxygen levels of receiving water, and may cause toxicity (depending on the nitrogen form) to aquatic organisms (Saeed and Sun 2012). The removal mechanisms for nitrogen in constructed wetlands are many and include volatilization, ammonification, nitrification/denitrification, and plant uptake and matrix adsorption; however, the major removal mechanism in most of the constructed wetlands is microbial nitrification/denitrification wherein ammonia is oxidized into nitrate by nitrifying bacteria in aerobic zones and nitrates are converted into dinitrogen gas by denitrifying bacteria in anoxic and anaerobic zones (UN-HABITAT 2008). Further, the detail mechanisms for nitrogen removal from wastewater have been reviewed by Saeed and Sun (2012).

### 14.4.2 Mechanisms for Phosphorous Removal from Wastewaters

Phosphorus (P) is an important nutrient that is critically needed for the normal functioning of ecosystems. Nevertheless, excess P and nitrogen (N) is the main cause of eutrophication (Vollenweider 1968; Schindler 1977; Tiessen 1995; Vohla et al. 2011). The mechanisms for phosphorus removal in constructed wetlands are adsorption, complexation and precipitation, storage, plant uptake, and microbial assimilation (Watson et al. 1989; UN-HABITAT 2008). Phosphorus removal in SSF-CWs is mainly related with the physical–chemical and hydrological properties of the filter media, because P is mainly sorbed by or precipitated in filter media; however, the removal of P in constructed treatment wetlands is a complicated and difficult to sustain treatment process (Faulkner and Richardson 1989; Kadlec and Knight 1996; Sakadevan and Bavor 1998; Vymazal et al. 2000; Vohla et al. 2011). The life expectancy of CWs for

**TABLE 14.2**
**Mechanisms of Pollutants Removal in Constructed Wetlands**

| Wastewater Pollutants | Negative Impact on Treatment Processes due to Pollutants Overloading | Removal Mechanisms |
|---|---|---|
| Organic matter (measured in terms of biochemical oxygen demand [BOD]) | Decomposition consumes $O_2$ in water column and hence leads to anaerobic condition, reversal of many pollutants removal processes | Sedimentation, biodegradation (aerobic/anaerobic) |
| Chemical oxygen demand (COD) | | Sedimentation, oxidation/reduction, biodegradation and rhizodegradation |
| Total suspended solids (TSSs) | | Sedimentation and filtration |
| Soluble organics | | Microbial degradation (aerobic/anaerobic) |
| Organic pollutants (OPs) such as pesticides and endocrine disrupting chemicals (EDCs) | Harmful effects on microbial population and hence reduces biological treatment processes | Sedimentation, adsorption, volatilization, photodegradation, chemical reactions, biodegradation (aerobic/anaerobic), phytodegradation, and rhizodegradation |
| Nitrogen | Eutrophication, algal bloom formation, potential toxic accumulation in water and sediments, dead algae increase BOD through decomposition, reversal of many pollutants removal processes | Sedimentation, matrix adsorption, ammonification followed by nitrification/denitrification, microbial and plant uptake, volatilization and anammox process |
| Phosphorous | | Sedimentation, filtration, matrix adsorption, and plant and microbial uptake |
| Heavy metals (HMs) | Kills or harm microbial biota, creates toxic effects in plants, reduced biological treatment processes and bioaccumulation | Sedimentation, filtration, adsorption, cation exchange, complexation, and phytoremediation, microbial oxidation/reduction (microbe-mediated removal) |
| Pathogens | Not available | Natural die-off, sedimentation, filtration, adsorption, UV degradation, predation (feeding by protozoa), and excretion of antibiotics from roots of macrophytes |

*Sources*: Cooper, P., et al., *Water Science and Technology*, 35(5), 215–221, 1996;Sundaravadivel, M., and Vigneswaran, S., *Critical Reviews in Environmental Science and Technology*, 31(4), 351–409, 2001; UN-HABITAT, *Constructed Wetlands Manual*, UN-HABITAT Water for Asian Cities Programme Nepal, Kathmandu, 2008.

phosphorous removal depends on the filter media that are used for the construction purpose (Sakadevan and Bavor 1998; Gruneberg and Kern 2001). Further, the detail mechanisms for phosphorous removal from wastewater using different filter media have been reviewed by many authors (Ballantine and Tanner 2010; Vohla et al. 2011).

### 14.4.3 Mechanisms for Sulfur Removal from Wastewaters

The physicochemical processes such as mineral precipitation and dissolution, biologically catalyzed redox reactions such as assimilatory and dissimilatory sulfate reduction, as well as oxidation/reduction and disproportionation are responsible for the sulfur transformations in CWs (Wu et al. 2013). The most important biotic reactions influencing sulfur cycling in CWs might be those catalyzed by dissimilatory sulfate-reducing bacteria (SRB), which reduce sulfate to sulfide using the energy generated from the transfer of electrons from organic substrates (Sturman et al. 2008; Wu et al. 2013). The produced sulfide can be transported by diffusion and water flow from anoxic to more aerobic zones and then be re-oxidized to elemental sulfur and/or sulfate through either abiotic processes or reactions catalyzed by sulfur-oxidizing bacteria facilitated by wetland plants due to the release of oxygen from their roots (Wiessner et al. 2010; Wu et al. 2013). Further, the concept of sulfur cycling in CWs, transformation dynamics of sulfur compounds, and their interactions with other element cycles (such as carbon, nitrogen, and phosphorus) and plants as well as importance for treatment efficiencies have been well reviewed by Wu et al. (2013).

## 14.5 TECHNIQUES TO STUDY THE MICROBIAL COMMUNITY STRUCTURE AND METABOLIC POTENTIAL IN CONSTRUCTED WETLANDS

CWs contain an array of physicochemical and microbiological processes for the treatment of industrial wastewaters, which are facilitated not only by wetland plants but also by microorganisms present in the rhizosphere of these plants. The microbial community in CWs mainly consists of autochthonous (indigenous: able to possess metabolically activity, survive and grow in wetland systems participating in purification processes) and allochthonous (foreign: including pathogens entering with wastewater usually do not survive or have any function in treatment wetland) (Vymazal 2005; Truu et al. 2009). The wastewater treatment and purification process in CWs mainly depends on the combined action between microbes and filter material followed by plants; however, the mineralization of organic matter is primarily carried out by microbes in both aerobic and anaerobic conditions (Truu et al. 2009). Literature is available on the removal/transformation processes including anammox for nitrogen, phosphorous, and sulfur in CWs (Oehl et al. 2004; Sundberg et al. 2007; Paredes et al. 2007). However, there are only a limited number of studies available on the microbial community investigations in CWs (Tietz et al. 2004; Nurk et al. 2005; Truu et al. 2005; Nicomrat et al. 2006; Sundberg et al. 2007; Tao et al. 2007). Therefore, further investigations are required to know the microbial

**TABLE 14.3**
**Techniques Used to Assess Microbial Density, Activity, and Diversity in Constructed Wetlands Used for Wastewater Treatment**

| Parameter Measured | Technique/Method | Merits | Demerits |
|---|---|---|---|
| *For assessment of microbial density in CW* | | | |
| Direct bacterial, protozoan, and fungal counts | Plate count | Applicable to a wide range of bacterial types on different media and does not require complex instrumentation | Error prone due to large number of dilutions required and method itself (CFU) and limited to several types of bacteria and fungi |
| | Most probable number (MPN) | Reduced standard deviation (SD) of the culture-based method | Large number of replicates required to narrow the confidence intervals, as few as 10% of soil bacteria able to grow |
| | Epifluorescence microscopy (DAPI, BacLight, SYBR, Sytox) | Accurate cell counts (especially with a specialized software coupled to the microscope) | Organic matter may interfere with staining and not suitable to study biofilm; hence they need to "detach" groups of bacteria from the biofilm |
| | Flow cytometry | Does not require many dilutions as compared to plate count or epifluorescence microscopy | Errors due to interferences associated with extracellular polymeric substances (EPSs) and other organic compounds in samples |
| Indirect biomass | DNA | Complete community snapshot | Inactive/dead cell DNA is also included |
| | Protein | Highly practical method, easy to estimate biomass, and does not require instrumentation | Compounds such as sulfide can interfere and reduce overall utility |
| | Phospholipids fatty acid (PLFA) | Highly accurate biomass estimation | Complex protocol |
| | Polysaccharide and EPS | Very simple to extract and well correlated with protein content | Recurrent overestimation of biomass |

*(Continued)*

**TABLE 14.3** ***(CONTINUED)***
**Techniques Used to Assess Microbial Density, Activity, and Diversity in Constructed Wetlands Used for Wastewater Treatment**

| Parameter Measured | Technique/Method | Merits | Demerits |
|---|---|---|---|
| | Chlorophyll-a | Very easy measurement and quick estimation | Applicable only to surface wetlands (presence of algae) and samples required to be preserved in the field |
| | N Biomass | Suitable to apply in soil biology | High variability in results and hence require replications |
| | Substrate-induced respiration (SIR) | Good quality estimation of biomass C in the sample | Better estimation rather than accurate measurement due to high standard deviation |
| | ATP | ATP values remain constant after freezing and relatively prolonged storage of samples | Complex procedure, extremely sensitive extraction technique, and nonselective for bacteria |
| *For assessment of microbial activity in CW* | | | |
| Potential activity | Respiration | Good estimation of degradation rate potential of CW | Sometimes overestimation of real value and plant effects is not considered |
| | Nitrification | Direct evidence for the presence of specific types of bacteria but a rough estimate of their density | Sometimes overestimation of real value and plant effects is not considered |
| | Anaerobic | Direct evidence for the presence of specific types of bacteria but a rough estimate of their density | Sometimes overestimation of real value and plant effects is not considered |
| | Denitrification (acethylene) | Direct evidence for the presence of specific types of bacteria but a rough estimate of their density | Sometimes overestimation of real value and plant effects is not considered |
| Anaerobic activity | Methane production | $CH_4$ production (a good indicator of anaerobic activity in CW) | Great variability generally observed |

*(Continued)*

**TABLE 14.3 *(CONTINUED)***
**Techniques Used to Assess Microbial Density, Activity, and Diversity in Constructed Wetlands Used for Wastewater Treatment**

| Parameter Measured | Technique/Method | Merits | Demerits |
|---|---|---|---|
| Enzymatic activity | Phosphatase, glucosidase (p-nitropheol) | Compatible with separate density measurements | Sometimes errors and large SD due to complex protocols |
| | Protease | Compatible with separate density measurements | Sometimes errors and large SD due to complex protocols |
| | Catalase | Compatible with separate density measurements | Sometimes errors and large SD due to complex protocols |
| | Urease | Compatible with separate density measurements | Sometimes errors and large SD due to complex protocols |
| | Cellulase | Compatible with separate density measurements | Sometimes errors and large SD due to complex protocols |
| | Ammonia oxidation | Compatible with separate density measurements | Sometimes errors and large SD due to complex protocols |
| | Denitrifying enzyme activity | Compatible with separate density measurements | Sometimes errors and large SD due to complex protocols |
| | Arylsulfatase | Compatible with separate density measurements | Sometimes errors and large SD due to complex protocols |
| | FDA | Correlation with other activity measurements such as potential respiration rate | Can be affected by very low redox conditions |
| Electron transport system (ETS) activity | Dehydrogenase (INT, TTC) | Relatively high precision and applicable to all respiration types | Can be affected by very low redox conditions |

*(Continued)*

**TABLE 14.3** ***(CONTINUED)***
**Techniques Used to Assess Microbial Density, Activity, and Diversity in Constructed Wetlands Used for Wastewater Treatment**

| Parameter Measured | Technique/Method | Merits | Demerits |
|---|---|---|---|
| Radioactive tracer | $^{14}C$ Leucine | High precision and can follow plant–microbe interactions | Difficult to apply at full-scale CW |
| | $^{14}C$ urea | High precision and can follow plant–microbe interactions | Difficult to apply at full-scale CW |
| | $^{15}N$ | High precision and can follow plant–microbe interactions | Difficult to apply at full-scale CW |
| | DNA precursor | High precision and can follow plant–microbe interactions | Difficult to apply at full-scale CW |
| | $^{32}S2^{-}$ | High precision and can follow plant–microbe interactions | Difficult to apply at full-scale CW |
| *For assessment of microbial activity in CW* | | | |
| Microscopy | Light microscopy | Frequent observation and good assessment of microbial community profile | Knowledge regarding identification required |
| | Environmental scanning and electron microscopy | Helps the understanding of biofilm formation on substrates | Difficult to obtain a representative sample |
| CLPP | Biolog E cop late | Accurate bacterial communities differentiation | Limited to relative comparisons between samples |
| Cell membrane of microorganisms | PLFA | Good comparison of microbial community change with time in the same system | Rough estimate and complex protocol |

*(Continued)*

**TABLE 14.3** *(CONTINUED)*
**Techniques Used to Assess Microbial Density, Activity, and Diversity in Constructed Wetlands Used for Wastewater Treatment**

| Parameter Measured | Technique/Method | Merits | Demerits |
|---|---|---|---|
| Genomic methods and semi-quantitative PCR | Ammonia-oxidizing bacteria | Accurately followed development of *Nitrosomas* sp. in the biofilm | Crude method that could be refined |
| Real-time PCR | Quantification | Successfully applied to quantify small communities | Limited by the need of culture required to compare |
| Amplified ribosomal DNA restriction analysis, ARDRA and PCR cloning | Phylogenic analysis | Successfully applied to determine microbial diversity in a CW | Coupling with a quantitative method is required |
| "Touchdown" PCR | Acidophilic microorganisms | Applicable to a wide range of acidophiles | Difficult to obtain a quantitative result and thus requires complementary tests |
| LH-PCR | Diversity analysis | Successfully applied to determine microbial diversity and relative abundance estimation | Further work required to assess microbial community structure |
| T-RFLP | Relative abundance and diversity | Similar in some aspects to PCR-DGGE | Similar in some aspects to PCR-DGGE |
| PCR-DGGE | Total bacteria community | Detailed phylogenic analysis | Difficult to summarize results especially when using different primer pairs |
| | Ammonia-oxidizing bacteria | Detailed phylogenic analysis | Difficult to summarize results especially when using different primer pairs |
| | Ammonium monooxygenase | Detailed phylogenic analysis | Difficult to summarize results especially when using different primer pairs |

*(Continued)*

**TABLE 14.3** ***(CONTINUED)***
**Techniques Used to Assess Microbial Density, Activity, and Diversity in Constructed Wetlands Used for Wastewater Treatment**

| Parameter Measured | Technique/Method | Merits | Demerits |
|---|---|---|---|
| | Methanotrophs | Observed species missed by culture-based method | Required a large sample |
| | Acidophilic microorganisms | Observation and identification of microbial community members and composition | Included previous enrichment steps may bias results |
| FISH probes | Most bacteria | Simultaneous estimation of diversity and abundance (according to the selected probes) and observation of fine architecture of biofilms | Sometimes nonspecific binding of probes to nontarget organisms or groups |
| | N-cycle bacteria | Simultaneous estimation of diversity and abundance (according to the selected probes) and observation of fine architecture of biofilms | Sometimes nonspecific binding of probes to nontarget organisms or groups |
| | S-cycle bacteria | Simultaneous estimation of diversity and abundance (according to the selected probes) and observation of fine architecture of biofilms | Sometimes nonspecific binding of probes to nontarget organisms or groups |
| | Methanotroph | Simultaneous estimation of diversity and abundance (according to the selected probes) and observation of fine architecture of biofilms | Sometimes nonspecific binding of probes to nontarget organisms or groups |
| | Acidophilic bacteria | Simultaneous estimation of diversity and abundance (according to the selected probes) and observation of fine architecture of biofilms | Sometimes nonspecific binding of probes to nontarget organisms or groups |

*Source*: Faulwetter, J.L., et al., *Ecological Engineering*, 35, 987–1004, 2009; Truu, M., et al., *Science of the Total Environment*, 407, 3958–3971, 2009.

community structure in CWs that helps to broaden our understanding regarding the roles of microbes in pollutants removal/transformation from wastewaters. There are many conventional methods applied to characterize the microbial communities in the environment, but they have advantages and disadvantages and often provide specific kinds of information (Jennifer L. Faulwetter et al. 2009; Truu et al. 2009). Hence, molecular techniques can be applied to characterize the microbial community structure and activity in CWs (Faulwetter et al. 2009; Truu et al. 2009). A list of various techniques used to characterize the microbial community structure and activity in CWs are summarized in Table 14.3.

## 14.6 APPLICATIONS OF CONSTRUCTED WETLANDS IN THE TREATMENT OF INDUSTRIAL WASTEWATERS

CWs have become a popular option for wastewater treatment because of high pollutants removal efficiency, low cost, water and nutrient reuse, and other ancillary benefits (Ghermandi et al. 2007; Kadlec and Wallace 2009; Chen 2011). CWs have been used to treat a variety of wastewaters. Some studies on the treatment of different types of wastewater using CWs are discussed below.

### 14.6.1 Constructed Wetlands for Tannery Wastewater

CWs have been successfully employed for the treatment of tannery wastewater. A number of studies are available on the treatment of tannery wastewater in constructed wetlands (Aguilar et al. 2007; Calheiros et al. 2007, 2008, 2009, 2012; Dotro et al. 2011; Saeed et al. 2012; Desta et al. 2014). Calheiros et al. (2007) have reported 41%–73% and 41%–58% removal for inlet COD (inlet organic loadings: 332–1602 kg $ha^{-1}$ $d^{-1}$) and $BOD_5$ (inlet organic loadings: 218–780 kg $ha^{-1}$ $d^{-1}$) with minimum nutrient removal from tannery wastewater in a subsurface horizontal flow (HF) constructed wetlands planted with *Canna indica, Typha latifolia, Phragmites australis, Stenotaphrum secundatum,* and *Iris pseudacorus,* two different hydraulic loading rates (HLRs), 3 and 6 $cmd^{-1}$ within 17 months in Portugal. Aguilar et al. (2008) have also reported removal efficiencies ranging from 96% to 98% for COD, 93% to 95% for $BOD_5$, 99% for total chromium, 92% to 94% for total Kjeldahl Nitrogen, and 99% for sulfide and 88%–92% for sulfate from tannery wastewater in a CW planted with *Scirpus americanus* and *Typha* sp. in Modjo town, Ethiopia. Calheiros et al. (2008) have also reported the mass removal of up to 652 kg $BOD_5$ $ha^{-1}$ $d^{-1}$ (inlet organic loadings: 1800 kg $BOD_5$ $ha^{-1}$ $d^{-1}$) and 1869 kg COD $ha^{-1}$ $d^{-1}$ (inlet organic loadings: 3849 kg COD $ha^{-1}$ $d^{-1}$) from tannery wastewater in a subsurface flow constructed wetlands planted with *T. latifolia*. Desta et al. (2014) have also reported removal efficiencies ranging from 94% to 96% for COD, 91% to 100% for $SO_4^{2-}$ and $S^{2-}$, 92% to 94% for $BOD_5$, 56% to 82% for total nitrogen, and 2% to 90% for $NH_3$-N from tannery wastewater in a CW planted with *P. australis* in Modjo town, Ethiopia. Calheiros et al. (2009) have reported 88% of $BOD_5$ (from an inlet of 420 to 1000 mg $L^{-1}$) and 92% of COD (from an inlet of 808 to 2449 mg $L^{-1}$) from tannery wastewater in a two-stage series of horizontal subsurface flow CW planted with *P. australis* and *T. latifolia* at hydraulic retention times (HRTs) of 2, 5, and 7 days, whereas the overall mass removals of up to 1294 kg COD $ha^{-1}$ $d^{-1}$ and 529 kg $BOD_5$ $ha^{-1}$ $d^{-1}$ were achieved for organic loadings ranging from 242 to 1925 kg COD $ha^{-1}$ $d^{-1}$ and 126 to 900 kg $BOD_5$ $ha^{-1}$ $d^{-1}$, respectively. Dotro et al. (2011) have reported up to 95%–99% removals of $BOD_5$ and 90%–99% removals of chromium from tannery wastewater in a laboratory-scale CW planted with *Typha* spp. within 184 days of operation in Memphis, TN. Saeed et al. (2012) have also reported the average percentage removals of 86%, 50%, 98%, 98%, and 87% for $NH_3$-N, $NO_3$-N, BOD, COD, and $PO_4$, respectively, in a hybrid CW consisting of a vertical flow (VF) wetland, followed by a HF and a VF wetland planted with common reed (*P. australis*) in Bangladesh. Further, Calheiros et al. (2012) have also reported removal efficiencies of 51% to 80% for COD (inlet: 68–425 mg $L^{-1}$) and 53% to 90% for $BOD_5$ (inlet: 16–220 mg $L^{-1}$) from tannery wastewater after secondary treatment in two series of horizontal subsurface flow CW planted with *Arundo donax* and *Sarcocornia fruticosa*, whereas 40%–93%, 31%–89%, and 41%–90% removal for total P, $NH_4^+$ for total Kjeldahl nitrogen were also reported.

### 14.6.2 Constructed Wetlands for Winery and Distillery Wastewater

There are many studies available on the treatment of distillery wastewater using CWs (Shepherd et al. 2001; Mulidzi 2007; Sohsalam and Sirianuntapiboon 2008; Serrano et al. 2011; Bama et al. 2013; de la Varga et al. 2013). Shepherd et al. (2001) have reported average percentage removals of 98%, 97%, 78.2%, 98.5%, 63.3%, 99.9%, 100%, and 77.9% for COD, TSS, nitrogen, sulfide, ortho-phosphate, volatile fatty acids, phenols, and tannins and lignins from high-strength winery wastewater in a pilot-scale subsurface-flow constructed wetland planted with cattails (*Typha dominicus*), bulrushes (*Scirpus acutus*), and some arrowheads (*Sagittaria latifolia*). Mulidzi (2007) have reported more than 90% COD removal from winery wastewater of an average COD of 14,000 mg/L in a CW planted with *Typha* spp., *Scirpus* spp., and *Phragmites communis* in South Africa. Serrano et al. (2011) have reported the average removal efficiencies of 86.8% for TSS, 73.3% for COD, 74.2% for $BOD_5$, 52.4% for TKN, 55.4% for NH3-N, and 17.4% for phosphates from winery wastewater in a hybrid CW consisting of vertical flow CW planted with *P. australis* and horizontal flow CW planted with *Juncus effusus* units at an average HLR of 19.5 $mmd^{-1}$ and average surface loading rates (SLRs) of 30.4 g $CODm^{-2}$ $d^{-1}$ and 18.4 g $BOD_5$ $m^{-2}$ $d^{-1}$ in Spain. de la Varga et al. (2013) have reported the average removal efficiencies of 74% for TSS, 60%–80% for COD and $BOD_5$, 6%–29% for nitrogen and phosphate, and 39% for polyphenols from winery wastewater in a horizontal flow CW planted with *Juncus effusus* at an average HLRs ranging from 7 to 93 mm/d and SLRs ranging from 4 to 85 g $COD/m^2{\cdot}d$, 2 to 49 g $BOD_5/m^2{\cdot}d$, and 0.5 to 6 g $TSS/m^2{\cdot}d$. Sohsalam and Sirianuntapiboon (2008) have reported the average removal efficiencies of 90%–93%, 88%–89%, 67%, 70%–76%, 77%–82%, 94%–95%, and 72%–77% for suspended solids, BOD, COD, total phosphorous, $NH^{4+}$, $NO^{3-}$, and molasses pigments from molasses-based distillery wastewater in a surface flow constructed wetland planted with *Cyperus involucratus, Typha angustifolia,* and *Thalia dealbata* at the organic loading rates of 612 kg $BOD_5$ $ha^{-1}$ day. Bama et al. (2013) have reported the average removal efficiencies of 50.14%, 48.42%,

45.5%, 56.3%, 35.3%, and 54% for electric conductivity, total dissolved solids, COD, BOD, nitrate, and phosphate, respectively, from distillery wastewater in a laboratory-scale constructed wetland planted with *Eichhornia* sp., *Lemna* sp., *Pistia* sp., and *Lemna minor.*

### 14.6.3 Constructed Wetlands for Pulp and Paper Mill Wastewater

There are very few studies available for the treatment of distillery wastewater using CWs (Abira et al. 2005; Abira 2008; Arivoli et al. 2015). Abira (2008) reported average removal efficiencies of 52%, 90%, 81%, and 73%–96% for total suspended solids, BOD, COD, and phenol, respectively, from pulp paper mill wastewater in a pilot-scale constructed treatment wetland planted with *Cyperus immensus, Phragmites mauritianus,* and *Typha domingensis* and *Cyperus papyrus*. Arivoli et al. (2015) have reported the average removal efficiencies of 72%, 80%, 60%, 70%, and 71% for Fe, Mn, Zn, Ni, and Cd, respectively, in vertical flow CW planted with *Typha angustifolia, Erianthus arundinaceus,* and *P. australis.* Abira et al. (2005) have reported average removal efficiencies for phenols were variable but reached 60% at 5-day HRT and 77% at 3-day HRT on average.

### 14.6.4 Constructed Wetlands for Petrochemical and Oil Refinery Wastewater

There are some studies available on the treatment of petrochemical and oil refinery wastewater using CWs (Aslam et al. 2007; Gillespie et al. 2008; Al-Baldawi et al. 2013; Murray-Gulde et al. 2013). Gillespie Jr. et al. (2000) have reported an average of 38% removal of the total recoverable, and 65% of the soluble Zn was removed from refinery wastewater in pilot-scale constructed wetland planted with *Scirpus californicus* with the 144-day experiment under normal operating conditions (0.3-m water depth, 24-h HRT). Aslam et al. (2007) have reported the average removal efficiencies of 51%–73%, 45%–78%, and 35%–83% for total suspended solids, COD, and BOD, respectively, from refinery wastewater in a compost-based VF constructed wetland planted with *Phragmites karka*, whereas they also reported the average removal efficiencies of 39%–56%, 33%–61%, and 35%–69% for total suspended solids, COD, and BOD, respectively, from refinery wastewater in a gravel-based VF constructed wetland planted with *P. karka.* Murray-Gulde et al. (2013) have reported the average removal efficiencies of 95% and 94% for conductivity and TDS, respectively, from oil field wastewater in a hybrid RO-constructed wetland planted with broadleaf cattail (*T. latifolia*) and giant bulrush (*S. californicus*). Al-Baldawi et al. (2013) have also reported the average removal efficiencies of 82%, 71%, and 67% for total petroleum hydrocarbon for diesel concentrations of 0.1%, 0.2%, and 0.25%, respectively, at the end of 72 days, whereas they also reported the average removal efficiencies of 100% and 75.4% for TSS and COD at the diesel concentration of 0.1% in a pilot-scale horizontal subsurface flow CW planted with *Scirpus grossus.*

### 14.6.5 Constructed Wetlands for Textile Wastewater

There are many studies available for the treatment of dye-containing textile wastewater using CWs (Mbuligwe 2005; Davies 2006; Bulc and Ojstrsek 2008; Davies et al. 2009; Ong et al. 2009; Saeed and Sun 2013; Sivakumar et al. 2013; Shehzadi et al. 2014; Dogdu and Yalcuk 2015). Mbuligwe (2005) reported the average removal efficiencies of 72%–77%, 68%–73%, and 53%–59% for color, COD, and sulfate, respectively, from textile wastewater in a pilot-scale engineered wetland system planted with *Typha* sp. and *Colocasia* sp. in Dar es Salaam city, Tanzania. Saeed and Sun (2013) have reported simultaneous removal of $BOD_5$ up to 74%–79% and ammonia up to 59%–66% from textile wastewater in the lab-scale hybrid wetland systems consisted of a VF and a HF wetland planted with *P. australis, Dracaena sanderiana,* and *Asplenium platyneuron*, and the wetland systems were operated under high hydraulic loading (HL) (566–5660 mm/d) and inorganic nitrogen (254–508 g $N/m^2$ d) and organics loadings (9840–19,680 g $COD/m^2$ d and 2154–4307 g $BOD_5/m^2$ d). Davies (2006) has reported simultaneous removals of color up to 99% and COD up to 93% during the aerobic degradation of acid orange 7 (AO7) in a VF constructed wetland planted with *Phragmites* sp., which was fed with 127 mg $L^{-1}$ of AO7 at hydraulic loads of 28, 40, 53, and 108 $lm^{-2}$ $day^{-1}$. Bulc and Ojstrsek (2008) have reported the average removal efficiencies of 84%, 65%, 89%, 52%, 87%, 88%, 80%, and 93% for COD, BOD, total organic carbon (TOC), total nitrogen, nitrogen organic, $NH_4$-N, sulfate, anionic surfactants, total suspended solids, and color, respectively, from dye-containing textile wastewater in a pilot-scale hybrid constructed wetland consisting of vertical and horizontal flow CW planted with *P. australis.* Dogdu and Yalcuk (2015) have reported an average removal in color up to 97% and COD up to 62% from synthetic textile wastewater containing commercial indigo dye in a VF constructed wetland planted with *C. indica* and *T. angustifolia.* Ong et al. (2009) have reported average removal efficiencies of 86% and 96% for COD and $NH_4$-N, respectively, in the aerated reactors, whereas the average removal efficiencies of 78%–82% and 41%–48% for COD and $NH_4$-N, respectively, from azo dye-containing wastewater in the nonaerated reactors of laboratory-scale up-flow constructed wetlands (UFCW) planted with *P. australis* and Manchurian wild rice. Sivakumar et al. (2013) have reported the average removal efficiencies of 87.2%, 90.2%, 82.6%, 86.8%, 78.5%, 91.3%, and 92.8% for EC, TDS, chloride, sulfate, phenols, BOD, and COD, respectively, from textile industry wastewater in a constructed wetland planted with *Eichhornia crassipes.* Shehzadi et al. (2014) have reported the average removal efficiencies of 79%, 77%, 59%, and 27% for COD, BOD, TDSs, and TSSs, respectively, from textile industry wastewater within 72 hours in a constructed wetland planted with *T. domingensis* inoculated with two endophytic bacterial strains, *Microbacterium arborescens* TYSI04 and *Bacillus pumilus* PIRI30. Davies et al. (2009) have reported the average removal efficiencies of approximately 68% ± 8%, 69% ± 8%, and 67% ± 4% for textile azo-dye, AO7, COD, and TOC, respectively, from AO7 dye-containing textile wastewater in a pilot-scale constructed wetland planted with *P. australis.*

## 14.7 MONITORING AND MANAGEMENT OF CONSTRUCTED WETLANDS

As CWs rely on the natural processes, regular monitoring is required for the operation and maintenance of the constructed wetland for efficient wastewater treatment, which includes the following (UN-HABITAT 2008):

1. Adjustment of water levels that may change due to leaks, clogged outlets, breached berms, storm water drainage, or other causes, because changes in water levels affect the hydraulic residence time, atmospheric oxygen diffusion into the water phase, and plant cover.
2. Maintenance of wastewater flow uniformly (both for inlet and outlet) that is extremely important to achieve the expected treatment performance and both inlet and outlet should be inspected routinely and regularly, and the debris that may clog the inlets and outlets must be cleaned.
3. Unwanted vegetation should be removed from the beds two to three times a year, and vegetation should be harvested from time to time to maintain the desired plant communities within the wetland system.
4. Odor control is also required as odor is a nuisance problem in properly loaded wetlands due to anaerobic conditions created by excessive BOD and ammonia loadings. Therefore, uniform distribution of wastewater is required into the bed to prevent odor.
5. Berms (walls) should be properly maintained and any earthen berm erosion or crack in the walls should be repaired as soon as it is noted; however, leaks around berms (walls) should be repaired by plugging, sealing, etc.

## 14.8 MERITS AND DEMERITS OF CONSTRUCTED WETLAND TECHNOLOGY

CWs have emerged as eco-friendly and sustainable wastewater treatment options and have been reported to treat a variety of wastewaters (Aguilar et al. 2007; Calheiros et al. 2009; Dotro et al. 2011; Saeed et al. 2012; Desta et al. 2014). CWs systems have distinct advantages for wastewater treatment: they are odorless, they require relatively low labor and low energy, they assist in the purification of water and recycling of atmospheric $CO_2$, and they support several food crops (Nelson et al. 2003). This technology requires little machinery and no chemicals and relies on natural ecological mechanisms (microbial and plant metabolism), maintenance requirements are minimized, and systems can be expected to have long operating lifetimes with proper maintenance (Nelson et al. 2003). However, the construction of treatment wetlands requires a large land area that is not readily available in the case of developing countries, and if available, requires high cost to install treatment wetlands. The development of CWs is more favorable in developing countries due to their more suitable tropical environments and easy availability of lands. Further research efforts are required to expand the scope and efficacy of treatment wetlands for wastewater treatment.

## 14.9 CONCLUSION

1. Industrial wastewater is considered as the major source of environmental pollution.
2. Physicochemical wastewater treatment methods are costly and require hazardous chemicals for wastewater treatment and also generate a large amount of sludge that acts as the secondary pollutant in the environment.
3. Secondary treatment methods like activated sludge treatment are also not very effective as they require the use of potential microbes to degrade and detoxify the organic and inorganic pollutants present in wastewaters.
4. Constricted wetlands can be eco-friendly treatment options for wastewater treatment because of their low cost.
5. Constructed wetlands offer many ecological benefits and support biodiversity.
6. Further, the knowledge of interactions among the plants, microbes, and substrates is required for better understanding of the underlying mechanisms for wastewater treatment and management.
7. Further research is required for the commercial applications of constructed wetlands for wastewater treatment.

## ACKNOWLEDGMENTS

The authors are extremely grateful to the "Science and Engineering Research Board" (SERB), Department of Science and Technology (DST), and Government of India (GOI), New Delhi, India, for financial support as "Major Research Project" (Grant No.: SB/EMEQ-357/2013) for this work, and the University Grant Commission (UGC) Fellowship received by Mr. Gaurav Saxena is also duly acknowledged.

## REFERENCES

Abira, M.A. 2008. *A pilot constructed treatment wetland for pulp and paper mill wastewater performance, processes and implications for the Nzoia river, Kenya* UNESCO-IHE PhD. Netherlands Wageningen University, UNESCO-IHE Institute for Water, Taylor & Francis/Balkema, The Netherlands.

Abira, M.A., Van Bruggen, J.J.A., and Denny, P. 2005. Potential of a tropical subsurface constructed wetland to remove phenol from pretreated pulp and paper mill wastewater. *Water Science and Technology* 51(9): 173–175.

Aguilar, J.R.P., Cabriales, J.J.P., and Vega, M.M. 2008. Identification and characterization of sulfur oxidizing bacteria in an artificial wetland that treats wastewater from a tannery. *Int. J. Phytoremediation* 10(5): 359–370.

Al-Baldawi I.A., Sheikh, A.S.R., Anuar, N., Suja, F., and Idris, M. 2013. Performance assessment of pilot horizontal sub-surface flow constructed wetlands for removal of diesel from wastewater by *Scirpus grossus*. *Water Science and Technology* 68(10): 2271–2278. doi: 10.2166/wst.2013.484

Andersson, J.L., Bastviken, S.K., and Tonderski, K.S. 2005. Free water surface wetlands for wastewater treatment in Sweden: Nitrogen and phosphorus removal. *Water Science and Technology* 51: 39–46.

Arivoli, A., Mohanraj, R., and Seenivasan, R. 2015. Application of vertical flow constructed wetland in treatment of heavy metals from pulp and paper industry wastewater. *Environmental Science and Pollution Research* 22: 13336–13343.

Arroyo, P., Ansola, G., and de Luis Calabuig, E. 2010. Effectiveness of a full-scale constructed wetland for the removal of metals from domestic wastewater, *Water, Air and Soil Pollution* 210: 473–481.

Aslam, M.M., Malikb, M., Baiga, M.A., Qazia, I.A., and Iqbal, J. 2007. Treatment performances of compost-based and gravel-based vertical flow wetlands operated identically for refinery wastewater treatment in Pakistan. *Ecological Engineering* 30: 34–42.

Babatunde, A.O., Zhao, Y.Q., O'Neill, M., and O'Sullivan, B. 2008. Constructed wetlands for environmental pollution control: A review of developments, research and practice in Ireland. *Environmental International* 34: 116–126.

Ballantine, D.J., and Tanner, C.C. 2010. Substrate and filter materials to enhance phosphorus removal in constructed wetlands treating diffuse farm runoff: A review. *New Zealand Journal of Agricultural Research* 53(1): 71–95.

Bama, P., Thushyanthy, M., Alvappillai, P., and Pirabhaharan, M. 2013. Evaluation of lab scale constructed wetlands to treat the toddy distillery effluent with different aquatic plants. *Archives of Applied Science Research* 5(5): 213–219.

Bitton, G. 2005. *Wastewater microbiology*, (3rd Edn.), Wiley-Liss, Hoboken, NJ, pp. 746.

Brix, H. 1994. Functions of macrophytes in constructed wetlands. *Water Science and Technology* 29: 71–78.

Brix, H. 1997. Do macrophytes play a role in constructed treatment wetlands? *Wat. Sci. Tech.* 35: 11–17.

Bulc, T.G., and Ojstrsek, A. 2008. The use of constructed wetland for dye-rich textile wastewater treatment. *Journal of Hazardous Material* 155(1–2): 76–82.

Calheiros C.S., Quitério, P.V., Silva, B., Crispim, L.F., Brix, H., Moura, S.C., et al. 2012. Use of constructed wetland systems with Arundo and Sarcocornia for polishing high salinity tannery wastewater. *Journal of Environmental Management* 95(1): 66–71.

Calheiros, C.S.C., Rangel, A.O.S.S., and Castro, P.M.L. 2007. Constructed wetland systems vegetated with different plants applied to the treatment of tannery wastewater. *Water Research* 41(8): 1790–1798.

Calheiros, C.S.C., Rangel, A.O.S.S., and Castro, P.M.L. 2008. Evaluation of different substrates to support the growth of *Typha latifolia* in constructed wetlands treating tannery wastewater over long-term operation. *Bioresource Technology* 99: 6866–6877.

Calheiros, C.S.C., Rangel, A.O.S.S., and Castro, P.M.L. 2009. Treatment of industrial wastewater with two-stage constructed wetlands planted with *Typha latifolia* and *Phragmites australis*. *Bioresource Technology* 100: 3205–3213.

Chandra R., Saxena G., and Kumar V. 2015. Phytoremediation of environmental pollutants: An eco-sustainable green technology to environmental management. In *Advances in biodegradation and bioremediation of industrial waste*, Chandra, R. (ed.), CRC Press, Boca Raton, FL, pp. 1–30.

Chen, H. 2011. Surface-flow constructed treatment wetlands for pollutant removal: Applications and perspectives. *Wetlands* 31: 805–814.

Chen, T.Y., Kao, C.M., Yeh, T.Y., Chien, H.Y., and Chao, A.C., 2006. Application of a constructed wetland for industrial wastewater treatment: A pilot-scale study. *Chemosphere* 64: 497–502.

Chen, Y., Wen, Y., Cheng, J., Xue, C.H., Yang, D.H., and Zhou, Q. 2011. Effects of dissolved oxygen on extracellular enzymes activities and transformation of carbon sources from plant biomass: Implications for denitrification in constructed wetlands. *Bioresource Technology* 102: 2433–2440.

Choudhary, A.K., Kumar, S., and Sharma, C. 2011. Constructed wetlands: An approach for wastewater treatment. *Elixir Pollution* 37: 3666–3672.

Cooper, P. 2009. What can we learn from old wetlands? Lessons that have been learned and some that may have been forgotten over the past 20 years. *Desalination* 246: 11–26.

Cooper, P., Smith, M., and Maynard, H. 1996. The design and performance of a nitrifying vertical flow reed bed treatment system. *Water Science and Technology* (Oxford) 35(5): 215–221.

Davies, L.C., Cabrita, G.J.M., Ferreira, R.A., Carias, C.C., Novais, J.M., and Martins-Dias, S. 2009. Integrated study of the role of Phragmites australis in azo-dye treatment in a constructed wetland: From pilot to molecular scale. *Ecological engineering* 35: 961–970.

Davies, L.C., Pedro, I.S., Novais, J.M., and Martins-Dias, S. 2006. Aerobic degradation of acid orange 7 in a vertical-flow constructed wetland. *Water Research* 40: 2055–2063.

Desta, A.F., Assefa, F., Leta, S., Stomeo, F., Wamalwa, M., Njahira, M., et al. 2014. Microbial community structure and diversity in an integrated system of anaerobic-aerobic reactors and a constructed wetland for the treatment of tannery wastewater in Modjo, Ethiopia. *PLoS One* 9(12): e115576.

Dogdu, G., and Yalcuk, A. 2015. Indigo dyeing wastewater treatment by eco-friendly constructed wetlands using different bedding media. *Desalination and Water Treatment* 57(32): 15007–15019.

Dotro, G., Larsen, D., and Palazolo, P. 2011. Treatment of chromium-bearing wastewaters with constructed wetlands. *Water and Environment Journal* 25: 241–249.

EPA. 2000. *Design manual: Constructed wetlands treatment of municipal wastewaters*. EPA/625/R-99/010. Office of Research and Development, Cincinnati, OH.

Faulkner, S.P., and Richardson C.J. 1989. Physical and chemical characteristics of freshwater wetland soils. In *Constructed wetlands for wastewater treatment: Municipal, industrial, and agricultural*, Hammer, D.A., (ed.), Lewis Publishers, Chelsea, MA, pp. 41.

Faulwetter, J.L., Gagnon, V., Sundberg, C., Chazarenc, F., Burr, M.D., Brisson, J., et al. 2009. Microbial processes influencing performance of treatment wetlands: A review. *Ecological Engineering* 35: 987–1004.

Foley, J.A., DeFries, R., Asner, G.P., Barford, C., Bonan, G., Carpenter, S.R., et al. 2005. Global consequences of land use. *Science* 309(5734): 570–574.

Ghermandi, A., Bixio, D., and Thoeye, C. 2007. The role of free water surface constructed wetlands as polishing step in municipal wastewater reclamation and reuse. *Science of the Total Environment* 380: 247–258.

Gillespie, W.B., Jr., Hawkins, B., Rodgers, J., Jr. Manuel, L., Cano, D., et al. 2008. Transfers and transformations of zinc in constructed wetlands: Mitigation of a refinery effluent. *Ecological Engineering* 14: 279–292.

Gruneberg, B., and Kern, J. 2001. Phosphorus retention capacity of iron-ore and blast furnace slag in subsurface flow constructed wetlands. *Water Science and Technology* 44: 69–75.

Hadad, H.R., Maine, M.A., and Bonetto, C.A. 2006. Macrophyte growth in a pilot-scale constructed wetland for industrial wastewater treatment. *Chemosphere* 63: 1744–1753.

IPCC. 2014. 2013 supplement to the 2006 IPCC guidelines for national greenhouse gas inventories: Wetlands. In *Intergovernmental panel on climate change*, Hiraishi, T., Krug, T., Tanabe, K., Srivastava, N., Baasansuren, J., Fukuda, M., and Troxler, T.G. (eds.), IPCC, Switzerland.

IWA. 2000. *Constructed wetland for pollution control. Processes, performance, design and operation*. IWA Publishing, London.

Jayaweera, M.W., Kasturiarachchi, J.C., Kularatne, R.K.A., and Wijeyekoon, S.L.J. 2008. Contribution of water hyacinth (*Eichhornia crassipes* (Mart.) Solms) grown under different nutrient conditions to Fe-removal mechanisms in constructed wetlands. *Journal of Environmental Management* 87: 450–460.

Jennifer, L.F., Vincent, G., Carina, S., Florent, C., Mark, D.B., Jacques, B., Anne, K.C., and Otto, R.S. 2009. Microbial processes influencing performance of treatment wetlands: A review. *Ecological Engineering* 35: 987–1004.

Jenssen, P.D., Krogstad, T., and Mahlum, T. 1993. *Potential use of constructed wetland for wastewater treatment in northern environment*. Department of Soil Science, Agricultural University of Norway, Norway. N-1432 As.

Kadlec, R., and Knight, R.L. 1996. *Treatment wetlands*. Lewis Publishers, Boca Raton, FL.

Kadlec, R.H., Knight, R.L., Vymazal, J., Brix, H., Cooper, P., and Haberl, R. 2000. *Constructed wetlands for pollution control: Processes, performance, design and operation. IWA specialist group on use of macrophytes in water pollution control*. IWA Publishing, London, UK, pp. 156.

Kadlec, R.H., and Wallace, S.D. 2009. *Treatment wetlands*, (2nd Edn.), Taylor and Francis Group, Boca Raton, FL.

Khan, S., Ahmad, I., Shah. M.T., Rehman S., and Khaliq, A. 2009. Use of constructed wetland for the removal of heavy metals from industrial wastewater. *Journal of Environmental Management* 90: 3451–3457.

Khan, S., Cao, Q., Chen, B., and Zhu, Y.G. 2006. Humic acids increase the phytoavailability of Cd and Pb to wheat plants cultivated in freshly spiked, contaminated soil. *Journal of Soil and Sediments* 6: 236–242.

Kivaisi, A.K. 2001. The potential for constructed wetlands for wastewater treatment and reuse in developing countries: A review. *Ecological Engineering* 16: 545–560.

Kosolapov, D.B., Kuschk, M.B., Vainshtein, A.V., Vatsourina, A., Wießner, M., and Kästner, R.A.M. 2004. Microbial processes of heavy metal removal from carbon-deficient effluents in constructed wetlands. *Engineering in Life Sciences* 4(5): 403–411.

Lee, C.G., Fletcher, T.D., and Sun, G. 2009. Nitrogen removal in constructed wetland systems. *Engineering in Life Science* 9(1): 11–22.

Lesage, E., Rousseau, D.P.L., Meers, E., Tack, F.M.G., and De Pauw, N., 2007. Accumulation of metals in a horizontal subsurface flow constructed wetland treating domestic wastewater in Flanders, Belgium. *Science of the Total Environment* 380: 102–115.

Libralato, G., Ghirardini, A.V., and Avezzu, F. 2012. To centralise or to decentralise: An overview of the most recent trends in wastewater treatment management. *Journal of Environmental Management* 94: 61–68.

Maine, M. A., Suñe, N., Hadad, H., Sánchez, G., and Bonetto, C. 2006. Nutrient and metal removal in a constructed wetland for wastewater treatment from a metallurgic industry. *Ecological Engineering* 26(4): 341–347.

Marchand, L., Mench, M., Jacob, D.L., and Otte, M.L. 2010. Metal and metalloid removal in constructed wetlands, with emphasis on the importance of plants and standardized measurements: a review. *Environmental Pollution* 158: 3447–3461.

Massoud, M.A., Tarhini, A., and Nasr, J.A. 2009. Decentralized approaches to wastewater treatment and management: Applicability in developing countries. *Journal of Environmental Management* 90: 652–659.

Mbuligwe, S.E. 2005. Comparative treatment of dye-rich wastewater in engineered wetland systems (EWSs) vegetated with different plants. *Water Res*. 39: 271–280.

Mitsch, W.J., and Gosselink, J.G. 2007. *Wetlands*, (4th Edn.), Wiley, NJ.

Mulidzi, A.R., 2007. Winery wastewater treatment by constructed wetlands and the use of treated wastewater for cash crop production. *Water Science and Technology* 52(2): 103–109.

Murray-Gulde, C., Heatley, J.E., Karanfil, T., Rodgers, J.H., and Jr, Myers, J.E. 2003. Performance of a hybrid reverse osmosis-constructed wetland treatment system for brackish oil field produced water. *Water Research* 37(3): 705–713.

Nelson, M., Alling, A., Dempster, W.F., van Thillo, M., and Allen, J. 2003. Advantages of using subsurface flow constructed wetlands for wastewater treatment in space applications: Ground-based Mars Base prototype. *Advances in Space Research* 31(7): 1799–1804.

Nicomrat, D., Dick, W.A., and Tuovinen, O.H. 2006. Assessment of the microbial community in a constructed wetland that receives acid coal mine drainage. *Microbial Ecology* 51: 83–89.

Nurk, K., Truu, J., Truu, M., and Mander, U. 2005. Microbial characteristics and nitrogen transformation in planted soil filter for domestic wastewater treatment. *Journal of Environmental Science and Health* 40: 1201–1214.

Oehl, F., Frossard, E., Fliessbach, A., Dubois, D., and Oberson A. 2004. Basal organic phosphorus mineralization in soils under different farming systems. *Soil Biology and Biochemistry* 36: 667–675.

Ong, S.A., Uchiyama, K., Inadama, D., and Yamagiwa, K. 2009. Simultaneous removal of color, organic compounds and nutrients in azo dye-containing wastewater using upflow constructed wetland. *Journal of Hazardous Materials* 165: 696–703.

Pacheco Aguilar, J.R., Peña Cabriales, J.J., and Maldonado Vega, M. 2008. Identification and characterization of sulfur-oxidizing bacteria in an artificial wetland that treats wastewater from a tannery. *International Journal of Phytoremediation* 10(5): 359–370.

Paredes, D., Kuschk, P., Mbwette, T.S.A., Stange, F., Muller, R.A., and Köser, H. 2007. New aspects of microbial nitrogen transformations in the context of wastewater treatment: A review. *Engineering and Life Science* 7(1): 13–25.

Ram Chandra., Ram Naresh Bharagava., Atya Kapley., and Hemant J. Purohit. 2012. Characterization of *Phragmites cummunis* rhizosphere bacterial communities and metabolic products during the two stage sequential treatment of post methanated distillery effluent by bacteria and wetland plants. *Bioresource Technology* 103: 78–86.

Reddy, K.R., Kadlec, R.H., Flaig, E., and Gale, P.M. 1999. Phosphorus retention in streams and wetlands: A review. *Critical Reviews in Environmental Science and Technology* 29(1): 83–146.

Saeed, T., Afrin, R., Muyeed, A.A., and Sun, G. 2012. Treatment of tannery wastewater in a pilot-scale hybrid constructed wetland system in Bangladesh *Chemosphere* 88: 1065–1073.

Saeed, T., and Sun, G. 2012. A review on nitrogen and organics removal mechanisms in subsurface flow constructed wetlands: Dependency on environmental parameters, operating conditions and supporting media. *Journal of Environmental Management* 112: 429–448.

Saeed, T., and Sun, G. 2013. A lab-scale study of constructed wetlands with sugarcane bagasse and sand media for the treatment of textile wastewater. *Bioresource Technology* 128: 438–444.

Sakadevan, K., and Bavor, H.J. 1998. Phosphate adsorption characteristics of soils, slags and zeolite to be used as substrates in constructed wetland systems. *Water Research* 32: 393–399.

Saxena, G., and Bharagava, R.N. 2015b. Persistent organic pollutants and bacterial communities present during the treatment of tannery wastewater. In *Environmental waste management*, Chandra, R. (ed.), CRC Press, Boca Raton, FL, pp. 217–247.

Saxena, G., and Bharagava, R.N. 2016b. Ram Chandra: Advances in biodegradation and bioremediation of industrial waste. *Clean Technology and Environmental Policy* 18: 979980.

Saxena, G., Bharagava, R.N., Kaithwas, G., and Raj, A. 2015a. Microbial indicators, pathogens and methods for their monitoring in water environment. *Journal of Water and Health* 13: 319–339.

Saxena, G., Chandra, R., and Bharagava, R.N. 2016a. Environmental pollution, toxicity profile and treatment approaches for tannery wastewater and its chemical pollutants. *Reviews of Environmental Contamination and Toxicology* 240: 31–69. doi: http://dx.doi.org/10.1007/398_2015_5009.

Schindler, D.W. 1977. Evolution of phosphorus limitation in lakes. *Science* 195: 260–262.

Serrano, L., de La Verga, D.., Ruiz, I., and Soto, M. 2011.Winery wastewater treatment in a hybrid constructed wetland. *Ecological Engineering* 37: 744–753.

Shehzadi, M., Afzal, M., Khan, M.U., Islam, E., Mobin, A., Anwar, S., et al. 2014. Enhanced degradation of textile effluent in constructed wetland system using Typha domingensis and textile effluent-degrading endophytic bacteria. *Water Research* 58: 152–159.

Shelef, O., Gross, A., and Rachmilevitch, S. 2013. Role of plants in a constructed wetland: Current and new perspectives. *Water* 5: 405–419.

Shepherd, H.L.,Grismer, M.E., and George, T. 2001. Treatment of high-strength winery wastewater using a subsurface-flow constructed wetland. *Water Environment Research* 73: 394–403.

Sivakumar, D., Shankar, D., Vijaya Prathima, A.J.R., and Valarmathi, M. 2013. Constructed wetland treatment of textile industry wastewater using aquatic macrophytes *International Journal of Environmental Sciences* (3)4: 1223–1232.

Snow, A., Ghaly, A.E., and Cote, R. 2008. Treatment of stormwater runoff and landfill leachates using a surface flow constructed wetland. *American Journal of Environmental Sciences* 4: 164–172.

Sohsalam, P., and Sirianuntapiboon, S., 2008. Feasibility of using constructed wetland treatment for molasses wastewater treatment. *Bioresource Technology* 99: 5610–5616.

Solano, M.L., Soriano, P., and Ciria, M.P. 2003. Constructed wetlands as a sustainable solution for wastewater treatment in small villages. *Biosystems Engineering* 87(1): 109–118.

Soukup, A.R.J., Williams, F.C., Cattell, R., and Krough, M.H. 1994. The function of a coastal wetland as an efficient remover of nutrients from sewage effluent: A case study. *Water Science and Technology* 29(4): 295–304.

Stottmeister, U., Wiener, A., Kuschk, P., Kappelmeyer, U., Kästner, M., Bederski, O., et al. 2003. Effects of plants and microorganisms in constructed wetlands for wastewater treatment. *Biotechnology Advances* 22: 93–117.

Sturman, P.J., Stein, O.R., Vymazal, J., and Kreopfelov L., 2008. Sulfur Cycling in constructed wetlands. In *Wastewater treatment, plant dynamics and management in constructed and natural wetlands*, Vymazal, J (ed.), Springer, Netherlands, pp. 329–344.

Sundaravadivel, M., and Vigneswaran, S. 2001. Constructed wetlands for wastewater treatment. *Critical Reviews in Environmental Science and Technology* 31(4): 351–409.

Sundberg, C., Tonderski, K., and Lindgren, P-E. 2007. Potential nitrification and denitrification and the corresponding composition of the bacterial communities in a compact constructed wetland treating landfill leachates. *Water Science and Technology* 56: 159–166.

Tao, W., Hall, K.J., and Ramey, W. 2007. Effects of influent strength on microorganisms in surface flow mesocosm wetlands. *Water Res.* 41: 4557–4565.

Tiessen, H., 1995. *Phosphorus in the global environment: Transfers, cycles, and management*, Wiley, New York.

Tietz, A., Kirschner, A., Langergraber, G., Sleytr, K., and Haberl, R. 2004. Characterisation of microbial biocoenosis in vertical subsurface flow constructed wetlands. *Science of The Total Environment* 380: 163–72.

Truu, J., Nurk, K., Juhanson, J., and Mander, U. 2005. Variation of microbiological parameters within planted soil filter for domestic wastewater treatment. *Journal of Environmental Health* 40: 1191–1200.

Truu, M., Juhanson, J., and Truu, J. 2009. Microbial biomass, activity and community composition in constructed wetlands. *Science of the Total Environment* 407: 3958–3971.

UN-HABITAT. 2008. *Constructed wetlands manual*. UN-HABITAT Water for Asian Cities Programme Nepal, Kathmandu.

Varga, D., Ruiz, I., and Soto, M. 2013. Winery wastewater treatment in subsurface constructed wetlands with different bed depths. *Water Air and Soil Pollution* 224: 1–13.

Vohla, C., Koiv, M., Bavor, H.J., Chazarenc, F., and Mander, U. 2011. Filter materials for phosphorus removal from wastewater in treatment wetlands: A review. *Ecological Engineering* 37: 70–89.

Vollenweider, R.A., 1968. *The scientific basis of lake and stream eutrophication with particular reference to phosphorus and nitrogen as eutrophication factors*. Technical Report DAS/C81/68, OECD, Paris, France.

Vymazal, J. 2005. Horizontal sub-surface flow and hybrid constructed wetlands systems for wastewater treatment. *Ecological Engineering* 25: 478–490.

Vymazal, J. 2007. Removal of nutrients in various types of constructed wetlands. *Science of the Total Environment* 380: 48–65.

Vymazal, J. 2008. Constructed wetlands, surface flow. In *Encyclopedia of Ecology*, Vol. 1. Jørgensen, S. E., Fath, B., (eds.), Elsevier BV, Amsterdam, The Netherlands.

Vymazal, J. 2010. Constructed Wetlands for Wastewater Treatment. *Water* 2(3): 530–549.

Vymazal, J. 2011. Constructed wetlands for wastewater treatment: Five decades of experience. *Environmental Science and Technology* 45: 61–69.

Vymazal, J., Kröpfelová, L., Švehla, J., and Štíchová, J. 2010. Can multiple harvest of above-ground biomass enhance removal of trace elements in constructed wetlands receiving municipal sewage? *Ecological Engineering* 36: 939–945.

Vymazal, J., Svehla, J., Kropfelova, L., and Chrastn, Y.V. 2007. Trace metals in *Phragmites australis* and *Phalaris arundinacea* growing in constructed and natural wetlands. *Science of the Total Environment* 380: 154–162.

Watson, J.T., Reed, S.C., Kadlec, R.H., Knight, R.L., and Whitehouse, A.E. 1989. Performance expectations and loading rates for constructed wetlands. In *Constructed wetlands for waste water treatment. Municipal, industrial and agricultural*, Hammer, D. (ed.), Lewis Publishers, Chelsea, MI, pp. 121–129.

Wetzel, R.G. 2001. Fundamental processes within natural and constructed wetland ecosystems: Short-term versus long-term objectives. *Water Science and Technology* 44: 1–8.

Whitney, D., Rossman, A., and Hayden, N. 2003. Evaluating an existing subsurface flow constructed wetland in Akumal, Mexico. *Ecological Engineering* 20: 105–111.

Wiessner, A., Rahman, K.Z., Kuschk, P., Kastner, M., and Jechorek, M., 2010. Dynamics of sulphur compounds in horizontal sub-surface flow laboratory-scale constructed wetlands treating artificial sewage. *Water Research* 44: 6175–6185.

Wu, S., Kuschk, P., Wiessner, A., Müller, J., Saad, R.A.B., and Dong, R. 2013. Sulphur transformations in constructed wetlands for wastewater treatment: A review. *Ecological Engineering* 52: 278–289.

Yang, Z., Zheng, S., Chen, J., and Sun, M. 2008. Purification of nitrate-rich agricultural runoff by a hydroponic system. *Bioresource Technology* 99: 8049–8053.

# 15 Biomedical Wastes
## *Environmental Threats and Management*

*Sahar Mohamed Soliman and Amel Ibrahim Ahmed*

## CONTENTS

15.1 Introduction ........ 428
15.2 Definition of Biomedical Waste ........ 428
15.3 Categories of Biomedical Waste ........ 428
15.3.1 Nonhazardous or "General Healthcare Waste" ........ 428
15.3.2 Biohazardous ........ 428
15.3.2.1 Sharp Wastes ........ 428
15.3.2.2 Infectious Wastes ........ 429
15.3.2.3 Pathological Wastes ........ 429
15.3.2.4 Pharmaceutical Waste Including Genotoxic Wastes ........ 429
15.4 Sources of Biomedical Wastes ........ 429
15.4.1 Major Sources ........ 429
15.4.2 Minor Sources ........ 430
15.5 Impacts of Biomedical Waste ........ 430
15.5.1 Persons at Risks ........ 430
15.5.2 Hazards from Biomedical Waste ........ 430
15.5.3 Environmental Impact of Biomedical Waste ........ 431
15.6 Management of Biomedical Waste ........ 431
15.6.1 Legalization and Policies ........ 432
15.6.2 Segregation, Packaging and Collection ........ 433
15.6.3 Storage and Transportation ........ 434
15.6.3.1 Interim Storage ........ 434
15.6.3.2 On-site Transportation ........ 434
15.6.3.3 Central Storage Inside the Facilities ........ 435
15.6.3.4 Off-site Transportation ........ 436
15.6.3.5 Treatment and Disposal Methods ........ 437
15.7 Future Prospects and Challenges ........ 440
15.8 Conclusion ........ 441
References ........ 441

## 15.1 INTRODUCTION

One of the essential human rights is to live in a clean environment that is free of pollution. Healthcare facilities are considered sources of environmental pollution due to the discarding of biomedical waste. Inadequate management of biomedical waste induces risk to human health and the environment. The management of biomedical waste is a serious concern and has become a major challenge in developing countries. Therefore, the disposal of biomedical waste must be carried out under a well-controlled system to control this hazardous waste (Singh 2001; Bokhoree et al. 2014).

## 15.2 DEFINITION OF BIOMEDICAL WASTE

*Healthcare waste* is a term used by the World Health Organization (WHO), and *biomedical waste* is an exchangeable term used to describe biological waste generated during the diagnosis, testing, treatment, research, or production of biological products for humans or animals. Biomedical wastes include different items, such as laboratory samples, culture dishes, swabs, glassware, body fluids, soiled or blood-soaked bandages, live vaccines, surgical gloves, surgical instruments, sharps, and body parts (Pullishery et al. 2016).

## 15.3 CATEGORIES OF BIOMEDICAL WASTE

Biomedical waste can be categorized into nonhazardous and biohazardous waste.

### 15.3.1 Nonhazardous or "General Healthcare Waste"

Nonhazardous waste is discarded waste that does not cause any specific biological, chemical, radioactive, or physical hazard. According to the WHO, 85% of the healthcare facilities' wastes are nonhazardous and as harmless as any other domestic waste. It is generated mostly from the administrative, kitchen, and housekeeping tasks at healthcare facilities. While more than half of all nonhazardous waste is generated from healthcare facilities is paper, cardboard, and plastics, while the rest comprises discarded food, metal, glass, textiles, plastics, and wood.

### 15.3.2 Biohazardous

The remaining 15% of healthcare facilities' wastes are "hazardous," which includes 10% sharps, along with infectious, pathological, and pharmaceutical waste, including genotoxic waste categories. Finally, the remaining 5% of healthcare waste includes chemical and radioactive hazardous waste.

#### 15.3.2.1 Sharp Wastes

Sharp waste items could cause cuts or puncture wounds, and include needles, hypodermic needles, intravenous or other needles; auto-disable syringes, syringes with attached needles, scalpels and other blades, knives, infusion sets, saws, broken glass, and pipettes.

The used or unused sharps should be handled as highly hazardous biomedical waste. They are always considered and treated as if they were potentially infected.

#### 15.3.2.2 Infectious Wastes

Infectious waste is material suspected to contain pathogens such as bacteria, viruses, parasites, or fungi in sufficient concentration to cause disease in susceptible hosts. This category includes waste such as dressings, bandages, swabs, gloves, masks, gowns, aprons, drapes, and other materials that are contaminated with blood or other body fluids. Laboratory cultures and stocks of infectious agents from laboratory work are highly infectious waste. Waste from autopsies, animal bodies, and other waste items that have been inoculated, infected, or in contact with highly infectious agents are highly infectious waste.

Waste from infected patients in isolation wards and discarded instruments or materials that have been in contact with persons or animals infected with highly infectious agents are also considered as infectious wastes.

#### 15.3.2.3 Pathological Wastes

Pathological waste consists of tissues, organs, body parts, blood, body fluids, excreta, and other waste from surgery and autopsies on patients with infectious diseases. Pathological waste includes also human fetuses and infected animal carcasses. Recognizable human or animal body parts are sometimes called anatomical waste. Pathological waste may include healthy body parts that have been removed during a medical procedure.

#### 15.3.2.4 Pharmaceutical Waste Including Genotoxic Wastes

Pharmaceutical waste in terms of biohazards waste includes contaminated pharmaceutical products, vaccines, and sera that are no longer required. This category also includes cytotoxic waste, pharmaceuticals that are expired, and items contaminated by or containing residues of these pharmaceuticals substances. Cytotoxic means substances that are toxic to the cell, cytotoxic wastes are chemotherapeutic or antineoplastic drugs. The principal substances in this category have the ability to kill or stop the growth of certain living cells and are used in chemotherapy of cancer. Cytotoxic drugs are mainly used in specialized departments, such as oncology and radiotherapy units, but their use is increasing in other hospital departments and in clinics outside hospitals. Genotoxic waste is highly hazardous and may have mutagenic (capable of inducing a genetic mutation), teratogenic (capable of causing defects in an embryo or fetus), or carcinogenic (cancer-causing) properties. Genotoxic waste may include certain cytostatic drugs.

## 15.4 SOURCES OF BIOMEDICAL WASTES

### 15.4.1 Major Sources

The major sources of biomedical waste are hospitals, including university hospitals, general hospitals, and district hospitals. Other healthcare facilities that generate biomedical waste are emergency medical care services, healthcare centers dispensaries, obstetric and maternity clinics, outpatient clinics, dialysis centers, long-term healthcare establishments and hospices, transfusion centers, military medical services, and prison hospitals or clinics. Biomedical wastes are also generated from related laboratories and research centers, medical and biomedical laboratories, veterinary colleges and animal research centers, biotechnology laboratories, medical research centers, mortuary, autopsy centers, testing centers, blood banks, blood collection services, and nursing homes for the elderly.

### 15.4.2 Minor Sources

Among the identified minor sources of biomedical wastes are small healthcare establishments that include first-aid posts and sick bays, physicians' offices, dental clinics, acupuncturists, and chiropractors. Other settings that generate biomedical waste include the specialized healthcare establishments and institutions with low waste generation, such as convalescent nursing homes, psychiatric hospitals, blood donation camps, vaccination centers, and disabled persons' institutions. In addition to these sources that are involving activities of intravenous or subcutaneous interventions, cosmetic ear-piercing and tattoo parlors, illicit drug users and needle exchanges, funeral services, ambulance services, and home treatment. Finally, animal houses and slaughterhouses are other sources for biomedical waste (Mathur et al. 2012).

## 15.5 IMPACTS OF BIOMEDICAL WASTE

Biomedical wastes are of great importance due to their potential environmental hazard and health problems. The waste generated in hospitals has the potential to spread infections and cause diseases and injury more than any other type of waste. In addition, it is a potential health hazard to those who are directly or indirectly related to the healthcare facility's environment. If biomedical waste is not managed properly, it also causes environmental, occupational, and public health hazards (Nema et al. 2011; Sefouhi et al. 2013).

### 15.5.1 Persons at Risks

All exposed personnel to biomedical waste are potentially at risk of being injured and/or catching infection. Those personnel include doctors, nurses, and workers in support services linked to healthcare facilities. Workers include those who are working in laundries, waste handling, waste disposal facilities, transportation services, sanitation tasks, and hospital maintenance activities. Moreover, patients who are receiving treatment in healthcare facilities and hospital visitors are at risk of exposure to biomedical waste's hazards. Also at risk is the general population and more specifically the children playing with items they can find in the waste outside the healthcare facilities when it is directly accessible to them.

### 15.5.2 Hazards from Biomedical Waste

The biomedical waste commonly induces needlestick injury that occurs due to handed recapping and the unsafe collection and disposal of sharps waste. Other circumstances that increase risk exposure of healthcare personnel to needlestick injuries are using unsafe sharps collection containers that are not puncture proof, inappropriate design and/or overflow of existing sharps container, and unprotected pits. The general population may be exposed to hazards of biomedical waste through the reuse of infectious syringes and dumping of sharps on an uncontrolled site where it can be easily accessed by the public, especially children and rag pickers. HIV, hepatitis B and C viruses are among the most common blood borne diseases transmitted by sharps.

Other negative health effects of biomedical waste are the effects of cytotoxic and genotoxic pharmaceutical waste. Exposure to this type of waste can induce local irritation effects after direct contact with skin or eyes. Exposure may also cause dizziness, nausea, headache, or dermatitis. Recurrent exposure to such substances may cause carcinogenic and mutagenic effects. Exposure to cytotoxic and genotoxic pharmaceutical waste occurs during handling or disposal processes. The main pathways of exposure are inhalation of dust or aerosols, absorption through the skin, and ingestion of food accidentally contaminated with cytotoxic waste. Exposure may also occur through contact with body fluids and secretions of patients undergoing chemotherapy (WHO 2014).

### 15.5.3 Environmental Impact of Biomedical Waste

Hazards from biomedical waste–treatment methods could occur due to flue gases from waste incinerators that are improperly operated or poorly maintained. Emissions from waste incinerators cause air pollution that directly impacts people living and working close to a treatment site. This exposure to emission particulates is associated with increased cardiovascular and respiratory mortality and morbidity. Incinerator ash contains a high concentration of heavy metals and other toxic items, such as mercury and cadmium, that disrupt the immune system, neurological system, lungs, and kidneys. In addition, the ash provides ideal conditions for the synthesis of dioxins and furans, because it is often exposed for a long time to a temperature range of 200°C–450°C. Dioxins, furans, and polycyclic aromatic hydrocarbons may also cause serious health effects. The burnt-out needles and glass will not be infected but can still cause physical injury (WHO 2014).

Other methods of treating biomedical waste such as autoclave and steam disinfection treatment methods can also pose potential hazards. The post-waste treatment water contains organic and inorganic contaminants that may be discharged into sewerage systems. Uncontrolled landfilling and dumping along water bodies causes water pollution and soil pollution. Landfill of biomedical wastes that contain pathogens and products of their treatment may be released into surface streams or groundwater (WHO 2014).

## 15.6 MANAGEMENT OF BIOMEDICAL WASTE

Inadequate management of biomedical waste is a serious concern in many developing countries due to the risks posed to human health and the environment. Proper management of biomedical waste is a crucial issue for maintaining human health and the environment (Sefouhi et al. 2013). The spectrum of proper management of biomedical waste ranged from effective legalization of handling and processing of biomedical waste to the proper selection of safe and appropriate methods of treatment (Health Care without Harm Asia 2007).

### 15.6.1 Legalization and Policies

Active governmental intervention is a cornerstone for improving effective biomedical waste management. Stating a national policy for biomedical waste management is the first step to sustain the healthcare waste management system including biomedical waste. A national policy should identify the needs and problems in the country, as well as take into account the relevant international agreements and conventions adopted nationally that govern public health, sustainable development, the environment, and safe management of hazardous waste. After preparation of a national policy, legislation supporting regulations should be developed. These regulations should describe what is expected from healthcare staff and explain the methods for their enforcement.

Planning for healthcare waste management at national, regional, or local levels should take into consideration the WHO core principles for achieving safe and sustainable management of healthcare waste. Planning should cover the following six objectives listed below:

1. Develop the legal and regulatory framework for healthcare waste management
2. Rationalize the waste-management practices within healthcare facilities
3. Develop specific financial investment and operational resources dedicated to waste management
4. Launch capacity building and training measures
5. Set up a monitoring plan
6. Reduce the pollution associated with waste management

Nations would be required to develop their own technical guideline that is intended to assist the implementation of legislation. Technical guidelines should be practical and directly applicable to local managers and staff. They should contain sufficient detail to ensure that safe practices and appropriate standards can be achieved. They should outline the legal framework to be met for the safe management of healthcare waste and how the guidance improves hospital hygiene and occupational health and safety. Technical guidelines address a broad range of relevant issues that describe the responsibilities of public health authorities, safe practices for waste minimization, separation, handling, storage, and transport of healthcare waste, and treatment and disposal methods.

Based on the national legalization, policies, and allied technical guidelines, each healthcare facility will implement its own plan to implement the national policies and plan (WHO 2014).

According to the size of the healthcare facility, the details of the established waste management plan would be described. In general, this document should include the knowledge and decisions for all waste management processes as well as the responsibilities of staff members (WHO 2014).

### 15.6.2 Segregation, Packaging and Collection

The first priority is to segregate wastes, preferably at the point of generation, into reusable and nonreusable, hazardous and nonhazardous components. Other important steps are the institution of a sharps management system, waste reduction, avoidance of hazardous substances whenever possible (e.g., PVC-containing products, mercury thermometers), assurance of worker safety, provision of secure methods of waste collection, transportation, installation of safe treatment and disposal mechanisms (Askarian et al. 2010).

The first producer of biomedical waste is the responsible person for ideal segregation of waste items. Segregation should take place in a medical area, at a bedside, in an operating theatre, or laboratory by nurses, physicians, and technicians. Biomedical waste is separated into two categories: used sharps and potentially infectious items. In the latter, the largest components are typically tubing, bandages, disposable medical items, swabs, and tissues. Pathological waste also may be segregated from other infectious waste according to the method of final disposal. Sharps waste should be placed directly into a sharps container. Pharmaceutical wastes should be segregated and collected separately and be clearly labeled with the type of toxic substance.

Packaging of biomedical waste should be followed throughout a country, and many countries have national legislation that prescribes the waste segregation categories to be used and a system of color coding for waste containers. Where there is no national legislation, WHO guidelines should be followed. Color-coding makes it easier for healthcare providers and workers to put waste items into the correct containers and to maintain segregation of the wastes during transport, storage, treatment, and disposal. Color-coding also provides a visual indication of the potential risk posed by the waste in that container (WHO 2014).

Waste containers have several shapes and sizes and may be made from different materials. Some waste containers are designed for automated systems that empty their contents into the waste-disposal system, wash and disinfect them mechanically. Other waste containers may be made out of reused plastic and metal containers. However, they should be strong, leak-proof, and lined with a sturdy plastic bag. The recommended thickness of bags for infectious waste is 70 μm. Plastics used for either containers or bags should be chlorine free. Not all plastic bags can withstand temperatures of 121°C, and some can melt during an autoclave process. Containers should have well-fitting lids, either removable by hand or preferably operated by a foot pedal. Both the container and the bag should be of the correct color for the waste they are intended to receive and should be labeled clearly. The containers should be large enough for the quantity of waste generated at that location during the period between collections.

Sharps containers may be disposable or designed for disinfection and reuse. Disposables are boxes made of plasticized cardboard or plastic; reusable designs are plastic or metal. Low-cost options include the reuse of plastic bottles or metal cans. If this is to be done, the original labels should be removed or obscured, and the containers should be clearly relabeled as "Sharps containers" (WHO 2014).

Both contaminated and uncontaminated sharps should be collected in a puncture-proof and impermeable container that is difficult to break open after closure.

Labeling of waste containers is used to identify the source, record the type and quantities of waste produced in each area, and allow problems with waste segregation to be traced back to a medical area. A simple approach is to attach a label to each filled container with the details of the medical area, date, and time of closure of the container, and the name of the person filling out the label. Using an international hazard symbol on each waste container is also recommended (WHO 2014).

Collection times of biomedical waste should be fixed and appropriate to the quantity of waste produced in each area of the healthcare facility. Waste bags and sharps containers should be filled to no more than three-quarters full. Once this level is reached, they should be sealed, ready for collection. Plastic bags should never be stapled but may be tied or sealed with a plastic tag or tie. Replacement bags or containers should be available at each waste-collection location so that filled ones can immediately be replaced (WHO 2014).

### 15.6.3 Storage and Transportation

Generally, biomedical wastes should be stored in a manner and location that are protected from weather and animals. Storage should not provide opportunity to be a breeding place or food source for insects or rodents. Biomedical wastes should be stored in an area that is secure, not accessible to the public. Biomedical waste should not be adjacent to supply storage areas or areas used for food preparation or consumption. The storage area should be refrigerated or is itself a refrigeration or freezer unit and should be clearly marked with the universal biohazard symbol.

#### 15.6.3.1 Interim Storage

It is recommended to find interim or short-term storage at medical areas to store biomedical waste. A utility room, which is designated for cleaning equipment, dirty linen, and waste, can be used for interim storage. This storage area is to ensure that the waste can be kept away from patients before removal, then collected conveniently, and transported to a central storage facility.

Another possibility for interim storage is a closed container stationed indoors, within or close to a medical area. A storage container used for biomedical waste should be clearly labeled and preferably lockable (WHO 2014).

#### 15.6.3.2 On-site Transportation

On-site transport should take place during less busy times whenever possible. Set routes should be used to prevent exposure to staff and patients and to minimize the passage of loaded carts through patient care and other clean areas. Depending on the design of the healthcare facility, the internal transport of waste should use separate floors, stairways, or elevators as far as possible. Regular transport routes and collection times should be fixed and reliable. Transport staff should wear adequate personal protective equipment, gloves, strong and closed shoes, overalls, and masks (WHO 2014).

Biomedical waste can be transported together with used sharps waste. Trolleys should be colored in the appropriate color code for infectious waste (yellow) and should be labeled with an "Infectious waste" sign. Biomedical waste and pharmaceutical waste should be transported separately in boxes to central storage sites (WHO 2014).

To avoid injuries and infection transmission, trolleys and carts should fulfill the following criteria:

- Be easy to load and unload
- Have no sharp edges that could damage waste bags or containers during loading and unloading
- Be easy to clean and, if enclosed, fitted with a drainage hole and plug
- Be labeled and dedicated to a particular waste type
- Be easy to push and pull
- Not be too high to avoid restricting the view of staff transporting waste
- Be secured with a lock (for hazardous waste)
- Be appropriately sized according to the volumes of waste generated at a healthcare facility

Spare trolleys should be available in case of breakdowns and maintenance. The vehicles should be cleaned and disinfected daily. All waste bag seals should be in place and intact at the end of transportation.

A waste route should follow the principle "from clean to dirty." Collection should start from the most hygienically sensitive medical areas (e.g., intensive care, dialysis, and theatres) and follow a fixed route around other medical areas and interim storage locations. The frequency of collection should be refined through experience to ensure that there are no overflowing waste containers at any time. Biologically active waste (e.g., infectious waste) must be collected at least daily (WHO 2014).

#### 15.6.3.3 Central Storage Inside the Facilities

Central storage areas are places within a healthcare facility, where different types of waste should be brought for safe retention until it is treated or collected for transport offsite (WHO 2014). The central storage areas should fulfill the following criteria:

- Have an impermeable, hard-standing floor with good drainage; the floor should be easy to clean and disinfect.
- Include the facility to keep biomedical waste separated from other waste types.
- Have a water supply for cleaning purposes.
- Have easy access for staff in charge of handling the waste.
- Be lockable to prevent access by unauthorized persons.
- Have easy access for waste-collection vehicles.
- Have protection from the sun.
- Be inaccessible to animals, insects, and birds.
- Have good lighting and at least passive ventilation.
- Not be situated in the proximity of fresh food stores and food preparation areas.
- Have a supply of cleaning equipment, protective clothing, and waste bags or containers located conveniently close to the storage area.
- Have a washing basin with running tap water and soap that is readily available for the staff.
- Clean regularly (at least once per week).
- Have spillage containment equipment.
- Be appropriate to the volumes of waste generated from each healthcare facility.

Special attention should be paid to storage of infectious, pathological waste and genotoxic and cytotoxic waste. Storage areas of infectious waste must be identified as infectious waste areas by using the biohazard sign. Floors and walls should be sealed or tiled to allow easy disinfection. The storage room should be connected to a special sewage system for infectious hospital wastewater. It is not permitted to compact untreated infectious waste or waste with a high content of blood or other body fluids destined for offsite disposal. Infectious waste should be kept cool or refrigerated at a temperature preferably no higher than 3°C–8°C if stored for more than a week, unless a refrigerated storage room is available. Storage times for infectious waste should not exceed 72 hours in winter, 48 hours in summer in a warm climate, and only 24 hours during the hot season. Pathological waste is considered as biologically active waste, and gas formation during storage should be expected. To minimize these possibilities, the storage places should have the same conditions as those for infectious and sharps wastes. In some cultures, body parts are passed to the family for ritual procedures or are buried in designated places. They should be placed in sealed bags to reduce infection risks before release to the public. Genotoxic waste is highly toxic and should be identified and stored carefully away from other healthcare waste in a designated secure location. It can be stored in the same manner as toxic chemical waste, although some cytotoxic waste may also carry a risk of infection (WHO 2014).

#### 15.6.3.4 Off-site Transportation

Off-site transport is the carriage of biomedical waste on the public streets away from a healthcare facility. Transporting biomedical waste should comply with national regulations, and with international agreements in case of transporting wastes across an international frontier for treatment. Certain requirements should be assured for the purpose of safe off-site transportation. These requirements include vehicle criteria, driver attributions, and documentation principles.

Essential requirements of the vehicle transporting biomedical waste include that it be road-worthy and labeled to indicate its load, and its payload must be secured to minimize the risk of accidents and spillages. Any vehicle used to transport healthcare waste should fulfill several design criteria in relation to size, design, securing system, internal structure, and labeling. The body of the vehicle should be of a suitable size commensurate with the design of the vehicle. There should be a bulkhead between the driver's cabin and the vehicle body, which is designed to retain the load if the vehicle is involved in a collision. There should be a suitable system for securing the load during transport. Empty plastic bags, suitable protective clothing, cleaning equipment, tools, and disinfectant, together with special kits for dealing with liquid spills, should be carried in a separate compartment in the vehicle. The internal finish of the vehicle should allow it to be steam-cleaned, and internal angles should be rounded to eliminate sharp edges to permit more thorough cleaning and prevent damage to waste containers. Labeling of the vehicle and containers should display the name and address of the waste carrier, the international hazard sign, and an emergency telephone number. Daily cleansing and disinfection of vehicles and transporting containers should be done mechanically after use by using soaps and detergents. A standard operating procedure for cleaning should be prepared and explained to cleaning staff. In addition, a schedule for preventive maintenance should be set up for all equipment and vehicles used in the transportation process (WHO 2014).

Drivers of vehicles carrying biomedical waste should have appropriate training about risks and handling of hazardous waste. Training should include issues that are related to relevant legal regulations, waste classifications and risks, safe handling of hazardous waste, labeling and documentation, as well as emergency and spillage procedures. In addition, drivers should be declared medically fit to drive vehicles. Vaccination against tetanus and hepatitis A and B is recommended, and vaccination and training details of staff should be recorded (WHO 2014).

Transport documentation that is commonly called a "consignment note" or "waste tracking note" should be prepared and carried by the driver. A consignment note should be designed to take into account the control system for waste transportation in operation within a country. The consignment note for a vehicle carrying a biomedical waste load should include basic information about waste classes, waste sources, pick-up date, destination, driver name, number of containers or volume, and receipt of load received from responsible person at pick-up areas. This information allows quick and effective countermeasures to be taken in the event of an accident or incident. Weight of waste is useful for commercial treatment and disposal operators who bill healthcare facilities for their waste services. On completion of a journey, the transporter should complete a consignment note and return it to the waste producer. Emergency response intervention cards should be kept inside the driver's cab and provide guidance on initial actions for fire crews (WHO 2014).

### 15.6.3.5 Treatment and Disposal Methods

Disposal refers to the final placement of treated waste, using a sanitary landfill or any other environmentally acceptable method of final storage (Ferdowsi et al. 2013; Singh et al. 2014).

#### *15.6.3.5.1 General Overview About Treatment Methods*

The treatment of biomedical waste is done in order to reduce the potential hazard posed by biomedical waste to human beings and the environment. To decide on which treatment methods of biomedical waste, certain factors should be considered. These factors involve consideration of waste characteristics, technology capabilities requirements, environmental safety factors, and costs. There are five basic processes for the treatment of biomedical waste, which are thermal, chemical, irradiation, biological, and mechanical treatment methods (WHO 2014).

*15.6.3.5.1.1 Thermal Processes* These processes use thermal energy to destroy pathogens that included the waste. Thermal processes have two designs: low heat and high heat. Low-heat thermal processes are those that use thermal energy at elevated temperatures high enough to destroy microorganisms but not sufficient to cause combustion or pyrolysis of the waste. Pyrolysis is the thermal degradation of a substance through the application of heat in the absence of oxygen. Pyrolysis is a special case of thermolysis and is commonly used for organic materials. It occurs at high temperatures but does not involve reactions with oxygen. In practice, it is difficult to have a completely oxygen-free atmosphere, so some oxidation takes place.

In general, low-heat thermal technologies operate between 100°C and 180°C. The low-heat processes take place in either moist or dry heat environments. Moist (or wet) thermal treatment involves the use of steam to disinfect waste and is commonly performed in an autoclave or steam-based treatment system. Microwave treatment is essentially a moist thermal process, because disinfection occurs through the action of moist heat (hot water and steam) generated by the microwave energy. Dry-heat processes use hot air without the addition of water or steam. In dry-heat systems, the waste is heated by conduction, convection, and/or thermal radiation using infrared or resistance heaters. The most commonly used dry heat technology treatment is incineration.

Incineration is one of the oldest methods for the treatment of medical wastes that does not require pretreatment. Incineration is a high-temperature, dry oxidation process that reduces organic and combustible waste to inorganic and incombustible matter and results in a significant reduction of waste volume and weight. High-heat thermal processes take place at temperatures from about 200°C to more than 1000°C. They involve the chemical and physical breakdown of organic material through the processes of combustion, pyrolysis, or gasification. The basic characteristics necessary for incineration include heating value above 2000 kcal/kg (8370 kJ/kg), calorific values within the regulatory and design requirements, content of combustible matter above 60%, content of noncombustible solids below 5%, content of noncombustible fines below 20%, and moisture content below 30% (Ferdowsi et al. 2013; Singh et al. 2014; WHO 2014).

*15.6.3.5.1.2 Chemical Processes* Chemical treatment methods use disinfectants such as dissolved chlorine dioxide, bleach (sodium hypochlorite), peracetic acid, lime solution, ozone gas, or dry inorganic chemicals such as calcium oxide powder. Chemical processes often involve shredding, grinding, or mixing to increase exposure of the waste to the chemical agent. In liquid systems, the waste may go through a dewatering section to remove and recycle the disinfectant. Besides chemical disinfectants, there are also encapsulating compounds that can solidify sharps, blood, or other body fluids within a solid matrix before disposal. Another example of a chemical process is a system that uses heated alkali to digest tissues, pathological waste, anatomical parts, and animal carcasses in heated stainless-steel tanks (WHO 2014).

*15.6.3.5.1.3 Irradiation Technologies* Irradiation treatment encompasses designs using irradiation from electron beams, cobalt-60, or ultraviolet sources. These technologies require shielding to prevent elevated occupational exposures to electromagnetic radiation. The pathogen destruction efficacy depends on the dose absorbed by the mass of waste. Electron beams are powerful enough to penetrate waste bags and containers. Germicidal ultraviolet radiation has been used to destroy airborne microorganisms as a supplement to other treatment technologies but is not able to penetrate closed waste bags (WHO 2014).

*15.6.3.5.1.4 Biological Treatment* Some biological treatment systems use enzymes to speed up the destruction of organic waste containing pathogens. Composting and vermiculture (digestion of organic wastes through the action of worms) are biological processes and have been used successfully to decompose hospital kitchen waste, as well as other organic digestible waste and placenta waste. The natural decomposition of pathological waste through burial is another example of a biological process (WHO 2014).

*15.6.3.5.1.5 Mechanical Processes* Mechanical treatment processes include several shredding, grinding, mixing, and compaction technologies that reduce waste volume, although they cannot destroy pathogens. Mechanical processes are not used alone in biomedical waste treatment processes but supplement other treatment methods. Mechanical destruction can be used to destroy needles and syringes. In the case of thermal or chemical treatment processes, mechanical devices such as shredders and mixers can also improve the rate of heat transfer or expose more surface area of waste to waste treatment. Unless shredders, mixers, and other mechanical devices are an integral part of a closed treatment system, they should not be used before disinfecting biomedical waste (WHO 2014).

#### *15.6.3.5.2 Treatment and Disposal Methods to Specific Biomedical Waste Categories*

*15.6.3.5.2.1 Sharps* Sharps are often collected in safety boxes and burned. Several nonburn methods have been developed in response to concerns about air pollution and the short life span of brick incinerators. The methods generally entail the following steps:

1. Using onsite mechanical needle cutters or electric needle destroyers
2. Shredding the treated plastic parts
3. Burying the metal pieces in sharp pits
4. Remelting the plastics for recycling

The sharps waste can be autoclaved, shredded, and then encapsulated in cement blocks that later become useful items such as hospital benches (WHO 2014).

*15.6.3.5.2.2 Pathological Waste* The treatment of pathological wastes may be bound by sociocultural, religious, and esthetic norms and practices. Burial in cemeteries or special burial sites and burning in crematoria or specially designed incinerators are two traditional options. A more recent option is alkaline digestion, especially for contaminated tissues and animal carcasses. Promession is a newer technology designed especially for human cadavers. In some countries, placenta waste is composted or buried in placenta pits designed to facilitate natural biological decomposition (WHO 2014).

*15.6.3.5.2.3 Cytotoxic and Genotoxic Waste* Cytotoxic and genotoxic waste should first be minimized by careful segregation, purchasing optimal drug quantities, using proper spill containment and cleanup procedures, and substituting environmentally persistent drugs with degradable drugs, where possible. They should be labeled and sorted using proper personal protective equipment before treatment. Cytotoxic waste is highly hazardous and should never be landfilled or discharged into the sewerage system. Disposal options include return to the original supplier, incineration at high temperatures, and chemical degradation in accordance with manufacturers' instructions.

Full destruction of all cytotoxic substances may require incineration temperatures up to 1200°C and a minimum gas residence time of 2 seconds in the second chamber. The incinerator should be equipped with gas cleaning equipment. Incineration at lower temperatures may release hazardous cytotoxic vapors into the atmosphere. Incineration in most municipal incinerators, in single-chamber incinerators, or by open-air burning, is inappropriate for the disposal of cytotoxic waste. Chemical degradation methods, which convert cytotoxic compounds into nontoxic/nongenotoxic compounds, can be used for drug residues and for cleaning contaminated urinals, spillages, and protective clothing. They are not appropriate for treating contaminated body fluids (WHO 2014).

#### *15.6.3.5.3 Final Disposal of Biomedical Waste*

Land disposal is required to remove the remaining materials of biomedical waste after minimization or treatment. In less developed areas, where a municipality or healthcare facility lacks the means to treat wastes before disposal, the direct use of a landfill is likely to be required for much of the material produced. The controlled landfilling is recommended for disposing of biomedical waste in engineered landfills under constraints of local regulations. A well-engineered landfill is designed to minimize contamination of soil, surface water, and groundwater; limit atmospheric releases and odors; block access to waste by pests and vectors; and prevent contact with the public (WHO 2014).

## 15.7 FUTURE PROSPECTS AND CHALLENGES

The future prospects of biomedical waste incorporate issues regarding changing pattern of diseases, environmental issues, and waste technology. The changing pattern of diseases could be summarized in the increasing number of new microbial strains that induce emerging human diseases. Therefore, the technology of treating biomedical waste should be upgraded continuously to overcome the challenges of new microbial strains. The periodical occurrence of pandemics as well as the mandatory use of vaccines consequently will increase the global burden of biomedical waste. Accordingly, a shortage of workforce personnel and required resources for managing the biomedical waste may occur. These obstacles can be reduced by considering the health status of waste workers and applying the central factory for the waste management system.

Gradual climatic change may affect the process of biomedical waste disposal. Waste disposal sites are often built on marginal sites, such as marshlands, flood plains, and coastal areas. These safe disposal sites will become vulnerable to flooding where average sea and river levels rise or more frequent extreme weather events inundate the land. This can be solved by decentralizing waste treatment and increasing storage capacity, as well as undertaking contingency or continuity planning at the facility and national levels. Fuel and power costs are predicted to rise, and power shortages may become more common. Planners should promote the adoption of low-energy technologies wherever possible.

Waste technology such as incinerators may induce environmental pollution, especially with the increased quantity of biomedical waste due to utilization of single-use instruments. Alternative environmentally friendly technology should be focused to reduce the negative effects of existing methods (Ghasemi et al. 2016).

Other challenges facing biomedical waste management include commitment to regulation and polices by facilities that generate biomedical waste and their partners in its treatment and disposal processes. The insufficient infrastructure that is required for safe biomedical waste management is a crucial issue that hinders the processes system. Prevention of exposure to hazards of biomedical waste would take place by improving regulation, emphasizing supervision and monitoring, as well as strengthening the infrastructure and supporting facilities of waste management. Poor training of healthcare workers on the safe handling of biomedical waste is an important issue and still requires executing efforts to improve the biomedical waste processing (Babu et al. 2009; Alzahrani 2013; Nie and Wu 2016).

## 15.8 CONCLUSION

Biomedical waste is generated mainly from healthcare facilities. This type of waste requires meticulous precautions to protect human beings and the environment from its dangerous influences. The undesirable effect of biomedical waste ranged from dermal irritation to the acquisition of serious communicable diseases and carcinogenic diseases. Segregation on generation point is the first and most important step in attaining the hazards elimination aim. A series of processes and standards should be followed to achieve the final disposing of biomedical waste in a safe manner.

Several challenges that debilitate the safe biomedical waste management may arise. Some of these challenges are related to pathogen mutations and pandemic occurrence, while others are related to the technologies used in treating biomedical waste as well as the increased burden of waste due to the advancement of single-use medical instruments. These challenges and others require actions to minimize and/or eliminate any challenge.

## REFERENCES

Alzahrani, D. 2013. Environmental legislation and the management of medical waste. People and the Planet 2013 Conference: Transforming the Future, 2–4 July, RMIT University, Melbourne, Australia. http://global-cities.info/news-events/conferences-forums/conferences-proceedings (accessed September 20, 2016).

Askarian, M., Heidarpoor, P., and Assadian, O. 2010. A total quality management approach to healthcare waste management in Namazi Hospital, Iran. *Waste Management* 30(11): 2321–2326.

Bokhoree, C., Beeharry, Y., Makoondlall-Chadee, T., Doobah, T., and Soomary, N. 2014. Assessment of environmental and health risks associated with the management of medical waste in Mauritius. *Procedia APCBEE* 9: 36–41.

Ferdowsi, A., Ferdosi, M., and Mehrani, M.J. 2013. Incineration or autoclave? A comparative study in Isfahan hospitals waste management system. *Materia Sociomedica* 25(1): 48–51.

Ghasemi, M.K., and Yusuff, R.B.M. 2016. Advantages and disadvantages of healthcare waste treatment and disposal alternatives: Malaysian scenario. *Polish Journal of Environmental Studies* 25(1): 17–25.

Health Care without Harm Asia. 2007. *Best practices in health care waste management: Examples from four Philippine hospitals*, Health Care without Harm Asia, Metro Manila, Philippines.

Mathur, P., Patan, S., and Shobhawat, A.S. 2012. Need of biomedical waste management system in hospitals—An emerging issue—A review. *Current World Environment* 7(1): 117–124.

Nema, A., Pathak, A., Bajaj, P., Singh, H., and Kumar, S. 2011. A case study: Biomedical waste management practices at city hospital in Himachal Pradesh. *Waste Management & Research* 29(6): 669–673.

Nie, L., and Wu, H. 2016. Investigation and analysis of rural medical waste management. *Journal of Environmental Protection* 7: 93–98.

Pullishery, F., Panchmal, G.S., Siddique, S., and Abraham, A. 2016. Awareness, knowledge, and practices on biomedical waste management among health care professionals in Mangalore: A cross sectional study. *IAIM* 3(1): 29–35.

Ramesh, B., Parande, A.K., Rajalakshmi, R., Suriyakala, P., and Volga, M. 2009. Management of biomedical waste in India and other countries: A review. *Journal of International Environmental Application & Science* 4(1): 65–78.

Sefouhi, L., Kalla, M., Bahmed, L., and Aouragh, L. 2013. The risk assessment for the health-care waste in the hospital of Batna city, Algeria. *International Journal of Environmental Science and Development* 4(4): 442–445.

Singh, R. 2001. Bio-medical wastes an emerging threat to environment. *JPAFMAT* 1: 3–5.

Singh, A., Singh, J., Singh, A.P., and Singh, R. 2014. Biomedical waste management and their possible health risks with controlling measures in Bareilly city, UP, India. *Octa Journal of Environmental Research* 2(4): 296–302.

World Health Organization (WHO). 2014. *Safe management of wastes from health-care activities*. 2nd edition. Y. Chartier et al. (eds.), World Health Organization, Geneva.

# Index


## A

Accumulation, root to shoot, 320
*Acetovibrio* sp., 292
*Achromobacter* sp., 70, 78, 189, 387
Acid dyes, 129
*Acinetobacter* sp., 247, 260
*Actinomadura* sp., 288
*Actinomycetes* sp., 157
Adsorption
  appropriate isotherms, 142–143
  colloid flotation, 136
  dye contamination treatment, 136, 141–146
  multicomponent adsorption, 144–146
  pesticides, 80–82
  remediation, toxic metals contamination, 216
  soil, pesticides biodegradation, 80–82
Advanced oxidation processes (AOPs)
  advantages/disadvantages, 146–147
  catalytic ozonation, 153
  electron transfers, 150
  electrophilic addition, 149–150
  fenton reaction, 153–154
  generation and interferences, 147–149
  heterogeous systems, 154–155
  homogeneous systems, radical generators, 150–154
  $H_2O_2$/UV, combined use, 151–152
  hydrogen atoms abstraction, 149
  hydroxyl radical, 147–150
  implementation limitations, 147
  $O_3/H_2O_2$, combined use, 152
  $O_3/H_2O_2$/UV, combined use, 152–153
  overview, 146
  ozone, 150–151
  photo-Fenton processes, 154
  principles of, 146
  radical–radical reactions, 150
  systems used in, 150–155
Advantages/disadvantages
  adsorption techniques, 146
  advanced oxidation processes, 146–147
  biological methods, 229–230
  bioremediation, 83, 119, 120
  constructed wetlands, 401, 419
  dye treatment technologies, 129, 133, 156
  excavation, 219
  *ex situ* bioremediation, 9, 16–18
  phytoremediation, 8–9, 325, 327
  phytostabilization, 317
  phytovolatilization, 229
  remediation, 229–230
  waste recycling, 363
Aerobic treatments, *see* Composting
*Aeromonas* sp., 157
*Agave tequilana,* 193
Agricultural field, 3, 64, 341
*Agrobacterium radiobacter,* 70
*Agrocybe semiorbicularis,* 72
Air, pesticide residues, 65
Air sparging method, 6, 245
*Alcaligenes* sp.
  dye contamination, 157
  lindane remediation, 189
  pesticide biodegradation, 70
  plant growth-promoting rhizobacteria, 260, 322
Algae, 156, 159, 223–225; *see also specific type of algae*
Alkaline digestion, 439
*Allescheriella* sp., 226
*Allium cepa*, 132
*Alternaria alternata*, 158
Aluminum, 376–377, 379, 381–382
*Alyssum bertolonii*, 260
*Amaranthus caudatus*, 66
*Anabaena* sp., 224
Animal husbandry waste, 341
AOPs, *see* Advanced oxidation processes
*Arabidopsis thaliana*, 307, 317, 323
*Arachis hypogaea*, 69
Ardystil syndrome, 132
"Arhus Protocol," 34
Arsenic toxicity, 36, 311–312
*Arthrobacter* sp.
  lindane mineralization, 191, 192
  pesticide biodegradation, 70, 72
  plant growth-promoting rhizobacteria, 260, 322
  surfactant mediated bioremediation, 86, 87
  tannery wastewater, 387
*Arundo donax*, 389, 416
Ash residues handling, 362
*Aspergillus* sp.
  biosorption, 84–85
  cellulase production, 288
  heavy metals bioremediation, 226
  pesticide biodegradation, 72
  solid-state fermentation, 283
  submerged fermentation, 287

tannery wastewater, 387
toxic metals removal, 222
*Asplenium platyneuron*, 418
*Astragalus sinicus*, 266
*Aureobasidium pullulans*, 222
*Auricularia auricula*, 72
Azo dyes
biological treatment, 155–156
complete destruction, 152
regulations, 27
tannery wastewater, 375–376
toxicity, 30–31
*Azospirillum* sp., 260, 322
*Azotobacter* sp., 190, 322

**B**

*Bacillus* sp.
biotransformation, role in, 112
cellulase production, 291–292
dye contamination, 157
lindane remediation, 189–190
pesticide biodegradation, 70, 76, 79
plant growth-promoting rhizobacteria, 260, 322
surfactant mediated bioremediation, 86, 87
tannery wastewater, 387
textile wastewater, 418
Bacterial approaches; *see also specific type of bacteria*
dye contamination, 156–157
industrial waste, 41
lindane contamination, 188–191
pesticide biodegradation, 72, 73
symbiotic relationship with plants, 13
toxic metals, 223, 225
Bacterial cellulases, 291–292
*Bacteroides* sp., 292, 359
Bag house filter, 362
*Barytelphusa cunicularis,* 69
*Basidiomycetousi* yeasts, 159
Berms and walls, textile wastewater, 419
Bioattenuation method, 5, 38
Bioaugmentation technique
bioreactive barrier, 117
integrated bioremediation strategies, 249
nonaqueous phase liquids, 243, 246–247
pesticides contamination, 83–84
*in situ* bioremediation, 38
soil bioremediation, 5, 11
uranium contamination, 112
Biobeating and pulp refining, 292–293
Biobed treatment system, 88–89
Biochar, 216
Biocrystallization, 115
Biodegradation, pesticides
adsorption, 80–82
bioavailability, 80–82
conjugation, 75
dehalogenation, 74–75
enzymes, role of, 72–73
microbes involved in, 72
nutrients availability, 76
overview, 72
oxidation, 74
oxygen availability, 76
pathway of, 74–75
pH, 76, 78–79
reduction, 74
in soil, 78–92
soil moisture content, 80
soil texture, 79–80
solubility of pesticides, 77–78
temperature, 76, 78–79
water, 75–78
Bioenergy production, 358–361, 405
Biohazardous wastes, 428; *see also* Biomedical wastes
Bioleaching mechanisms, 114
Biological approaches
biomedical waste, 438–439
dye contamination, 155–160
toxic metals contamination, 221–230
"Biological index," 33
Biological stains, 129
Biomass composition, 295
Biomedical wastes
biohazardous, 428
categories, 428–429
central storage areas, 435–436
challenges, 440–441
collection, 433–434
defined, 428
disposal methods, 437–440
environmental impact of, 431
final disposal, 440
future prospects, 440–441
"general healthcare waste," 428
genetoxic waste, 439
genotoxic wastes, 429
hazards from, 430–431
impacts of, 430–431
infectious wastes, 429
interim storage, 434
introduction, 428
irradiation technologies, 438
legalization and policies, 432
management of, 431–440
mechanical processes, 439
nonhazardous, 428

off-site transportation, 436–437
on-site transportation, 434–435
packaging, 433–434
pathological wastes, 429, 439
persons at risk, 430
pharmaceutical wastes, 429
segregation, 433–434
sharps, 428–429, 439
sources of, 429–430
storage, 434–440
thermal processes, 437–438
transportation, 434–440
treatment, 437–440

Biomineralization
pesticide biodegradation, 72
uranium contamination, 109–110, 115, 121

Biopiles technique, 10
Biopolymers, 16, 109, 211, 278
Bioprecipitation, 115
Bioreactive barriers, 117, 122
Bioreduction, 109, 115, 118, 122
Bioremediation approaches; *see also specific contamination*
air sparging, 6
bioattenuation, 5
bioaugmentation, 5
biopiles, 10
bioprecipitation, 115
bioreactive barrier, 117
bioreactors, 10–11, 112
bioreduction, 115
biosorption, 114–115
biosparging, 6, 112
biostimulation, 5
bioventing, 6, 112
challenges, 44
chelation, 117–118
composting, 9–10
defined, 2
engineered *in situ* techniques, 111
enhancement, 13–16
environmental pollutants, 37–43
*ex situ* bioremediation, 9–11, 16–18
*ex situ* techniques, 112
factors affecting process, 11–13
future perspectives, 44
genetic engineering, 13–16
industrial waste, 37–43
injection wells, 118–119
intrinsic bioremediation, 111
introduction, 2–3
landfarming, 9
lindane contamination, 188–194
mechanisms, 114–119
merits and demerits, 16–18
metals and metalloids, 112–113
microbial processes, 113–114
microorganisms, 112–113, 385–386
nano-biotechnology, role of, 16
overview, 3–4
phytoremediation, 6–9
principles of, 110–111
radionuclides, 112–114
remediation, toxic metals contamination, 220–222
*in situ* techniques, 4–9, 16–18, 111
slurry-phase treatment, 18, 112
solid-phase treatment, 18, 112
successful outcome, 230
types of, 4–11, 111–112
uranium contamination, 109–119

Biosorption, 84–85, 114–115, 156
Biosparging, 6, 84, 112; *see also* Sparging
Biostimulation method, 3–5, 38
Biotechnology divisions, 3
Biotic systems, pesticide residues, 65–66
Biotransformation, role of microorganisms, 112–113
Bioventing techniques, 6, 112, 244–245
*Bjerkandera adusta,* 158, 193
Boiler/steam turbine, 362
*Bordetella petrii,* 78
*Bordetella* sp., 70
*Botryosphaeria rhodina,* 226
Bound pesticide residues, 80–81
*Brachiaria decumbens,* 389
*Brassicaceae* family, 266
*Brassica* sp., 69, 259, 317
*Brevibacterium* sp., 191, 387
Bubbles, *see* Flotation
*Burkholderia* sp.
bioaugmentation, 84
pesticide biodegradation, 70, 72
plant growth-promoting rhizobacteria, 260, 322

Burning, solid waste, 346
*Butyrivibrio fibrosolvens,* 359


## C


Cadmium
carcinogenic properties, 384
tannery wastewater, 377, 379–380, 383
toxicity, 34–35, 310

*Calothrix* sp., 224
*Candida* sp., 158, 193, 247, 288
*Canna generalis,* 247
*Canna indica,* 389, 415, 418
Capping systems, 220

*Capsicum frutescens,* 66
Casmalia Resources superfund site, 82
Catalytic ozonation, 153
Cationic dyes, 129
*Cellulomonas* sp., 291–292
Cellulosomes, 276, 278, 282, 292
Central storage, biomedical waste, 435–436
*Cephalosporium aphidicola,* 226
*Ceriporiopsis subvermispora,* 158
Chakrabarty, Ananda, 13
*Chara canescens,* 317
Charge neutralization, 137, 138–139
Chelation, uranium contamination, 117–118
*Chelatobacter heintzii,* 70
Chemotaxis, 5
Chitosan, 141
*Chlamydomonas* sp., 85
*Chlorella pyrenoidosa,* 159
*Chlorella vulgaris,* 159
Chloride toxicity, 381
Chlorinated phenols (CPs), 29–30
*Chlorlla* sp., 85
Chromium, 35–36, 312–313, 383–384
*Chryseobacterium* sp., 261
*Cicer arietinum,* 194
*Cintractia sorghi,* 193
*Citrobacter* sp., 76, 189
*Cladophora fascicularis,* 224
*Cladosporium resinae,* 222
*Clavibacter* sp., 191
Clean technologies, 3
Climatic change, biomedical waste disposal, 440
*Clostridium* sp.
  anerobic bacteria degrading cellulose, 359
  cellulase production, 291
  cellulosomes, 282
  enzymatic reduction, uranium, 119
  lindane remediation, 188
Coagulation–flocculation, 137–141
Collection, biomedical waste, 433–434
Collection systems, solid waste, 342
Colloid flotation, adsorbing, 136
*Colocasia* sp., 418
Combustion, 345, 354, 362; *see also* Incineration
Co-metabolism, 11
Commercial waste, 341
Competitive interactions, adsorbates, 143–144
Composting
  biomedical waste, 438
  industrial wastes, 37–38
  pesticide contamination, 85
  soil bioremediation, 9–10
  solid waste, 345, 354, 356–358
  toxic metals contamination, 221–222
*Conidiobolus* sp., 193
Conjugating enzymes, 73
Conjugation, pesticides biodegradation, 75
"Consignment note," 437
Consortia conditions, 11
Constructed wetlands (CWs)
  characteristics, 399–403
  distillery wastewater, 416–417
  hybrid type, 402–403
  industrial wastewater applications, 415–418
  introduction, 398–399
  management and monitoring, 419
  mechanisms of action, 406–408
  merits and demerits, 419
  metabolic potential, 408–415
  microbes role, wastewater treatment, 405
  microbial community structure, 408–415
  natural wetlands, 399–400
  nitrogen removal, 406
  nonaqueous phase liquids, 248
  oil refinery wastewater, 417
  overview, 8
  petrochemical wastewater, 417
  phosphorous removal, 406, 408
  plant role, wastewater treatment, 403–405
  pulp and paper mill wastewater, 417
  subsurface flow type, 402
  sulfur removal, 408
  surface flow type, 400–402
  tannery wastewater, 415–416
  textile wastewater, 418
  types, 399–403
  uranium contamination, 120–121
  winery wastewater, 416–417
Construction waste, 341
"Controlled disequilibrium," 6
Conversion, lignocellulose into bio-ethanol, 295–296
*Corchorus olitorius,* 66
*Coriolus versicolor,* 72
Corrosive solid waste, 349
*Corynebacterium* sp., 86, 191, 247
*Cratylia argentea,* 66
*Cunninghamella elegans,* 226
*Cupriavidus metallidurans,* 227
CW, *see* Constructed wetlands
*Cyanothece* sp., 224
*Cyathus bulleri,* 192
*Cyperus immensus,* 417
*Cyperus involucratus,* 416
*Cyperus papyrus,* 417
*Cyprinus carpio,* 32
*Cytisus striatus,* 194
*Cytophaga* sp., 292
Cytotoxic waste, 429, 431, 436, 439–440

## D

*Debaryomyces polymorphus*, 158
*Dehalobacter* sp., 189
*Dehalobacte* sp., 76
Dehalogenation, 74–75
Dehydrohalogenase enzyme, 73
Deinking process, 293–294
*Deinococcus geothemalis*, 227
*Deinococcus radiodurans*, 16, 228
Demerits, *see* Advantages/disadvantages
Demolition waste, 341
Destabilization, dye contamination, 137–140
*Desulfobacter* sp., 76
*Desulfococcus* sp., 76, 189
*Desulfosporosinus* sp., 119
*Desulfotomaculum* sp., 264
*Desulfovibrio* sp.
  enzymatic reduction, uranium, 119
  heavy metal and microbe interaction, 264
  lindane remediation, 189, 191
  tannery wastewater, 387
*Dichomitus squalens*, 72, 158
*Dioscorea alata*, 66
Direct dyes, 129
Dirty Dozen POPs, 34
Disadvantages, *see* Advantages/disadvantages
Disinfectants, biomedical waste, 438
Dispersed-air flotation, 135, 136
Disperse dyes, 129
Dissimilatory metal reduction, 227
Dissolved-air flotation, 135, 136
Dissolved organic matter (DOM), 87–88
Distillery industry, 32, 39, 416–417
DLVO theory, 138
Double-layer compression mechanism, 137, 138–139
*Dracaena sanderiana*, 418
Drinking water
  aluminum, 376
  cadmium, 380
  chlorinated phenols, 29
  lead, 377
  pesticides, 62–63
  sodium chloride, 378
Dyes contamination
  acid dyes, 129
  adsorption, 141–146
  advanced oxidation process, 146–155
  advantages/disadvantages, 146–147
  basic dyes, 129
  biodegradation, 156–161
  biological treatment approaches, 155–160
  coagulation–flocculation, 137–141
  direct dyes, 129
  disperse dyes, 129
  dyes biodegradation, 160–161
  ecotoxicological effects, 131–132
  equalization, 134–135
  fate in environment, 130–131
  flotation, 135–136
  health hazards, 131–132
  homogenization, 134–135
  introduction, 128–129
  mordant dyes, 130
  multicomponent adsorption, 144–146
  nature, characteristics, classification, 129–130
  physiochemical treatment approaches, 133–155
  reactive dyes, 130
  removal of dyes, 143, 150
  solvent dyes, 130
  sources of, 131
  sulfur dyes, 130
  transport in environment, 130–131
  vat dyes, 130
  wastewater treatment approaches, 132–160
Dye/Textile wastewater, 140–141

## E

Earthworm assisted approaches, 85
Ecotoxicological effects, dyes, 131–132
EDCs, *see* Endocrine disrupting chemicals
*Eichhornia* sp., 417–418
Electro-bioremediation technology, 41
Electrokinetics, 41, 217–218
Electrolytic flotation, 136
Electron transfers, 150
Electrophilic addition, 149–150
Emergent aquatic vegetation (EAV) system, 401
Endocrine disrupting chemicals (EDCs), 28–29
Engineered techniques, 13, 111, 221
Enhanced bioremediation, 13–16
Enhanced biostimulation strategies, 244–246
*Enterobacter* sp., 260, 322, 387
*Enterococcus casseliflavus,* 387–388
Environmental pollutants; *see also specific contamination*
  arsenic, 36
  azo dye, 30–31
  bioremediation approaches, 37–43
  cadmium, 34–35
  chlorinated phenols, 29–30
  chromium, 35–36
  endocrine disrupting chemicals, 28–29
  industrial wastes, 27–37
  lead, 36–37
  melanoidins, 32
  mercury, 37

organic pollutants, 27–34
persistent organic pollutants, 33–34
pesticides, 32–33
petroleum hydrocarbons, 31–32
phenol, 28
Enzymes, 72–73, 119–120, 160
Equalization, 134–135
*Erianthus arundinaceus,* 417
*Erwinia* sp., 260
Erythrosine, 132
*Escherichia* sp.
dye contamination, 157
heavy metals bioremediation, 227
lindane remediation, 189
pesticide biodegradation, 76
tannery wastewater, 387
toxic metals removal, 225
Ethanol, 275, 295–296
*Eurydice* sp., 387
Excavation, 219
Exhaust gas cleaning, 362
Explosives, concentration reduction, 10
Exposure, solid waste, 349–350
*Ex situ* techniques; *see also specific contamination*
bioremediation process, factors affecting, 11–12
costs, 38
pesticides, soil and water, 83–89
slurry-phase treatment, 18
solid-phase treatment, 18
toxic metals contamination, 221–222
types, 9–11
uranium contamination, 112

## F

Fate, 66–67, 81, 130–131
Fermentation industry, 32
Filtration, 215–216
Final disposal/destruction
biomedical waste, 440
bioremediation processes, 104
reactive dyes and azo dye, 152
solid waste, 352
First-order kinetics, 251
*Flammulina velutipes,* 72
*Flavobacterium* sp., 189, 191, 260
Flocculation, *see* Coagulation-flocculation
Flotation, 135–136
Flow equalization, 134–135
Foods and food products
aluminum contamination, 376
cadmium, 377
dyes in, 129, 131
lead, 377
lindane contamination, 184
pesticide contamination, 66, 68–69
solid waste, 339–440
toxic metals contamination, 214
Forced aeration, 357
Free-floating aquatic vegetation (FFAV) system, 401
Freundlich isotherms, 143
Fumigants, 33; *see also* Pesticides
*Funalia trogii,* 222
Fungi; *see also specific type*
cellulase enzyme, 288–291
conjugation, 74
dye contamination, 156–158
extracellular oxidative enzymes, 192–193
lindane contamination, 192–193
pesticide biodegradation, 72, 73
toxic metals contamination, 222–223
toxic metals removal, 223
Fungicides, 33
*Fusarium* sp., 193
Future prospects and directions
biomedical wastes management, 440–441
bioremediation, 44
microbial cellulase enzyme, 297
phytoremediation, metal-contaminated soils, 325–327
solid waste, 364–365

## G

*Galactomyces geotrichum*, 158
*Ganoderma* sp., 70, 193, 222
"General healthcare waste," 428
Genetic engineering
enhancement, bioremediation, 13–16
heavy metal-contaminated soils, 266–267
microbes, dyes biodegradation, 160–161
modified plants approach, 323–325, 326
Genetoxic waste, 439
Genomics, current technologies, 13
Genotoxic wastes, 429, 431, 436, 439–440
*Geobacter* sp., 119, 120, 227
*Geotrichum candidum,* 158
Germicidal ultraviolet radiation, 438
*Gloeocapsa pleurocapsoides,* 157
*Glycine max,* 69, 194
Government initiatives, lindane, 194–196
Greenhouse gases, landfills, 344–347

## H

Healthcare waste, 428
Heavy metals; *see also* Metals and metalloids
bacterial-assisted phytoremediation, 42–43
genetically engineered strategies, 266–267

internal physiological mechanisms, 264–266
introduction, 257–258
microbe and heavy metal interaction, 264
origin, 212–213
phytoremediation technologies, 258–260
plant, soil, and microbe interaction impact, 261–263
plant growth-promoting rhizobacteria, 260–261
*Helianthus* sp., 259, 307
Herbicides, 33
Heterogeous systems, 154–155
Homogeneous systems, radical generators, 150–154
Homogenization, 134–135
Horizontal flow (HF), constructed wetlands, 402–403
$H_2O_2$/UV, combined use, 151–152
Human health hazards
aluminum, 379, 381–382
arsenic, 36, 312
azo dye, 30, 31
cadmium, 34, 310, 379–381, 383
carcinogenic properties, 383–385
cardiovascular system, 379
chloride, 381
chlorinated phenols, 30
chromium, 35, 312–313
digestive system, 382
dyes, 128, 131–132
erythrosine, 132
heavy metals, 308, 309
hypertension, 379
lead, 37, 311, 380
lindane contamination, 185–187
mercury, 37, 311
nervous system, 381–382
persistent organic pollutants, 33, 350
pesticides, 33, 62, 66, 67–68, 70–71
petroleum hydrocarbons, 31–32
phenol toxicity, 28
pregnancy, exposure during, 383
skeletal effects, 383
solid waste, 348–351
sulfide, 378–379, 381, 382
tannery wastewater, 376, 378–385
toxic metals contamination, 213–214
uranium contamination, 107–109
urinary system, 379
zinc toxicity, 382
*Humicola insolens*, 291
Hybrid constructed wetlands, 402–403
Hydrocarbon-contamination, 10
*Hydrodictyon* sp., 224
Hydrogen atoms abstraction, 149
Hydrolysis of cellulose, 295
Hydrolytic enzymes, 72
Hydroponics, 221
Hydroxyl radical, 147–150
*Hymenoscyphus ericae*, 226
Hyperaccumulators
continuous phytoextraction, 316
future prospects, 327
heavy metals, 320–322, 325
phytoremediation, 307, 325
rhizofiltration, 229, 259
toxic metals, 104, 260, 267
*Hypholoma fasciculare*, 72
*Hypocrea jecorina*, 291


## I


Ideal adsorbed solution theory (IAST), 144
Ignitable solid waste, 349
Immobilization technologies, 218
Incineration, 361–362, 438, 440; *see also* Combustion
Industrial fields, clean technologies, 3
Industrial wastes
arsenic, 36
azo dye, 30–31
bacterial-assisted phytoremediation, 41, 42–43
bioremediation, 37–43
cadmium, 34–35
chlorinated phenols, 29–30
chromium, 35–36
composting, 37–38
electro-bioremediation, 41
endocrine disrupting chemicals, 28–29
environmental pollutants, toxicity, 27–37
inorganic pollutants, 34–37
introduction, 24–25
lead, 36–37
melanoidins, 32
mercury, 37
nature and characteristics, 25–27
organic pollutants, 27–34
persistent organic pollutants, 33–34
pesticides, 32–33
petroleum hydrocarbons, 31–32
phenol toxicity, 28
phytoremediation, 38–41
solid waste, 341
wastewater, constructed wetlands, 415–418
Infectious biomedical wastes, management, 429
Injection wells, uranium contamination, 118–119
Inorganic pollutants, industrial wastes, 34–37
Insecticides, 33, 60; *see also* Pesticides

*In situ* techniques; *see also specific contamination*
advantages/disadvantages, 16–18
bioremediation process, factors affecting, 11–12
costs, 38
pesticides contamination, 83–89
soil bioremediation, 4–9
toxic metals contamination, 221
uranium contamination, 111
Insoluble nonionic dyes, 129
Institutional waste, 341
Integrated solid waste management (ISWM), 353, 354–355, 364
Integrated strategies, NAPLs, 249
Intentional stimulation, *see* Biostimulation
Interferences, hydroxyl radical, 147–149
Interim storage, biomedical waste, 434
Internal physiological mechanisms, 264–266
Interparticle bridging, 137, 139–140
Intrinsic bioremediation, 13, 83, 111, 221
Inverting mechanism, 282
Ion flotation, 136
*Iris pseudacorus,* 389, 415–416
*Irpex lacteus,* 158, 193
Irradiation technologies, 438
Irrigation, 9, 35, 373, 398
*Ischnoderma resinosum,* 158
*Issatchenkia* sp., 158, 159

## J

*Jatropha curcas*, 194
*Juncus effusus*, 416

## K

Kinetic models, NAPLs, 250–252
*Klebsiella* sp.
heavy metals bioremediation, 226, 228
lindane remediation, 194
pesticide biodegradation, 70, 72
plant growth-promoting rhizobacteria, 322
*Kocuria rhizophila*, 194

## L

*Lactobacillus* sp., 79, 190
*Lactuca sativa,* 194
Landfarming, 9; *see also* Biopiles technique
Landfilling
biomedical waste, 431
greenhouse gases, 344–347
solid waste, 343, 354, 363–364
Land forming, 221
Langmuir isotherms, 143, 144, 216
*Lathyrus sativum,* 194
Leachate contamination, 343–344
Leaching, soil contamination, 219
Lead
carcinogenic properties, 385
pregnancy, exposure during, 383
tannery wastewater, 377, 380, 383
toxicity, 36–37, 311
Leather, *see* Tannery industries
Legalization, biomedical waste, 432
*Lemna* sp., 417
*Lentinus Tigris,* 193
Leucomalachite green dye, 132
Lignocellulosic waste degradation into ethanol, 295–296
Lindane contamination
acute health effects, 186–187
applications, 181–183
bacteria, role of, 188–191
bioremediation approaches, 188–194
cancer hazard, 187
chemical properties, 179–180
chronic health effects, 187
commercial production, 180
fungal extracellular oxidative enzymes, 192–193
government initiatives, 194–196
health hazards, 186–187
introduction, 178
lindane-producing countries, 182
long-term effects, 187
mineralization, 191–192
overview, 178–183
physical properties, 179–180
physiological effects, 185–186
plant–microbe association, role of, 193–194
private initiatives, 194–196
reproductive hazard, 187
sources, 183–184
structure of, 179
toxicological effects, 185–187
*Litoria freycineti*, 69
*Lumbricus terrestris*, 85
*Lycopersicum esculentum*, 66

## M

Malachite green dye, 132
Management strategies, biomedical wastes
central storage areas, 435–436
collection, 433–434
final disposal, 440

interim storage, 434
legalization and policies, 432
off-site transportation, 436–437
on-site transportation, 434–435
overview, 431
packaging, 433–434
segregation, 433–434
storage, 434–440
transportation, 434–440
treatment and disposal methods, 437–440
Management strategies, solid waste
bioenergy production, 358–361
challenges, 364–365
combustion, 354
composting, 354, 356–358
disposal, 354
future prospects, 364–365
incineration, 361–362
landfilling, 354, 363–364
overview, 352–353
plan development, 354–356
processing, 356–364
recycling, 354, 362–363
waste prevention, 353–354
Materials recovery facilities (MRFs), 347–348
"Mauveine," 131
Mechanisms of action
bioremediation, 225–228
constructed wetlands, 406–408
microbial cellulase enzyme, 280–282
phytoremediation, 318–321
uranium contamination, 114–119
*Medicago sativa*, 194
Melanoidins, 32, 39
Membrane technology, 215–216
Mercury, 37, 310–311
Merits, *see* Advantages/disadvantages
*Mesorhizobium huakuii*, 266
Metabolic potential, 408–415
Metals and metalloids; *see also* Heavy metals
composting, inability to detoxify, 10
defined, 212
electrolytic flotation, 136
phytoremediation, 320–321
sequestration/volatilization, 321
solubilization, soil matrix, 319–320
uptake, plant root, 320
uranium contamination, 112–113
*Methanobacterium* sp., 360
*Methanobrevibacter* sp., 360
*Methanosaeta* sp., 360
*Methanosarcina* sp., 360
*Methanothermobacter* sp., 360
*Methanotrix* sp., 360
Micelle formation, 85–86
*Microbacterium* sp., 190, 194, 418
Microbes; *see also specific type*
chromium, 35
constructed wetlands, 408–415
dye contamination, 160–161
heavy metal-contaminated soils, 261–264
industrial wastewaters, 39
lindane bioremediation, 193–194
NAPLs-polluted sites, 246
pesticides contamination, 70, 72, 83
role in constructed wetlands, 405
seeding strategies, 250
uranium contamination, 113–114
wastewater treatment, 405
Microbial cellulase enzyme
bacterial cellulases, 291–292
carbon sources, 288
cellulolytic microorganisms, 288–294
chemical composition, 288
classification, 277–278
concentration of, 288
factors affecting, 287–288
fungal cellulases, 288–291
future prospects, 297
introduction, 274–275
lignocellulosic waste degradation into ethanol, 294–296
mechanism of action, 280–282
mode of action, 278–280
nitrogen sources, 288
occurrence of, 275–276
overview, 275
pH, 288
physiological role of, 275–276
problems and issues, 296
production of, 282–287
pulp and paper industry application, 292–294
solid-state fermentation, 283, 284–285
structure, 275–276
submerged fermentation, 283, 286, 287
temperature, 287
Microbial growth kinetics, 250–252
*Microbiospora* sp., 292
*Micrococcus* sp., 190, 225, 292, 387
*Micromonospora* sp., 191, 192, 359
*Micropterus salmoides,* 132
Mineralization, lindane bioremediation, 191–192
Mixotrophy, 15
Mode of action, cellulase enzyme, 278–280
Moisture, soil content, 80
Molecular mechanisms, metals contamination, 227–228
Monitoring, 5, 10, 64–65, 419
Monod's kinetics, 251
Mordant dyes, 130

*Moringa oleifera,* 141
"Muck," lindane contamination, 178, 188, 190
Multicomponent adsorption, 144–146
Multielectron acceptors/microbial seeding strategies, 250
Municipal solid waste (MSW), 339–341, 357; *see also* Solid waste
*Mycobacterium* sp., 72, 191
Mycoremediation, 222

## N

Nano-biotechnology, role of, 16
Natural attenuation, 3
Natural sources, metal-contaminated soils, 308–309
Natural wetlands, 399–400; *see also* Constructed wetlands
Negative interactions, 263
*Neocosmospora vasinfecta,* 226
*Neurospora* sp., 288
*Nicotiana* sp., 259, 323
*Nocardia* sp., 86, 191
*Nocardioides* sp., 191
*Nodularia* sp., 224
Nonaqueous phase liquids (NAPLs)
  bioaugmentation, 246–247
  constructed wetland treatment, 248
  engineered bioremediation techniques, 243–244
  enhanced biostimulation strateges, 244–246
  integrated strategies, 249
  introduction, 241–242
  kinetics models, 250–252
  microbial seeding strategies, 250
  movement in subsurfaces, 242
  multielectron acceptors strategies, 250
  overview, 243–244
  pesticide residues, 82
  plant-assisted strategies, 247–248
  recommendations, 252
Nonhazardous biomedical wastes, 428
*Nostoc* sp., 224
*Novosphingobium lindaniclasticum,* 191
Nutrients, 76

## O

Odor control, textile wastewater, 419
*Oedogonium* sp., 224
Off-site transportation, 436–437
Oil refinery wastewater, 417
Olfactory paralysis, 379
On-site transportation, 434–435
Open burning, solid waste, 346
Organic amendment enhanced bioremediation, 87–88
Organic pollutants, industrial wastes
  azo dye, 30–31
  chlorinated phenols, 29–30
  endocrine disrupting chemicals, 28–29
  melanoidins, 32
  persistent organic pollutants, 33–34
  pesticides, 32–33
  petroleum hydrocarbons, 31–32
  phenol, 28
  toxicity, 27–28
Orthokinetic flocculation, 137, 140
*Oscillatoria* sp., 159, 224
Outcome success, 230
Oxidation, 72, 74
Oxygen, 76
Ozone, 150–151

## P

Packaging, biomedical waste, 433–434
*Pandoraea* sp., 190
Passive aeration, 357
Pathological biomedical wastes, 429, 439
Pathways, pesticides biodegradation, 74–75
*Penicillium* sp., 85, 222, 283, 288
*Pennisetum purpureum,* 389
Perikinetic flocculation, 137, 140
Perkin, William Henry, 131
Persistent organic pollutants (POPs), 33–34, 178, 196, 350
Pesticides
  adsorption, 80–82
  air, 65
  bioaugmentation, 83–84
  bioavailability, 80–82
  biobed, 88–89
  biodegradation, 72–82
  bioremediation, 82–89
  biosorption, 84–85
  biosparging, 84
  biotic system, 65–66
  classifications, 58
  composting, 85
  conjugation, 75
  dehalogenation, 74–75
  earthworm assisted bioremediation, 85
  environmental pollutants, toxicity, 32–33
  enzymes, role of, 72–73
  *ex situ* techniques, 83–89
  fate of, 66–67
  human health hazards, 70–71

introduction, 58–59
microbes, 70, 72
microbial catalyst, 83
moisture content, 80
nutrients availability, 76
organic amendment enhanced bioremediation, 87–88
overview, 72
oxidation, 74
oxygen availability, 76
pathway of, 74–75
pH, 76, 78–79
plants, 68–69
reduction, 74
residues in environment, 62–67
sediment, 63–65
*in situ* techniques, 83–89
soil, 63–65, 78–92
solubility of pesticides, 77–78
surfactant mediated bioremediation, 85–87
temperature, 76, 78–79
texture of soil, 79–80
toxicological impact, 67–71
types, 58
in water, 75–78
water resources, 62–63
worldwide use, 59–61
Petrochemical wastewater, 417
Petroleum hydrocarbons, 31–32
PGPR, *see* Plant growth-promoting rhizobacteria
*Phanerochaete* sp., 72, 157, 192
Pharmaceutical biomedical wastes, 429
Phenol toxicity, 28
pH factor
cellulase enzyme production, 288
electrokinetic remediation, 218
enhanced biostimulation strategies, 246
pesticides degradation, 76, 78–79
*Phlebia* sp., 158, 226
*Phormidium ceylanicum,* 157
Phosphorous removal, constructed wetlands, 406, 408
*Phragmites* sp.
constructed wetlands, 389, 400
distillery wastewater, 416
petrochemical and oil refinery wastewater, 417
pulp and paper mill wastewater, 417
tannery wastewater, 415–416
textile wastewater, 418
winery wastewater, 416
Phthalates, 27, 371–372, 374–375; *see also* Endocrine disrupting chemicals
*Phycomycetes* sp., 193
Physical separation, 220
Physicochemical remediation techniques, 214–220
Physicochemical treatment approaches
adsorbing colloid flotation, 136
adsorption, 141–146
advanced oxidation process, 146–155
advantages/disadvantages, 146–147
appropriate isotherms, 142–143
capping, 220
catalytic ozonation, 153
coagulation–flocculation, 137–141
destabilization mechanisms, 138–140
dissolved-air flotation, 136
dye/textile wastewater, 140–141
electrolytic flotation, 136
electron transfers, 150
electrophilic addition, 149–150
equalization, 134–135
excavation, 219
fenton reaction, 153–154
flotation, 135–136
generation and interferences, 147–149
heterogeous systems, 154–155
homogeneous systems, 150–154
homogenization, 134–135
hydrogen atoms abstraction, 149
hydroxyl radical, 147–150
immobilization, 218
ion flotation, 136
limitations, 200
multicomponent adsorption, 144–146
ozone, 150–151
photo-fenton processes, 154
physical separation, 220
precipitate flotation, 136
radical–radical reactions, 150
remediation, 214–218
soil leaching, 219
solidification, 219
stabilization, 219
systems used in, 150–155
vitrification, 218–219
Physio-bio-chemical mechanism, 226–227
Phytoaccumulation, heavy metals, 259
Phytodegradation, 7, 39–40, 317–318
Phytoextraction
applications, 7, 8, 40
metals contamination, 228, 259, 315–316
overview, 38
Phytoimmobilization, 259
Phytoremediation
applications, 40
heavy metal-contaminated soils, 258–260
industrial waste, 38–41
soil bioremediation, 6–9
toxic metals contamination, 228–229

Phytoremediation, metal-contaminated soils
- accumulation, root to shoot, 320
- advantages/disadvantages, 325, 327
- anthropogenic sources of contamination, 309
- arsenic toxicity, 311–312
- cadmium toxicity, 310
- chromium toxicity, 312–313
- future prospects, 325–327
- genetically modified plants approach, 323–325, 326
- introduction, 306–307
- lead toxicity, 311
- mercury toxicity, 310–311
- metal detoxification and chelation, 320–321
- metal sequestration/volatilization, 321
- metal solubilization, soil matrix, 319–320
- metal uptake, plant root, 320
- natural sources of contamination, 308–309
- phytodegradation, 317–318
- phytoextraction, 315–316
- phytostabilization, 317
- phytovolatilization, 317
- plant growth-promoting rhizobacteria, 322–323
- plant types used, 321–322
- rhizodegradation, 318
- rhizofiltration, 316
- sources of contamination, 308–313
- toxic effects, heavy metals, 309–313
- types of, 313–318

Phytosiderophores, 118

Phytostabilization
- applications, 7, 8, 40
- heavy metals contamination, 259
- metals contamination, 228–229, 317
- overview, 38

Phytotechnologies, 38–41; *see also specific type*

Phytotransformation, 7

Phytovolatilization
- applications, contaminated sites, 7, 8
- heavy metal-contaminated soils, 259
- metal-contaminated soils, 317
- overview, 39
- remediation, toxic metals contamination, 229

*Piromyces* sp., 288

*Pistia* sp., 417

Planning and development, 354–356, 432

Plant growth-promoting rhizobacteria (PGPR)
- heavy metal-contaminated soils, 260–261
- metal-contaminated soils, 322–323
- phytoremediation, 41, 42–43, 307

Plants; *see also* Constructed wetlands; Phytoremediation
- bioremediation approaches, lindane, 193–194
- chromium, 35
- genetic engineering of, 3
- heavy metal-contaminated soils, 261–263
- lindane bioremediation, 193–194
- metal-contaminated soils, 321–322
- metal uptake, roots, 320
- nonaqueous phase liquids, 247–248
- pesticides, toxicity in, 68–69
- phytosiderophores, 118
- soil and microbe interaction impact, 261–263
- symbiotic relationship with bacteria, 13
- wastewater treatment, constructed wetlands, 403–405

"Plasmids," genetic engineering, 13

*Pleurotus* sp., 72, 158, 192–193, 226

Policies, biomedical wastes, 432

*Poliporus ciliatus,* 193

*Pontibacter* sp., 191

POPs, *see* Persistent organic pollutants

Positive interactions, 262–263

Precipitate flotation, 136

Pre-polymerized coagulants, 139

Prevention, solid waste, 353–354

Private initiatives, lindane, 194–196

Production, cellulase enzyme, 283–288

Promession, 439

Proteus vulgaris, 74

*Pseudoarthrobacter* sp., 194

*Pseudomonas* sp.
- biotransformation, role in, 112
- cellulase production, 292
- dehalogenation, 74
- dye contamination, 157
- genetic engineering, 13
- heavy metals bioremediation, 227–228
- lindane, 178, 189–190, 194
- pesticide biodegradation, 70, 72, 79, 80
- plant growth-promoting rhizobacteria, 260, 322
- surfactant mediated bioremediation, 86, 87
- tannery wastewater, 387
- toxic metals removal, 225

Pseudo solubilization, 86

*Pseudoxanthomonas indica,* 191

*Pseudozyma* sp., 158, 193

*Pteris vittata,* 321

Pulp and paper industry application
- biobeating and pulp refining, 292–293
- cellulase enzyme, application, 292–294
- constructed wetlands, 417
- deinking process, 293–294
- pulp brightness, 294
- pulp drainage rate, 293

*Pycnoporus cinnabarinus*, 158

*Pyrobaculum* sp., 119

## Q


Quantity, solid waste, 340–341


## R


Radical–radical reactions, 150
Radiological toxicity, 108
Radionuclides, 112–114
*Ralstonia* sp., 70, 227
*Raphanus sativus,* 66
Reactive dyes, 130, 152
Reactive solid waste, 349
Recombinant DNA techniques, 14–15, 160
Recycling, solid waste, 354, 362–363
Reduction, pesticides, 74
Reductive enzymes, 72–73
Regulations, industrial wastes, 25, 27
Remediation, toxic metals contamination
  absorption, 215
  adsorption, 216
  advantages/disadvantages, 229–230
  algae bioremediation, 224–225
  bacterial bioremediation, 225
  biological approaches, 220–229
  bioremediation, 220–222
  capping, 220
  electrokinetic remediation, 217–218
  excavation, 219
  *ex situ* bioremediation techniques, 221–222
  filtration, 215–216
  fungal bioremediation, 222–223
  immobilization, 218
  limitations, 220
  mechanisms of, 225–228
  membrane technology, 215–216
  physical separation, 220
  physicochemical techniques, 214–220
  phytoremediation, 228–229
  remediation techniques, 214–218
  *in situ* bioremediation techniques, 221
  soil leaching, 219
  soil treatment processes, 218–220
  solidification, 219
  stabilization, 219
  vitrification, 218–219
Removal of dyes, 135–136
Reproductive hazards, lindane, 187
Residential waste, 341
Residues, 62–67; *see also* Pesticides
Respiratory system, 378
Retaining mechanism, 280–282
Rhizobacteria, 41
*Rhizobium* sp., 261
*Rhizoclonium* sp., 224
Rhizodegradation (rhizosphere degradation), 7, 39, 318
Rhizofiltration
  applications, contaminated sites, 7, 8
  heavy metal-contaminated soils, 259–260
  metal-contaminated soils, 316
  overview, 38
*Rhizopus* sp., 85, 222, 288
Rhizovolatilization, 259
*Rhodanobacter lindanclasticus,* 189
*Rhodobacter sphaeroides,* 76
*Rhodococcus* sp.
  bioaugmentation, 84
  lindane mineralization, 191
  lindane remediation, 194
  pesticide biodegradation, 72
  surfactant mediated bioremediation, 86
*Rhodotorula* sp., 193
Rodenticides, 33
Rotterdam Convention, 196
*Ruminococcus* sp., 292, 359


## S


*Saccharomonospora* sp., 288
*Saccharomyces cerevisiae*, 158, 226
*Saccoglossus kowalevskii*, 191
*Sagittaria latifolia*, 416
*Salmonella typhimurium*, 132
Salt, 378
*Sarcocornia fruticosa*, 389, 416
*Scenedesmus* sp., 85
*Scirpus* sp.
  distillery wastewater, 416
  petrochemical and oil refinery wastewater, 417
  tannery wastewater, 415
  winery wastewater, 416
*Secale* sp., 259
Segregation, biomedical waste, 433–434
*Selenastrum capricornulum,* 85
Semivolatile organic compounds (SVOCs), 10
Separation techniques, 220
Sequestration, 266, 321, 345
*Serratia* sp., 70, 322
Service waste, 341
*Setia quisquiliarum,* 190
Sharps, biomedical waste, 428–429, 433, 434, 439
*Shewanella* sp., 119, 120
Slurry-phase treatment, 10, 18, 112, 221
Smoluchowski theory, 140
Snail shells, 71, 141
Sodium chloride, tannery wastewater, 378
Soil
  heavy metal-contaminated soils, 261–263
  leaching, removal and soil isolation, 219

moisture content, 80
pesticides biodegradation, 78–82
pesticides bioremediation, 82–89
residues of pesticides, 63–65
texture, 79–80
*Solanum melongena*, 69
Solidification, 219
Solid-phase treatment, 11, 18, 112, 221
Solid-state fermentation (SSF), 282, 283–285
Solid waste
bioenergy production, 358–361
challenges, 364–365
characteristics, 341–342
collection systems, 342
combustion, 354
composting, 354, 356–358
disposal, 354
environmental threats and pollution from, 343–351
exposure impacts, 349–350
functional elements, 351–352
future prospects, 364–365
greenhouse gas emissions, landfills, 344–347
groundwater contamination, 343–344
health impacts, 348–351
incineration, 361–362
introduction, 338–339
issues, 351–352
landfilling, 354, 363–364
management of, 351–352
materials recovery facilities, 347–348
municipal types, 339–341
plan development, 354–356
processing, 356–364
quantity of, 340–341
recommendations, 365
recycling, 354, 362–363
sources of generation, 341–342
strategies, 352–356
toxicity, 350–351
waste prevention, 353–354
Solubility, pesticides, 77–78
Solvent dyes, 130
Sources of contamination
arsenic, 36
azo dye, 30
biomedical waste management, 429–430
cadmium, 34
chlorinated phenols, 29
chromium, 35
inorganic pollutants, 34
lead, 36–37
lindane contamination, 183–184
melanoidins, 32
mercury, 37
persistent organic pollutants, 33
pesticides, 64, 65
petroleum hydrocarbons, 31
phytoremediation technologies, metal-contaminated soils, 308–313
solid waste generation, 341–342, 353–354
toxic metals contamination, 213
uranium, 105–107
Sparging, 6; *see also* Biosparging
Special waste, 341
*Sphingobium* sp., 189–190, 191
*Sphingomonadaceae* family, 190
*Sphingomonas* sp., 70, 194
*Spinacia* sp., 259
*Spirogyra* sp., 224
*Spirulina,* 224, 389
*Sporocytophaga* sp., 292
Stabilization, toxic metals contamination, 219
*Stachybotrys* sp., 226
*Staphylococcus* sp., 76, 194
*Stenotaphrum secundatum,* 389, 415
*Stenotrophomonas maltophilia,* 72
*Stereum hirsutum,* 72, 193
Stochastic molecular modeling (SMM), 279
Stockholm Convention, 34, 178, 196
Storage, biomedical waste, 434–440
*Streptomyces* sp.
biotransformation, role in, 112
cellulase production, 288, 292
lindane mineralization, 191–192
lindane remediation, 194
surfactant mediated bioremediation, 86
tannery wastewater, 387
toxic metals removal, 225
Structure, microbial cellulase enzyme, 275–276
Submerged aquatic vegetation (SAV) system, 401
Submerged fermentation, 283, 286, 287
Subsurface flow, constructed wetlands, 402
Subsurfaces, NAPLs, 242
Successful outcome, 230
"Sudden sniffing death," 32
Sulfide, tannery wastewater, 378–379, 380–382
Sulfur removal, 130, 408
Surface flow, constructed wetlands, 400–402
Surfactant mediated bioremediation, 85–87
"Sweep flocculation," 137
*Synechococcus* sp., 224


## T


*Talinum triangulare*, 66
Tannery industries
aluminum, 376–377, 379, 381–382
cadmium, 377, 379, 383–385
carcinogenic properties, 383–385

cardiovascular system, 379
chloride toxicity, 381
chlorophenolic compounds, 375
chromium, 35, 374–375, 383–384
constructed wetlands, 415–416
digestive system, 382
dyes, 375–376
hypertension, 379
lead, 377, 380, 383, 385
major pollutants, 373–378
microbes, 39
nervous system, 381–382
nitrogen, 374
oils and grease, 378
pharmacological aspects, 378–385
phenolic compounds, 375
phthalates, 375
pregnancy, exposure during, 383
regulations, 25, 27
respiratory system, 378–379
salts, 378
skeletal effects, 383
sodium chloride, 378
solids, 374
sulfide, 378–382
urinary system, 379
zinc, 377
zinc toxicity, 382
Temperature, 76, 78–79, 245, 287
*Terrabacter* sp., 70
Textile industries
constructed wetlands, 418
dye contamination, 132, 151, 152
microbes, 39
*Thalia dealbata*, 416
Thermal processes, 437–438
*Thermobifida* sp., 292
*Thermomonospora* sp., 292
Thermophilic treatments, *see* Composting
*Thermus* sp., 119
*Thlaspi caerulescens*, 260, 266
Toxic metals contamination
arsenic, 311–312
biological approaches, 220–229
cadmium, 310
chromium, 312–313
defined, 211
health hazards, 213–214, 309–313
introduction, 210–211
lead, 311
mercury, 310–311
nomenclature, 212–213
origin, heavy metals, 212–213
physicochemical remediation techniques, 214–218
physicochemical soil treatment processes, 218–220
remediation approaches, 214–230
sources of, 213
toxicological effects, 213–214
Toxic metals removal, 224
Toxic solid waste, 349
Tracking note, waste, 437
*Trametes* sp., 158, 192, 222
Transgenic plants, 326; *see also* Genetically engineered strategies
Transportation, biomedical waste, 434–437
Transport in environment, dyes, 130–131
*Trichoderma* sp.
cellulase production, 288, 291
pulp and paper industry, 293, 294, 296
solid-state fermentation, 283
submerged fermentation, 283
*Trichosporon* sp., 159
*Typha* sp.
constructed wetlands, 389, 400
distillery wastewater, 416
petrochemical and oil refinery wastewater, 417
pulp and paper mill wastewater, 417
tannery wastewater, 415–416
textile wastewater, 418
winery, 416


## U


Ultrafiltration, 215–216
Underground storage tank sites, 5
Uranium contamination
advantages/disadvantages, 119
animal health, 109
anthropogenic sources, 107
bioremediation, 109–119
chemical toxicity, 107–108
constructed wetlands, 120–121
enzymatic reduction, 119–120
general properties, 105
health hazards, 107–109
human health, 109
introduction, 104
natural sources of, 105–106
radiological toxicity, 108
sources of, 105–107
toxicological effects, 107–109


## V


Vat dyes, 130
Vertical flow (VF), constructed wetlands, 402–403
*Verticillium terrestre,* 226

Vitrification, 218–219
Volatile organic compounds (VOCs), 6, 10, 267, 351

## W

Walls and berms, textile wastewater, 419
Wastes, 4, 435; *see also specific type*
"Waste tracking note," 437
Wastewater treatment approaches, dye contamination
  adsorbing colloid flotation, 136
  adsorption, 141–146
  advanced oxidation process, 146–155
  algae biodegradation of dyes, 159
  bacterial biodegradation of dyes, 156–157
  biological treatment approaches, 155–160
  coagulation–flocculation, 137–141
  dissolved-air flotation, 136
  dyes biodegradation, 156–157
  electrolytic flotation, 136
  enzymatic biodegradation of dyes, 160
  equalization, 134–135
  flotation, 135–136
  fungal biodegradation of dyes, 157–158
  homogenization, 134–135
  ion flotation, 136
  overview, 132–134
  physiochemical treatment approaches, 133–155
  precipitate flotation, 136
  yeast biodegradation of dyes, 158–159
Well injection, 118–119
Wetlands, *see* Constructed wetlands
Windrow composting, 10
Winery wastewater, 416–417
*Withania somnifera,* 194
Worldwide use, pesticides, 59–61

## X

*Xanthomonas* sp., 190

## Y

*Yarrowia lipolytica*, 158
Yeasts, 158–159; *see also specific type*

## Z

*Zea mays,* 66, 194, 259
Zero-order kinetics, 251
Zinc toxicity, 382